STUDENT SOLUTIONS MANUAL

BEGINNING AND INTERMEDIATE ALGEBRA
FIFTH EDITION

John Tobey
North Shore Community College
Danvers, Massachusetts

Jeffrey Slater
North Shore Community College
Danvers, Massachusetts

Jamie Blair
Orange Coast College
Costa Mesa, California

Jennifer Crawford
Normandale Community College
Bloomington, Minnesota

PEARSON

Boston Columbus Indianapolis New York San Francisco
Amsterdam Cape Town Dubai London Madrid Milan Munich Paris Montréal Toronto
Delhi Mexico City São Paulo Sydney Hong Kong Seoul Singapore Taipei Tokyo

Contents

Copyright © 2017, 2013 by Pearson Education, Inc.
Publishing as Pearson, 501 Boylston Street, Boston, MA 02116.

ISBN-13: 978-0-13-418888-1
ISBN-10: 0-13-418888-8

6

www.pearsonhighered.com

Chapter 0

0.1 Exercises

1. $\dfrac{12}{13}$: numerator = 12

3. When two or more numbers are multiplied, each number that is multiplied is called a factor. In 2×3, 2 and 3 are factors.

5. $2\dfrac{2}{3}$

7. $\dfrac{9}{15} = \dfrac{3 \times 3}{5 \times 3} = \dfrac{3}{5}$

9. $\dfrac{12}{36} = \dfrac{1 \times 12}{3 \times 12} = \dfrac{1}{3}$

11. $\dfrac{60}{12} = \dfrac{12 \times 5}{12 \times 1} = 5$

13. $\dfrac{24}{36} = \dfrac{12 \times 2}{12 \times 3} = \dfrac{2}{3}$

15. $\dfrac{30}{85} = \dfrac{5 \times 6}{5 \times 17} = \dfrac{6}{17}$

17. $\dfrac{42}{54} = \dfrac{6 \times 7}{6 \times 9} = \dfrac{7}{9}$

19. $\dfrac{17}{6} = 6\overline{)17} = 2\dfrac{5}{6}$
$\phantom{\dfrac{17}{6} = 6)}\underline{12}$
$\phantom{\dfrac{17}{6} = 6)17}5$

21. $\dfrac{47}{5} = 5\overline{)47} = 9\dfrac{2}{5}$
$\phantom{\dfrac{47}{5} = 5)}\underline{45}$
$\phantom{\dfrac{47}{5} = 5)47}2$

23. $\dfrac{38}{7} = 7\overline{)38} = 5\dfrac{3}{7}$
$\phantom{\dfrac{38}{7} = 7)}\underline{35}$
$\phantom{\dfrac{38}{7} = 7)38}3$

25. $\dfrac{41}{2} = 2\overline{)41} = 20\dfrac{1}{2}$
$\phantom{\dfrac{41}{2} = 2)}\underline{4}$
$\phantom{\dfrac{41}{2} = 2)4}01$
$\phantom{\dfrac{41}{2} = 2)4}\underline{0}$
$\phantom{\dfrac{41}{2} = 2)44}1$

27. $\dfrac{32}{5} = 5\overline{)32} = 6\dfrac{2}{5}$
$\phantom{\dfrac{32}{5} = 5)}\underline{30}$
$\phantom{\dfrac{32}{5} = 5)3}2$

29. $\dfrac{111}{9} = 9\overline{)111} = 12\dfrac{3}{9} = 12\dfrac{1}{3}$
$\phantom{\dfrac{111}{9} = 9)}\underline{9}$
$\phantom{\dfrac{111}{9} = 9)}21$
$\phantom{\dfrac{111}{9} = 9)}\underline{18}$
$\phantom{\dfrac{111}{9} = 9)1}3$

31. $3\dfrac{1}{5} = \dfrac{(3 \times 5) + 1}{5} = \dfrac{15 + 1}{5} = \dfrac{16}{5}$

33. $6\dfrac{3}{5} = \dfrac{(6 \times 5) + 3}{5} = \dfrac{30 + 3}{5} = \dfrac{33}{5}$

35. $1\dfrac{2}{9} = \dfrac{(1 \times 9) + 2}{9} = \dfrac{9 + 2}{9} = \dfrac{11}{9}$

37. $8\dfrac{3}{7} = \dfrac{(8 \times 7) + 3}{7} = \dfrac{56 + 3}{7} = \dfrac{59}{7}$

39. $24\dfrac{1}{4} = \dfrac{(24 \times 4) + 1}{4} = \dfrac{96 + 1}{4} = \dfrac{97}{4}$

41. $\dfrac{72}{9} = 72 \div 9$

8
$9\overline{)72}$
$\underline{72}$
0

$\dfrac{72}{9} = 8$

43. $\dfrac{3}{8} = \dfrac{?}{64} \Rightarrow \dfrac{3 \times 8}{8 \times 8} = \dfrac{24}{64}$

45. $\dfrac{3}{5} = \dfrac{?}{35} \Rightarrow \dfrac{3 \times 7}{5 \times 7} = \dfrac{21}{35}$

47. $\dfrac{4}{13} = \dfrac{?}{39} \Rightarrow \dfrac{4 \times 3}{13 \times 3} = \dfrac{12}{39}$

49. $\dfrac{3}{7} = \dfrac{?}{49} \Rightarrow \dfrac{3 \times 7}{7 \times 7} = \dfrac{21}{49}$

51. $\dfrac{3}{4} = \dfrac{?}{20} \Rightarrow \dfrac{3 \times 5}{4 \times 5} = \dfrac{15}{20}$

53. $\dfrac{35}{40} = \dfrac{?}{80} \Rightarrow \dfrac{35 \times 2}{40 \times 2} = \dfrac{70}{80}$

55. $\dfrac{799}{34} = 23\dfrac{17}{34} = 23\dfrac{1}{2}$

$$\begin{array}{r} 23 \\ 34\overline{)799} \\ 68 \\ \hline 119 \\ 102 \\ \hline 17 \end{array}$$

57. $\dfrac{13{,}200}{64{,}000} = \dfrac{400 \times 33}{400 \times 160} = \dfrac{33}{160}$

59. $\dfrac{1}{1+1+2} = \dfrac{1}{4}$ is nuts.

61. $\dfrac{1+1}{1+1+2} = \dfrac{2}{4} = \dfrac{1}{2}$ is not sunflower seeds.

63. $\dfrac{12}{8+12} = \dfrac{12}{20} = \dfrac{3 \times 4}{5 \times 4} = \dfrac{3}{5}$

$\dfrac{3}{5}$ of the students enrolled in 2014 were female.

Quick Quiz 0.1

1. $\dfrac{84}{92} = \dfrac{21 \times 4}{23 \times 4} = \dfrac{21}{23}$

2. $6\dfrac{9}{11} = \dfrac{(6 \times 11) + 9}{11} = \dfrac{66 + 9}{11} = \dfrac{75}{11}$

3. $\dfrac{103}{21} = 4\dfrac{19}{21}$

$$\begin{array}{r} 4 \\ 21\overline{)103} \\ 84 \\ \hline 19 \end{array}$$

4. Answers may vary. Possible solution: First, multiply the whole number by the denominator. Then, add this to the numerator. The result is the new numerator. The denominator does not change.

0.2 Exercises

1. Answers may vary. A sample answer is: 8 is the LCD of $\dfrac{3}{4}$ and $\dfrac{5}{8}$ because 8 is exactly divisible by 4.

3. $\dfrac{4}{9}$ and $\dfrac{5}{12}$
$9 = 3 \cdot 3$
$12 = 2 \cdot 2 \cdot 3$
$\text{LCD} = 2 \cdot 2 \cdot 3 \cdot 3 = 36$

5. $\dfrac{7}{10}$ and $\dfrac{1}{4}$
$10 = 2 \cdot 5$
$4 = 2 \cdot 2$
$\text{LCD} = 2 \cdot 2 \cdot 5 = 20$

7. $\dfrac{5}{18}$ and $\dfrac{7}{54}$
$18 = 2 \cdot 3 \cdot 3$
$54 = 2 \cdot 3 \cdot 3 \cdot 3$
$\text{LCD} = 2 \cdot 3 \cdot 3 \cdot 3 = 54$

9. $\dfrac{1}{15}$ and $\dfrac{4}{21}$
$15 = 3 \cdot 5$
$21 = 3 \cdot 7$
$\text{LCD} = 3 \cdot 5 \cdot 7 = 105$

11. $\dfrac{17}{40}$ and $\dfrac{13}{60}$
$40 = 2 \cdot 2 \cdot 2 \cdot 5$
$60 = 2 \cdot 2 \cdot 3 \cdot 5$
$\text{LCD} = 2 \cdot 2 \cdot 2 \cdot 3 \cdot 5 = 120$

13. $\dfrac{2}{5}, \dfrac{3}{8},$ and $\dfrac{5}{12}$

$5 = 5$
$8 = 2 \cdot 2 \cdot 2$
$12 = 2 \cdot 2 \cdot 3$
$\text{LCD} = 2 \cdot 2 \cdot 2 \cdot 3 \cdot 5 = 120$

15. $\dfrac{5}{6}, \dfrac{9}{14},$ and $\dfrac{17}{26}$

$6 = 2 \cdot 3$
$14 = 2 \cdot 7$
$26 = 2 \cdot 13$
$\text{LCD} = 2 \cdot 3 \cdot 7 \cdot 13 = 546$

17. $\dfrac{1}{2}, \dfrac{1}{18}$ and $\dfrac{13}{30}$

$2 = 2$
$18 = 2 \cdot 3 \cdot 3$
$30 = 2 \cdot 3 \cdot 5$
$\text{LCD} = 2 \cdot 3 \cdot 3 \cdot 5 = 90$

19. $\dfrac{3}{8} + \dfrac{2}{8} = \dfrac{3+2}{8} = \dfrac{5}{8}$

21. $\dfrac{5}{14} - \dfrac{1}{14} = \dfrac{5-1}{14} = \dfrac{4}{14} = \dfrac{2}{7}$

23. $\dfrac{5}{12} + \dfrac{5}{8} = \dfrac{5\times2}{12\times2} + \dfrac{5\times3}{8\times3} = \dfrac{10}{24} + \dfrac{15}{24} = \dfrac{25}{24}$ or $1\dfrac{1}{24}$

25. $\dfrac{5}{7} - \dfrac{2}{9} = \dfrac{5\times9}{7\times9} - \dfrac{2\times7}{9\times7} = \dfrac{45}{63} - \dfrac{14}{63} = \dfrac{31}{63}$

27. $\dfrac{1}{3} + \dfrac{2}{5} = \dfrac{1\times5}{3\times5} + \dfrac{2\times3}{5\times3} = \dfrac{5}{15} + \dfrac{6}{15} = \dfrac{11}{15}$

29. $\dfrac{5}{9} + \dfrac{5}{12} = \dfrac{5\times4}{9\times4} + \dfrac{5\times3}{12\times3} = \dfrac{20}{36} + \dfrac{15}{36} = \dfrac{35}{36}$

31. $\dfrac{11}{15} - \dfrac{31}{45} = \dfrac{11\times3}{15\times3} - \dfrac{31}{45} = \dfrac{33}{45} - \dfrac{31}{45} = \dfrac{2}{45}$

33. $\dfrac{16}{24} - \dfrac{1}{6} = \dfrac{16}{24} - \dfrac{1\times4}{6\times4} = \dfrac{16}{24} - \dfrac{4}{24} = \dfrac{12}{24} = \dfrac{1}{2}$

35. $\dfrac{3}{8} + \dfrac{4}{7} = \dfrac{3\times7}{8\times7} + \dfrac{4\times8}{7\times8} = \dfrac{21}{56} + \dfrac{32}{56} = \dfrac{53}{56}$

37. $\dfrac{2}{3} + \dfrac{7}{12} + \dfrac{1}{4} = \dfrac{8}{12} + \dfrac{7}{12} + \dfrac{3}{12} = \dfrac{18}{12} = \dfrac{3}{2}$ or $1\dfrac{1}{2}$

39.
$$\dfrac{5}{30} + \dfrac{3}{40} + \dfrac{1}{8} = \dfrac{5\times4}{30\times4} + \dfrac{3\times3}{40\times3} + \dfrac{1\times15}{8\times15}$$
$$= \dfrac{20}{120} + \dfrac{9}{120} + \dfrac{15}{120}$$
$$= \dfrac{44}{120}$$
$$= \dfrac{11}{30}$$

41.
$$\dfrac{1}{3} + \dfrac{1}{12} - \dfrac{1}{6} = \dfrac{1\times4}{3\times4} + \dfrac{1}{12} - \dfrac{1\times2}{6\times2}$$
$$= \dfrac{4}{12} + \dfrac{1}{12} - \dfrac{2}{12}$$
$$= \dfrac{3}{12}$$
$$= \dfrac{1}{4}$$

43.
$$\dfrac{5}{36} + \dfrac{7}{9} - \dfrac{5}{12} = \dfrac{5}{36} + \dfrac{7\times4}{9\times4} - \dfrac{5\times3}{12\times3}$$
$$= \dfrac{5}{36} + \dfrac{28}{36} - \dfrac{15}{36}$$
$$= \dfrac{18}{36}$$
$$= \dfrac{1}{2}$$

45. $4\dfrac{1}{3} + 3\dfrac{2}{5} = \dfrac{13}{3} + \dfrac{17}{5} = \dfrac{65}{15} + \dfrac{51}{15} = \dfrac{116}{15} = 7\dfrac{11}{15}$

47. $1\dfrac{5}{24} + \dfrac{5}{18} = \dfrac{29}{24} + \dfrac{5}{18} = \dfrac{87}{72} + \dfrac{20}{72} = \dfrac{107}{72} = 1\dfrac{35}{72}$

49. $7\dfrac{1}{6} - 2\dfrac{1}{4} = \dfrac{43}{6} - \dfrac{9}{4} = \dfrac{86}{12} - \dfrac{27}{12} = \dfrac{59}{12} = 4\dfrac{11}{12}$

51. $8\dfrac{5}{7} - 2\dfrac{1}{4} = \dfrac{61}{7} - \dfrac{9}{4} = \dfrac{244}{28} - \dfrac{63}{28} = \dfrac{181}{28} = 6\dfrac{13}{28}$

53. $2\dfrac{1}{8} + 3\dfrac{2}{3} = \dfrac{17}{8} + \dfrac{11}{3} = \dfrac{51}{24} + \dfrac{88}{24} = \dfrac{139}{24} = 5\dfrac{19}{24}$

55. $11\dfrac{1}{7} - 6\dfrac{5}{7} = \dfrac{78}{7} - \dfrac{47}{7} = \dfrac{31}{7} = 4\dfrac{3}{7}$

57. $3\dfrac{5}{12} + 5\dfrac{7}{12} = \dfrac{41}{12} + \dfrac{67}{12} = \dfrac{108}{12} = 9$

59. $\dfrac{7}{8} + \dfrac{1}{12} = \dfrac{42}{48} + \dfrac{4}{48} = \dfrac{46}{48} = \dfrac{23}{24}$

61. $3\dfrac{3}{16}+4\dfrac{3}{8}=\dfrac{51}{16}+\dfrac{35}{8}=\dfrac{51}{16}+\dfrac{70}{16}=\dfrac{121}{16}=7\dfrac{9}{16}$

63. $\dfrac{16}{21}-\dfrac{2}{7}=\dfrac{16}{21}-\dfrac{6}{21}=\dfrac{10}{21}$

65. $5\dfrac{1}{5}-2\dfrac{1}{2}=\dfrac{26}{5}-\dfrac{5}{2}=\dfrac{52}{10}-\dfrac{25}{10}=\dfrac{27}{10}=2\dfrac{7}{10}$

67. $25\dfrac{2}{3}-6\dfrac{1}{7}=\dfrac{77}{3}-\dfrac{43}{7}=\dfrac{539}{21}-\dfrac{129}{21}=\dfrac{410}{21}=19\dfrac{11}{21}$

69. $1\dfrac{1}{6}+\dfrac{3}{8}=\dfrac{7}{6}+\dfrac{3}{8}=\dfrac{56}{48}+\dfrac{18}{48}=\dfrac{74}{48}=\dfrac{37}{24}=1\dfrac{13}{24}$

71. $8\dfrac{1}{4}+3\dfrac{5}{6}=\dfrac{33}{4}+\dfrac{23}{6}$

$\qquad\qquad=\dfrac{198}{24}+\dfrac{92}{24}$

$\qquad\qquad=\dfrac{290}{24}$

$\qquad\qquad=\dfrac{145}{12}$

$\qquad\qquad=12\dfrac{1}{12}$

73. $36-2\dfrac{4}{7}=\dfrac{36}{1}-\dfrac{18}{7}=\dfrac{252}{7}-\dfrac{18}{7}=\dfrac{234}{7}=33\dfrac{3}{7}$

75. $8\dfrac{1}{4}+10\dfrac{2}{3}+5\dfrac{3}{4}=\dfrac{33}{4}+\dfrac{32}{3}+\dfrac{23}{4}$

$\qquad\qquad\qquad=\dfrac{99}{12}+\dfrac{128}{12}+\dfrac{69}{12}$

$\qquad\qquad\qquad=\dfrac{296}{12}$

$\qquad\qquad\qquad=24\dfrac{8}{12}$

$\qquad\qquad\qquad=24\dfrac{2}{3}$

Their total distance was $24\dfrac{2}{3}$ miles.

77. $15\dfrac{1}{2}-3\dfrac{2}{3}-9\dfrac{1}{4}=\dfrac{31}{2}-\dfrac{11}{3}-\dfrac{37}{4}$

$\qquad\qquad\qquad=\dfrac{186}{12}-\dfrac{44}{12}-\dfrac{111}{12}$

$\qquad\qquad\qquad=\dfrac{31}{12}$

$\qquad\qquad\qquad=2\dfrac{7}{12}$

There will be $2\dfrac{7}{12}$ hours available to wash the front windows.

79. $A=2+\dfrac{1}{2}+3\dfrac{1}{2}+\dfrac{1}{2}+3\dfrac{1}{2}+\dfrac{1}{2}+1\dfrac{1}{2}=12$ in.

$\quad B=\dfrac{1}{2}+4\dfrac{5}{8}+\dfrac{1}{2}+4\dfrac{5}{8}+\dfrac{1}{2}+4\dfrac{5}{8}+\dfrac{1}{2}=15\dfrac{7}{8}$ in.

81. $2\dfrac{1}{2}-\dfrac{7}{8}=\dfrac{5}{2}-\dfrac{7}{8}=\dfrac{20}{8}-\dfrac{7}{8}=\dfrac{13}{8}=1\dfrac{5}{8}$

The mower blade must be lowered $1\dfrac{5}{8}$ inches.

Cumulative Review

83. $\dfrac{36}{44}=\dfrac{9\times4}{1\times4}=\dfrac{9}{11}$

84. $26\dfrac{3}{5}=\dfrac{26\times5+3}{5}=\dfrac{133}{5}$

Quick Quiz 0.2

1. $\dfrac{3}{4}+\dfrac{1}{2}+\dfrac{5}{12}=\dfrac{9}{12}+\dfrac{6}{12}+\dfrac{5}{12}=\dfrac{20}{12}=\dfrac{5}{3}$ or $1\dfrac{2}{3}$

2. $2\dfrac{3}{5}+4\dfrac{14}{15}=2\dfrac{9}{15}+4\dfrac{14}{15}=6\dfrac{23}{15}=7\dfrac{8}{15}$

3. $6\dfrac{1}{9}-3\dfrac{5}{6}=\dfrac{55}{9}-\dfrac{23}{6}=\dfrac{110}{18}-\dfrac{69}{18}=\dfrac{41}{18}=2\dfrac{5}{18}$

4. Answers may vary. Possible solution:
Write each denominator as the product of prime factors. The LCD is a product containing each different factor. If a factor occurs more than once in any one denominator, the LCD will contain that factor repeated the greatest number of times that it occurs in any one denominator.

0.3 Exercises

1. First change each mixed number to an improper fraction. Look for a common factor in the numerator and denominator to divide by; if one is found, perform the division. Multiply the numerators. Multiply the denominators.

3. $\dfrac{28}{5}\times\dfrac{6}{35}=\dfrac{7\cdot4}{5}\times\dfrac{6}{7\cdot5}=\dfrac{24}{25}$

5. $\dfrac{17}{18}\times\dfrac{3}{5}=\dfrac{17}{6\cdot3}\times\dfrac{3}{5}=\dfrac{17}{30}$

7. $\dfrac{4}{5}\times\dfrac{3}{10}=\dfrac{2\cdot2}{5}\times\dfrac{3}{2\cdot5}=\dfrac{6}{25}$

9. $\dfrac{24}{25}\times\dfrac{5}{2}=\dfrac{12\cdot2}{5\cdot5}\times\dfrac{5}{2}=\dfrac{12}{5}$ or $2\dfrac{2}{5}$

11. $\dfrac{7}{12}\times\dfrac{8}{28}=\dfrac{7}{2\cdot2\cdot3}\times\dfrac{2\cdot4}{4\cdot7}=\dfrac{1}{6}$

13. $\dfrac{6}{35}\times5=\dfrac{6}{5\cdot7}\times\dfrac{5}{1}=\dfrac{6}{7}$

15. $9\times\dfrac{2}{5}=\dfrac{9}{1}\times\dfrac{2}{5}=\dfrac{18}{5}$ or $3\dfrac{3}{5}$

17. $\dfrac{8}{5}\div\dfrac{8}{3}=\dfrac{8}{5}\times\dfrac{3}{8}=\dfrac{3}{5}$

19. $\dfrac{3}{7}\div3=\dfrac{3}{7}\div\dfrac{3}{1}=\dfrac{3}{7}\times\dfrac{1}{3}=\dfrac{1}{7}$

21. $10\div\dfrac{5}{7}=\dfrac{10}{1}\times\dfrac{7}{5}=\dfrac{2\cdot5}{1}\times\dfrac{7}{5}=14$

23. $\dfrac{6}{14}\div\dfrac{3}{8}=\dfrac{6}{14}\times\dfrac{8}{3}=\dfrac{2\cdot3}{2\cdot7}\times\dfrac{8}{3}=\dfrac{8}{7}$ or $1\dfrac{1}{7}$

25. $\dfrac{7}{24}\div\dfrac{9}{8}=\dfrac{7}{24}\times\dfrac{8}{9}=\dfrac{7}{3\cdot8}\times\dfrac{8}{9}=\dfrac{7}{27}$

27. $\dfrac{\frac{7}{8}}{\frac{3}{4}}=\dfrac{7}{8}\div\dfrac{3}{4}=\dfrac{7}{8}\times\dfrac{4}{3}=\dfrac{7}{2\cdot4}\times\dfrac{2\cdot2}{3}=\dfrac{14}{12}=\dfrac{7}{6}=1\dfrac{1}{6}$

29. $\dfrac{\frac{5}{6}}{\frac{7}{9}}=\dfrac{5}{6}\div\dfrac{7}{9}=\dfrac{5}{6}\times\dfrac{9}{7}=\dfrac{5}{2\cdot3}\times\dfrac{3\cdot3}{7}=\dfrac{15}{14}=1\dfrac{1}{14}$

31. $1\dfrac{3}{7}\div6\dfrac{1}{4}=\dfrac{10}{7}\div\dfrac{25}{4}=\dfrac{10}{7}\times\dfrac{4}{25}=\dfrac{2\cdot5}{7}\times\dfrac{4}{5\cdot5}=\dfrac{8}{35}$

33. $3\dfrac{1}{3}\div2\dfrac{1}{2}=\dfrac{10}{3}\div\dfrac{5}{2}=\dfrac{2\cdot5}{3}\times\dfrac{2}{5}=\dfrac{4}{3}=1\dfrac{1}{3}$

35. $6\dfrac{1}{2}\div\dfrac{3}{4}=\dfrac{13}{2}\div\dfrac{3}{4}=\dfrac{13}{2}\times\dfrac{4}{3}=\dfrac{13}{2}\times\dfrac{2\cdot2}{3}=\dfrac{26}{3}=8\dfrac{2}{3}$

37. $\dfrac{15}{2\frac{2}{5}}=15\div2\dfrac{2}{5}$

    $=\dfrac{15}{1}\div\dfrac{12}{5}$

    $=\dfrac{15}{1}\times\dfrac{5}{12}$

    $=\dfrac{3\cdot5}{1}\times\dfrac{5}{3\cdot4}$

    $=\dfrac{25}{4}$

    $=6\dfrac{1}{4}$

39. $\dfrac{\frac{2}{3}}{1\frac{1}{4}}=\dfrac{\frac{2}{3}}{\frac{5}{4}}=\dfrac{2}{3}\div\dfrac{5}{4}=\dfrac{2}{3}\times\dfrac{4}{5}=\dfrac{8}{15}$

41. $\dfrac{4}{7}\times\dfrac{21}{2}=\dfrac{2\cdot2}{7}\times\dfrac{7\cdot3}{2}=6$

43. $\dfrac{5}{14}\div\dfrac{2}{7}=\dfrac{5}{14}\times\dfrac{7}{2}=\dfrac{5}{2\cdot7}\times\dfrac{7}{2}=\dfrac{5}{4}$ or $1\dfrac{1}{4}$

45. $10\dfrac{3}{7}\times5\dfrac{1}{4}=\dfrac{73}{7}\times\dfrac{21}{4}=\dfrac{73}{7}\times\dfrac{3\cdot7}{4}=\dfrac{219}{4}=54\dfrac{3}{4}$

47. $25\div\dfrac{5}{8}=\dfrac{25}{1}\div\dfrac{5}{8}=\dfrac{25}{1}\times\dfrac{8}{5}=\dfrac{5\cdot5}{1}\times\dfrac{8}{5}=40$

49. $6\times4\dfrac{2}{3}=\dfrac{6}{1}\times\dfrac{14}{3}=\dfrac{2\cdot3}{1}\times\dfrac{14}{3}=28$

51. $2\dfrac{1}{2}\times\dfrac{1}{10}\times\dfrac{3}{4}=\dfrac{5}{2}\times\dfrac{1}{5\cdot2}\times\dfrac{3}{4}=\dfrac{3}{16}$

53. a. $\dfrac{1}{15}\times\dfrac{25}{21}=\dfrac{1}{3\cdot5}\times\dfrac{5\cdot5}{21}=\dfrac{5}{63}$

   b. $\dfrac{1}{15}\div\dfrac{25}{21}=\dfrac{1}{15}\times\dfrac{21}{25}=\dfrac{1}{3\cdot5}\times\dfrac{3\cdot7}{5\cdot5}=\dfrac{7}{125}$

55. a. $\dfrac{2}{3} \div \dfrac{12}{21} = \dfrac{2}{3} \times \dfrac{21}{12} = \dfrac{2}{3} \times \dfrac{3 \cdot 7}{2 \cdot 6} = \dfrac{7}{6}$ or $1\dfrac{1}{6}$

    b. $\dfrac{2}{3} \times \dfrac{12}{21} = \dfrac{2}{3} \times \dfrac{3 \cdot 4}{21} = \dfrac{8}{21}$

57. $\quad 71\dfrac{1}{2} \div 2\dfrac{3}{4} = \dfrac{143}{2} \div \dfrac{11}{4}$

$\qquad\qquad\qquad = \dfrac{143}{2} \times \dfrac{4}{11}$

$\qquad\qquad\qquad = \dfrac{11 \cdot 13}{2} \times \dfrac{2 \cdot 2}{11}$

$\qquad\qquad\qquad = 26$

The material can make 26 shirts.

59. $\quad 11\dfrac{1}{3} \times 12 = \dfrac{34}{3} \times \dfrac{12}{1} = \dfrac{34}{3} \times \dfrac{3 \cdot 4}{1} = 136$

The area of the window is 136 square feet.

Cumulative Review

61. $\quad \dfrac{11}{15} = \dfrac{?}{75} \Rightarrow \dfrac{11 \times 5}{15 \times 5} = \dfrac{55}{75}$

62. $\quad \dfrac{7}{9} = \dfrac{?}{63} \Rightarrow \dfrac{7 \times 7}{9 \times 7} = \dfrac{49}{63}$

Quick Quiz 0.3

1. $\quad \dfrac{7}{15} \times \dfrac{25}{14} = \dfrac{7}{5 \cdot 3} \times \dfrac{5 \cdot 5}{7 \cdot 2} = \dfrac{5}{6}$

2. $\quad 3\dfrac{1}{4} \times 4\dfrac{1}{2} = \dfrac{13}{4} \times \dfrac{9}{2} = \dfrac{117}{8} = 14\dfrac{5}{8}$

3. $\quad 3\dfrac{3}{10} \div 2\dfrac{1}{2} = \dfrac{33}{10} \div \dfrac{5}{2}$

$\qquad\qquad\qquad = \dfrac{33}{10} \times \dfrac{2}{5}$

$\qquad\qquad\qquad = \dfrac{33}{2 \cdot 5} \times \dfrac{2}{5}$

$\qquad\qquad\qquad = \dfrac{33}{25}$

$\qquad\qquad\qquad = 1\dfrac{8}{25}$

4. Answers may vary. Possible solution:
Change each mixed number to an improper
fraction. Invert the second fraction and multiply
the result by the first fraction. Simplify.

0.4 Exercises

1. A decimal is another way of writing a fraction
whose denominator is <u>10, 100, 1000, 10,000,
and so on.</u>

3. When dividing 7432.9 by 1000 we move the
decimal point <u>3</u> places to the <u>left</u>.

5. $\dfrac{7}{8} = 8\overline{)7.000} = 0.875$ with long division:

$$\begin{array}{r} 0.875 \\ 8\overline{)7.000} \\ \underline{6\ 4} \\ 60 \\ \underline{56} \\ 40 \\ \underline{40} \\ 0 \end{array}$$

7. $\dfrac{3}{15} = 15\overline{)3.0} = 0.2$

$$\begin{array}{r} 0.2 \\ 15\overline{)3.0} \\ \underline{3.0} \\ 0 \end{array}$$

9. $\dfrac{7}{11} = 11\overline{)7.000} = 0.\overline{63}$

$$\begin{array}{r} 0.63 \\ 11\overline{)7.000} \\ \underline{6\ 6} \\ 40 \\ \underline{33} \\ 7 \end{array}$$

11. $\quad 0.8 = \dfrac{8}{10} = \dfrac{4}{5}$

13. $\quad 0.25 = \dfrac{25}{100} = \dfrac{1}{4}$

15. $\quad 0.625 = \dfrac{625}{1000} = \dfrac{5}{8}$

17. $\quad 0.06 = \dfrac{6}{100} = \dfrac{3}{50}$

19. $\quad 3.4 = \dfrac{34}{10} = \dfrac{17 \times 2}{5 \times 2} = \dfrac{17}{5}$ or $3\dfrac{2}{5}$

21. $\quad 5.5 = \dfrac{55}{10} = \dfrac{5 \times 11}{5 \times 2} = \dfrac{11}{2}$ or $5\dfrac{1}{2}$

23.
$$\begin{array}{r} 1.71 \\ +\,0.38 \\ \hline 2.09 \end{array}$$

25.
$$\begin{array}{r} 2.50 \\ 3.42 \\ +\,4.90 \\ \hline 10.82 \end{array}$$

27.
$$\begin{array}{r} 46.030 \\ 215.100 \\ +\,\ \ 0.078 \\ \hline 261.208 \end{array}$$

29.
$$\begin{array}{r} 147.18 \\ -\,15.39 \\ \hline 131.79 \end{array}$$

31.
$$\begin{array}{r} 6.0054 \\ -\,2.0257 \\ \hline 3.9797 \end{array}$$

33.
$$\begin{array}{r} 125.43 \\ -\,\ \ 2.80 \\ \hline 122.63 \end{array}$$

35.
$$\begin{array}{r} 7.21 \\ \times\ \ 4.2 \\ \hline 1\,442 \\ 28\,84 \\ \hline 30.282 \end{array}$$

37.
$$\begin{array}{r} 0.04 \\ \times\ \ 0.08 \\ \hline 0.0032 \end{array}$$

39.
$$\begin{array}{r} 4.23 \\ \times\ 0.025 \\ \hline 2115 \\ 846 \\ \hline 0.10575 \end{array}$$

41.
$$\begin{array}{r} 58,200 \\ \times\,0.0015 \\ \hline 29\,1000 \\ 58\,\ 200 \\ \hline 87.3000 \end{array}$$
or 87.3

43.
$$\begin{array}{r} 0.0565 \\ 64\,\overline{)3.6160} \\ 3\,20 \\ \hline 416 \\ 384 \\ \hline 320 \\ 320 \\ \hline 0 \end{array}$$

45.
$$\begin{array}{r} 2.64 \\ 3.02\wedge\,\overline{)7.97\wedge28} \\ 6\,04 \\ \hline 1\,93\ \ 2 \\ 1\,81\ \ 2 \\ \hline 12\ \ 08 \\ 12\ \ 08 \\ \hline 0 \end{array}$$

47.
$$\begin{array}{r} 261.5 \\ 0.002\wedge\,\overline{)0.523\wedge0} \\ 4 \\ \hline 12 \\ 12 \\ \hline 3 \\ 2 \\ \hline 1\ \ 0 \\ 1\ \ 0 \\ \hline 0 \end{array}$$

49.
$$\begin{array}{r} 0.508 \\ 0.06\wedge\,\overline{)0.03\wedge048} \\ 3\ \ 0 \\ \hline 48 \\ 48 \\ \hline 0 \end{array}$$

51. $3.45 \times 1000 = 3450$

53. $0.76 \div 100 = 0.0076$

55. $7.36 \times 10,000 = 73,600$

57. $73,892 \div 100,000 = 0.73892$

59. $0.1498 \times 100 = 14.98$

61. $1.931 \div 100 = 0.01931$

63.
$$\begin{array}{r} 54.8 \\ \times\,0.15 \\ \hline 2\ 740 \\ 5\ 48 \\ \hline 8.220 \text{ or } 8.22 \end{array}$$

65.
$$\begin{array}{r} 13.75 \\ 2.55 \\ +\ \ 0.078 \\ \hline 16.378 \end{array}$$

67.
$$\begin{array}{r} 2.12 \\ 0.027\wedge\overline{)0.057\wedge 24} \\ \underline{54} \\ 3\ 2 \\ \underline{2\ 7} \\ 54 \\ \underline{54} \\ 0 \end{array}$$

69. $0.7683 \times 1000 = 768.3$

71.
$$\begin{array}{r} 56.37 \\ -\ \ 4.29 \\ \hline 52.08 \end{array}$$

73. $153.7 \div 100 = 1.537$

75.
$$\begin{array}{r} 0.4732 \\ \times\ \ \ \ 5.5 \\ \hline 23660 \\ 23660 \\ \hline 2.60260 \end{array}$$
The measured data is 2.6026 L.

77.
$$\begin{array}{r} 20.5 \\ 9\overline{)185.0} = 20.\overline{5} \\ \underline{18} \\ 05 \\ \underline{0} \\ 5\ 0 \\ \underline{4\ 5} \\ 5 \end{array}$$
He will need 21 hours to achieve his goal.
$9 \times 21 = 189$
He will exceed his goal by $4 each week.

Cumulative Review

79. $3\dfrac{1}{2} \div 5\dfrac{1}{4} = \dfrac{7}{2} \div \dfrac{21}{4} = \dfrac{7}{2} \cdot \dfrac{4}{21} = \dfrac{1 \cdot 2}{1 \cdot 3} = \dfrac{2}{3}$

80. $\dfrac{3}{8} \cdot \dfrac{12}{27} = \dfrac{3}{2 \cdot 4} \cdot \dfrac{3 \cdot 4}{3 \cdot 3 \cdot 3} = \dfrac{1}{6}$

81. $\dfrac{12}{25} + \dfrac{9}{20} = \dfrac{48}{100} + \dfrac{45}{100} = \dfrac{48+45}{100} = \dfrac{93}{100}$

82. $1\dfrac{3}{5} - \dfrac{1}{2} = \dfrac{8}{5} - \dfrac{1}{2} = \dfrac{16}{10} - \dfrac{5}{10} = \dfrac{16-5}{10} = \dfrac{11}{10} = 1\dfrac{1}{10}$

Quick Quiz 0.4

1.
$$\begin{array}{r} 8.0567 \\ -\ 2.3489 \\ \hline 5.7078 \end{array}$$

2.
$$\begin{array}{r} 58.7 \\ \times\,0.06 \\ \hline 3.522 \end{array}$$

3.
$$\begin{array}{r} 28.8 \\ 0.16\wedge\overline{)4.60\wedge 8} \\ \underline{3\ 2} \\ 1\ 40 \\ \underline{1\ 28} \\ 12\ \ 8 \\ \underline{12\ \ 8} \\ 0 \end{array}$$

4. Answers may vary. Possible solution:
Move the decimal point over 4 places to the right on the divisor ($0.0035 \Rightarrow 35$). Then move the decimal point the same number of places on the dividend ($0.252 \Rightarrow 2520$).

Use Math to Save Money

1. $100, $300, $2000, $8000, $8000, $8000, $12,000

2. $3 \times \$25 + \$50 + \$200 + 2 \times \$20$
$= \$75 + \$50 + \$200 + \$40$
$= \$365$

3. The three smallest debts are $100 loan, $300 loan, and $2000 car loan.

4. $20 each, or $40

5. $2000 - $400 = $1600

$\dfrac{\$1600}{\$240} \approx 7$ (rounded to the nearest whole number)

It will take them 7 more months to pay off the third smallest debt.

6. Answers will vary.

How Am I Doing? Sections 0.1–0.4
(Available online through MyMathLab.)

1. $\dfrac{15}{55} = \dfrac{3 \cdot 5}{11 \cdot 5} = \dfrac{3}{11}$

2. $\dfrac{46}{115} = \dfrac{2 \cdot 23}{5 \cdot 23} = \dfrac{2}{5}$

3. $\dfrac{15}{4} = 4\overline{)15} = 3\dfrac{3}{4}$
$\phantom{\dfrac{15}{4} = 4\overline{)}}\underline{12}$
$\phantom{\dfrac{15}{4} = 4\overline{)1}}3$

4. $4\dfrac{5}{7} = \dfrac{4 \cdot 7 + 5}{7} = \dfrac{28 + 5}{7} = \dfrac{33}{7}$

5. $\dfrac{4}{5} = \dfrac{?}{30} \Rightarrow \dfrac{4 \times 6}{5 \times 6} = \dfrac{24}{30}$

6. $\dfrac{5}{9} = \dfrac{?}{81} \Rightarrow \dfrac{5 \times 9}{9 \times 9} = \dfrac{45}{81}$

7. $8 = 2 \cdot 2 \cdot 2$
$6 = 2 \cdot 3$
$15 = 3 \cdot 5$
$LCD = 2 \cdot 2 \cdot 2 \cdot 3 \cdot 5 = 120$

8. $\dfrac{3}{7} + \dfrac{2}{7} = \dfrac{3 + 2}{7} = \dfrac{5}{7}$

9. $\dfrac{5}{14} + \dfrac{2}{21} = \dfrac{5 \cdot 3}{14 \cdot 3} + \dfrac{2 \cdot 2}{21 \cdot 2} = \dfrac{15}{42} + \dfrac{4}{42} = \dfrac{19}{42}$

10. $2\dfrac{3}{4} + 5\dfrac{2}{3} = \dfrac{11}{4} + \dfrac{17}{3}$
$= \dfrac{11 \cdot 3}{4 \cdot 3} + \dfrac{17 \cdot 4}{3 \cdot 4}$
$= \dfrac{33}{12} + \dfrac{68}{12}$
$= \dfrac{101}{12}$
$= 8\dfrac{5}{12}$

11. $\dfrac{17}{18} - \dfrac{5}{9} = \dfrac{17}{18} - \dfrac{5 \cdot 2}{9 \cdot 2} = \dfrac{17}{18} - \dfrac{10}{18} = \dfrac{7}{18}$

12. $\dfrac{6}{7} - \dfrac{2}{3} = \dfrac{6 \cdot 3}{7 \cdot 3} - \dfrac{2 \cdot 7}{3 \cdot 7} = \dfrac{18}{21} - \dfrac{14}{21} = \dfrac{4}{21}$

13. $3\dfrac{1}{5} - 1\dfrac{3}{8} = \dfrac{16}{5} - \dfrac{11}{8}$
$= \dfrac{16 \cdot 8}{5 \cdot 8} - \dfrac{11 \cdot 5}{8 \cdot 5}$
$= \dfrac{128}{40} - \dfrac{55}{40}$
$= \dfrac{73}{40}$
$= 1\dfrac{33}{40}$

14. $\dfrac{25}{7} \times \dfrac{14}{45} = \dfrac{5 \cdot 5}{7} \times \dfrac{2 \cdot 7}{5 \cdot 9} = \dfrac{10}{9} = 1\dfrac{1}{9}$

15. $12 \times 3\dfrac{1}{2} = \dfrac{12}{1} \times \dfrac{7}{2} = \dfrac{6 \cdot 2}{1} \times \dfrac{7}{2} = 42$

16. $4 \div \dfrac{8}{7} = \dfrac{4}{1} \cdot \dfrac{7}{8} = \dfrac{4}{1} \times \dfrac{7}{2 \cdot 4} = \dfrac{7}{2} = 3\dfrac{1}{2}$

17. $2\dfrac{1}{3} \div 3\dfrac{1}{4} = \dfrac{7}{3} \div \dfrac{13}{4} = \dfrac{7}{3} \times \dfrac{4}{13} = \dfrac{28}{39}$

18. $\dfrac{3}{8} = 8\overline{)3.000} = 0.375$

$\dfrac{3}{8} = 0.375$
$\phantom{\dfrac{3}{8} = 8)}\dfrac{2\ 4}{}$
$\phantom{\dfrac{3}{8} = 8)xx}60$
$\phantom{\dfrac{3}{8} = 8)xx}\dfrac{56}{}$
$\phantom{\dfrac{3}{8} = 8)xxx}40$
$\phantom{\dfrac{3}{8} = 8)xxx}\dfrac{40}{}$
$\phantom{\dfrac{3}{8} = 8)xxxx}0$

19. $\dfrac{5}{6} = 6\overline{)}\begin{array}{r}0.833\\5.000\end{array} = 0.8\overline{3}$

$$\begin{array}{r}4\,8\\\hline 20\\18\\\hline 20\\18\\\hline 2\end{array}$$

20. $\dfrac{3}{200} = 200\overline{)}\begin{array}{r}0.015\\3.000\end{array} = 0.015$

$$\begin{array}{r}2\,00\\\hline 1\,000\\1\,000\\\hline 0\end{array}$$

21. $\begin{array}{r}15.230\\3.600\\+\;\;0.821\\\hline 19.651\end{array}$

22. $\begin{array}{r}3.28\\\times\;\;0.63\\\hline 984\\1\,968\\\hline 2.0664\end{array}$

23. $\dfrac{3.015}{6.7} = 6.7\wedge\overline{)}\begin{array}{r}0.45\\3.0\wedge 15\end{array} = 0.45$

$$\begin{array}{r}2\,6\;\;8\\\hline 3\;\;35\\3\;\;35\\\hline 0\end{array}$$

24. $\begin{array}{r}12.130\\-\;\;9.884\\\hline 2.246\end{array}$

0.5 Exercises

1. Answers may vary. Sample answers follow:
   19% means 19 out of 100 parts. Percent means per 100. 19% is really a fraction with a denominator of 100. In this case it would be $\dfrac{19}{100}$.

3. $0.79 = 79\%$

5. $0.568 = 56.8\%$

7. $0.076 = 7.6\%$

9. $2.39 = 239\%$

11. $3.6 = 360\%$

13. $3.672 = 367.2\%$

15. $3\% = 0.03$

17. $0.4\% = 0.004$

19. $250\% = 2.5$

21. $7.4\% = 0.074$

23. $0.52\% = 0.0052$

25. $100\% = 1.00$ or 1

27. 8% of 65
    $0.08 \times 65 = 5.2$

29. 10% of 130
    $0.10 \times 130 = 13$

31. 112% of 65
    $1.12 \times 65 = 72.8$

33. 36 is what percent of 24?
    $\dfrac{36}{24} = 1.50 = 150\%$

35. What percent of 340 is 17?
    $\dfrac{17}{340} = \dfrac{1}{20} = 0.05 = 5\%$

37. 30 is what percent of 500?
    $\dfrac{30}{500} = \dfrac{3}{50} = 0.06 = 6\%$

39. 80 is what percent of 200?
    $\dfrac{80}{200} = \dfrac{2}{5} = 0.4 = 40\%$

41. $\dfrac{68}{80} = 0.85 = 85\%$
    His grade was 85%.

43. $0.15 \times 32.80 = 4.92$
    $32.80 + 4.92 = 37.72$
    The tip is $4.92, and the total bill is $37.72.

45. $\dfrac{380}{1850} \approx 0.21 = 21\%$

About 21% of the budget is for food.

47. $1.5\% \times 36,000 = 0.015 \times 36,000 = 540$
They can expect 540 gifts to be exchanged.

49. a. 3.8% of $780,000 = 0.038 \times 780,000$
$= 29,640$
His sales commission was $29,640.

b. $450 \times 12 + 29,640 = 5400 + 29,640 = 35,040$
His annual salary was $35,040.

51. $586 \times 421 \approx 600 \times 400 = 240,000$

53. $3547 \times 4693 \approx 4000 \times 5000 = 20,000,000$

55. $14 + 73 + 80 + 21 + 56 \approx 10 + 70 + 80 + 20 + 60$
$= 240$

57. $41 \overline{)829,346} \approx 40 \overline{)800,000}^{\,20,000}$

59. $\dfrac{2714}{31,500} \approx \dfrac{3000}{30,000} = 0.1$

61. 17% of $21,365.85 \approx 0.20 \times \$20,000 \approx \$4000$

63. $4 \times 22 \times 82 \approx 4 \times 20 \times 80 = 6400$
The store receives approximately $6400 each hour.

65. $\dfrac{117.7}{3.8} \approx \dfrac{100}{4} = 25$
The car achieves about 25 miles per gallon.

Cumulative Review

67. Distance: $69,229.5 - 68,459.5 = 770$
$\dfrac{770}{35} = 22$
His car achieved 22 miles per gallon.

68. Total $= 4.6 + 4.5 + 2.9 = 12$
Average $= \dfrac{12}{3} = 4.0$ inches per month.

Quick Quiz 0.5

1. 114% of $85 = 1.14 \times 85 = 96.9$

2. 63 is what percent of 420?
$\dfrac{63}{420} = 0.15 = 15\%$

3. $34,987 \overline{)567,238} \approx 30,000 \overline{)600,000}^{\,20}$
$\underline{600\ 00}$
00

4. Answers may vary. Possible solution:
Move the decimal point two places to the right and add the % symbol.

0.6 Exercises

1. Area $= 7\dfrac{1}{3} \times 4\dfrac{1}{3} = \dfrac{22}{3} \times \dfrac{13}{3} = \dfrac{286}{9}$

Square feet $= \dfrac{286}{9} \times \dfrac{9}{1} = 286$

Cost $= 286 \times 4.50 = 1287$
The total cost is $1287.

3. a. Perimeter $= 2 \times 14\dfrac{1}{2} + 2 \times 23\dfrac{1}{2}$
$= 2 \times \dfrac{29}{2} + 2 \times \dfrac{47}{2}$
$= 29 + 47$
$= 76$
They need 76 feet of fencing.

b. Per foot cost $= \$11.20 \times 76 = \$851.20$
90 feet package $= \$859$
He should buy the cut-to-order fencing.
He will save $859 - \$851.20 = \$7.80$.

5. Increase $1\dfrac{1}{3} \times 1\dfrac{1}{3} = \dfrac{4}{3} \times \dfrac{4}{3} = \dfrac{16}{9}$

Jog: $\dfrac{16}{9} \times 1\dfrac{1}{2} = \dfrac{16}{9} \times \dfrac{3}{2} = \dfrac{48}{18} = 2\dfrac{2}{3}$ mi

Walk: $\dfrac{16}{9} \times 1\dfrac{3}{4} = \dfrac{16}{9} \times \dfrac{7}{4} = \dfrac{28}{9} = 3\dfrac{1}{9}$ mi

Rest: $\dfrac{16}{9} \times 2\dfrac{1}{2} = \dfrac{16}{9} \times \dfrac{5}{2} = \dfrac{40}{9} = 4\dfrac{4}{9}$ min

Walk: $\dfrac{16}{9} \times 1 = \dfrac{16}{9} = 1\dfrac{7}{9}$ mi

7. Betty will have a more demanding schedule on day 3 because Melinda increases each activity by $\frac{2}{3}$ by day 3 and Betty increases each activity by $\frac{7}{9}$ by day 3.

9. Increase: $1\frac{1}{2} \times \frac{1}{3} = \frac{3}{2} \times \frac{1}{3} = \frac{1}{2}$

Day 7: $1\frac{1}{2} + \left(6 \times \frac{1}{2}\right) = \frac{3}{2} + \frac{6}{2} = \frac{9}{2} = 4\frac{1}{2}$

She will jog $4\frac{1}{2}$ miles on day 7.

11. a. $\dfrac{6,300,000 \text{ lb/week}}{7 \text{ days/week}} = 900,000 \text{ lb/day}$

In one day, 900,000 pounds of Snickers Bars are produced.

b. $\dfrac{900,000}{6,300,000} = \dfrac{1}{7} \approx 0.1429 = 14.29\%$

About 14.29% of the weekly total is produced in one day.

13. a. Percent available = $100\% - 29\% = 71\%$
$0.71 \times 68,500 = 48,635$
They have \$48,635 available.

b. $0.29 \times 48,635 = 14,104.15$
\$14,104.15 is budgeted for food.

15. $\dfrac{138.97 + 67.76 + 5.18}{1150} = \dfrac{211.91}{1150} \approx 0.18$

About 18% of Fred's gross pay is deducted for federal, state, and local taxes.

17. $\dfrac{790.47}{1150} \approx 0.69 = 69\%$

He actually takes home 69% of his gross pay.

Quick Quiz 0.6

1. $\dfrac{525}{1.5} = 350$ stones

It will require 350 stones to pave the courtyard.

2. 4% of $345,000 = 0.04 \times 345,000 = 13,800$

$\dfrac{13,800}{16,200 + 13,800} = \dfrac{13,800}{30,000} = 0.46 = 46\%$

Her commission was 46% of her total income.

3. $2(40) + 2(95) = 80 + 190 = 270$
$4.50 \times 270 = 1215$
The fence will cost \$1215.

4. Answers may vary. Possible solution: Multiply the number of kilometers by 0.62 to obtain miles.

Career Exploration Problems

1. 90% of the first \$826 = $0.9(\$826) = \$743.40$
32% of earnings between \$826 and \$4980
$= 0.32(\$4200 - \$826)$
$= 0.32(\$3374)$
$= \$1079.68$
She has no earnings over \$4980.
Monthly benefit = $\$743.40 + \$1079.68$
$\qquad\qquad\qquad = \$1823.08$
Cora's monthly benefit will be \$1823.08.

2. $\dfrac{\text{monthly benefit}}{\text{monthly earnings}} = \dfrac{\$1823.08}{\$4200} \approx 0.434$

Cora's benefit is about 43.4% of her pre-retirement earnings.

3. a. The reduction at age 66 is 6.7%.
$100\% - 6.7\% = 93.3\%$ or 0.933
$0.933(\$1823.08) \approx \$1700.93$
Cora's monthly benefit will be \$1700.93.
$\$1823.08 - \$1700.93 = \$122.15$
The total by which it would be reduced is \$122.15.

b. The reduction at age 65 is 13.3%.
$100\% - 13.3\% = 86.7\%$ or 0.867
$0.867(\$1823.08) \approx \$1580.61$
Cora's monthly benefit will be \$1580.61.
$\$1823.08 - \$1580.61 = \$242.47$
The total by which it would be reduced is \$242.47.

You Try It

1. a. $\dfrac{18}{27} = \dfrac{9 \times 2}{9 \times 3} = \dfrac{2}{3}$

b. $\dfrac{45}{60} = \dfrac{15 \times 3}{15 \times 4} = \dfrac{3}{4}$

c. $\dfrac{34}{85} = \dfrac{17 \times 2}{17 \times 5} = \dfrac{2}{5}$

2. a. $\dfrac{21}{5} = 5\overline{)21}^{\,4} = 4\dfrac{1}{5}$
$\phantom{\dfrac{21}{5} = 5)}\underline{20}$
$\phantom{\dfrac{21}{5} = 5)}\,1$

b. $\dfrac{37}{7} = 7\overline{)37}^{\,5} = 5\dfrac{2}{7}$
$\phantom{\dfrac{37}{7} = 7)}\underline{35}$
$\phantom{\dfrac{37}{7} = 7)}\,2$

3. a. $2\dfrac{2}{5} = \dfrac{2\times 5 + 2}{5} = \dfrac{10+2}{5} = \dfrac{12}{5}$

b. $6\dfrac{1}{9} = \dfrac{6\times 9 + 1}{9} = \dfrac{54+1}{9} = \dfrac{55}{9}$

4. $\dfrac{3}{8} = \dfrac{?}{40} \Rightarrow \dfrac{3\times 5}{8\times 5} = \dfrac{15}{40}$

5. a. $10 = 2\times 5$
$12 = 2\times 2\times 3$
$\text{LCD} = 2\times 2\times 3\times 5 = 60$

b. $25 = 5\times 5$
$20 = 2\times 2\times 5$
$\text{LCD} = 2\times 2\times 5\times 5 = 100$

6. a. $\dfrac{1}{2} + \dfrac{1}{9} = \dfrac{1\times 9}{2\times 9} + \dfrac{1\times 2}{9\times 2} = \dfrac{9}{18} + \dfrac{2}{18} = \dfrac{11}{18}$

b. $\dfrac{19}{24} - \dfrac{3}{8} = \dfrac{19}{24} - \dfrac{3\times 3}{8\times 3} = \dfrac{19}{24} - \dfrac{9}{24} = \dfrac{10}{24} = \dfrac{5}{12}$

7. a. $2\dfrac{2}{3} + 3\dfrac{1}{9} = \dfrac{8}{3} + \dfrac{28}{9} = \dfrac{24}{9} + \dfrac{28}{9} = \dfrac{52}{9} = 5\dfrac{7}{9}$

b. $3\dfrac{5}{6} - 1\dfrac{1}{3} = \dfrac{23}{6} - \dfrac{4}{3} = \dfrac{23}{6} - \dfrac{8}{6} = \dfrac{15}{6} = \dfrac{5}{2} = 2\dfrac{1}{2}$

8. a. $\dfrac{5}{11} \times \dfrac{2}{3} = \dfrac{5\times 2}{11\times 3} = \dfrac{10}{33}$

b. $\dfrac{7}{10} \times \dfrac{15}{21} = \dfrac{7}{2\cdot 5} \times \dfrac{3\cdot 5}{3\cdot 7} = \dfrac{1}{2}$

c. $6 \times \dfrac{3}{5} = \dfrac{6}{1} \times \dfrac{3}{5} = \dfrac{18}{5}$ or $3\dfrac{3}{5}$

9. a. $\dfrac{3}{14} \div \dfrac{2}{5} = \dfrac{3}{14} \times \dfrac{5}{2} = \dfrac{15}{28}$

b. $\dfrac{7}{12} \div \dfrac{7}{5} = \dfrac{7}{12} \times \dfrac{5}{7} = \dfrac{5}{12}$

10. a. $1\dfrac{5}{6} \times 2\dfrac{1}{4} = \dfrac{11}{6} \times \dfrac{9}{4} = \dfrac{99}{24} = \dfrac{33}{8}$ or $4\dfrac{1}{8}$

b. $3\dfrac{1}{2} \div 1\dfrac{3}{4} = \dfrac{7}{2} \div \dfrac{7}{4} = \dfrac{7}{2} \times \dfrac{4}{7} = \dfrac{7}{2} \times \dfrac{2\cdot 2}{7} = \dfrac{2}{1} = 2$

11. $\dfrac{7}{8} = 8\overline{)7.000}^{\,0.875} = 0.875$
$\phantom{\dfrac{7}{8} = 8)}\underline{6\ 4}$
$\phantom{\dfrac{7}{8} = 8)7.}\,60$
$\phantom{\dfrac{7}{8} = 8)7.}\underline{56}$
$\phantom{\dfrac{7}{8} = 8)7.0}\,40$
$\phantom{\dfrac{7}{8} = 8)7.0}\underline{40}$
$\phantom{\dfrac{7}{8} = 8)7.00}\,0$

12. a. $0.29 = \dfrac{29}{100}$

b. $0.175 = \dfrac{175}{1000} = \dfrac{7}{40}$

13. a. $\begin{array}{r} 2.338 \\ +\ 6.195 \\ \hline 8.533 \end{array}$

b. $\begin{array}{r} 6.00 \\ -\ 2.54 \\ \hline 3.46 \end{array}$

14. a. $\begin{array}{r} 1.5 \\ \times\ 0.9 \\ \hline 1.35 \end{array}$

b. $\begin{array}{r} 5.12 \\ \times\ 0.67 \\ \hline 3584 \\ 3\ 072 \\ \hline 3.4304 \end{array}$

15.

$$0.5_\wedge \overline{)9.2_\wedge 5}$$ quotient 18.5

$$\begin{array}{r} 18\,.\,5 \\ 0.5_\wedge\overline{)9.2_\wedge 5} \\ \underline{5} \\ 4\,2 \\ \underline{4\,0} \\ 2\ 5 \\ \underline{2\ 5} \\ 0 \end{array}$$

16. a. $0.52 = 52\%$

   b. $0.008 = 0.8\%$

   c. $1.86 = 186\%$

   d. $0.077 = 7.7\%$

   e. $0.0009 = 0.09\%$

17. a. $28\% = 0.28$

   b. $7.42\% = 0.0742$

   c. $165\% = 1.65$

   d. $0.25\% = 0.0025$

18. 15% of 92
$0.15 \times 92 = 13.8$

19. a. What percent of 12 is 10?
$$\frac{10}{12} = \frac{5}{6} \approx 0.833 = 83.3\%$$

   b. 50 is what percent of 40?
$$\frac{50}{40} = \frac{5}{4} = 1.25 = 125\%$$

20. Area $= 27 \times 11.5 \approx 25 \times 12 = 300$
The room is approximately 300 ft^2.

21. 15 ft = 5 yd
$21\frac{3}{4}$ ft $= 7\frac{1}{4}$ yd
Area $= 5 \times 7\frac{1}{4} = \frac{5}{1} \times \frac{29}{4} = \frac{145}{4} \text{ yd}^2$
Cost $= 4.25 \times \frac{145}{4} \approx 154.06$
The tile will cost \$154.06.

Chapter 0 Review Problems

1. $\dfrac{36}{48} = \dfrac{12 \times 3}{12 \times 4} = \dfrac{3}{4}$

2. $\dfrac{15}{50} = \dfrac{5 \times 3}{5 \times 10} = \dfrac{3}{10}$

3. $\dfrac{36}{82} = \dfrac{18 \times 2}{41 \times 2} = \dfrac{18}{41}$

4. $\dfrac{18}{30} = \dfrac{6 \times 3}{6 \times 5} = \dfrac{3}{5}$

5. $7\dfrac{1}{8} = \dfrac{7 \times 8 + 1}{8} = \dfrac{56 + 1}{8} = \dfrac{57}{8}$

6. $\dfrac{34}{5} = 5\overline{)34} = 6\dfrac{4}{5}$
$$\begin{array}{r} 6 \\ 5\overline{)34} \\ \underline{30} \\ 4 \end{array}$$

7. $\dfrac{80}{3} = 3\overline{)80} = 26\dfrac{2}{3}$
$$\begin{array}{r} 26 \\ 3\overline{)80} \\ \underline{6} \\ 20 \\ \underline{18} \\ 2 \end{array}$$

8. $\dfrac{5}{8} = \dfrac{?}{24} \Rightarrow \dfrac{5 \cdot 3}{8 \cdot 3} = \dfrac{15}{24}$

9. $\dfrac{1}{7} = \dfrac{?}{35} \Rightarrow \dfrac{1 \cdot 5}{7 \cdot 5} = \dfrac{5}{35}$

10. $\dfrac{3}{5} = \dfrac{?}{75} \Rightarrow \dfrac{3 \times 15}{5 \times 15} = \dfrac{45}{75}$

11. $\dfrac{2}{5} = \dfrac{?}{55} \Rightarrow \dfrac{2 \times 11}{5 \times 11} = \dfrac{22}{55}$

12. $\dfrac{3}{5} + \dfrac{1}{4} = \dfrac{3 \times 4}{5 \times 4} + \dfrac{1 \times 5}{4 \times 5} = \dfrac{12}{20} + \dfrac{5}{20} = \dfrac{17}{20}$

13. $\dfrac{7}{12} + \dfrac{5}{8} = \dfrac{7 \times 2}{12 \times 2} + \dfrac{5 \times 3}{8 \times 3} = \dfrac{14}{24} + \dfrac{15}{24} = \dfrac{29}{24} = 1\dfrac{5}{24}$

14. $\dfrac{7}{20} - \dfrac{1}{12} = \dfrac{21}{60} - \dfrac{5}{60} = \dfrac{16}{60} = \dfrac{4}{15}$

15. $\dfrac{7}{10} - \dfrac{4}{15} = \dfrac{7 \cdot 3}{10 \cdot 3} - \dfrac{4 \cdot 2}{5 \cdot 2} = \dfrac{21}{30} - \dfrac{8}{30} = \dfrac{13}{30}$

16. $3\dfrac{1}{6} + 2\dfrac{3}{5} = \dfrac{19}{6} + \dfrac{13}{5} = \dfrac{95}{30} + \dfrac{78}{30} = \dfrac{173}{30} = 5\dfrac{23}{30}$

17. $2\dfrac{7}{10} + 3\dfrac{3}{4} = \dfrac{27}{10} + \dfrac{15}{4} = \dfrac{54}{20} + \dfrac{75}{20} = \dfrac{129}{20} = 6\dfrac{9}{20}$

18. $6\dfrac{2}{9} - 3\dfrac{5}{12} = \dfrac{56}{9} - \dfrac{41}{12}$

$\qquad = \dfrac{56 \cdot 4}{9 \cdot 4} - \dfrac{41 \cdot 3}{12 \cdot 3}$

$\qquad = \dfrac{224}{36} - \dfrac{123}{36}$

$\qquad = \dfrac{101}{36}$

$\qquad = 2\dfrac{29}{36}$

19. $3\dfrac{1}{15} - 1\dfrac{3}{20} = \dfrac{46}{15} - \dfrac{23}{20}$

$\qquad = \dfrac{184}{60} - \dfrac{69}{60}$

$\qquad = \dfrac{115}{60}$

$\qquad = 1\dfrac{55}{60}$

$\qquad = 1\dfrac{11}{12}$

20. $6 \times \dfrac{5}{11} = \dfrac{6}{1} \times \dfrac{5}{11} = \dfrac{30}{11} = 2\dfrac{8}{11}$

21. $2\dfrac{1}{3} \times 4\dfrac{1}{2} = \dfrac{7}{3} \times \dfrac{9}{2} = \dfrac{7}{3} \times \dfrac{3 \cdot 3}{2} = \dfrac{21}{2} = 10\dfrac{1}{2}$

22. $16 \times 3\dfrac{1}{8} = \dfrac{16}{1} \times \dfrac{25}{8} = \dfrac{8 \cdot 2}{1} \times \dfrac{25}{8} = 50$

23. $\dfrac{4}{7} \times 5 = \dfrac{4}{7} \times \dfrac{5}{1} = \dfrac{20}{7} = 2\dfrac{6}{7}$

24. $\dfrac{3}{8} \div 6 = \dfrac{3}{8} \times \dfrac{1}{6} = \dfrac{3}{8} \times \dfrac{1}{2 \cdot 3} = \dfrac{1}{16}$

25. $\dfrac{\frac{8}{3}}{\frac{5}{9}} = \dfrac{8}{3} \div \dfrac{5}{9} = \dfrac{8}{3} \times \dfrac{9}{5} = \dfrac{8}{3} \times \dfrac{3 \cdot 3}{5} = \dfrac{24}{5} = 4\dfrac{4}{5}$

26. $\dfrac{15}{16} \div 6\dfrac{1}{4} = \dfrac{15}{16} \div \dfrac{25}{4} = \dfrac{15}{16} \times \dfrac{4}{25} = \dfrac{3}{4} \times \dfrac{1}{5} = \dfrac{3}{20}$

27. $2\dfrac{6}{7} \div \dfrac{10}{21} = \dfrac{20}{7} \times \dfrac{21}{10} = \dfrac{2 \cdot 10}{7} \times \dfrac{3 \cdot 7}{10} = 6$

28.
$$\begin{array}{r} 1.634 \\ 3.007 \\ + \ 2.560 \\ \hline 7.201 \end{array}$$

29.
$$\begin{array}{r} 24.831 \\ - \ 17.094 \\ \hline 7.737 \end{array}$$

30.
$$\begin{array}{r} 47.251 \\ - \ 17.690 \\ \hline 29.561 \end{array}$$

31.
$$\begin{array}{r} 1.900 \\ 2.530 \\ + \ 0.006 \\ \hline 4.436 \end{array}$$

32.
$$\begin{array}{r} 5.35 \\ \times \ \ 0.007 \\ \hline 0.03745 \end{array}$$

33. $362.341 \times 1000 = 362{,}341$

34. $2.6 \times 0.03 \times 1.02 = 0.07956$

35. $2.51 \times 100 \times 0.05 = 125.5$

36. $71.32 \div 1000 = 0.07132$

37. $0.523 \div 0.4 = 1.3075$

38.
$$0.015_\wedge \overline{)1.350_\wedge} \begin{array}{c} 90 \\ \end{array}$$
$$\begin{array}{r} 1\ 35 \\ \hline 00 \end{array}$$

Chapter 0: *Prealgebra Review*

SSM: Beginning and Intermediate Algebra

39. $\dfrac{4.186}{2.3} = 2.3 \wedge \overline{)4.1 \wedge 86} = 1.82$

$$\begin{array}{r} 1.82 \\ \hline 4.1\wedge 86 \\ 2\ 3 \\ \hline 1\ 8\ \ 8 \\ 1\ 8\ \ 4 \\ \hline 4\ 6 \\ 4\ 6 \\ \hline 0 \end{array}$$

40. $\dfrac{3}{8} = 8\overline{)3.000} = 0.375$

$$\begin{array}{r} 0.375 \\ \hline 3.000 \\ 2\ 4 \\ \hline 6\ 0 \\ 5\ 6 \\ \hline 4\ 0 \\ 4\ 0 \\ \hline 0 \end{array}$$

41. $0.36 = \dfrac{36}{100} = \dfrac{9}{25}$

42. $1.4\% = 0.014$

43. $36.1\% = 0.361$

44. $0.02\% = 0.0002$

45. $125.3\% = 1.253$

46. $0.0025 = 0.25\%$

47. $0.325 = 32.5\%$

48. $0.9 = 90\%$

49. $0.1 = 10\%$

50. 30% of 400
$0.30 \times 400 = 120$

51. 7.2% of 55
$0.072 \times 55 = 3.96$

52. 76 is what percent of 80?
$\dfrac{76}{80} = 0.95 = 95\%$

53. $\dfrac{750}{1250} = 0.6 = 60\%$

54. 80% of 16,850
$0.8 \times 16,850 = 13,480$
13,480 students have cell phones.

55. $\dfrac{720}{960} = 0.75 = 75\%$
75% of the class had a math deficiency.

56. $234,897 \times 1,936,112 \approx 200,000 \times 2,000,000$
$\qquad\qquad = 400,000,000,000$

57. $357 + 923 + 768 + 417 \approx 400 + 900 + 800 + 400$
$\qquad\qquad\qquad = 2500$

58. $634,318 - 284,000 \approx 600,000 - 300,000$
$\qquad\qquad\qquad \approx 300,000$

59. $21\dfrac{1}{5} - 8\dfrac{4}{5} - 1\dfrac{2}{3} \approx 20 - 9 - 2 = 9$

60. 18% of $56,297 \approx 0.2 \times 60,000 = \$12,000$

61. $12,482 \div 389 \approx 10,000 \div 400 = 25$

62. $38.5 \times 8.35 \approx 40 \times 8 = \$320$
Her estimated salary is $320.

63. $\dfrac{3875}{5} \approx \dfrac{4000}{5} = \$800$
Each family owes about $800.

64. Maximum distance $= 7\dfrac{2}{3} \times 240$
$\qquad\qquad\qquad = \dfrac{23}{3} \times 240$
$\qquad\qquad\qquad = 1840$ miles
The plane can fly 1840 miles.
Longest trip $= 80\% \times 1840$
$\qquad\qquad = 0.8 \times 1840$
$\qquad\qquad = 1472$ miles
The longest trip he would take is 1472 miles.

65. Distance at maximum speed:
$6\dfrac{1}{4} \times 240 = \dfrac{25}{4} \times 240 = 1500$
The plane can fly 1500 miles.
$0.70 \times 1500 = 1050$
The longest trip he would take is 1050 miles.

66. $Area = 12\frac{1}{2} \times 9\frac{2}{3}$

$= \frac{25}{2} \times \frac{29}{3}$

$= 120\frac{5}{6}$ sq feet

$= \frac{120\frac{5}{6}}{9}$

$= 13\frac{23}{54}$ sq yards

$Cost = 26.00 \times 13\frac{23}{54} = 26 \times \frac{725}{54} = \$349.07$

The total cost is \$349.07.

67. 8% of 5785

$0.08 \times 5785 = \$462.80$

His commission was \$462.80.

68. Amount paid $= 225 \times 4 \times 12 = \$10,800$

Interest $= 10,800 - 9214.50 = \$1585.50$

They paid \$1585.50 more than the car loan.

69. Regular pay $= 7.50 \times 40 = 300$

O/T hours $= 49 - 40 = 9$

O/T pay rate $= 1.5 \times 7.5 = 11.25$

O/T pay $= 11.25 \times 9 = 101.25$

Total pay $= 300 + 101.25 = \$401.25$

He was paid a total of \$401.25.

How Am I Doing? Chapter 0 Test

1. $\frac{16}{18} = \frac{2 \times 8}{2 \times 9} = \frac{8}{9}$

2. $\frac{48}{36} = \frac{4 \times 12}{3 \times 12} = \frac{4}{3}$

3. $6\frac{3}{7} = \frac{6 \times 7 + 3}{7} = \frac{45}{7}$

4. $\frac{105}{9} = 9)\overline{105} = 11\frac{6}{9} = 11\frac{2}{3}$

$\quad\quad\quad\;\; \dfrac{11}{}$
$\quad\quad\quad \dfrac{9}{15}$
$\quad\quad\quad\;\; \dfrac{9}{6}$

5. $\frac{2}{3} + \frac{5}{6} + \frac{3}{8} = \frac{2 \cdot 8}{3 \cdot 8} + \frac{5 \cdot 4}{6 \cdot 4} + \frac{3 \cdot 3}{8 \cdot 3}$

$= \frac{16}{24} + \frac{20}{24} + \frac{9}{24}$

$= \frac{45}{24}$

$= \frac{15}{8}$ or $1\frac{7}{8}$

6. $1\frac{1}{8} + 3\frac{3}{4} = \frac{9}{8} + \frac{15}{4} = \frac{9}{8} + \frac{30}{8} = \frac{9 + 30}{8} = \frac{39}{8} = 4\frac{7}{8}$

7. $3\frac{2}{3} - 2\frac{5}{6} = \frac{11}{3} - \frac{17}{6} = \frac{22}{6} - \frac{17}{6} = \frac{5}{6}$

8. $\frac{5}{7} \times \frac{28}{15} = \frac{1}{1} \times \frac{4}{3} = \frac{4}{3}$ or $1\frac{1}{3}$

9. $\frac{7}{4} \div \frac{1}{2} = \frac{7}{4} \times \frac{2}{1} = \frac{7}{2} \times \frac{1}{1} = \frac{7}{2}$ or $3\frac{1}{2}$

10. $5\frac{3}{8} \div 2\frac{3}{4} = \frac{43}{8} \div \frac{11}{4}$

$= \frac{43}{2 \cdot 4} \times \frac{4}{11}$

$= \frac{43}{22}$

$= 1\frac{21}{22}$

11. $2\frac{1}{2} \times 3\frac{1}{4} = \frac{5}{2} \times \frac{13}{4} = \frac{65}{8} = 8\frac{1}{8}$

12. $\frac{\frac{7}{8}}{\frac{1}{4}} = \frac{7}{8} \div \frac{1}{4} = \frac{7}{8} \times \frac{4}{1} = \frac{7}{2} \times \frac{1}{1} = \frac{7}{2}$ or $3\frac{1}{2}$

13.
$$\begin{array}{r} 1.60 \\ 3.24 \\ + 9.80 \\ \hline 14.64 \end{array}$$

14.
$$\begin{array}{r} 7.0046 \\ - 3.0149 \\ \hline 3.9897 \end{array}$$

15.
$$\begin{array}{r} 32.8 \\ \times\; 0.04 \\ \hline 1.312 \end{array}$$

16. $0.07385 \times 1000 = 73.85$

17.

$$
0.056 \!\wedge\! \overline{)12.880\wedge} \quad \begin{array}{r} 230 \\ \hline 11\ 2 \\ \hline 1\ 68 \\ 1\ 68 \\ \hline 0 \end{array}
$$

18. $26{,}325.9 \div 100 = 263.259$

19. $0.073 = 7.3\%$

20. $196.5\% = 1.965$

21. What is 3.5% of 180?
$0.035 \times 180 = 6.3$

22. 39 is what percent of 650?
$\dfrac{39}{650} = 0.06 = 6\%$

23. $4 \div \dfrac{2}{9} = \dfrac{4}{1} \times \dfrac{9}{2} = \dfrac{2}{1} \times \dfrac{9}{1} = 18$
18 chips are in the stack.

24. $52{,}344\overline{)4{,}678{,}987} \approx 50{,}000\overline{)5{,}000{,}000}$ with quotient 100

25. $285.36 + 311.85 + 113.6 \approx 300 + 300 + 100$
$= 700$

26. Commission: $(0.03)(870{,}000) = 26{,}100$
Total income: $26{,}100 + 14{,}000 = 40{,}100$
Percent: $\dfrac{26{,}100}{40{,}100} \approx 0.65 = 65\%$
Her commission was 65% of her total income.

27. $210 \div 3\dfrac{1}{2} = \dfrac{210}{1} \div \dfrac{7}{2}$
$= \dfrac{210}{1} \times \dfrac{2}{7}$
$= \dfrac{30 \times 7}{1} \times \dfrac{2}{7}$
$= 60$
They will need 60 tiles.

Chapter 1

	Number	Int	RaN	IR	ReN
1.	23	X	X		X
3.	π			X	X
5.	$-6.666\ldots$		X		X
7.	$-2.3434\ldots$		X		X
9.	$\sqrt{2}$			X	X

11. 20,000 leagues under the sea $\Rightarrow -20{,}000$

13. Lost $37\frac{1}{2} \Rightarrow -37\frac{1}{2}$

15. Rise $7° \Rightarrow +7$

17. Additive inverse of 8 is -8.

19. Opposite of -2.73 is 2.73.

21. $|-1.3| = 1.3$

23. $\left|\dfrac{5}{6}\right| = \dfrac{5}{6}$

25. $-8 + (-7) = -15$

27. $-20 + (-30) = -50$

29. $-\dfrac{7}{20} + \dfrac{13}{20} = \dfrac{6}{20} = \dfrac{3}{10}$

31. $-\dfrac{2}{13} + \left(-\dfrac{5}{13}\right) = -\dfrac{7}{13}$

33. $-\dfrac{2}{5} + \dfrac{3}{7} = -\dfrac{14}{35} + \dfrac{15}{35} = \dfrac{1}{35}$

35. $-10.3 + (-8.9) = -19.2$

37. $0.6 + (-0.2) = 0.4$

39. $-5.26 + (-8.9) = -14.16$

41. $-8 + 5 + (-3) = -3 + (-3) = -6$

43. $-2 + (-8) + 10 = -10 + 10 = 0$

45. $-\dfrac{3}{10} + \dfrac{3}{4} = -\dfrac{6}{20} + \dfrac{15}{20} = \dfrac{9}{20}$

47. $-14 + 9 + (-3) = -5 + (-3) = -8$

49. $8 + (-11) = -3$

51. $-83 + 142 = 59$

53. $-\dfrac{4}{9} + \dfrac{5}{6} = -\dfrac{8}{18} + \dfrac{15}{18} = \dfrac{7}{18}$

55. $-\dfrac{1}{10} + \dfrac{1}{2} = -\dfrac{1}{10} + \dfrac{5}{10} = \dfrac{4}{10} = \dfrac{2}{5}$

57. $5.18 + (-7.39) = -2.21$

59. $4 + (-8) + 16 = -4 + 16 = 12$

61. $26 + (-19) + 12 + (-31) = 7 + 12 + (-31)$
$= 19 + (-31)$
$= -12$

63. $17.85 + (-2.06) + 0.15 = 15.79 + 0.15 = 15.94$

65. Profit $= 214 - 47 = 167$
Her profit was $167.

67. $-2300 + (-1500) = -3800$
He owed $-$3800.

69. $-15 + 3 + 21 = -12 + 21 = 9$
His total was a 9-yard gain.

71. $8000 + (-3000) + (-1500) = 5000 + (-1500)$
$= 3500$
The new population was 3500.

73. $30 + 14 + (-16) = 44 + (-16) = 28$
The total earnings were $28,000,000.

75. $-13 + ? = 5$
$-13 + 18 = 5$
$? = 18$

Cumulative Review

77. $\dfrac{15}{16} + \dfrac{1}{4} = \dfrac{15}{16} + \dfrac{4}{16} = \dfrac{19}{16}$ or $1\dfrac{3}{16}$

78. $\dfrac{3}{7} \times \dfrac{14}{9} = \dfrac{3 \cdot 14}{7 \cdot 9} = \dfrac{3 \cdot 7 \cdot 2}{7 \cdot 3 \cdot 3} = \dfrac{2}{3}$

79. $\dfrac{2}{15} - \dfrac{1}{20} = \dfrac{8}{60} - \dfrac{3}{60} = \dfrac{5}{60} = \dfrac{1}{12}$

80. $2\dfrac{1}{2} \div 3\dfrac{2}{5} = \dfrac{5}{2} \div \dfrac{17}{5} = \dfrac{5}{2} \cdot \dfrac{5}{17} = \dfrac{25}{34}$

81. $0.72 + 0.8 = 1.52$

82. $1.63 - 0.98 = 0.65$

83. $\begin{array}{r} 1.63 \\ \times\ 0.7 \\ \hline 1.141 \end{array}$

84. $0.208 \div 0.8 = 0.26$

Quick Quiz 1.1

1. $-18 + (-16) = -34$

2. $-2.7 + 8.6 + (-5.4) = 5.9 + (-5.4) = 0.5$

3. $-\dfrac{5}{6} + \dfrac{7}{24} = -\dfrac{20}{24} + \dfrac{7}{24} = -\dfrac{13}{24}$

4. Answers may vary. Possible solution: Adding a negative number to a negative number will always result in a number further in the negative direction. However, adding numbers of opposite sign could result in a negative number if the absolute value of the negative number is larger than that of the positive number, or a positive number, if the absolute value of the positive number is greater than that of the negative number, or 0 if their absolute values are equal.

1.2 Exercises

1. First change subtracting −3 to adding a positive three. Then use the rules for addition of two real numbers with different signs. Thus, $-8 - (-3) = -8 + 3 = -5$

3. $27 - 49 = 27 + (-49) = -22$

5. $19 - 23 = 19 + (-23) = -4$

7. $-14 - (-3) = -14 + 3 = -11$

9. $-52 - (-60) = -52 + 60 = 8$

11. $0 - (-5) = 0 + 5 = 5$

13. $-18 - (-18) = -18 + 18 = 0$

15. $-17 - (-20) = -17 + 20 = 3$

17. $\dfrac{2}{5} - \dfrac{4}{5} = \dfrac{2}{5} + \left(-\dfrac{4}{5}\right) = -\dfrac{2}{5}$

19. $\dfrac{3}{4} - \left(-\dfrac{3}{5}\right) = \dfrac{15}{20} + \left(+\dfrac{12}{20}\right) = \dfrac{27}{20}$ or $1\dfrac{7}{20}$

21. $-\dfrac{3}{4} - \dfrac{5}{6} = -\dfrac{9}{12} + \left(-\dfrac{10}{12}\right) = -\dfrac{19}{12}$ or $-1\dfrac{7}{12}$

23. $-0.6 - 0.3 = -0.6 + (-0.3) = -0.9$

25. $2.64 - (-1.83) = 2.64 + 1.83 = 4.47$

27. $\dfrac{3}{5} - 4 = \dfrac{3}{5} + \left(-\dfrac{20}{5}\right) = -\dfrac{17}{5}$ or $-3\dfrac{2}{5}$

29. $-\dfrac{2}{3} - 4 = -\dfrac{2}{3} + \left(-\dfrac{12}{3}\right) = -\dfrac{14}{3}$ or $-4\dfrac{2}{3}$

31. $34 - 87 = 34 + (-87) = -53$

33. $-25 - 48 = -25 + (-48) = -73$

35. $2.3 - (-4.8) = 2.3 + 4.8 = 7.1$

37. $8 - \left(-\dfrac{3}{4}\right) = 8 + \dfrac{3}{4} = 8\dfrac{3}{4}$ or $\dfrac{35}{4}$

39. $\dfrac{5}{6} - 7 = \dfrac{5}{6} + \left(-\dfrac{42}{6}\right) = -\dfrac{37}{6}$ or $-6\dfrac{1}{6}$

41. $-\dfrac{2}{7} - \dfrac{4}{5} = -\dfrac{10}{35} + \left(-\dfrac{28}{35}\right) = -\dfrac{38}{35}$ or $-1\dfrac{3}{35}$

43. $-135 - (-126.5) = -135 + 126.5 = -8.5$

45. $\dfrac{1}{5} - 6 = \dfrac{1}{5} + \left(-\dfrac{30}{5}\right) = -\dfrac{29}{5}$ or $-5\dfrac{4}{5}$

47. $4.5 - (-1.56) = 4.5 + 1.56 = 6.06$

49. $-3 - 2.047 = -3 + (-2.047) = -5.047$

51. $-2 - (-9) = -2 + 9 = 7$

53. $-35 - 13 = -35 + (-13) = -48$

55. $7+(-6)-3 = 7+(-6)+(-3) = 1+(-3) = -2$

57. $-13+12-(-1) = -13+12+1 = -1+1 = 0$

59. $16+(-20)-(-15)-1 = 16+(-20)+15+(-1)$
$$= -4+15+(-1)$$
$$= 11+(-1)$$
$$= 10$$

61. $-7.8-(-5.2)+3.7 = -7.8+5.2+3.7$
$$= -2.6+3.7$$
$$= 1.1$$

63. $600-300+200-(-126)$
$$= 600+(-300)+200+126$$
$$= 300+200+126$$
$$= 500+126$$
$$= 626$$
The helicopter is 626 feet above the submarine.

Cumulative Review

65. $-37+16 = -21$

66. $-37+(-14) = -51$

67. $-3+(-6)+(-10) = -9+(-10) = -19$

68. $-5+20 = 15$
The afternoon temperature was 15°F.

69. $\left(\dfrac{4}{5}\right)\left(8\dfrac{1}{3}\right) = \left(\dfrac{4}{5}\right)\left(\dfrac{25}{3}\right) = \dfrac{20}{3} = 6\dfrac{2}{3}$

There were $6\dfrac{2}{3}$ miles covered in snow.

Quick Quiz 1.2

1. $-8-(-15) = -8+15 = 7$

2. $-1.3-0.6 = -1.3+(-0.6) = -1.9$

3. $\dfrac{5}{8}-\left(-\dfrac{2}{7}\right) = \dfrac{5}{8}+\dfrac{2}{7} = \dfrac{35}{56}+\dfrac{16}{56} = \dfrac{51}{56}$

4. Answers may vary. Possible solution:
The result could be positive, if the absolute value of the number subtracted is greater than the other number $[-2-(-3) = -2+3 = 1]$ or the result could be zero if the two numbers are the same $[-2-(-2) = -2+2 = 0]$, or the result could be negative if the absolute value of the number being subtracted is less than the other number $[-2-(-1) = -2+1 = -1]$.

1.3 Exercises

1. To multiply two real numbers, multiply the absolute values. The sign of the result is positive if both numbers have the same sign, but negative if the two numbers have different signs.

3. $8(-5) = -40$

5. $0(-12) = 0$

7. $14(3.5) = 49$

9. $(-1.32)(-0.2) = 0.264$

11. $1.8(-2.5) = -4.5$

13. $\left(\dfrac{3}{8}\right)(-4) = \left(\dfrac{3}{8}\right)\left(-\dfrac{4}{1}\right) = -\dfrac{3}{2}$ or $-1\dfrac{1}{2}$

15. $\left(-\dfrac{3}{5}\right)\left(-\dfrac{15}{11}\right) = \dfrac{9}{11}$

17. $\left(\dfrac{12}{13}\right)\left(\dfrac{-5}{24}\right) = -\dfrac{5}{26}$

19. $0 \div (-9) = 0$

21. $-48 \div (-8) = \dfrac{-48}{-8} = 6$

23. $-120 \div (-8) = 15$

25. $156 \div (-13) = -12$

27. $-9.1 \div 0.07 = -\dfrac{9.1}{0.07} = -130$

29. $0.54 \div (-0.9) = \dfrac{0.54}{-0.9} = -0.6$

31. $-6.3 \div 7 = \dfrac{-6.3}{7} = -0.9$

33. $\left(-\dfrac{1}{5}\right) \div \left(\dfrac{2}{3}\right) = \left(-\dfrac{1}{5}\right)\left(\dfrac{3}{2}\right) = -\dfrac{3}{10}$

35. $\left(-\dfrac{5}{7}\right) \div \left(-\dfrac{3}{28}\right) = \dfrac{5}{7}\cdot\dfrac{28}{3} = \dfrac{20}{3}$ or $6\dfrac{2}{3}$

37. $-\dfrac{7}{12} \div \left(-\dfrac{5}{6}\right) = \left(\dfrac{7}{12}\right)\left(\dfrac{6}{5}\right) = \dfrac{7}{10}$

39. $\dfrac{-6}{-\frac{3}{7}} = \left(-\dfrac{6}{1}\right)\left(-\dfrac{7}{3}\right) = 14$

41. $\dfrac{-\frac{2}{3}}{\frac{8}{15}} = \left(-\dfrac{2}{3}\right)\left(\dfrac{15}{8}\right) = -\dfrac{5}{4}$ or $-1\dfrac{1}{4}$

43. $\dfrac{\frac{8}{3}}{-4} = \dfrac{8}{3} \cdot \left(-\dfrac{1}{4}\right) = -\dfrac{8}{12} = -\dfrac{2}{3}$

45. $(-1)(-2)(-3)(4) = -(1)(2)(3)(4) = -24$

47. $-2(4)(3)(-1)(-3) = -72$

49. $(-3)(0)\left(\dfrac{1}{3}\right)(-4)(2) = 0$

51. $25(-0.04)(-0.3)(-1) = -1(-0.3)(-1)$
$\qquad\qquad\qquad\qquad = 0.3(-1)$
$\qquad\qquad\qquad\qquad = -0.3$

53. $\left(-\dfrac{4}{5}\right)\left(-\dfrac{6}{7}\right)\left(-\dfrac{1}{3}\right) = -\left(\dfrac{24}{35}\right)\left(\dfrac{1}{3}\right) = -\dfrac{8}{35}$

55. $\left(-\dfrac{3}{4}\right)\left(-\dfrac{7}{15}\right)\left(-\dfrac{8}{21}\right)\left(-\dfrac{5}{9}\right)$
$= +\left(\dfrac{3}{4}\right)\left(\dfrac{7}{3\cdot5}\right)\left(\dfrac{4\cdot2}{3\cdot7}\right)\left(\dfrac{5}{9}\right)$
$= \dfrac{2}{27}$

57. $-36 \div (-4) = \dfrac{-36}{-4} = 9$

59. $12 + (-8) = 4$

61. $8 - (-9) = 8 + 9 = 17$

63. $6(-12) = -72$

65. $-37 \div 37 = \dfrac{-37}{37} = -1$

67. $17.60 \div 4 = \dfrac{17.60}{4} = 4.40$
He gave \$4.40 to each person and to himself.

69. $\dfrac{14{,}136}{60} = 235.60$
Her monthly bill is \$235.60.

71. $5(4) = +20$, gained 20 yards in small gains.

73. $-10(7) = -70$, lost 70 yards in medium losses.

75. Total $= -70 + 90 = 20$, gained 20 yards

Cumulative Review

76. $-17.4 + 8.31 + 2.40 = -9.09 + 2.40 = -6.69$

77. $-\dfrac{3}{4} + \left(-\dfrac{2}{3}\right) + \left(-\dfrac{5}{12}\right) = -\dfrac{9}{12} + \left(-\dfrac{8}{12}\right) + \left(-\dfrac{5}{12}\right)$
$\qquad\qquad\qquad\qquad\qquad = -\dfrac{17}{12} + \left(-\dfrac{5}{12}\right)$
$\qquad\qquad\qquad\qquad\qquad = -\dfrac{22}{12}$
$\qquad\qquad\qquad\qquad\qquad = -\dfrac{11}{6}$ or $-1\dfrac{5}{6}$

78. $-47 - (-32) = -47 + 32 = -15$

79. $-37 - 51 = -37 + (-51) = -88$

Quick Quiz 1.3

1. $\left(-\dfrac{3}{8}\right)(5) = -\dfrac{3}{8} \cdot \dfrac{5}{1} = -\dfrac{15}{8}$ or $-1\dfrac{7}{8}$

2. $-4(3)(-5)(-2) = -12(-5)(-2) = 60(-2) = -120$

3. $-2.4 \div (-0.6) = \dfrac{-2.4}{-0.6} = 4$

4. Answers may vary. Possible solution: Multiplying an even number of negative numbers results in a positive number, whereas multiplying an odd number of negative numbers results in a negative number.

1.4 Exercises

1. The base is 3 and the exponent is 4. Thus you multiply $(3)(3)(3)(3) = 81$.

3. The answer is negative. When you raise a negative number to an odd power the result is always negative.

5. If you have parentheses surrounding the -2, then the base is -2 and the exponent is 4. The result is 16. Without parentheses, the base is 2. You evaluate to obtain 16 and then take the opposite of 16, which is -16. Thus, $(-2)^4 = 16$, but $-2^4 = -16$.

7. $(6) \cdot (6) \cdot (6) \cdot (6) \cdot (6) = 6^5$

9. $(w)(w) = w^2$

11. $(p) \cdot (p) \cdot (p) \cdot (p) = p^4$

13. $(3q) \cdot (3q) \cdot (3q) = (3q)^3 \text{ or } 3^3 q^3$

15. $3^3 = 27$

17. $3^4 = 81$

19. $6^3 = 216$

21. $(-3)^3 = (-3)(-3)(-3) = -27$

23. $(-4)^2 = (-4)(-4) = 16$

25. $-5^2 = -(5)(5) = -25$

27. $\left(\dfrac{1}{4}\right)^2 = \left(\dfrac{1}{4}\right)\left(\dfrac{1}{4}\right) = \dfrac{1}{16}$

29. $\left(\dfrac{2}{5}\right)^3 = \left(\dfrac{2}{5}\right)\left(\dfrac{2}{5}\right)\left(\dfrac{2}{5}\right) = \dfrac{8}{125}$

31. $(2.1)^2 = (2.1)(2.1) = 4.41$

33. $(0.2)^5 = (0.2)(0.2)(0.2)(0.2)(0.2) = 0.00032$

35. $(-16)^2 = (-16)(-16) = 256$

37. $-16^2 = -(16)(16) = -256$

39. $5^3 + 6^2 = 125 + 36 = 161$

41. $10^2 - 11^2 = 100 - 121 = -21$

43. $(-4)^2 - (12)^2 = 16 - 144 = -128$

45. $2^5 - (-3)^2 = 32 - 9 = 23$

47. $(-5)^3 (-2)^3 = (-125)(-8) = 1000$

Cumulative Review

48. $\begin{aligned}(-11) + (-13) + 6 + (-9) + 8 &= -24 + 6 + (-9) + 8 \\ &= -18 + (-9) + 8 \\ &= -27 + 8 \\ &= -19\end{aligned}$

49. $\begin{aligned}\dfrac{3}{4} \div \left(-\dfrac{9}{20}\right) &= \left(\dfrac{3}{4}\right)\left(-\dfrac{20}{9}\right) \\ &= \left(\dfrac{3}{4}\right)\left(-\dfrac{4 \cdot 5}{3 \cdot 3}\right) \\ &= -\dfrac{5}{3} \text{ or } -1\dfrac{2}{3}\end{aligned}$

50. $-17 - (-9) = -17 + 9 = -8$

51. $(-2.1)(-1.2) = 2.52$

52. 6% of $1600 = 0.06 \times 1600 = 96$
$1600 + 96 = 1696$
She has $1696 at the end of the year.

Quick Quiz 1.4

1. $(-4)^4 = (-4)(-4)(-4)(-4) = 256$

2. $(1.8)^2 = (1.8)(1.8) = 3.24$

3. $\left(\dfrac{3}{4}\right)^3 = \left(\dfrac{3}{4}\right)\left(\dfrac{3}{4}\right)\left(\dfrac{3}{4}\right) = \dfrac{27}{64}$

4. Answers may vary. Possible solution:
If you have parentheses surrounding the -2, then the base is -2 and the exponent is 6. The result is 64. If you do not have parentheses, then the base is 2. You evaluate to obtain 64 and then take the opposite of 64, which is -64. Thus, $(-2)^6 = 64$ and $-2^6 = -64$.

1.5 Exercises

1. $3(4) + 6(5)$

3. a. $3(4) + 6(5) = 12 + 6(5) = 18(5) = 90$

   b. $3(4) + 6(5) = 12 + 30 = 42$

5. $\begin{aligned}(7 - 9)^2 \div 2 \times 5 &= (-2)^2 \div 2 \times 5 \\ &= 4 \div 2 \times 5 \\ &= 2 \times 5 \\ &= 10\end{aligned}$

7. $9 + 4(5 + 2 - 8) = 9 + 4(-1) = 9 + (-4) = 5$

9. $8 - 2^3 \cdot 5 + 3 = 8 - 8 \cdot 5 + 3$
$\qquad = 8 - 8 \cdot 5 + 3$
$\qquad = 8 - 40 + 3$
$\qquad = -29$

11. $4 + 42 \div 3 \cdot 2 - 8 = 4 + 14 \cdot 2 - 8$
$\qquad = 4 + 28 - 8$
$\qquad = 32 - 8$
$\qquad = 24$

13. $3 \cdot 5 + 7 \cdot 3 - 5 \cdot 3 = 15 + 21 - 15 = 21$

15. $8 - 5(2)^3 \div (-8) = 8 - 5(8) \div (-8)$
$\qquad = 8 - 40 \div (-8)$
$\qquad = 8 + 5$
$\qquad = 13$

17. $3(5 - 7)^2 - 6(3) = 3(-2)^2 - 6(3)$
$\qquad = 3(4) - 6(3)$
$\qquad = 12 - 18$
$\qquad = -6$

19. $5 \cdot 6 - (3 - 5)^2 + 8 \cdot 2 = 5 \cdot 6 - (-2)^2 + 8 \cdot 2$
$\qquad = 5 \cdot 6 - 4 + 8 \cdot 2$
$\qquad = 30 - 4 + 16$
$\qquad = 42$

21. $\dfrac{1}{2} \div \dfrac{2}{3} + 6 \cdot \dfrac{1}{4} = \dfrac{1}{2} \cdot \dfrac{3}{2} + \dfrac{6}{1} \cdot \dfrac{1}{4}$
$\qquad = \dfrac{3}{4} + \dfrac{6}{4}$
$\qquad = \dfrac{9}{4}$ or $2\dfrac{1}{4}$

23. $0.8 + 0.3(0.6 - 0.2)^2 = 0.8 + 0.3(0.4)^2$
$\qquad = 0.8 + 0.3(0.16)$
$\qquad = 0.8 + 0.048$
$\qquad = 0.848$

25. $\dfrac{3}{8}\left(-\dfrac{1}{6}\right) - \dfrac{7}{8} + \dfrac{1}{2} = -\dfrac{3}{48} - \dfrac{7}{8} + \dfrac{1}{2}$
$\qquad = -\dfrac{1}{16} - \dfrac{14}{16} + \dfrac{8}{16}$
$\qquad = -\dfrac{15}{16} + \dfrac{8}{16}$
$\qquad = -\dfrac{7}{16}$

27. $(3 - 7)^2 \div 8 + 3 = (-4)^2 \div 8 + 3$
$\qquad = 16 \div 8 + 3$
$\qquad = 2 + 3$
$\qquad = 5$

29. $\left(\dfrac{3}{4}\right)^2 (-16) + \dfrac{4}{5} \div \dfrac{-8}{25} = \dfrac{9}{16}(-16) + \dfrac{4}{5} \div \dfrac{-8}{25}$
$\qquad = -9 + \dfrac{4}{5} \div \dfrac{-8}{25}$
$\qquad = -9 + \dfrac{4}{5}\left(\dfrac{25}{-8}\right)$
$\qquad = -9 + \left(\dfrac{5}{-2}\right)$
$\qquad = -\dfrac{18}{2} + \left(-\dfrac{5}{2}\right)$
$\qquad = -\dfrac{23}{2}$ or $-11\dfrac{1}{2}$

31. $-2.6 - (-1.8)(2.3) + (4.1)^2$
$\qquad = -2.6 - (-1.8)(2.3) + 16.81$
$\qquad = -2.6 + 4.14 + 16.81$
$\qquad = 18.35$

33. $\left(\dfrac{2}{3}\right)^2 + \dfrac{1}{2} - \left(\dfrac{1}{3} - \dfrac{3}{4}\right) + \left(-\dfrac{1}{2}\right)^2$
$\qquad = \left(\dfrac{2}{3}\right)^2 + \dfrac{1}{2} - \left(-\dfrac{5}{12}\right) + \left(-\dfrac{1}{2}\right)^2$
$\qquad = \dfrac{4}{9} + \dfrac{1}{2} + \dfrac{5}{12} + \left(\dfrac{1}{4}\right)$
$\qquad = \dfrac{16}{36} + \dfrac{18}{36} + \dfrac{15}{36} + \dfrac{9}{36}$
$\qquad = \dfrac{58}{36}$
$\qquad = \dfrac{29}{18}$ or $1\dfrac{11}{18}$

35. $1(-2) + 5(-1) + 10(0) + 2(+1)$

37. $1(-2) + 5(-1) + 10(0) + 2(+1)$
$\qquad = -2 + 5(-1) + 10(0) + 2(+1)$
$\qquad = 3(-1) + 10(0) + 2(+1)$
$\qquad = -3 + 10(0) + 2(+1)$
$\qquad = 7(0) + 2(+1)$
$\qquad = 0 + 2(+1)$
$\qquad = 2(+1)$
$\qquad = +2$ or 2 above par

Cumulative Review

39. $(0.5)^3 = (0.5)(0.5)(0.5) = 0.125$

40. $-\dfrac{3}{4} - \dfrac{5}{6} = -\dfrac{9}{12} - \dfrac{10}{12} = -\dfrac{19}{12} = -1\dfrac{7}{12}$

41. $-1^{20} = -1$

42. $3\dfrac{3}{5} \div 6\dfrac{1}{4} = \dfrac{18}{5} \div \dfrac{25}{4} = \dfrac{18}{5} \cdot \dfrac{4}{25} = \dfrac{72}{125}$

Quick Quiz 1.5

1. $7 - 3^4 + 2 - 5 = 7 - 81 + 2 - 5$
$$= -74 + 2 - 5$$
$$= -72 - 5$$
$$= -77$$

2. $(0.3)^2 - 4.2(-4) + 0.07$
$$= 0.09 - 4.2(-4) + 0.07$$
$$= 0.09 + 16.8 + 0.07$$
$$= 16.89 + 0.07$$
$$= 16.96$$

3. $(7 - 9)^4 + 22 \div (-2) + 6 = (-2)^4 + 22 \div (-2) + 6$
$$= (16) + 22 \div (-2) + 6$$
$$= 16 + (-11) + 6$$
$$= 5 + 6$$
$$= 11$$

4. Answers may vary. Possible solution:
Evaluate within the parentheses. Then raise that result to the power of 3. Then divide. Finally, evaluate addition and subtraction from left to right.

Use Math to Save Money

1. $(\$2500 \times 0.05) \times 12 = \$1500$

2. $(\$2500 \times 0.15) \times 12 = \$4500$

3. $\$3000 + \$450 = \$3450$

4. $\$2500 \times 0.05 = \$125$
$\$3450 \div \$125 = 27.6 \approx 28$ months
He will need to save for 28 months or 2 years, 4 months.

5. $\$2500 \times 0.10 = \$250$
$\$3450 \div \$250 = 13.8 \approx 14$ months
He will need to save for 14 months or 1 year, 2 months.

6. $\left[2500 + \left(\dfrac{5800}{12}\right)\right] \times 0.05 = \$149.17$ per month

7. $\left[2500 + \left(\dfrac{5800}{12}\right)\right] \times 0.20 = \$596.67$ per month

8. Answers will vary.

9. Answers will vary.

10. Answers will vary.

How Am I Doing? Sections 1.1–1.5
(Available online through MyMathLab.)

1. $3 + (-12) = -9$

2. $-\dfrac{5}{6} + \left(-\dfrac{7}{8}\right) = -\dfrac{20}{24} + \left(-\dfrac{21}{24}\right) = \dfrac{-41}{24} = -1\dfrac{17}{24}$

3. $\begin{array}{r} 0.34 \\ +\ 0.90 \\ \hline 1.24 \end{array}$

4. $-3.5 + 9 + 2.3 + (-3) = 5.5 + 2.3 + (-3)$
$$= 7.8 + (-3)$$
$$= 4.8$$

5. $-23 - (-34) = -23 + 34 = 11$

6. $-\dfrac{1}{6} - \dfrac{4}{5} = -\dfrac{1}{6} + \left(-\dfrac{4}{5}\right) = -\dfrac{5}{30} + \left(-\dfrac{24}{30}\right) = -\dfrac{29}{30}$

7. $4.5 - (-7.8) = 4.5 + 7.8 = 12.3$

8. $-4 - (-5) + 9 = -4 + 5 + 9 = 10$

9. $(-3)(-8)(2)(-2) = 24(2)(-2) = 48(-2) = -96$

10. $\left(-\dfrac{6}{11}\right)\left(-\dfrac{5}{3}\right) = \dfrac{10}{11}$

11. $-0.072 \div 0.08 = \dfrac{-0.072}{0.08} = -0.9$

12. $\dfrac{5}{8} \div \left(-\dfrac{17}{16}\right) = \left(\dfrac{5}{8}\right) \cdot \left(-\dfrac{16}{17}\right) = -\dfrac{10}{17}$

13. $(0.7)^3 = (0.7)(0.7)(0.7) = 0.343$

14. $(-4)^4 = (-4)(-4)(-4)(-4) = 256$

15. $-2^8 = -(2)(2)(2)(2)(2)(2)(2)(2) = -256$

16. $\left(\dfrac{2}{3}\right)^3 = \left(\dfrac{2}{3}\right)\left(\dfrac{2}{3}\right)\left(\dfrac{2}{3}\right) = \dfrac{8}{27}$

17. $-3^3 + 3^4 = -27 + 81 = 54$

18. $\begin{aligned} 20 - 12 \div 3 - 8(-1) &= 20 - 4 - 8(-1) \\ &= 20 - 4 + 8 \\ &= 16 + 8 \\ &= 24 \end{aligned}$

19. $\begin{aligned} 15 + 3 - 2 + (-6) &= 18 + (-2) + (-6) \\ &= 16 + (-6) \\ &= 10 \end{aligned}$

20. $\begin{aligned} (9 - 13)^2 + 15 \div (-3) &= (-4)^2 + 15 \div (-3) \\ &= 16 + 15 \div (-3) \\ &= 16 + (-5) \\ &= 11 \end{aligned}$

21. $\begin{aligned} &-0.12 \div 0.6 + (-3)(1.2) - (-0.5) \\ &= -0.2 + (-3)(1.2) + 0.5 \\ &= -0.2 + (-3.6) + 0.5 \\ &= -3.8 + 0.5 \\ &= -3.3 \end{aligned}$

22. $\begin{aligned} &\left(\dfrac{3}{4}\right)\left(-\dfrac{2}{5}\right) + \left(-\dfrac{1}{2}\right)\left(\dfrac{4}{5}\right) + \left(\dfrac{1}{2}\right)^2 \\ &= \left(\dfrac{3}{4}\right)\left(-\dfrac{2}{5}\right) + \left(-\dfrac{1}{2}\right)\left(\dfrac{4}{5}\right) + \dfrac{1}{4} \\ &= -\dfrac{3}{10} + \left(-\dfrac{1}{2}\right)\left(\dfrac{4}{5}\right) + \dfrac{1}{4} \\ &= -\dfrac{3}{10} + \left(-\dfrac{2}{5}\right) + \dfrac{1}{4} \\ &= -\dfrac{6}{20} + \left(-\dfrac{8}{20}\right) + \dfrac{5}{20} \\ &= -\dfrac{14}{20} + \dfrac{5}{20} \\ &= -\dfrac{9}{20} \end{aligned}$

1.6 Exercises

1. A <u>variable</u> is a symbol used to represent an unknown number.

3. We are multiplying 4 by x by x. Since we know from the definition of exponents that x multiplied by x is x^2, this gives us an answer of $4x^2$.

5. Yes, $a(b - c)$ can be written as $a[b + (-c)]$.
$3(10 - 2) = (3 \times 10) - (3 \times 2)$
$\qquad 3 \times 8 = 30 - 6$
$\qquad\quad 24 = 24$

7. $5(2x - 5y) = 5(2x) + 5(-5y) = 10x - 25y$

9. $-2(4a - 3b) = -2(4a) + (-2)(-3b) = -8a + 6b$

11. $3(3x + y) = 3(3x) + 3(y) = 9x + 3y$

13. $8(-m - 3n) = 8(-m) + 8(-3n) = -8m - 24n$

15. $-(x - 3y) = (-1)(x) + (-1)(-3y) = -x + 3y$

17. $\begin{aligned} -9(9x - 5y + 8) &= (-9)(9x) + (-9)(-5y) + (-9)(8) \\ &= -81x + 45y - 72 \end{aligned}$

19. $\begin{aligned} 2(-5x + y - 6) &= 2(-5x) + 2(y) + 2(-6) \\ &= -10x + 2y - 12 \end{aligned}$

21. $\begin{aligned} &\dfrac{5}{6}(12x^2 - 24x + 18) \\ &= \left(\dfrac{5}{6}\right)(12x^2) + \left(\dfrac{5}{6}\right)(-24x) + \left(\dfrac{5}{6}\right)(18) \\ &= 10x^2 - 20x + 15 \end{aligned}$

23. $\begin{aligned} \dfrac{x}{5}(x + 10y - 4) &= \dfrac{x}{5}(x) + \dfrac{x}{5}(10y) + \dfrac{x}{5}(-4) \\ &= \dfrac{x^2}{5} + 2xy - \dfrac{4x}{5} \end{aligned}$

25. $\begin{aligned} &5x(x + 2y + z) \\ &= 5x(x) + 5x(2y) + 5x(z) \\ &= 5x^2 + 10xy + 5xz \end{aligned}$

27. $\begin{aligned} (-4.5x + 5)(-3) &= (-4.5x)(-3) + (5)(-3) \\ &= 13.5x - 15 \end{aligned}$

29. $\begin{aligned} (6x + y - 1)(3x) &= 6x(3x) + y(3x) - 1(3x) \\ &= 18x^2 + 3xy - 3x \end{aligned}$

31. $\begin{aligned} (3x + 2y - 1)(-xy) &= 3x(-xy) + 2y(-xy) - 1(-xy) \\ &= -3x^2y - 2xy^2 + xy \end{aligned}$

33. $(-a-2b+4)5ab$
$= (-a)(5ab)+(-2b)(5ab)+4(5ab)$
$= -5a^2b-10ab^2+20ab$

35. $\dfrac{1}{4}(8a^2-16a-5) = \dfrac{1}{4}(8a^2)+\dfrac{1}{4}(-16a)+\dfrac{1}{4}(-5)$
$\qquad\qquad\qquad\quad = 2a^2-4a-\dfrac{5}{4}$

37. $-0.3x(-1.2x^2-0.3x+0.5)$
$= -0.3x(-1.2x^2)+(-0.3x)(-0.3x)+(-0.3x)(0.5)$
$= 0.36x^3+0.09x^2-0.15x$

39. $0.4q(-3.3q^2-0.7r-10)$
$= 0.4q(-3.3q^2)+0.4q(-0.7r)+0.4q(-10)$
$= -1.32q^3-0.28qr-4q$

41. $800(5x+14y) = 800(5x)+800(14y)$
$\qquad\qquad\qquad = 4000x+11{,}200y$

The area is $(4000x+11{,}200y)$ ft^2.

43. $4x(3000-2y) = 4x(3000)+4x(-2y)$
$\qquad\qquad\qquad = 12{,}000x-8xy$

The area is $(12{,}000x-8xy)$ ft^2.

Cumulative Review

44. $-18+(-20)+36+(-14) = -38+36+(-14)$
$\qquad\qquad\qquad\qquad\qquad = -2+(-14)$
$\qquad\qquad\qquad\qquad\qquad = -16$

45. $(-2)^6 = (-2)(-2)(-2)(-2)(-2)(-2) = 64$

46. $-27-(-41) = -27+41 = 14$

47. $25 \div 5(2)+(-6) = 5(2)+(-6) = 10+(-6) = 4$

48. $(12-10)^2+(-3)(-2) = (2)^2+(-3)(-2)$
$\qquad\qquad\qquad\qquad\quad = 4+(-3)(-2)$
$\qquad\qquad\qquad\qquad\quad = 4+6$
$\qquad\qquad\qquad\qquad\quad = 10$

Quick Quiz 1.6

1. $5(-3a-7b) = 5(-3a)+5(-7b) = -15a-35b$

2. $-2x(x-4y+8)$
$= -2x(x)+(-2x)(-4y)+(-2x)(8)$
$= -2x^2+8xy-16x$

3. $-3ab(4a-5b-9)$
$= -3ab(4a)+(-3ab)(-5b)+(-3ab)(-9)$
$= -12a^2b+15ab^2+27ab$

4. Answers may vary. Possible solution:
Distribute $\left(-\dfrac{3}{7}\right)$ to each term within the parentheses.
$\left(-\dfrac{3}{7}\right)(21x^2-14x+3)$
$= \left(-\dfrac{3}{7}\right)(21x^2)+\left(-\dfrac{3}{7}\right)(-14x)+\left(-\dfrac{3}{7}\right)(3)$
$= -9x^2+6x-\dfrac{9}{7}$

1.7 Exercises

1. A term is a number, a variable, a product, or a quotient of numbers and variables.

3. The two terms $5x$ and $-8x$ are like terms because they both have the variable x with the exponent of one.

5. The only like terms are $7xy$ and $-14xy$ because the other two have different exponents even though they have the same variables.

7. $-16x^2-15x^2 = (-16-15)x^2 = -31x^2$

9. $5a^3-7a^2+a^3 = 5a^3+1a^3-7a^2$
$\qquad\qquad\qquad\quad = (5+1)a^3-7a^2$
$\qquad\qquad\qquad\quad = 6a^3-7a^2$

11. $3x+2y-8x-7y = (3-8)x+(2-7)y = -5x-5y$

13. $1.3x-2.6y+5.8x-0.9y$
$= (1.3+5.8)x+(-2.6-0.9)y$
$= 7.1x-3.5y$

15. $1.6x-2.8y-3.6x-5.9y$
$= (1.6-3.6)x+(-2.8-5.9)y$
$= -2x-8.7y$

17. $3p-4q+2p+3+5q-21$
$= (3+2)p+(-4+5)q+3-21$
$= 5p+q-18$

19. $2ab+5bc-6ac-2ab = (2-2)ab+5bc-6ac$
$\qquad\qquad\qquad\qquad\qquad = 5bc-6ac$

21. $2x^2 - 3x - 5 - 7x + 8 - x^2$
$= (2-1)x^2 + (-3-7)x - 5 + 8$
$= x^2 - 10x + 3$

23. $2y^2 - 8y + 9 - 12y^2 - 8y + 3$
$= (2-12)y^2 + (-8-8)y + 9 + 3$
$= -10y^2 - 16y + 12$

25. $\dfrac{1}{3}x - \dfrac{2}{3}y - \dfrac{2}{5}x + \dfrac{4}{7}y$
$= \left(\dfrac{1}{3} - \dfrac{2}{5}\right)x + \left(-\dfrac{2}{3} + \dfrac{4}{7}\right)y$
$= \left(\dfrac{5}{15} - \dfrac{6}{15}\right)x + \left(-\dfrac{14}{21} + \dfrac{12}{21}\right)y$
$= -\dfrac{1}{15}x - \dfrac{2}{21}y$

27. $\dfrac{3}{4}a^2 - \dfrac{1}{3}b - \dfrac{1}{5}a^2 - \dfrac{1}{2}b$
$= \left(\dfrac{3}{4} - \dfrac{1}{5}\right)a^2 + \left(-\dfrac{1}{3} - \dfrac{1}{2}\right)b$
$= \left(\dfrac{15}{20} - \dfrac{4}{20}\right)a^2 + \left(-\dfrac{2}{6} - \dfrac{3}{6}\right)b$
$= \dfrac{11}{20}a^2 - \dfrac{5}{6}b$

29. $3rs - 8r + s - 5rs + 10r - s$
$= (3-5)rs + (-8+10)r + (1-1)s$
$= -2rs + 2r$

31. $4xy + \dfrac{5}{4}x^2y + \dfrac{3}{4}xy + \dfrac{3}{4}x^2y$
$= \left(4 + \dfrac{3}{4}\right)xy + \left(\dfrac{5}{4} + \dfrac{3}{4}\right)x^2y$
$= \left(\dfrac{16}{4} + \dfrac{3}{4}\right)xy + \left(\dfrac{5}{4} + \dfrac{3}{4}\right)x^2y$
$= \dfrac{19}{4}xy + \dfrac{8}{4}x^2y$
$= \dfrac{19}{4}xy + 2x^2y$

33. $5(2a - b) - 3(5b - 6a) = 10a - 5b - 15b + 18a$
$= 28a - 20b$

35. $-3b(5a - 3b) + 4(-3ab - 5b^2)$
$= -15ab + 9b^2 - 12ab - 20b^2$
$= -27ab - 11b^2$

37. $6(c - 2d^2) - 2(4c - d^2) = 6c - 12d^2 - 8c + 2d^2$
$= (6-8)c + (-12+2)d^2$
$= -2c - 10d^2$

39. $3(4 - x) - 2(-4 - 7x) = 12 - 3x + 8 + 14x$
$= 11x + 20$

41. $3a + 2b + 4a + 7b = 7a + 9b$
He needs $(7a + 9b)$ to enclose the pool.

43. $2(5x - 10) + 2(2x + 6) = 10x - 20 + 4x + 12$
$= 14x - 8$
The perimeter is $(14x - 8)$ feet.

Cumulative Review

45. $-\dfrac{3}{4} - \dfrac{1}{3} = -\dfrac{9}{12} - \dfrac{4}{12} = -\dfrac{13}{12}$ or $-1\dfrac{1}{12}$

46. $\left(\dfrac{2}{3}\right)\left(-\dfrac{9}{16}\right) = -\dfrac{18}{48} = -\dfrac{3 \cdot 6}{8 \cdot 6} = -\dfrac{3}{8}$

47. $\dfrac{4}{5} + \left(-\dfrac{1}{25}\right) + \left(-\dfrac{3}{10}\right) = \dfrac{40}{50} + \left(-\dfrac{2}{50}\right) + \left(-\dfrac{15}{50}\right)$
$= \dfrac{23}{50}$

48. $\left(\dfrac{5}{7}\right) \div \left(-\dfrac{14}{3}\right) = \left(\dfrac{5}{7}\right)\left(-\dfrac{3}{14}\right) = -\dfrac{15}{98}$

Quick Quiz 1.7

1. $3xy - \dfrac{2}{3}x^2y - \dfrac{5}{6}xy + \dfrac{7}{3}x^2y$
$= \left(3 - \dfrac{5}{6}\right)xy + \left(-\dfrac{2}{3} + \dfrac{7}{3}\right)x^2y$
$= \dfrac{13}{6}xy + \dfrac{5}{3}x^2y$

2. $8.2a^2b + 5.5ab^2 - 7.6a^2b - 9.9ab^2$
$= (8.2 - 7.6)a^2b + (5.5 - 9.9)ab^2$
$= 0.6a^2b - 4.4ab^2$

3. $2(3x - 5y) - 2(-7x - 4y) = 6x - 10y + 14x + 8y$
$= (6 + 14)x + (-10 + 8)y$
$= 20x - 2y$

4. Answers may vary. Possible solution:
Use the distributive property to remove the parentheses. Then simplify by combining like terms.

$1.2(3.5x - 2.2y) - 4.5(2.0x + 1.5y)$
$= 4.2x - 2.64y - 9x - 6.75y$
$= -4.8x - 9.39y$

1.8 Exercises

1. If $x = 4$, then
$-3x + 5 = -3(4) + 5 = -12 + 5 = -7.$

3. If $y = -10$, then
$\dfrac{2}{5}y - 8 = \dfrac{2}{5}(-10) - 8 = -4 - 8 = -12.$

5. If $x = \dfrac{1}{2}$, then
$5x + 10 = 5\left(\dfrac{1}{2}\right) + 10 = \dfrac{5}{2} + 10 = \dfrac{25}{2}$ or $12\dfrac{1}{2}.$

7. If $x = 7$, then $2 - 4x = 2 - 4(7) = 2 - 28 = -26.$

9. If $x = 2.4$, then
$3.5 - 2x = 3.5 - 2(2.4) = 3.5 - 4.8 = -1.3.$

11. If $x = -\dfrac{3}{4}$, then
$9x + 13 = 9\left(-\dfrac{3}{4}\right) + 13 = -\dfrac{27}{4} + \dfrac{52}{4} = \dfrac{25}{4}$ or $6\dfrac{1}{4}.$

13. If $x = -2$, then
$x^2 - 3x = (-2)^2 - 3(-2) = 4 + 6 = 10.$

15. If $y = -1$, then $5y^2 = 5(-1)^2 = 5(1) = 5.$

17. If $x = 2$, then $-3x^3 = -3(2)^3 = -3(8) = -24.$

19. If $x = -2$, then $-5x^2 = -5(-2)^2 = -5(4) = -20.$

21. If $x = -3$, then $2x^2 + 3x = 2(-3)^2 + 3(-3)$
$= 2(9) - 9$
$= 18 - 9$
$= 9.$

23. If $x = 3$, then
$(2x)^2 + x = [2(3)]^2 + 3 = [6]^2 + 3 = 36 + 3 = 39.$

25. If $x = -2$, then
$2 - (-x)^2 = 2 - [-(-2)]^2 = 2 - (2)^2 = 2 - 4 = -2.$

27. If $a = -2$, then $10a + (4a)^2 = 10(-2) + [4(-2)]^2$
$= 10(-2) + (-8)^2$
$= -10 + 64$
$= 44.$

29. If $x = \dfrac{1}{2}$, then $4x^2 - 6x = 4\left(\dfrac{1}{2}\right)^2 - 6\left(\dfrac{1}{2}\right)$
$= 4\left(\dfrac{1}{4}\right) - 6\left(\dfrac{1}{2}\right)$
$= 1 - 3$
$= -2.$

31. If $x = 2$ and $y = 5$, then
$x^3 - 7y + 3 = (2)^3 - 7(5) + 3$
$= 8 - 7(5) + 3$
$= 8 - 35 + 3$
$= -24.$

33. If $a = -4$ and $b = \dfrac{2}{3}$, then
$\dfrac{1}{2}a^2 - 3b + 9 = \dfrac{1}{2}(-4)^2 - 3\left(\dfrac{2}{3}\right) + 9$
$= \dfrac{1}{2}(16) - 2 + 9$
$= 8 - 2 + 9$
$= 15.$

35. If $r = -1$ and $s = 3$, then
$2r^2 + 3s^2 - rs = 2(-1)^2 + 3(3)^2 - (-1)(3)$
$= 2(1) + 3(9) - (-3)$
$= 2 + 27 + 3$
$= 32.$

37. If $a = 5$, $b = 9$, and $c = -1$, then
$a^3 + 2abc - 3c^2 = 5^3 + 2(5)(9)(-1) - 3(-1)^2$
$= 125 - 90 - 3(1)$
$= 125 - 90 - 3$
$= 32.$

39. If $a = -1$ and $b = -2$, then
$\dfrac{a^2 + ab}{3b} = \dfrac{(-1)^2 + (-1)(-2)}{3(-2)} = \dfrac{1 + 2}{-6} = -\dfrac{1}{2}.$

41. $A = ab$, $b = 22$, $a = 16$
$A = (22)(16) = 352$
The area is 352 square feet.

43. $A = s^2$

Increase $= A_{new} - A_{old}$
$$= (3.2)^2 - (3)^2$$
$$= 10.24 - 9$$
$$= 1.24$$

The area is increased by 1.24 square centimeters.

45. $A = \frac{1}{2}a(b_1 + b_2)$, $a = 4$, $b_1 = 9$, $b_2 = 7$

$$A = \frac{1}{2}(4)(9 + 7)$$
$$= \frac{4(16)}{2}$$
$$= 32$$

The area is 32 square inches.

47. $A = \frac{1}{2}ab$, $a = 400$, $b = 280$

$$A = \frac{1}{2}(400)(280) = 56,000$$

The area is 56,000 square feet.

49. $A = \pi r^2$, $r = 3$

$$A \approx (3.14)(3)^2$$
$$\approx (3.14)(9)$$
$$\approx 28.26$$

The area is approximately 28.26 square feet.

51. $C = \frac{5}{9}(F - 32)$, $F = -109.3$

$$C = \frac{5}{9}(-109.3 - 32)$$
$$C = \frac{5}{9}(-141.3)$$
$$C = 5(-15.7)$$
$$C = -78.5$$

The temperature is $-78.5°C$.

53. $A = \frac{1}{2}ab$, $a = 20$, $b = 12$

$$A = \frac{1}{2}(20)(12) = 120$$

Cost $= 19.50(120) = 2340$
The cost is $2340.

55. $F = \frac{9}{5}C + 32$

$$F = \frac{9}{5}(-238) + 32 = -428.4 + 32 = -396.4$$
$$F = \frac{9}{5}(123) + 32 = 221.4 + 32 = 253.4$$

The coldest temperature was $-396.4°F$.
The warmest temperature was $253.4°F$.

Cumulative Review

57. $(-2)^4 - 4 \div 2 - (-2) = 16 - 4 \div 2 - (-2)$
$$= 16 - 2 + 2$$
$$= 16$$

58. $3(x - 2y) - (x^2 - y) - (x - y)$
$$= 3x - 6y - x^2 + y - x + y$$
$$= -x^2 + 2x - 4y$$

Quick Quiz 1.8

1. If $x = -2$, then
$$2x^2 - 4x - 14 = 2(-2)^2 - 4(-2) - 14$$
$$= 2(4) - 4(-2) - 14$$
$$= 8 + 8 - 14$$
$$= 2.$$

2. If $a = \frac{1}{2}$ and $b = -\frac{1}{3}$, then
$$5a - 6b = 5\left(\frac{1}{2}\right) - 6\left(-\frac{1}{3}\right)$$
$$= \frac{5}{2} + 2$$
$$= \frac{5}{2} + \frac{4}{2}$$
$$= \frac{9}{2} \text{ or } 4\frac{1}{2}.$$

3. If $x = -2$ and $y = 3$, then
$$x^3 + 2x^2y + 5y + 2$$
$$= (-2)^3 + 2(-2)^2(3) + 5(3) + 2$$
$$= -8 + 24 + 15 + 2$$
$$= 33.$$

4. Answers may vary. Possible solution:

The formula for the area of a circle is $A = \pi r^2$. Therefore, first find the radius from the diameter $\left(r = \dfrac{d}{2} \right)$. Use $\pi \approx 3.14$.

$$A \approx 3.14(6 \text{ m})^2 \approx 3.14(36 \text{ m}^2) \approx 113.04 \text{ m}^2$$

1.9 Exercises

1. $-3x - 2y = -(3x + 2y)$

3. To simplify expressions with grouping symbols, we use the <u>distributive</u> property.

5. $8x - 4(x - 3y) = 8x - 4x + 12y = 4x + 12y$

7. $\begin{aligned} 5(c - 3d) - (3c + d) &= 5c - 15d - 3c - d \\ &= (5 - 3)c + (-15 - 1)d \\ &= 2c - 16d \end{aligned}$

9. $\begin{aligned} -3(x + 3y) + 2(2x + y) &= -3x - 9y + 4x + 2y \\ &= x - 7y \end{aligned}$

11. $\begin{aligned} 2x[4x^2 - 2(x - 3)] &= 2x[4x^2 - 2x + 6] \\ &= 8x^3 - 4x^2 + 12x \end{aligned}$

13. $\begin{aligned} 2[5(x + y) - 2(3x - 4y)] &= 2[5x + 5y - 6x + 8y] \\ &= 2(-x + 13y) \\ &= -2x + 26y \end{aligned}$

15. $\begin{aligned} [10 - 4(x - 2y)] + 3(2x + y) \\ = 10 - 4x + 8y + 6x + 3y \\ = (-4 + 6)x + (8 + 3)y + 10 \\ = 2x + 11y + 10 \end{aligned}$

17. $\begin{aligned} 5[3a - 2a(3a + 6b) + 6a^2] \\ = 5[3a - 6a^2 - 12ab + 6a^2] \\ = 5[3a - 12ab] \\ = 15a - 60ab \end{aligned}$

19. $\begin{aligned} 6a(2a^2 - 3a - 4) - a(a - 2) \\ = 12a^3 - 18a^2 - 24a - a^2 + 2a \\ = 12a^3 - 19a^2 - 22a \end{aligned}$

21. $\begin{aligned} 3a^2 - 4[2b - 3b(b + 2)] &= 3a^2 - 4(2b - 3b^2 - 6b) \\ &= 3a^2 - 4(-4b - 3b^2) \\ &= 3a^2 + 16b + 12b^2 \end{aligned}$

23. $\begin{aligned} 5b + \{-[3a + 2(5a - 2b)] - 1\} \\ = 5b + [-3a - 2(5a - 2b)] - 1 \\ = 5b + (-3a - 10a + 4b) - 1 \\ = 5b - 13a + 4b - 1 \\ = -13a + 9b - 1 \end{aligned}$

25. $\begin{aligned} 2\{3x^2 + 5[2x - (3 - x)]\} &= 2\{3x^2 + 5[2x - 3 + x]\} \\ &= 2\{3x^2 + 5[3x - 3]\} \\ &= 2\{3x^2 + 15x - 15\} \\ &= 6x^2 + 30x - 30 \end{aligned}$

27. $\begin{aligned} -4\{3a^2 - 2[4a^2 - (b + a^2)]\} \\ = -4[3a^2 - 2(4a^2 - b - a^2)] \\ = -4[3a^2 - 2(3a^2 - b)] \\ = -4(3a^2 - 6a^2 + 2b) \\ = -4(-3a^2 + 2b) \\ = 12a^2 - 8b \end{aligned}$

Cumulative Review

29. If $C = 1064.18$, then
$$\begin{aligned} F &= 1.8C + 32 \\ &= 1.8(1064.18) + 32 \\ &= 1915.524 + 32 \\ &= 1947.524 \end{aligned}$$
The melting point is $1947.52°$F.

30. $A = \pi r^2$, $r = 380$
$$\begin{aligned} A &\approx 3.14(380)^2 \\ &\approx 3.14(144,400) \\ &\approx 453,416 \end{aligned}$$
The area is approximately 453,416 square feet.

31. $k = 0.45p$
If $p = 120$, $k = 0.45(120) = 54$.
If $p = 150$, $k = 0.45(150) = 67.5$.
Great Danes weigh on average from 54 to 67.5 kg.

32. $k = 0.45p$
If $p = 9$, $k = 0.45(9) = 4.05$.
If $p = 14$, $k = 0.45(14) = 6.3$.
Miniature Pinschers weigh on average 4.05 to 6.3 kg.

Quick Quiz 1.9

1. $\begin{aligned} 2[3x - 2(5x + y)] &= 2(3x - 10x - 2y) \\ &= 2(-7x - 2y) \\ &= -14x - 4y \end{aligned}$

2. $3[x - 3(x + 4) + 5y] = 3(x - 3x - 12 + 5y)$
$$= 3(-2x + 5y - 12)$$
$$= -6x + 15y - 36$$

3. $-4\{2a + 2[2ab - b(1 - a)]\}$
$$= -4\{2a + 2[2ab - b + ab]\}$$
$$= -4\{2a + 2[3ab - b]\}$$
$$= -4\{2a + 6ab - 2b\}$$
$$= -8a - 24ab + 8b$$

4. Answers may vary. Possible solution:
First use the distributive property to remove the parentheses. Then simplify within the square brackets. Next use the distributive property to remove grouping symbols, combining like terms after each step.
$3\{2 - 3[4x - 2(x + 3) + 5x]\}$
$= 3\{2 - 3[4x - 2x - 6 + 5x]\}$
$= 3\{2 - 3[7x - 6]\}$
$= 3\{2 - 21x + 18\}$
$= 3\{20 - 21x\}$
$= 60 - 63x$

Career Exploration Problems

1. $P = I \times E = 12 \text{ A} \times 1.8 \text{ V} = 21.6 \text{ W}$
The power loss is 21.6 W per conductor. Since there are two conductors, the total power loss is $2 \times 21.6 \text{ W} = 43.2 \text{ W}$.

2. $E = 3\% \times 240 \text{ V} = 0.03 \times 240 \text{ V} = 7.2 \text{ V}$
$P = I \times E = 24 \text{ A} \times 7.2 \text{ V} = 172.8 \text{ W}$
The power loss is 172.8 W.

3. $I = \dfrac{E}{R} = \dfrac{120 \text{ V}}{200 \text{ ohms}} = 0.6 \text{ A}$
The amount of current flow is 0.6 A.

4. $V = L \times W \times H = 4 \text{ in.} \times 4 \text{ in.} \times 1\frac{1}{2} \text{ in.} = 24 \text{ in.}^3$

The volume of one box is 24 in.^{3}.

volume of one box $\times 7$ boxes $= 24 \text{ in.}^3 \times 7$
$$= 168 \text{ in.}^3$$
The total volume is 168 in.^{3}.

You Try It

1. a. $|5| = 5$

   b. $|-1| = 1$

   c. $|0.5| = 0.5$

   d. $\left|-\dfrac{1}{4}\right| = \dfrac{1}{4}$

   e. $|-4.57| = 4.57$

2. $-10 + (-4) = -14$

3. a. $(-5) + 11 = 6$

   b. $5 + (-11) = -6$

4. $-4 + 1 + (-8) + 12 + (-3) + 5$
$$= -3 + (-8) + 12 + (-3) + 5$$
$$= -11 + 12 + (-3) + 5$$
$$= 1 + (-3) + 5$$
$$= -2 + 5$$
$$= 3$$

5. $-8 - (-7) = -8 + 7 = -1$

6. a. $9(-6) = -54$

   b. $24 \div (-3) = -8$

   c. $-48 \div (-8) = 6$

   d. $-3(-7) = 21$

7. a. $3^4 = 3 \cdot 3 \cdot 3 \cdot 3 = 81$

   b. $1.5^2 = 1.5 \cdot 1.5 = 2.25$

   c. $\left(\dfrac{1}{2}\right)^4 = \left(\dfrac{1}{2}\right)\left(\dfrac{1}{2}\right)\left(\dfrac{1}{2}\right)\left(\dfrac{1}{2}\right) = \dfrac{1}{16}$

8. a. $(-2)^3 = (-2)(-2)(-2) = -8$

   b. $(-4)^4 = (-4)(-4)(-4)(-4) = 256$

9. $4^2 + 2(6 - 3)^3 - (5 - 2)^2 \div 3$
$$= 4^2 + 2(3)^3 - (3)^2 \div 3$$
$$= 16 + 2(27) - 9 \div 3$$
$$= 16 + 54 - 9 \div 3$$
$$= 16 + 54 - 3$$
$$= 67$$

10. a. $4(2a - 3) = 4(2a) + 4(-3) = 8a - 12$

   b. $-5(5x - 1) = -5(5x) + (-5)(-1) = -25x + 5$

11. $9a^2 - 10a + 3ab + 7a - 12a^2 + 5ab$
$$= (9-12)a^2 + (-10+7)a + (3+5)ab$$
$$= -3a^2 - 3a + 8ab$$

12. If $x = 4$ and $y = -1$, then
$$6x^2 - xy + 3y^2 = 6(4)^2 - (4)(-1) + 3(-1)^2$$
$$= 6(16) - (4)(-1) + 3(1)$$
$$= 96 - (-4) + 3$$
$$= 96 + 4 + 3$$
$$= 103$$

13. $A = \dfrac{1}{2}a(b_1 + b_2)$
$$= \dfrac{1}{2}(50)(40+60)$$
$$= 25(100)$$
$$= 2500$$
The area is 2500 square feet.

14. $4\{9x - [2(x+3) - 8]\} = 4\{9x - [2x + 6 - 8]\}$
$$= 4\{9x - [2x - 2]\}$$
$$= 4\{9x - 2x + 2\}$$
$$= 4\{7x + 2\}$$
$$= 28x + 8$$

Chapter 1 Review Problems

1. $-6 + (-2) = -8$

2. $-12 + 7.8 = -4.2$

3. $5 + (-2) + (-12) = 3 + (-12) = -9$

4. $3.7 + (-1.8) = 1.9$

5. $\dfrac{1}{2} + \left(-\dfrac{5}{6}\right) = \dfrac{3}{6} + \left(-\dfrac{5}{6}\right) = -\dfrac{2}{6} = -\dfrac{1}{3}$

6. $-\dfrac{3}{11} + \left(-\dfrac{1}{22}\right) = -\dfrac{6}{22} + \left(-\dfrac{1}{22}\right) = -\dfrac{7}{22}$

7. $\dfrac{3}{4} + \left(-\dfrac{1}{12}\right) + \left(-\dfrac{1}{2}\right) = \dfrac{9}{12} + \left(-\dfrac{1}{12}\right) + \left(-\dfrac{6}{12}\right)$
$$= \dfrac{2}{12}$$
$$= \dfrac{1}{6}$$

8. $\dfrac{2}{15} + \dfrac{1}{6} + \left(-\dfrac{4}{5}\right) = \dfrac{4}{30} + \dfrac{5}{30} + \left(-\dfrac{24}{30}\right)$
$$= -\dfrac{15}{30}$$
$$= -\dfrac{1}{2}$$

9. $5 - (-3) = 5 + 3 = 8$

10. $-2 - (-15) = -2 + 15 = 13$

11. $-30 - (+3) = -30 + (-3) = -33$

12. $8 - (-1.2) = 8 + 1.2 = 9.2$

13. $-\dfrac{7}{8} + \left(-\dfrac{3}{4}\right) = -\dfrac{7}{8} + \left(-\dfrac{6}{8}\right) = -\dfrac{13}{8}$ or $-1\dfrac{5}{8}$

14. $-\dfrac{3}{8} + \dfrac{5}{6} = -\dfrac{9}{24} + \dfrac{20}{24} = \dfrac{11}{24}$

15. $-20.8 - 1.9 = -20.8 + (-1.9) = -22.7$

16. $-151 - (-63) = -151 + 63 = -88$

17. $87 \div (-29) = -3$

18. $-10.4 \div (-0.8) = 13$

19. $\dfrac{-24}{-\frac{3}{4}} = -24 \div \left(-\dfrac{3}{4}\right) = \left(\dfrac{-24}{1}\right)\left(-\dfrac{4}{3}\right) = 32$

20. $-\dfrac{2}{3} \div \left(-\dfrac{4}{5}\right) = -\dfrac{2}{3} \cdot \left(-\dfrac{5}{4}\right) = \dfrac{10}{12} = \dfrac{5}{6}$

21. $\dfrac{5}{7} \div \left(-\dfrac{5}{25}\right) = \dfrac{5}{7} \cdot \left(-\dfrac{25}{5}\right) = -\dfrac{25}{7}$ or $-3\dfrac{4}{7}$

22. $-6(3)(4) = (-18)(4) = -72$

23. $-1(-4)(-3)(-5) = 4(-3)(-5) = -12(-5) = 60$

24. $(-5)\left(-\dfrac{1}{2}\right)(4)(-3) = \left(\dfrac{5}{2}\right)(4)(-3) = 10(-3) = -30$

25. $(-3)^5 = (-3)(-3)(-3)(-3)(-3) = -243$

26. $(-2)^6 = (-2)(-2)(-2)(-2)(-2)(-2) = 64$

27. $(-5)^4 = (-5)(-5)(-5)(-5) = 625$

28. $\left(-\dfrac{2}{3}\right)^3 = \left(-\dfrac{2}{3}\right)\left(-\dfrac{2}{3}\right)\left(-\dfrac{2}{3}\right) = -\dfrac{8}{27}$

29. $-9^2 = -(9)(9) = -81$

30. $(0.6)^2 = (0.6)(0.6) = 0.36$

31. $\left(\dfrac{5}{6}\right)^2 = \left(\dfrac{5}{6}\right)\left(\dfrac{5}{6}\right) = \dfrac{25}{36}$

32. $\left(\dfrac{3}{4}\right)^3 = \left(\dfrac{3}{4}\right)\left(\dfrac{3}{4}\right)\left(\dfrac{3}{4}\right) = \dfrac{27}{64}$

33. $\begin{aligned}(5)(-4)+(3)(-2)^3 &= (5)(-4)+(3)(-8)\\ &= -20+(-24)\\ &= -44\end{aligned}$

34. $\begin{aligned}8 \div 0.4 + 0.1 \times (0.2)^2 &= 8 \div 0.4 + 0.1 \times 0.04\\ &= 20 + 0.1 \times 0.04\\ &= 20 + 0.004\\ &= 20.004\end{aligned}$

35. $\begin{aligned}(3-6)^2 + (-12) \div (-3)(-2)\\ = (-3)^2 + (-12) \div (-3)(-2)\\ = 9 + (-12) \div (-3)(-2)\\ = 9 + 4(-2)\\ = 9 - 8\\ = 1\end{aligned}$

36. $7(-3x+y) = 7(-3x) + 7(y) = -21x + 7y$

37. $\begin{aligned}3x(6-x+3y) &= 3x(6) + 3x(-x) + 3x(3y)\\ &= 18x - 3x^2 + 9xy\end{aligned}$

38. $\begin{aligned}&-(7x^2 - 3x + 11)\\ &= -1(7x^2) + (-1)(-3x) + (-1)(11)\\ &= -7x^2 + 3x - 11\end{aligned}$

39. $\begin{aligned}&(2xy + x - y)(-3y^2)\\ &= (2xy)(-3y^2) + x(-3y^2) - y(-3y^2)\\ &= -6xy^3 - 3xy^2 + 3y^3\end{aligned}$

40. $\begin{aligned}&3a^2b - 2bc + 6bc^2 - 8a^2b - 6bc^2 + 5bc\\ &= (3-8)a^2b + (-2+5)bc + (6-6)bc^2\\ &= -5a^2b + 3bc\end{aligned}$

41. $9x + 11y - 12x - 15y = -3x - 4y$

42. $\begin{aligned}&4x^2 - 13x + 7 - 9x^2 - 22x - 16\\ &= (4-9)x^2 + (-13-22)x + 7 - 16\\ &= -5x^2 - 35x - 9\end{aligned}$

43. $\begin{aligned}&-x + \dfrac{1}{2} + 14x^2 - 7x - 1 - 4x^2\\ &= (14-4)x^2 + (-7-1)x + \dfrac{1}{2} - 1\\ &= 10x^2 - 8x - \dfrac{1}{2}\end{aligned}$

44. If $x = -7$, then
$7x - 6 = 7(-7) - 6 = -49 - 6 = -55.$

45. If $x = 8$, then $7 - \dfrac{3}{4}x = 7 - \dfrac{3}{4}(8) = 7 - 6 = 1.$

46. If $x = -3$, then
$\begin{aligned}x^2 + 3x - 4 &= (-3)^2 + 3(-3) - 4\\ &= 9 - 9 - 4\\ &= -4.\end{aligned}$

47. If $x = 3$, then
$-x^2 + 5x - 9 = -3^2 + 5(3) - 9 = -9 + 15 - 9 = -3.$

48. If $x = -1$, then
$\begin{aligned}2x^3 - x^2 + 6x + 9 &= 2(-1)^3 - (-1)^2 + 6(-1) + 9\\ &= -2 - 1 - 6 + 9\\ &= 0.\end{aligned}$

49. If $a = -1$, $b = 5$, and $c = -2$, then
$b^2 - 4ac = (5)^2 - 4(-1)(-2) = 25 - 8 = 17.$

50. If $m = -4$, $M = 15$, $G = -1$, and $r = -2$, then
$\dfrac{mMG}{r^2} = \dfrac{-4(15)(-1)}{(-2)^2} = \dfrac{60}{4} = 15.$

51. If $p = 6000$, $r = 18\%$, and $t = \dfrac{3}{4}$, then

$I = prt = 6000(0.18)\left(\dfrac{3}{4}\right) = 810.$

The interest is $810.

52. $F = \dfrac{9C + 160}{5}$

$\quad = \dfrac{9(20) + 160}{5}$

$\quad = \dfrac{180 + 160}{5}$

$\quad = \dfrac{340}{5}$

$\quad = 68$

$F = \dfrac{9C + 160}{5}$

$\quad = \dfrac{9(25) + 160}{5}$

$\quad = \dfrac{225 + 160}{5}$

$\quad = \dfrac{385}{5}$

$\quad = 77$

The range is 68°F to 77°F.

53. If $r = 4$, $A = \pi r^2$

$\quad \approx 3.14(4)^2$

$\quad \approx 3.14(16)$

$\quad \approx 50.24$ square feet.

Cost = (50.24 sq ft)($1.50/sq ft) = $75.36

The total cost is $75.36.

54. $P = 180S - R - C$

$\quad = 180(56) - 300 - 1200$

$\quad = 10,080 - 300 - 1200$

$\quad = 8580$

The daily profit is $8580.

55. $A = \dfrac{1}{2}a(b_1 + b_2)$, $a = 200$, $b_1 = 700$, $b_2 = 300$

$A = \dfrac{1}{2}(200)(700 + 300)$

$\quad = 100(1000)$

$\quad = 100,000$

Cost = 2(100,000) = 200,000

The area is $100,000$ ft^2 and the cost is $200,000.

56. $A = \dfrac{1}{2}ab$, $a = 3.8$, $b = 5.5$

$A = \dfrac{1}{2}(3.8)(5.5) = 10.45$

Cost = 66(10.45) = 689.70

The area is 10.45 ft^2 and the cost is $689.70.

57. $5x - 7(x - 6) = 5x - 7x + 42 = -2x + 42$

58. $3(x - 2) - 4(5x + 3) = 3x - 6 - 20x - 12$

$\quad\quad\quad\quad\quad\quad\quad\quad\quad = -17x - 18$

59. $2[3 - (4 - 5x)] = 2(3 - 4 + 5x)$

$\quad\quad\quad\quad\quad\quad\quad = 2(-1 + 5x)$

$\quad\quad\quad\quad\quad\quad\quad = -2 + 10x$

60. $-3x[x + 3(x - 7)] = -3x(x + 3x - 21)$

$\quad\quad\quad\quad\quad\quad\quad\quad\quad = -3x(4x - 21)$

$\quad\quad\quad\quad\quad\quad\quad\quad\quad = -12x^2 + 63x$

61. $2xy^3 - 6x^3 y - 4x^2 y^2 + 3(xy^3 - 2x^2 y - 3x^2 y^2)$

$= 2xy^3 - 6x^3 y - 4x^2 y^2 + 3xy^3 - 6x^2 y - 9x^2 y^2$

$= (2 + 3)xy^3 - 6x^3 y + (-4 - 9)x^2 y^2 - 6x^2 y$

$= 5xy^3 - 6x^3 y - 13x^2 y^2 - 6x^2 y$

62. $-5(x + 2y - 7) + 3x(2 - 5y)$

$= -5x - 10y + 35 + 6x - 15xy$

$= x - 10y + 35 - 15xy$

63. $-(a + 3b) + 5[2a - b - 2(4a - b)]$

$= -(a + 3b) + 5(2a - b - 8a + 2b)$

$= -(a + 3b) + 5(-6a + b)$

$= -a - 3b - 30a + 5b$

$= -31a + 2b$

64. $-5\{2a - [5a - b(3 + 2a)]\}$

$= -5\{2a - [5a - 3b - 2ab]\}$

$= -5\{2a - 5a + 3b + 2ab\}$

$= -5\{-3a + 3b + 2ab\}$

$= 15a - 15b - 10ab$

65. $-3\{2x - [x - 3y(x - 2y)]\}$

$= -3[2x - (x - 3xy + 6y^2)]$

$= -3(2x - x + 3xy - 6y^2)$

$= -3(x + 3xy - 6y^2)$

$= -3x - 9xy + 18y^2$

66. $2\{3x + 2[x + 2y(x - 4)]\}$

$= 2[3x + 2(x + 2xy - 8y)]$

$= 2(3x + 2x + 4xy - 16y)$

$= 2(5x + 4xy - 16y)$

$= 10x + 8xy - 32y$

67. $-6.3 + 4 = -2.3$

68. $4 + (-8) + 12 = -4 + 12 = 8$

69. $-\dfrac{2}{3} - \dfrac{4}{5} = -\dfrac{10}{15} + \left(-\dfrac{12}{15}\right) = -\dfrac{22}{15}$ or $-1\dfrac{7}{15}$

70. $-\dfrac{7}{8} - \left(-\dfrac{3}{4}\right) = -\dfrac{7}{8} + \dfrac{6}{8} = -\dfrac{1}{8}$

71. $3 - (-4) + (-8) = 3 + 4 + (-8) = 7 + (-8) = -1$

72. $-1.1 - (-0.2) + 0.4 = -1.1 + 0.2 + 0.4$
$\qquad\qquad\qquad\qquad\quad = -0.9 + 0.4$
$\qquad\qquad\qquad\qquad\quad = -0.5$

73. $\left(-\dfrac{9}{10}\right)\left(-2\dfrac{1}{4}\right) = \left(-\dfrac{9}{10}\right)\left(-\dfrac{9}{4}\right) = \dfrac{81}{40}$ or $2\dfrac{1}{40}$

74. $3.6 \div (-0.45) = -8$

75. $-14.4 \div (-0.06) = 240$

76. $(-8.2)(3.1) = -25.42$

77. $400 + 1000 - 800 = 1400 - 800 = 600$
Her score was $600.

78. $(-0.3)^4 = (-0.3)(-0.3)(-0.3)(-0.3) = 0.0081$

79. $-0.5^4 = -(0.5)(0.5)(0.5)(0.5) = -0.0625$

80. $9(5) - 5(2)^3 + 5 = 9(5) - 5(8) + 5$
$\qquad\qquad\qquad\quad = 45 - 5(8) + 5$
$\qquad\qquad\qquad\quad = 45 - 40 + 5$
$\qquad\qquad\qquad\quad = 5 + 5$
$\qquad\qquad\qquad\quad = 10$

81. $3.8x - 0.2y - 8.7x + 4.3y$
$\quad = (3.8 - 8.7)x + (-0.2 + 4.3)y$
$\quad = -4.9x + 4.1y$

82. If $p = -2$ and $q = 3$, then
$\dfrac{2p + q}{3q} = \dfrac{2(-2) + 3}{3(3)} = \dfrac{-4 + 3}{9} = -\dfrac{1}{9}.$

83. If $s = -3$ and $t = -2$, then
$\dfrac{4s - 7t}{s} = \dfrac{4(-3) - 7(-2)}{-3} = \dfrac{-12 + 14}{-3} = -\dfrac{2}{3}.$

84. $F = \dfrac{9}{5}C + 32, \ C = 38.6$

$F = \dfrac{9}{5}(38.6) + 32$
$F = 69.48 + 32$
$F = 101.48°$
Your dog does not have a fever; in fact, its temperature is below normal.

85. $-7(x - 3y^2 + 4) + 3y(4 - 6y)$
$\quad = -7x + 21y^2 - 28 + 12y - 18y^2$
$\quad = -7x + 3y^2 + 12y - 28$

86. $-2\{6x - 3[7y - 2y(3 - x)]\}$
$\quad = -2\{6x - 3[7y - 6y + 2xy]\}$
$\quad = -2\{6x - 3[y + 2xy]\}$
$\quad = -2\{6x - 3y - 6xy\}$
$\quad = -12x + 6y + 12xy$

How Am I Doing? Chapter 1 Test

1. $-2.5 + 6.3 + (-4.1) = 3.8 + (-4.1) = -0.3$

2. $-5 - (-7) = -5 + 7 = 2$

3. $\left(-\dfrac{2}{3}\right)(7) = -\dfrac{14}{3} = -4\dfrac{2}{3}$

4. $-5(-2)(7)(-1) = -(10)(7)(1) = -(70)(1) = -70$

5. $-12 \div (-3) = 4$

6. $-1.8 \div (0.6) = -3$

7. $(-4)^3 = (-4)(-4)(-4) = -64$

8. $(1.6)^2 = (1.6)(1.6) = 2.56$

9. $\left(\dfrac{2}{3}\right)^4 = \left(\dfrac{2}{3}\right)\left(\dfrac{2}{3}\right)\left(\dfrac{2}{3}\right)\left(\dfrac{2}{3}\right) = \dfrac{16}{81}$

10. $(0.2)^2 - (2.1)(-3) + 0.46$
$\quad = 0.04 - (2.1)(-3) + 0.46$
$\quad = 0.04 - (-6.3) + 0.46$
$\quad = 0.04 + 6.3 + 0.46$
$\quad = 6.34 + 0.46$
$\quad = 6.8$

11. $3(4-6)^3 + 12 \div (-4) + 2$
$= 3(-2)^3 + 12 \div (-4) + 2$
$= 3(-8) + 12 \div (-4) + 2$
$= -24 + (-3) + 2$
$= -25$

12. $-5x(x + 2y - 7) = -5x(x) - 5x(2y) - 5x(-7)$
$ = -5x^2 - 10xy + 35x$

13. $-2ab^2(-3a - 2b + 7ab)$
$= -2ab^2(-3a) - 2ab^2(-2b) - 2ab^2(7ab)$
$= 6a^2b^2 + 4ab^3 - 14a^2b^3$

14. $6ab - \dfrac{1}{2}a^2b + \dfrac{3}{2}ab + \dfrac{5}{2}a^2b$
$= \left(6 + \dfrac{3}{2}\right)ab + \left(-\dfrac{1}{2} + \dfrac{5}{2}\right)a^2b$
$= \left(\dfrac{12}{2} + \dfrac{3}{2}\right)ab + \dfrac{4}{2}a^2b$
$= \dfrac{15}{2}ab + 2a^2b$

15. $2.3x^2y - 8.1xy^2 + 3.4xy^2 - 4.1x^2y$
$= (2.3 - 4.1)x^2y + (-8.1 + 3.4)xy^2$
$= -1.8x^2y - 4.7xy^2$

16. $3(2-a) - 4(-6 - 2a) = 6 - 3a + 24 + 8a$
$ = 5a + 30$

17. $5(3x - 2y) - (x + 6y) = 15x - 10y - x - 6y$
$ = (15 - 1)x + (-10 - 6)y$
$ = 14x - 16y$

18. If $x = 3$ and $y = -4$, then
$x^3 - 3x^2y + 2y - 5 = 3^3 - 3(3)^2(-4) + 2(-4) - 5$
$ = 27 - 3(9)(-4) - 8 - 5$
$ = 27 + 108 - 8 - 5$
$ = 122.$

19. If $x = -3$, then
$3x^2 - 7x - 11 = 3(-3)^2 - 7(-3) - 11$
$ = 3(9) - 7(-3) - 11$
$ = 27 + 21 - 11$
$ = 37.$

20. If $a = \dfrac{1}{3}$ and $b = -\dfrac{1}{2}$, then
$2a - 3b = 2\left(\dfrac{1}{3}\right) - 3\left(-\dfrac{1}{2}\right)$
$ = \dfrac{2}{3} + \dfrac{3}{2}$
$ = \dfrac{4}{6} + \dfrac{9}{6}$
$ = \dfrac{13}{6} \text{ or } 2\dfrac{1}{6}.$

21. $k = 1.61r = 1.61(60) = 96.6$
You are traveling at 96.6 kilometers per hour.

22. $A = \dfrac{1}{2}a(b_1 + b_2),\ a = 120,\ b_1 = 200,\ b_2 = 180$
$A = \dfrac{1}{2}(120)(200 + 180)$
$ = 60(380)$
$ = 22,800$
The area is 22,800 square feet.

23. $A = \dfrac{1}{2}ab,\ a = 6.8,\ b = 8.5$
$A = \dfrac{1}{2}(6.8)(8.5) = 28.9$
Cost $= 0.80(28.9) = 23.12$
The cost is \$23.12.

24. $A = 60 \times 10 = 600$
$600 \text{ sq ft} \times \dfrac{1 \text{ can}}{200 \text{ sq ft}} = 3 \text{ cans}$
You should buy 3 cans.

25. $3[x - 2y(x + 2y) - 3y^2] = 3[x - 2xy - 4y^2 - 3y^2]$
$ = 3[x - 2xy - 7y^2]$
$ = 3x - 6xy - 21y^2$

26. $-3\{a + b[3a - b(1 - a)]\} = -3[a + b(3a - b + ab)]$
$\phantom{-3\{a + b[3a - b(1 - a)]\}} = -3(a + 3ab - b^2 + ab^2)$
$\phantom{-3\{a + b[3a - b(1 - a)]\}} = -3a - 9ab + 3b^2 - 3ab^2$

Chapter 2

2.1 Exercises

1. When we use the <u>equals</u> sign, we indicate two expressions are <u>equal</u> in value.

3. The <u>solution</u> of an equation is a value of the variable that makes the equation true.

5. Answers may vary. A sample answer is to isolate the variable x.

7.
$$x + 14 = 21$$
$$x + 14 + (-14) = 21 + (-14)$$
$$x = 7$$
Check: $7 + 14 \overset{?}{=} 21$
$$21 = 21 \checkmark$$

9.
$$20 = 9 + x$$
$$20 + (-9) = 9 + (-9) + x$$
$$11 = x$$
Check: $20 \overset{?}{=} 9 + 11$
$$20 = 20 \checkmark$$

11.
$$x - 3 = 14$$
$$x - 3 + 3 = 14 + 3$$
$$x = 17$$
Check: $17 - 3 \overset{?}{=} 14$
$$14 = 14 \checkmark$$

13.
$$0 = x + 5$$
$$0 + (-5) = x + 5 + (-5)$$
$$-5 = x$$
Check: $0 \overset{?}{=} -5 + 5$
$$0 = 0 \checkmark$$

15.
$$x - 6 = -19$$
$$x - 6 + 6 = -19 + 6$$
$$x = -13$$
Check: $-13 - 6 \overset{?}{=} -19$
$$-19 = -19 \checkmark$$

17.
$$-12 + x = 50$$
$$-12 + 12 + x = 50 + 12$$
$$x = 62$$
Check: $-12 + 62 \overset{?}{=} 50$
$$50 = 50 \checkmark$$

19.
$$3 + 5 = x - 7$$
$$8 = x - 7$$
$$8 + 7 = x - 7 + 7$$
$$15 = x$$
Check: $3 + 5 \overset{?}{=} 15 - 7$
$$8 = 8 \checkmark$$

21.
$$32 - 17 = x - 6$$
$$15 = x - 6$$
$$15 + 6 = x - 6 + 6$$
$$21 = x$$
Check: $32 - 17 \overset{?}{=} 21 - 6$
$$15 = 15 \checkmark$$

23.
$$4 + 8 + x = 6 + 6$$
$$12 + x = 12$$
$$12 + (-12) + x = 12 + (-12)$$
$$x = 0$$
Check: $4 + 8 + 0 \overset{?}{=} 6 + 6$
$$12 = 12 \checkmark$$

25.
$$18 - 7 + x = 7 + 9 - 5$$
$$11 + x = 11$$
$$11 + (-11) + x = 11 + (-11)$$
$$x = 0$$
Check: $18 - 7 + 0 \overset{?}{=} 7 + 9 - 5$
$$11 = 11 \checkmark$$

27.
$$-12 + x - 3 = 15 - 18 + 9$$
$$-15 + x = 6$$
$$-15 + 15 + x = 6 + 15$$
$$x = 21$$
Check: $-12 + 21 - 3 \overset{?}{=} 15 - 18 + 9$
$$6 = 6 \checkmark$$

29. $-7 + x = 2$, $x \overset{?}{=} 5$
$$-7 + 5 \overset{?}{=} 2$$
$$-2 \neq 2$$
$x = 5$ is not the solution.
$$-7 + x = 2$$
$$-7 + 7 + x = 2 + 7$$
$$x = 9$$

31. $-11 + 5 = x + 8$, $x \overset{?}{=} -6$
$$-11 + 5 \overset{?}{=} -6 + 8$$
$$-6 \neq 2$$
$x = -6$ is not the solution.

$$-11+5 = x+8$$
$$-6 = x+8$$
$$-6+(-8) = x+8+(-8)$$
$$-14 = x$$

33. $x-23 = -56, \ x \overset{?}{=} -33$
$$-33-23 \overset{?}{=} -56$$
$$-56 = -56$$
$x = -33$ is the solution.

35. $15-3+20 = x-3, \ x \overset{?}{=} 35$
$$15-3+20 \overset{?}{=} 35-3$$
$$32 = 32$$
$x = 35$ is the solution.

37. $2.5+x = 0.7$
$$2.5+(-2.5)+x = 0.7+(-2.5)$$
$$x = -1.8$$

39. $12.5+x-8.2 = 4.9$
$$x+4.3 = 4.9$$
$$x+4.3+(-4.3) = 4.9+(-4.3)$$
$$x = 0.6$$

41. $x-\dfrac{1}{4} = \dfrac{3}{4}$
$$x-\dfrac{1}{4}+\dfrac{1}{4} = \dfrac{3}{4}+\dfrac{1}{4}$$
$$x = \dfrac{4}{4}$$
$$x = 1$$

43. $\dfrac{2}{3}+x = \dfrac{1}{6}+\dfrac{1}{4}$
$$\dfrac{8}{12}+x = \dfrac{2}{12}+\dfrac{3}{12}$$
$$\dfrac{8}{12}+x = \dfrac{5}{12}$$
$$\dfrac{8}{12}+\left(-\dfrac{8}{12}\right)+x = \dfrac{5}{12}+\left(-\dfrac{8}{12}\right)$$
$$x = -\dfrac{3}{12}$$
$$x = -\dfrac{1}{4}$$

45. $3+x = -12+8$
$$3+x = -4$$
$$3+(-3)+x = -4+(-3)$$
$$x = -7$$

47. $5\dfrac{1}{6}+x = 8$
$$\dfrac{31}{6}+x = \dfrac{48}{6}$$
$$\dfrac{31}{6}+\left(-\dfrac{31}{6}\right)+x = \dfrac{48}{6}+\left(-\dfrac{31}{6}\right)$$
$$x = \dfrac{17}{6} \text{ or } 2\dfrac{5}{6}$$

49. $\dfrac{3}{14}-\dfrac{2}{7} = x-\dfrac{1}{2}$
$$\dfrac{3}{14}-\dfrac{4}{14} = x-\dfrac{1}{2}$$
$$-\dfrac{1}{14} = x-\dfrac{1}{2}$$
$$-\dfrac{1}{14}+\dfrac{1}{2} = x-\dfrac{1}{2}+\dfrac{1}{2}$$
$$-\dfrac{1}{14}+\dfrac{7}{14} = x$$
$$\dfrac{6}{14} = x$$
$$\dfrac{3}{7} = x$$

51. $1.6+x-3.2 = -2+5.6$
$$x-1.6 = 3.6$$
$$x-1.6+1.6 = 3.6+1.6$$
$$x = 5.2$$

53. $x-18.225 = 1.975$
$$x-18.225+18.225 = 1.975+18.225$$
$$x = 20.2$$

Cumulative Review

55. $x+3y-5x-7y+2x = (1-5+2)x+(3-7)y$
$$= -2x-4y$$

56. $y^2+y-12-3y^2-5y+16$
$$= (1-3)y^2+(1-5)y-12+16$$
$$= -2y^2-4y+4$$

Quick Quiz 2.1

1. $x-4.7 = 9.6$
$$x-4.7+4.7 = 9.6+4.7$$
$$x = 14.3$$

2. $-8.6 + x = -12.1$
$-8.6 + 8.6 + x = -12.1 + 8.6$
$x = -3.5$

3. $3 - 12 + 7 = 8 + x - 2$
$-2 = 6 + x$
$-2 - 6 = 6 - 6 + x$
$-8 = x$

4. Answers may vary. Possible solution: Substitute 3.8 for x in the equation. Simplify. If the resultant equation is true, $x = 3.8$ is the solution. If the resultant equation is not true, $x = 3.8$ is not the solution.

2.2 Exercises

1. To solve the equation $6x = -24$, divide each side of the equation by <u>6</u>.

3. To solve the equation $\frac{1}{7}x = -2$, multiply each side of the equation by <u>7</u>.

5. $\frac{1}{8}x = 6$
$8\left(\frac{1}{8}x\right) = 8(6)$
$x = 48$
Check: $\frac{1}{8}(48) \overset{?}{=} 6$
$6 = 6$ ✔

7. $\frac{1}{2}x = -15$
$2\left(\frac{1}{2}x\right) = 2(-15)$
$x = -30$
Check: $\frac{1}{2}(-30) \overset{?}{=} -15$
$-15 = -15$ ✔

9. $\frac{x}{5} = 16$
$5\left(\frac{x}{5}\right) = 5(16)$
$x = 80$
Check: $\frac{80}{5} \overset{?}{=} 16$
$16 = 16$ ✔

11. $-3 = \frac{x}{5}$
$5(-3) = 5\left(\frac{x}{5}\right)$
$-15 = x$
Check: $-3 \overset{?}{=} \frac{-15}{5}$
$-3 = -3$ ✔

13. $13x = 52$
$\frac{13x}{13} = \frac{52}{13}$
$x = 4$
Check: $13(4) \overset{?}{=} 52$
$52 = 52$ ✔

15. $56 = 7x$
$\frac{56}{7} = \frac{7x}{7}$
$8 = x$
Check: $56 \overset{?}{=} 7(8)$
$56 = 56$ ✔

17. $-16 = 6x$
$\frac{-16}{6} = \frac{6x}{6}$
$-\frac{8}{3} = x$
Check: $-16 \overset{?}{=} 6\left(-\frac{8}{3}\right)$
$-16 = -16$ ✔

19. $1.5x = 75$
$\frac{1.5x}{1.5} = \frac{75}{1.5}$
$x = 50$
Check: $1.5(50) \overset{?}{=} 75$
$75 = 75$ ✔

21. $-15 = -x$
$\frac{-15}{-1} = \frac{-x}{-1}$
$15 = x$
Check: $-15 \overset{?}{=} (-1)(15)$
$-15 = -15$ ✔

23. $-112 = 16x$

$$\frac{-112}{16} = \frac{16x}{16}$$

$$-7 = x$$

Check: $-112 \overset{?}{=} 16(-7)$

$$-112 = -112 \checkmark$$

25. $0.4x = 0.08$

$$\frac{0.4x}{0.4} = \frac{0.08}{0.4}$$

$$x = 0.2$$

Check: $(0.4)(0.2) \overset{?}{=} 0.08$

$$0.08 = 0.08 \checkmark$$

27. $-3.9x = -15.6$

$$\frac{-3.9x}{-3.9} = \frac{-15.6}{-3.9}$$

$$x = 4$$

Check: $-3.9(4) \overset{?}{=} -15.6$

$$-15.6 = -15.6 \checkmark$$

29. $-3x = 21,\ x \overset{?}{=} 7$

$$-3(7) \overset{?}{=} 21$$

$$-21 \neq 21$$

$x = 7$ is not the solution.

$$-3x = 21$$

$$\frac{-3x}{-3} = \frac{21}{-3}$$

$$x = -7$$

31. $-x = 15,\ x \overset{?}{=} -15$

$$-(-15) \overset{?}{=} 15$$

$$15 = 15$$

$x = -15$ is the solution.

33. $7y = -0.21$

$$\frac{7y}{7} = \frac{-0.21}{7}$$

$$y = -0.03$$

35. $-56 = -21t$

$$\frac{-56}{-21} = \frac{-21t}{-21}$$

$$\frac{8}{3} = t$$

37. $4.6y = -3.22$

$$\frac{4.6y}{4.6} = \frac{-3.22}{4.6}$$

$$y = -0.7$$

39. $4x + 3x = 21$

$$7x = 21$$

$$\frac{7x}{7} = \frac{21}{7}$$

$$x = 3$$

41. $2x - 7x = 20$

$$-5x = 20$$

$$\frac{-5x}{-5} = \frac{20}{-5}$$

$$x = -4$$

43. $\dfrac{1}{4}x = -9$

$$4\left(\frac{1}{4}x\right) = 4(-9)$$

$$x = -36$$

45. $12 - 19 = -7x$

$$-7 = -7x$$

$$\frac{-7}{-7} = \frac{-7x}{-7}$$

$$1 = x$$

47. $8m = -14 + 30$

$$8m = 16$$

$$\frac{8m}{8} = \frac{16}{8}$$

$$m = 2$$

49. $\dfrac{3}{4}x = 63$

$$\frac{4}{3}\left(\frac{3}{4}x\right) = \frac{4}{3}(63)$$

$$x = 84$$

51. $-2.5133x = 26.38965$

$$\frac{-2.5133x}{-2.5133} = \frac{26.38965}{-2.5133}$$

$$x = -10.5$$

Cumulative Review

53. $-3y(2x + y) + 5(3xy - y^2)$

$$= -6xy - 3y^2 + 15xy - 5y^2$$

$$= (-6 + 15)xy + (-3 - 5)y^2$$

$$= 9xy - 8y^2$$

54. $-\{2(x-3) + 3[x-(2x-5)]\}$
$= -\{2(x-3) + 3[x-2x+5]\}$
$= -\{2(x-3) + 3[-x+5]\}$
$= -\{2x-6-3x+15\}$
$= -\{-x+9\}$
$= x-9$

55. Find 25% of 30.
25% of $30 = 0.25 \times 30 = 7.5$
The whale will lose 7.5 tons.
$30 - 7.5 = 22.5$
The whale will weigh 22.5 tons.

56. Find 35% of 20.
35% of $20 = 0.35 \times 20 = 7$
The number of earthquakes is expected to increase by 7.
$20 + 7 = 27$
A total of 27 earthquakes can be expected.

Quick Quiz 2.2

1. $2.5x = -95$
$\dfrac{2.5x}{2.5} = \dfrac{-95}{2.5}$
$x = -38$

2. $-3.9x = -54.6$
$\dfrac{-3.9x}{-3.9x} = \dfrac{-54.6}{-3.9}$
$x = 14$

3. $7x - 12x = 60$
$-5x = 60$
$\dfrac{-5x}{-5} = \dfrac{60}{-5}$
$x = -12$

4. Answers may vary. Possible solution:
Change $36\dfrac{2}{3}$ to an improper fraction. Substitute that value for x in the equation. Simplify. If the resultant equation is true, $x = 36\dfrac{2}{3}$ is the solution.

2.3 Exercises

1. $3x + 23 = 50$
$3x + 23 - 23 = 50 - 23$
$3x = 27$
$\dfrac{3x}{3} = \dfrac{27}{3}$
$x = 9$
Check: $3(9) + 23 \overset{?}{=} 50$
$27 + 23 \overset{?}{=} 50$
$50 = 50$ ✓

3. $4x - 11 = 13$
$4x - 11 + 11 = 13 + 11$
$4x = 24$
$\dfrac{4x}{4} = -\dfrac{24}{4}$
$x = 6$
Check: $4(6) - 11 \overset{?}{=} 13$
$13 = 13$ ✓

5. $7x - 18 = -46$
$7x - 18 + 18 = -46 + 18$
$7x = -28$
$\dfrac{7x}{7} = \dfrac{-28}{7}$
$x = -4$
Check: $7(-4) - 18 \overset{?}{=} -46$
$-28 - 18 \overset{?}{=} -46$
$-46 = -46$ ✓

7. $-4x + 17 = -35$
$-4x + 17 + (-17) = -35 + (-17)$
$-4x = -52$
$\dfrac{-4x}{-4} = \dfrac{-52}{-4}$
$x = 13$
Check: $-4(13) + 17 \overset{?}{=} -35$
$-52 + 17 \overset{?}{=} -35$
$-35 = -35$ ✓

9. $2x + 3.2 = 9.4$
$2x + 3.2 + (-3.2) = 9.4 + (-3.2)$
$2x = 6.2$
$\dfrac{2x}{2} = \dfrac{6.2}{2}$
$x = 3.1$
Check: $2(3.1) + 3.2 \overset{?}{=} 9.4$
$6.2 + 3.2 \overset{?}{=} 9.4$
$9.4 = 9.4$ ✓

11. $\dfrac{1}{4}x + 6 = 13$

$\dfrac{1}{4}x + 6 - 6 = 13 - 6$

$\dfrac{1}{4}x = 7$

$4\left(\dfrac{1}{4}x\right) = 4(7)$

$x = 28$

Check: $\dfrac{1}{4}(28) + 6 \stackrel{?}{=} 13$

$7 + 6 \stackrel{?}{=} 13$

$13 = 13$ ✓

13. $\dfrac{1}{3}x + 5 = -4$

$\dfrac{1}{3}x + 5 + (-5) = -4 + (-5)$

$\dfrac{1}{3}x = -9$

$3\left(\dfrac{1}{3}x\right) = 3(-9)$

$x = -27$

Check: $\dfrac{1}{3}(-27) + 5 \stackrel{?}{=} -4$

$-9 + 5 \stackrel{?}{=} -4$

$-4 = -4$ ✓

15. $8x = 48 + 2x$

$8x + (-2x) = 48 + 2x + (-2x)$

$6x = 48$

$\dfrac{6x}{6} = \dfrac{48}{6}$

$x = 8$

Check: $8(8) \stackrel{?}{=} 48 + 2(8)$

$64 \stackrel{?}{=} 48 + 16$

$64 = 64$ ✓

17. $-6x = -27 + 3x$

$-6x + (-3x) = -27 + 3x + (-3x)$

$-9x = -27$

$\dfrac{-9x}{-9} = \dfrac{-27}{-9}$

$x = 3$

Check: $-6(3) \stackrel{?}{=} -27 + 3(3)$

$-18 \stackrel{?}{=} -27 + 9$

$-18 = -18$ ✓

19. $44 - 2x = 6x$

$44 - 2x + 2x = 6x + 2x$

$44 = 8x$

$\dfrac{44}{8} = \dfrac{8x}{8}$

$\dfrac{11}{2} = x$ or $x = 5.5$

Check: $44 - 2\left(\dfrac{11}{2}\right) \stackrel{?}{=} 6\left(\dfrac{11}{2}\right)$

$44 - 11 \stackrel{?}{=} 33$

$33 = 33$ ✓

21. $54 - 2x = -8x$

$54 - 2x + 2x = -8x + 2x$

$54 = -6x$

$\dfrac{54}{-6} = \dfrac{-6x}{-6}$

$-9 = x$

Check: $54 - 2(-9) \stackrel{?}{=} -8(-9)$

$54 + 18 \stackrel{?}{=} 72$

$72 = 72$ ✓

23. $2y + 3y = 12 - y,\ y \stackrel{?}{=} 2$

$2(2) + 3(2) \stackrel{?}{=} 12 - 2$

$4 + 6 \stackrel{?}{=} 10$

$10 = 10$

$y = 2$ is the solution.

25. $7x + 6 - 3x = 2x - 5 + x,\ x \stackrel{?}{=} 11$

$7(11) + 6 - 3(11) \stackrel{?}{=} 2(11) - 5 + 11$

$77 + 6 - 33 \stackrel{?}{=} 22 - 5 + 11$

$50 \neq 28$

$x = 11$ is not the solution.

$7x + 6 - 3x = 2x - 5 + x$

$4x + 6 = 3x - 5$

$4x + (-3x) + 6 = 3x + (-3x) - 5$

$x + 6 = -5$

$x + 6 + (-6) = -5 + (-6)$

$x = -11$

27. $14 - 2x = -5x + 11$

$14 - 2x + 5x = -5x + 5x + 11$

$14 + 3x = 11$

$3x + 14 + (-14) = 11 + (-14)$

$3x = -3$

$\dfrac{3x}{3} = \dfrac{-3}{3}$

$x = -1$

29.
$$x - 6 = 8 - x$$
$$x + x - 6 = 8 - x + x$$
$$2x - 6 = 8$$
$$2x - 6 + 6 = 8 + 6$$
$$2x = 14$$
$$\frac{2x}{2} = \frac{14}{2}$$
$$x = 7$$

31.
$$0.6y + 0.8 = 0.1 - 0.1y$$
$$0.6y + 0.1y + 0.8 = 0.1 - 0.1y + 0.1y$$
$$0.7y + 0.8 = 0.1$$
$$0.7y + 0.8 - 0.8 = 0.1 - 0.8$$
$$0.7y = -0.7$$
$$\frac{0.7y}{0.7} = \frac{-0.7}{0.7}$$
$$y = -1$$

33.
$$5x - 9 = 3x + 23$$
$$5x + (-3x) - 9 = 3x + (-3x) + 23$$
$$2x - 9 = 23$$
$$2x - 9 + 9 = 23 + 9$$
$$2x = 32$$
$$\frac{2x}{2} = \frac{32}{2}$$
$$x = 16$$

35. $-3 + 10y + 6 = 15 + 12y - 18$
Left
$$10y + 3 = 12y - 3$$
$$10y + (-12y) + 3 = 12y + (-12y) - 3$$
$$-2y + 3 = -3$$
$$-2y + 3 + (-3) = -3 + (-3)$$
$$-2y = -6$$
$$\frac{-2y}{-2} = \frac{-6}{-2}$$
$$y = 3$$
Right
$$10y + 3 = 12y - 3$$
$$10y + (-10y) + 3 = 12y + (-10y) - 3$$
$$3 = 2y - 3$$
$$3 + 3 = 2y - 3 + 3$$
$$6 = 2y$$
$$\frac{6}{2} = \frac{2y}{2}$$
$$3 = y$$
Neither approach is better.

37.
$$5(x + 3) = 35$$
$$5x + 15 = 35$$
$$5x + 15 - 15 = 35 - 15$$
$$5x = 20$$
$$\frac{5x}{5} = \frac{20}{5}$$
$$x = 4$$
Check: $5(4 + 3) \overset{?}{=} 35$
$$5(7) \overset{?}{=} 35$$
$$35 = 35 \checkmark$$

39. $5(4x - 3) + 8 = -2$
$$20x - 15 + 8 = -2$$
$$20x - 7 = -2$$
$$20x - 7 + 7 = -2 + 7$$
$$20x = 5$$
$$\frac{20x}{20} = \frac{5}{20}$$
$$x = \frac{1}{4}$$
Check: $5[4\left(\dfrac{1}{4}\right) - 3] + 8 \overset{?}{=} -2$
$$5(1 - 3) + 8 \overset{?}{=} -2$$
$$5(-2) + 8 \overset{?}{=} -2$$
$$-10 + 8 \overset{?}{=} -2$$
$$-2 = -2 \checkmark$$

41. $7x - 3(5 - x) = 10$
$$7x - 15 + 3x = 10$$
$$10x - 15 = 10$$
$$10x = 25$$
$$\frac{10x}{10} = \frac{25}{10}$$
$$x = \frac{5}{2}$$
Check: $7\left(\dfrac{5}{2}\right) - 3\left[5 - \left(\dfrac{5}{2}\right)\right] \overset{?}{=} 10$
$$\frac{35}{2} - 3\left(\frac{5}{2}\right) \overset{?}{=} 10$$
$$\frac{35}{2} - \frac{15}{2} \overset{?}{=} 10$$
$$10 = 10 \checkmark$$

43.
$$0.5x - 0.3(2 - x) = 4.6$$
$$0.5x - 0.6 - 0.3x = 4.6$$
$$0.8x - 0.6 = 4.6$$
$$0.8x - 0.6 + 0.6 = 4.6 + 0.6$$
$$0.8x = 5.2$$
$$\frac{0.8x}{0.8} = \frac{5.2}{0.8}$$
$$x = 6.5$$
Check: $0.5(6.5) - 0.3(2 - 6.5) \overset{?}{=} 4.6$
$$0.5(6.5) - 0.3(-4.5) \overset{?}{=} 4.6$$
$$3.25 + 1.35 \overset{?}{=} 4.6$$
$$4.6 = 4.6 \checkmark$$

45.
$$4(a - 3) + 2 = 2(a - 5)$$
$$4a - 12 + 2 = 2a - 10$$
$$4a - 10 = 2a - 10$$
$$4a - 2a - 10 = 2a - 2a - 10$$
$$2a - 10 = -10$$
$$2a - 10 + 10 = -10 + 10$$
$$2a = 0$$
$$\frac{2a}{2} = \frac{0}{2}$$
$$a = 0$$
Check: $4(0 - 3) + 2 \overset{?}{=} 2(0 - 5)$
$$4(-3) + 2 \overset{?}{=} 2(-5)$$
$$-12 + 2 \overset{?}{=} -10$$
$$-10 = -10 \checkmark$$

47.
$$-2(x + 3) + 4 = 3(x + 4) + 2$$
$$-2x - 6 + 4 = 3x + 12 + 2$$
$$-2x - 2 = 3x + 14$$
$$-2x + (-3x) - 2 = 3x + (-3x) + 14$$
$$-5x - 2 = 14$$
$$-5x - 2 + 2 = 14 + 2$$
$$-5x = 16$$
$$\frac{-5x}{-5} = \frac{16}{-5}$$
$$x = -\frac{16}{5}$$
Check: $-2\left[\left(-\dfrac{16}{5}\right) + 3\right] + 4 \overset{?}{=} 3\left[\left(-\dfrac{16}{5}\right) + 4\right] + 2$
$$-2\left(-\frac{1}{5}\right) + 4 \overset{?}{=} 3\left(\frac{4}{5}\right) + 2$$
$$\frac{2}{5} + 4 \overset{?}{=} \frac{12}{5} + 2$$
$$\frac{22}{5} = \frac{22}{5} \checkmark$$

49.
$$-3(y - 3y) + 4 = -4(3y - y) + 6 + 13y$$
$$-3(-2y) + 4 = -4(2y) + 6 + 13y$$
$$6y + 4 = -8y + 6 + 13y$$
$$6y + 4 = 5y + 6$$
$$6y - 5y + 4 = 5y - 5y + 6$$
$$y + 4 = 6$$
$$y + 4 - 4 = 6 - 4$$
$$y = 2$$
Check:
$$-3[2 - 3(2)] + 4 \overset{?}{=} -4[3(2) - 2] + 6 + 13(2)$$
$$-3(2 - 6) + 4 \overset{?}{=} -4(6 - 2) + 6 + 26$$
$$-3(-4) + 4 \overset{?}{=} -4(4) + 32$$
$$12 + 4 \overset{?}{=} -16 + 32$$
$$16 = 16 \checkmark$$

51.
$$5.7x + 3 = 4.2x - 3$$
$$5.7x - 4.2x + 3 = 4.2x - 4.2x - 3$$
$$1.5x + 3 = -3$$
$$1.5x + 3 - 3 = -3 - 3$$
$$1.5x = -6$$
$$\frac{1.5x}{1.5} = \frac{-6}{1.5}$$
$$x = -4$$

53.
$$5z + 7 - 2z = 32 - 2z$$
$$3z + 7 = 32 - 2z$$
$$3z + 2z + 7 = 32 - 2z + 2z$$
$$5z + 7 = 32$$
$$5z + 7 - 7 = 32 - 7$$
$$5z = 25$$
$$\frac{5z}{5} = \frac{25}{5}$$
$$z = 5$$

55.
$$-0.3a + 1.4 = -1.2 - 0.7a$$
$$-0.3a + 0.7a + 1.4 = -1.2 - 0.7a + 0.7a$$
$$0.4a + 1.4 = -1.2$$
$$0.4a + 1.4 + (-1.4) = -1.2 + (-1.4)$$
$$0.4a = -2.6$$
$$\frac{0.4a}{0.4} = \frac{-2.6}{0.4}$$
$$a = -6.5$$

57.
$$6x + 8 - 3x = 11 - 12x - 13$$
$$3x + 8 = -12x - 2$$
$$3x + 12x + 8 = -12x + 12x - 2$$
$$15x + 8 = -2$$
$$15x + 8 - 8 = -2 - 8$$
$$15x = -10$$
$$\frac{15x}{15} = \frac{-10}{15}$$
$$x = -\frac{2}{3}$$

59.
$$-3.5x + 1.3 = -2.7x + 1.5$$
$$-3.5x + 3.5x + 1.3 = -2.7x + 3.5x + 1.5$$
$$1.3 = 0.8x + 1.5$$
$$1.3 - 1.5 = 0.8x + 1.5 - 1.5$$
$$-0.2 = 0.8x$$
$$\frac{-0.2}{0.8} = \frac{0.8x}{0.8}$$
$$-0.25 = x$$

61.
$$5(4 + x) = 3(3x - 1) - 9$$
$$20 + 5x = 9x - 3 - 9$$
$$20 + 5x = 9x - 12$$
$$20 + 5x - 5x = 9x - 5x - 12$$
$$20 = 4x - 12$$
$$20 + 12 = 4x - 12 + 12$$
$$32 = 4x$$
$$\frac{32}{4} = \frac{4x}{4}$$
$$8 = x$$

63.
$$-1.7x + 4.4 + 5x = 0.3x - 0.1$$
$$3.3x + 4.4 = 0.3x - 0.1$$
$$3.3x - 0.3x + 4.4 = 0.3x - 0.3x - 0.1$$
$$3x + 4.4 = -0.1$$
$$3x + 4.4 - 4.4 = -0.1 - 4.4$$
$$3x = -4.5$$
$$\frac{3x}{3} = \frac{-4.5}{3}$$
$$x = -1.5$$

Cumulative Review

65. $(-6)(-8) + (-3)(2) = 48 - 6 = 42$

66.
$$(-3)^3 + (-20) \div 2 = -27 + (-20) \div 2$$
$$= -27 + (-10)$$
$$= -37$$

67. $5 + (2 - 6)^2 = 5 + (-4)^2 = 5 + 16 = 21$

68. We multiply and then add.
$35 \times \$9.11 = \$318.85$
$16 \times \$22.70 = \$363.20$
$5 \times \$100.46 = \$502.30$
$\$318.85 + \$363.20 + \$502.30 = \$1184.35$
The market value was \$1184.35 on May 1, 2015.

69. a. 30% of \$899 = $0.30 \times \$899 = \$269.70$
$\$899 - \$269.70 = \$629.30$
With a total discount of 30%, the sale price
is \$629.30.

b. 20% of \$899 = $0.20 \times \$899 = \$179.80$
$\$899 - \$179.80 = \$719.20$
The price after the 20% discount is \$719.20.
10% of \$719.20 = $0.10 \times \$719.20 = \$71.92$
$\$719.20 - \$71.92 = \$647.28$
The sale price after both discounts is
\$647.28.

Quick Quiz 2.3

1.
$$7x - 6 = -4x - 10$$
$$7x + 4x - 6 = -4x + 4x - 10$$
$$11x - 6 = -10$$
$$11x - 6 + 6 = -10 + 6$$
$$11x = -4$$
$$\frac{11x}{11} = \frac{-4}{11}$$
$$x = -\frac{4}{11}$$

2.
$$-3x + 6.2 = -5.8$$
$$-3x + 6.2 - 6.2 = -5.8 - 6.2$$
$$-3x = -12$$
$$\frac{-3x}{-3} = \frac{-12}{-3}$$
$$x = 4$$

3.
$$2(3x - 2) = 4(5x + 3)$$
$$6x - 4 = 20x + 12$$
$$6x - 6x - 4 = 20x - 6x + 12$$
$$-4 = 14x + 12$$
$$-4 - 12 = 14x + 12 - 12$$
$$-16 = 14x$$
$$\frac{-16}{14} = \frac{14x}{14}$$
$$-\frac{8}{7} = x$$

4. Answers may vary. Possible solution:
Use the distributive property to remove parentheses. Combine like terms on the left side of the equation. Move the variable terms to the left side of the equation and the constants to the right side of the equation. Simplify.

2.4 Exercises

1.
$$\frac{1}{6}x + \frac{2}{3} = -\frac{1}{2}$$
$$6\left(\frac{1}{6}x\right) + 6\left(\frac{2}{3}\right) = 6\left(-\frac{1}{2}\right)$$
$$x + 4 = -3$$
$$x + 4 + (-4) = -3 + (-4)$$
$$x = -7$$

Check: $\frac{1}{6}(-7) + \frac{2}{3} \overset{?}{=} -\frac{1}{2}$
$$-\frac{7}{6} + \frac{2}{3} \overset{?}{=} -\frac{1}{2}$$
$$-\frac{7}{6} + \frac{4}{6} \overset{?}{=} -\frac{1}{2}$$
$$-\frac{3}{6} \overset{?}{=} -\frac{1}{2}$$
$$-\frac{1}{2} = -\frac{1}{2} \checkmark$$

3.
$$\frac{2}{3}x = \frac{1}{15}x + \frac{3}{5}$$
$$15\left(\frac{2}{3}x\right) = 15\left(\frac{1}{15}x\right) + 15\left(\frac{3}{5}\right)$$
$$10x = x + 9$$
$$10x + (-x) = x + (-x) + 9$$
$$9x = 9$$
$$\frac{9x}{9} = \frac{9}{9}$$
$$x = 1$$

Check: $\frac{2}{3}(1) \overset{?}{=} \frac{1}{15}(1) + \frac{3}{5}$
$$\frac{2}{3} \overset{?}{=} \frac{1}{15} + \frac{9}{15}$$
$$\frac{2}{3} \overset{?}{=} \frac{10}{15}$$
$$\frac{2}{3} = \frac{2}{3} \checkmark$$

5.
$$\frac{x}{2} + \frac{x}{5} = \frac{7}{10}$$
$$10\left(\frac{x}{2}\right) + 10\left(\frac{x}{5}\right) = 10\left(\frac{7}{10}\right)$$
$$5x + 2x = 7$$
$$7x = 7$$
$$\frac{7x}{7} = \frac{7}{7}$$
$$x = 1$$

Check: $\frac{1}{2} + \frac{1}{5} \overset{?}{=} \frac{7}{10}$
$$\frac{5}{10} + \frac{2}{10} \overset{?}{=} \frac{7}{10}$$
$$\frac{7}{10} = \frac{7}{10} \checkmark$$

7.
$$5 - \frac{1}{3}x = \frac{1}{12}x$$
$$12(5) - 12\left(\frac{1}{3}x\right) = 12\left(\frac{1}{12}x\right)$$
$$60 - 4x = x$$
$$60 - 4x + 4x = x + 4x$$
$$60 = 5x$$
$$\frac{60}{5} = \frac{5x}{5}$$
$$12 = x$$

Check: $5 - \frac{1}{3}(12) \overset{?}{=} \frac{1}{12}(12)$
$$5 - 4 \overset{?}{=} 1$$
$$1 = 1 \checkmark$$

9.
$$2 + \frac{y}{2} = \frac{3y}{4} - 3$$
$$4(2) + 4\left(\frac{y}{2}\right) = 4\left(\frac{3y}{4}\right) - 4(3)$$
$$8 + 2y = 3y - 12$$
$$8 = y - 12$$
$$20 = y$$

Check: $2 + \left(\frac{20}{2}\right) \overset{?}{=} \frac{3(20)}{4} - 3$
$$2 + 10 \overset{?}{=} 15 - 3$$
$$12 = 12 \checkmark$$

11.
$$\frac{x-3}{5} = 1 - \frac{x}{3}$$
$$15\left(\frac{x-3}{5}\right) = 15(1) - 15\left(\frac{x}{3}\right)$$
$$3(x-3) = 15 - 5x$$
$$3x - 9 = 15 - 5x$$
$$3x + 5x - 9 = 15 - 5x + 5x$$
$$8x - 9 = 15$$
$$8x - 9 + 9 = 15 + 9$$
$$8x = 24$$
$$\frac{8x}{8} = \frac{24}{8}$$
$$x = 3$$
Check: $\dfrac{3-3}{5} \stackrel{?}{=} 1 - \dfrac{3}{3}$
$$\frac{0}{5} \stackrel{?}{=} 1 - 1$$
$$0 = 0 \checkmark$$

13.
$$\frac{x+3}{4} = \frac{x}{2} + \frac{1}{6}$$
$$12\left(\frac{x+3}{4}\right) = 12\left(\frac{x}{2}\right) + 12\left(\frac{1}{6}\right)$$
$$3(x+3) = 6x + 2$$
$$3x + 9 = 6x + 2$$
$$3x + (-6x) + 9 = 6x + (-6x) + 2$$
$$-3x + 9 = 2$$
$$-3x + 9 + (-9) = 2 + (-9)$$
$$-3x = -7$$
$$\frac{-3x}{-3} = \frac{-7}{-3}$$
$$x = \frac{7}{3}$$
Check: $\dfrac{\frac{7}{3}+3}{4} \stackrel{?}{=} \dfrac{\frac{7}{3}}{2} + \dfrac{1}{6}$
$$\frac{\frac{7}{3}+\frac{9}{3}}{4} \stackrel{?}{=} \frac{7}{3} \cdot \frac{1}{2} + \frac{1}{6}$$
$$\frac{16}{3} \cdot \frac{1}{4} \stackrel{?}{=} \frac{7}{6} + \frac{1}{6}$$
$$\frac{4}{3} \stackrel{?}{=} \frac{8}{6}$$
$$\frac{4}{3} = \frac{4}{3} \checkmark$$

15.
$$0.6x + 5.9 = 3.8$$
$$10(0.6x) + 10(5.9) = 10(3.8)$$
$$6x + 59 = 38$$
$$6x + 59 - 59 = 38 - 59$$
$$6x = -21$$
$$\frac{6x}{6} = \frac{-21}{6}$$
$$x = -\frac{7}{2} = -3.5$$
Check: $0.6(-3.5) + 5.9 \stackrel{?}{=} 3.8$
$$-2.1 + 5.9 \stackrel{?}{=} 3.8$$
$$3.8 = 3.8$$

17. $\dfrac{1}{2}(y-2) + 2 = \dfrac{3}{8}(3y-4)$, $y \stackrel{?}{=} 4$
$$\frac{1}{2}(4-2) + 2 \stackrel{?}{=} \frac{3}{8}(3 \cdot 4 - 4)$$
$$1 + 2 \stackrel{?}{=} \frac{3}{8}(8)$$
$$3 = 3$$
Yes, $y = 4$ is a solution.

19. $\dfrac{1}{2}\left(y - \dfrac{1}{5}\right) = \dfrac{1}{5}(y+2)$, $y \stackrel{?}{=} \dfrac{5}{8}$
$$\frac{1}{2}\left(\frac{5}{8} - \frac{1}{5}\right) \stackrel{?}{=} \frac{1}{5}\left(\frac{5}{8} + 2\right)$$
$$\frac{1}{2}\left(\frac{17}{40}\right) \stackrel{?}{=} \frac{1}{5}\left(\frac{21}{8}\right)$$
$$\frac{17}{80} \neq \frac{42}{80}$$
No, $y = \dfrac{5}{8}$ is not a solution.

21.
$$\frac{3}{4}(3x+1) = 2(3-2x) + 1$$
$$\frac{9}{4}x + \frac{3}{4} = 6 - 4x + 1$$
$$\frac{9}{4}x + \frac{3}{4} = -4x + 7$$
$$4\left(\frac{9}{4}x\right) + 4\left(\frac{3}{4}\right) = 4(-4x) + 4(7)$$
$$9x + 3 = -16x + 28$$
$$9x + 16x + 3 = -16x + 16x + 28$$
$$25x + 3 = 28$$
$$25x + 3 - 3 = 28 - 3$$
$$25x = 25$$
$$\frac{25x}{25} = \frac{25}{25}$$
$$x = 1$$

23.
$$2(x-2) = \frac{2}{5}(3x+1) + 2$$
$$2x - 4 = \frac{6}{5}x + \frac{2}{5} + 2$$
$$2x - 4 = \frac{6}{5}x + \frac{12}{5}$$
$$5(2x) - 5(4) = 5\left(\frac{6}{5}x\right) + 5\left(\frac{12}{5}\right)$$
$$10x - 20 = 6x + 12$$
$$10x - 6x - 20 = 6x - 6x + 12$$
$$4x - 20 = 12$$
$$4x - 20 + 20 = 12 + 20$$
$$4x = 32$$
$$\frac{4x}{4} = \frac{32}{4}$$
$$x = 8$$

25.
$$0.3x - 0.2(3 - 5x) = -0.5(x - 6)$$
$$0.3x - 0.6 + x = -0.5x + 3$$
$$1.3x - 0.6 = -0.5x + 3$$
$$10(1.3x) - 10(0.6) = 10(-0.5x) + 10(3)$$
$$13x - 6 = -5x + 30$$
$$13x + 5x - 6 = -5x + 5x + 30$$
$$18x - 6 = 30$$
$$18x - 6 + 6 = 30 + 6$$
$$18x = 36$$
$$\frac{18x}{18} = \frac{36}{18}$$
$$x = 2$$

27.
$$-8(0.1x + 0.4) - 0.9 = -0.1$$
$$-0.8x - 3.2 - 0.9 = -0.1$$
$$-0.8x - 4.1 = -0.1$$
$$10(-0.8x) - 10(4.1) = 10(-0.1)$$
$$-8x - 41 = -1$$
$$-8x - 41 + 41 = -1 + 41$$
$$-8x = 40$$
$$\frac{-8x}{-8} = \frac{40}{-8}$$
$$x = -5$$

29.
$$\frac{1}{3}(y+2) = 3y - 5(y-2)$$
$$\frac{1}{3}y + \frac{2}{3} = 3y - 5y + 10$$
$$\frac{1}{3}y + \frac{2}{3} = -2y + 10$$
$$3\left(\frac{1}{3}y\right) + 3\left(\frac{2}{3}\right) = 3(-2y) + 3(10)$$
$$y + 2 = -6y + 30$$
$$y + 6y + 2 = -6y + 6y + 30$$
$$7y + 2 = 30$$
$$7y + 2 - 2 = 30 - 2$$
$$7y = 28$$
$$y = 4$$

31.
$$\frac{1+2x}{5} + \frac{4-x}{3} = \frac{1}{15}$$
$$15\left(\frac{1+2x}{5}\right) + 15\left(\frac{4-x}{3}\right) = 15\left(\frac{1}{15}\right)$$
$$3(1+2x) + 5(4-x) = 1$$
$$3 + 6x + 20 - 5x = 1$$
$$x + 23 = 1$$
$$x + 23 - 23 = 1 - 23$$
$$x = -22$$

33.
$$\frac{3}{4}(x-2) + \frac{3}{5} = \frac{1}{5}(x+1)$$
$$\frac{3}{4}x - \frac{3}{2} + \frac{3}{5} = \frac{1}{5}x + \frac{1}{5}$$
$$\frac{3}{4}x - \frac{9}{10} = \frac{1}{5}x + \frac{1}{5}$$
$$20\left(\frac{3}{4}x\right) - 20\left(\frac{9}{10}\right) = 20\left(\frac{1}{5}x\right) + 20\left(\frac{1}{5}\right)$$
$$15x - 18 = 4x + 4$$
$$15x - 18 - 4x = 4x + 4 - 4x$$
$$11x - 18 = 4$$
$$11x - 18 + 18 = 4 + 18$$
$$11x = 22$$
$$\frac{11x}{11} = \frac{22}{11}$$
$$x = 2$$

35.
$$\frac{1}{3}(x-2) = 3x - 2(x-1) + \frac{16}{3}$$
$$\frac{1}{3}x - \frac{2}{3} = 3x - 2x + 2 + \frac{16}{3}$$
$$\frac{1}{3}x - \frac{2}{3} = x + \frac{22}{3}$$
$$3\left(\frac{1}{3}x\right) - 3\left(\frac{2}{3}\right) = 3(x) + 3\left(\frac{22}{3}\right)$$
$$x - 2 = 3x + 22$$
$$x - 3x - 2 = 3x - 3x + 22$$
$$-2x - 2 = 22$$
$$-2x - 2 + 2 = 22 + 2$$
$$-2x = 24$$
$$\frac{-2x}{-2} = \frac{24}{-2}$$
$$x = -12$$

37.
$$\frac{4}{5}x - \frac{2}{3} = \frac{3x+1}{2}$$
$$30\left(\frac{4}{5}x\right) - 30\left(\frac{2}{3}\right) = 30\left(\frac{3x+1}{2}\right)$$
$$24x - 20 = 15(3x+1)$$
$$24x - 20 = 45x + 15$$
$$24x + (-45x) - 20 = 45x + (-45x) + 15$$
$$-21x - 20 = 15$$
$$-21x - 20 + 20 = 15 + 20$$
$$-21x = 35$$
$$\frac{-21x}{-21} = \frac{35}{-21}$$
$$x = -\frac{5}{3}$$

39.
$$0.3x - 0.2(5x-1) = -0.4(x+2)$$
$$0.3x - 1.0x + 0.2 = -0.4x - 0.8$$
$$10(0.3x) - 10(1.0x) + 10(0.2) = 10(-0.4x) - 10(0.8)$$
$$3x - 10x + 2 = -4x - 8$$
$$-7x + 2 = -4x - 8$$
$$-7x + 4x + 2 = -4x + 4x - 8$$
$$-3x + 2 = -8$$
$$-3x + 2 - 2 = -8 - 2$$
$$-3x = -10$$
$$\frac{-3x}{-3} = \frac{-10}{-3}$$
$$x = \frac{10}{3}$$

41.
$$-1 + 5(x-2) = 12x + 3 - 7x$$
$$-1 + 5x - 10 = 5x + 3$$
$$5x - 11 = 5x + 3$$
$$5x - 5x - 11 = 5x - 5x + 3$$
$$-11 = 3, \text{ no solution}$$

43.
$$9(x+3) - 6 = 24 - 2x - 3 + 11x$$
$$9x + 27 - 6 = 9x + 21$$
$$9x + 21 = 9x + 21$$
$$9x - 9x + 21 = 9x - 9x + 21$$
$$21 = 21$$
Infinite number of solutions

45.
$$-3(4x-1) = 5(2x-1) + 8$$
$$-12x + 3 = 10x - 5 + 8$$
$$-12x + 3 = 10x + 3$$
$$-12x - 10x + 3 = 10x - 10x + 3$$
$$-22x + 3 = 3$$
$$-22x + 3 - 3 = 3 - 3$$
$$-22x = 0$$
$$\frac{-22x}{-22} = \frac{0}{-22}$$
$$x = 0$$

47.
$$3(4x+1) - 2x = 2(5x-3)$$
$$12x + 3 - 2x = 10x - 6$$
$$10x + 3 = 10x - 6$$
$$10x + (-10x) + 3 = 10x + (-10x) - 6$$
$$3 = -6, \text{ no solution}$$

Cumulative Review

49.
$$\left(-3\frac{1}{4}\right)\left(5\frac{1}{3}\right) = \left(-\frac{13}{4}\right)\left(\frac{16}{3}\right)$$
$$= -\frac{13 \cdot \cancel{4} \cdot 4}{\cancel{4} \cdot 3}$$
$$= -\frac{52}{3} \text{ or } -17\frac{1}{3}$$

50.
$$5\frac{1}{2} \div 1\frac{1}{4} = \frac{11}{2} \div \frac{5}{4}$$
$$= \frac{11}{2} \cdot \frac{4}{5}$$
$$= \frac{11 \cdot \cancel{2} \cdot 2}{\cancel{2} \cdot 5}$$
$$= \frac{22}{5} \text{ or } 4\frac{2}{5}$$

51. 30% of 440 = 0.30 × 440 = 132
440 + 132 = 572
30% of 750 = 0.3 × 750 = 225
750 + 225 = 975
The weight range for females is
572 – 975 grams.

52. Find the area of the seating area.

$\text{Area} = \dfrac{1}{2}a(b_1 + b_2)$

$= \dfrac{1}{2}(200)(150 + 88)$

$= 100(238)$

$= 23{,}800 \text{ ft}^2$

Find the area required for each seat.

$\text{Area} = L \cdot W = 2.5 \cdot 3 = 7.5 \text{ ft}^2$

Now divide.

$23{,}800 \div 7.5 \approx 3173$

The auditorium will hold approximately
3173 seats.

Quick Quiz 2.4

1.

$$\dfrac{3}{4}x + \dfrac{5}{12} = \dfrac{1}{3}x - \dfrac{1}{6}$$

$$12\left(\dfrac{3}{4}x\right) + 12\left(\dfrac{5}{12}\right) = 12\left(\dfrac{1}{3}x\right) - 12\left(\dfrac{1}{6}\right)$$

$$9x + 5 = 4x - 2$$

$$9x - 4x + 5 = 4x - 4x - 2$$

$$5x + 5 = -2$$

$$5x + 5 - 5 = -2 - 5$$

$$5x = -7$$

$$\dfrac{5x}{5} = \dfrac{-7}{5}$$

$$x = -\dfrac{7}{5} \text{ or } -1.4$$

2.

$$\dfrac{2}{3}x - \dfrac{3}{5} + \dfrac{7}{5}x + \dfrac{1}{3} = 1$$

$$15\left(\dfrac{2}{3}x\right) - 15\left(\dfrac{3}{5}\right) + 15\left(\dfrac{7}{5}x\right) + 15\left(\dfrac{1}{3}\right) = 15(1)$$

$$10x - 9 + 21x + 5 = 15$$

$$31x - 4 = 15$$

$$31x - 4 + 4 = 15 + 4$$

$$31x = 19$$

$$\dfrac{31x}{31} = \dfrac{19}{31}$$

$$x = \dfrac{19}{31}$$

3.

$$\dfrac{2}{3}(x + 2) + \dfrac{1}{4} = \dfrac{1}{2}(5 - 3x)$$

$$\dfrac{2}{3}x + \dfrac{4}{3} + \dfrac{1}{4} = \dfrac{5}{2} - \dfrac{3}{2}x$$

$$\dfrac{2}{3}x + \dfrac{19}{12} = \dfrac{5}{2} - \dfrac{3}{2}x$$

$$12\left(\dfrac{2}{3}x\right) + 12\left(\dfrac{19}{12}\right) = 12\left(\dfrac{5}{2}\right) - 12\left(\dfrac{3}{2}x\right)$$

$$8x + 19 = 30 - 18x$$

$$8x + 18x + 19 = 30 - 18x + 18x$$

$$26x + 19 = 30$$

$$26x + 19 - 19 = 30 - 19$$

$$26x = 11$$

$$\dfrac{26x}{26} = \dfrac{11}{26}$$

$$x = \dfrac{11}{26}$$

4. Answers may vary. Possible solution:
Multiply both sides of the equation by the LCD,
12. Add or subtract terms on both sides of the
equation to get all terms containing *x* on one side
of the equation. Add or subtract a constant value
on both sides of the equation to get all terms not
containing *x* on the other side of the equation.
Divide both sides by the coefficient of *x*, and
simplify the solution if necessary. Finally, check
the solution.

Use Math to Save Money

1. Shell: $4.55
ARCO: $4.43 + $0.45 = $4.88

2. Shell: 3($4.55) = $13.65
ARCO: 3($4.43) + $0.45 = $13.29 + $0.45
= $13.74

3. Shell: 4($4.55) = $18.20
ARCO: 4($4.43) + $0.45 = $17.72 + $0.45
= $18.17

4. Shell: 10($4.55) = $45.50
ARCO: 10($4.43) + $0.45 = $44.30 + $0.45
= $44.75

5. $4.55x = 4.43x + 0.45$
$0.12x = 0.45$
$x = 3.75$
The price is the same for 3.75 gallons of gas.

6. For less than four gallons, the Shell station is
less expensive.

7. For more than four gallons, the ARCO station is less expensive.

8. Answers will vary.

9. Answers will vary.

10. Answers will vary.

How Am I Doing? Sections 2.1–2.4
(Available online through MyMathLab.)

1. $5 - 8 + x = -12$
$-3 + x = -12$
$-3 + 3 + x = -12 + 3$
$x = -9$

2. $-2.8 + x = 4.7$
$-2.8 + 2.8 + x = 4.7 + 2.8$
$x = 7.5 \text{ or } 7\frac{1}{2}$

3. $-45 = -5x$
$\frac{-45}{-5} = \frac{-5x}{-5}$
$9 = x$

4. $12x - 6x = -48$
$6x = -48$
$\frac{6x}{6} = \frac{-48}{6}$
$x = -8$

5. $-1.2x + 3.5 = 2.7$
$-1.2x + 3.5 - 3.5 = 2.7 - 3.5$
$-1.2x = -0.8$
$\frac{-1.2x}{-1.2} = \frac{-0.8}{-1.2}$
$x = \frac{2}{3}$

6. $-14x + 9 = 2x + 7$
$-14x - 2x + 9 = 2x - 2x + 7$
$-16x + 9 = 7$
$-16x + 9 - 9 = 7 - 9$
$-16x = -2$
$\frac{-16x}{-16} = \frac{-2}{-16}$
$x = \frac{1}{8}$

7. $14x + 2(7 - 2x) = 20$
$14x + 14 - 4x = 20$
$10x + 14 = 20$
$10x + 14 - 14 = 20 - 14$
$10x = 6$
$\frac{10x}{10} = \frac{6}{10}$
$x = \frac{3}{5}$

8. $0.5(1.2x - 3.4) = -1.4x + 5.8$
$0.6x - 1.7 = -1.4x + 5.8$
$0.6x + 1.4x - 1.7 = -1.4x + 1.4x + 5.8$
$2x - 1.7 = 5.8$
$2x - 1.7 + 1.7 = 5.8 + 1.7$
$2x = 7.5$
$\frac{2x}{2} = \frac{7.5}{2}$
$x = 3.75 \text{ or } 3\frac{3}{4}$

9. $3(x + 6) = -2(4x - 1) + x$
$3x + 18 = -8x + 2 + x$
$3x + 18 = -7x + 2$
$3x + 7x + 18 = -7x + 7x + 2$
$10x + 18 = 2$
$10x + 18 - 18 = 2 - 18$
$10x = -16$
$\frac{10x}{10} = -\frac{16}{10}$
$x = -\frac{8}{5}$

10. $\frac{x}{3} + \frac{x}{4} = \frac{5}{6}$
$12\left(\frac{x}{3}\right) + 12\left(\frac{x}{4}\right) = 12\left(\frac{5}{6}\right)$
$4x + 3x = 10$
$7x = 10$
$\frac{7x}{7} = \frac{10}{7}$
$x = \frac{10}{7}$

11.
$$\frac{1}{4}(x+3) = 4x - 2(x-3)$$
$$\frac{1}{4}x + \frac{3}{4} = 4x - 2x + 6$$
$$\frac{1}{4}x + \frac{3}{4} = 2x + 6$$
$$4\left(\frac{1}{4}x\right) + 4\left(\frac{3}{4}\right) = 4(2x) + 4(6)$$
$$x + 3 = 8x + 24$$
$$x - x + 3 = 8x - x + 24$$
$$3 = 7x + 24$$
$$3 - 24 = 7x + 24 - 24$$
$$-21 = 7x$$
$$\frac{-21}{7} = \frac{7x}{7}$$
$$-3 = x$$

12.
$$\frac{1}{2}(x-1) + 2 = 3(2x-1)$$
$$\frac{1}{2}x - \frac{1}{2} + 2 = 6x - 3$$
$$\frac{1}{2}x + \frac{3}{2} = 6x - 3$$
$$2\left(\frac{1}{2}x\right) + 2\left(\frac{3}{2}\right) = 2(6x) - 2(3)$$
$$x + 3 = 12x - 6$$
$$x - x + 3 = 12x - x - 6$$
$$3 = 11x - 6$$
$$3 + 6 = -11x - 6 + 6$$
$$9 = 11x$$
$$\frac{9}{11} = \frac{11x}{11}$$
$$\frac{9}{11} = x$$

13.
$$\frac{1}{7}(7x-14) - 2 = \frac{1}{3}(x-2)$$
$$x - 2 - 2 = \frac{1}{3}x - \frac{2}{3}$$
$$x - 4 = \frac{1}{3}x - \frac{2}{3}$$
$$3(x) - 3(4) = 3\left(\frac{1}{3}x\right) - 3\left(\frac{2}{3}\right)$$
$$3x - 12 = x - 2$$
$$3x - x - 12 = x - x - 2$$
$$2x - 12 = -2$$
$$2x - 12 + 12 = -2 + 12$$
$$2x = 10$$
$$\frac{2x}{2} = \frac{10}{2}$$
$$x = 5$$

14.
$$0.2(x-3) = 4(0.2x - 0.1)$$
$$0.2x - 0.6 = 0.8x - 0.4$$
$$10(0.2x) - 10(0.6) = 10(0.8x) - 10(0.4)$$
$$2x - 6 = 8x - 4$$
$$2x - 2x - 6 = 8x - 2x - 4$$
$$-6 = 6x - 4$$
$$-6 + 4 = 6x - 4 + 4$$
$$-2 = 6x$$
$$\frac{-2}{6} = \frac{6x}{6}$$
$$-\frac{1}{3} = x$$

2.5 Exercises

1. eleven more than a number: $x + 11$

3. twelve less than a number: $x - 12$

5. one-eighth of a quantity: $\frac{1}{8}x$ or $\frac{x}{8}$

7. twice a quantity: $2x$

9. three more than half of a number: $3 + \frac{1}{2}x$

11. double a quantity increased by nine: $2x + 9$

13. one-third of the sum of a number and seven: $\frac{1}{3}(x+7)$

15. one-third of a number reduced by twice the same

number: $\dfrac{1}{3}x - 2x$

17. five times a quantity decreased by eleven:
$5x - 11$

19. Since the value of the IBM stock is being compared to the value of the AT&T stock, we let the variable represent the value of the AT&T stock.
x = value of a share of AT&T stock
The value of a share of IBM stock is $74.50 more than the value of a share of AT&T stock.
$x + 74.50$ = value of a share of IBM stock

21. Since the length of the rectangle is being compared to the width, we let the variable represent the width of a rectangle.
w = width of a rectangle
The length is 7 inches more than twice the width.
$2w + 7$ = length of rectangle

23. Since the numbers of boxes of cookies sold by Sarah and Imelda are being compared to the number sold by Keiko, we let the variable represent the number of boxes sold by Keiko.
x = number of boxes of cookies sold by Keiko
The number of boxes sold by Sarah was 43 fewer than the number sold by Keiko.
$x - 43$ = number of boxes of cookies sold by Sarah
The number of boxes sold by Imelda was 53 more than the number sold by Keiko.
$x + 53$ = number of boxes of cookies sold by Imelda

25. Since the measures of the first and third angles are being compared to the measure of the second angle, we let the variable represent the measure of the second angle.
s = measure of the second angle
The measure of the first angle is 25 degrees less than the measure of the second angle.
$s - 25$ = measure of the first angle
The measure of the third angle is triple the measure of the second.
$3s$ = measure of the third angle

27. Since the exports of Japan are being compared to the exports of Canada, we let the variable represent the value of the exports of Canada.
v = value of exports of Canada
The value of the exports of Japan was twice the value of the exports of Canada.
$2v$ = value of exports of Japan

29. Since the price of the All Star Concert tickets is being compared to the price of the Summer on the Beach Concert tickets, we let the variable represent the price of the Summer on the Beach Concert tickets.
p = price of the Summer on the Beach Concert tickets
The price of the all Star Concert tickets was one-half the price of the Summer on the Beach Concert tickets.
$\dfrac{1}{2}p$ = the price of the all Star Concert tickets

31. x = number of men aged 16 to 24
Since $302 - 220 = 82$, 82 more men aged 25 to 34 rented kayaks than men aged 16 to 24.
$x + 82$ = number of men aged 25 to 34
Since $220 - 195 = 25$, 25 fewer men aged 35 to 44 rented kayaks than men aged 16 to 24.
$x - 25$ = number of men aged 35 to 44
Since $220 - 110 = 110$, 110 fewer men aged 45+ rented kayaks than men aged 16 to 24.
$x - 110$ = number of men aged 45+

Cumulative Review

33.
$$x + \frac{1}{2}(x - 3) = 9$$
$$x + \frac{1}{2}x - \frac{3}{2} = 9$$
$$\frac{3}{2}x - \frac{3}{2} = 9$$
$$2\left(\frac{3}{2}x\right) - 2\left(\frac{3}{2}\right) = 2(9)$$
$$3x - 3 = 18$$
$$3x - 3 + 3 = 18 + 3$$
$$3x = 21$$
$$\frac{3x}{3} = \frac{21}{3}$$
$$x = 7$$

34. $\dfrac{3}{5}x - 3(x-1) = 9$

$$\dfrac{3}{5}x - 3x + 3 = 9$$

$$-\dfrac{12}{5}x + 3 = 9$$

$$5\left(-\dfrac{12}{5}x\right) + 5(3) = 5(9)$$

$$-12x + 15 = 45$$

$$-12x + 15 - 15 = 45 - 15$$

$$-12x = 30$$

$$\dfrac{-12x}{-12} = \dfrac{30}{-12}$$

$$x = -\dfrac{5}{2} \text{ or } -2\dfrac{1}{2}$$

Quick Quiz 2.5

1. Ten greater than a number: $x + 10$ or $10 + x$

2. Five less than double a number: $2x - 5$

3. Since the measures of the first and third angles are being compared to the measure of the second angle, we let the variable represent the measure of the second angle.
   x = the measure of the second angle
   The measure of the first angle is 15 degrees more than the measure of the second angle.
   $x + 15$ = the measure of the first angle
   The measure of the third angle is five times the measure of the second angle.
   $5x$ = the measure of the third angle

4. Answers may vary. Possible solution: "one-third of the sum" means you multiply $\dfrac{1}{3}$ times the sum. The sum will be a quantity in parentheses because the $\dfrac{1}{3}$ is multiplied by the whole sum;
   $\dfrac{1}{3}(x + 7)$.

2.6 Exercises

1. x = the number
   $$x - 543 = 718$$
   $$x - 543 + 543 = 718 + 543$$
   $$x = 1261$$
   The number is 1261.
   Check:
   Does 1261 minus 543 give 718?
   $$1261 - 543 \overset{?}{=} 718$$
   $$718 = 718 \checkmark$$

3. x = the number
   $$\dfrac{x}{8} = 296$$
   $$8\left(\dfrac{x}{8}\right) = 8(296)$$
   $$x = 2368$$
   The number is 2368.
   Check:
   Is 2368 divided by 8 equal to 296?
   $$\dfrac{2368}{8} \overset{?}{=} 296$$
   $$296 = 296 \checkmark$$

5. x = the number
   $$x + 17 = 199$$
   $$x + 17 - 17 = 199 - 17$$
   $$x = 182$$
   The number is 182.
   Check:
   Is 17 greater than 182 equal to 199?
   $$182 + 17 \overset{?}{=} 199$$
   $$199 = 199 \checkmark$$

7. x = the number
   $$2x + 7 = 93$$
   $$2x + 7 - 7 = 93 - 7$$
   $$2x = 86$$
   $$\dfrac{2x}{2} = \dfrac{86}{2}$$
   $$x = 43$$
   The number is 43.
   Check:
   When 43 is doubled and then increased by 7, is the result 93?
   $$2(43) + 7 \overset{?}{=} 93$$
   $$86 + 7 \overset{?}{=} 93$$
   $$93 = 93 \checkmark$$

9. x = the number

$$18 - \frac{2}{3}x = 12$$

$$3(18) - 3\left(\frac{2}{3}x\right) = 3(12)$$

$$54 - 2x = 36$$

$$54 - 2x - 54 = 36 - 54$$

$$-2x = -18$$

$$\frac{-2x}{-2} = \frac{-18}{-2}$$

$$x = 9$$

The number is 9.
Check:
When 18 is reduced by two-thirds of 9, is the result 12?

$$18 - \frac{2}{3}(9) \stackrel{?}{=} 12$$

$$18 - 6 \stackrel{?}{=} 12$$

$$12 = 12 \checkmark$$

11. x = the number

$$3x - 8 = 5x$$

$$3x - 8 - 3x = 5x - 3x$$

$$-8 = 2x$$

$$\frac{-8}{2} = \frac{2x}{2}$$

$$-4 = x$$

The number is -4.
Check:
Is 8 less than triple -4 the same as 5 times -4?

$$3(-4) - 8 \stackrel{?}{=} 5(-4)$$

$$-12 - 8 \stackrel{?}{=} -20$$

$$-20 = -20 \checkmark$$

13. x = the number

$$x + \frac{1}{2}x + \frac{1}{3}x = 22$$

$$6x + 6\left(\frac{1}{2}x\right) + 6\left(\frac{1}{3}x\right) = 6(22)$$

$$6x + 3x + 2x = 132$$

$$11x = 132$$

$$\frac{11x}{11} = \frac{132}{11}$$

$$x = 12$$

The number is 12.
Check:
When 12, half of 12, and one-third of 12 are added, is the result 22?

$$12 + \frac{1}{2}(12) + \frac{1}{3}(12) \stackrel{?}{=} 22$$

$$12 + 6 + 4 \stackrel{?}{=} 22$$

$$22 = 22 \checkmark$$

15. x = number of red cases
$3x$ = number of black cases

$$3x = 120$$

$$\frac{3x}{3} = \frac{120}{3}$$

$$x = 40$$

Lester should order 40 red tablet cases.

17. Let x = the measure of the equal angles.

$$x + x + 36 = 180$$

$$2x + 36 = 180$$

$$2x + 36 - 36 = 180 - 36$$

$$2x = 144$$

$$\frac{2x}{2} = \frac{144}{2}$$

$$x = 72$$

The two equal angles each measure 72°.

19. x = measure of 2nd angle
$2x$ = measure of 1st angle
$2(2x) + 19 = 4x + 19$ = measure of 3rd angle

$$x + 2x + 4x + 19 = 180$$

$$7x + 19 = 180$$

$$7x = 161$$

$$x = 23$$

$2x = 46$
$4x + 19 = 111$
1st angle measures 46°.
2nd angle measures 23°.
3rd angle measures 111°.

21. x = number of pieces of jewelry

$$2(38) + 49 + 11.50x = 171$$

$$76 + 49 + 11.50x = 171$$

$$125 + 11.50x = 171$$

$$125 + 11.50x - 125 = 171 - 125$$

$$11.50x = 46$$

$$\frac{11.50x}{11.50} = \frac{46}{11.50}$$

$$x = 4$$

She could buy 4 pieces of jewelry.

23. x = running time for last meet

$$\frac{11.7+11.6+12+12.1+11.9+x}{6}=11.8$$

$$\frac{59.3+x}{6}=11.8$$

$$59.3+x=70.8$$

$$x=11.5$$

Her running time in the last meet was 11.5 seconds.

25. Nell: $r=12$, $t=2.5$

$d=rt=12(2.5)=30$

Kristin: $r=14$, $t=2.5$

$d=rt=14(2.5)=35$

Distance apart = $35-30=5$

They will be 5 miles apart.

27. r = rate

Highway route: $5r=320$

$$r=64$$

Highway route at 64 mph.

Mountain route: $6r=312$

$$r=52$$

Mountain route at 52 mph.

Difference: $64-52=12$ mph

The highway route was 12 mph faster than the mountain route.

29. x = score on the final lab

$$\frac{84+81+93+89+89+94+94+x+x+x}{10}=90$$

$$\frac{624+3x}{10}=90$$

$$624+3x=900$$

$$3x=276$$

$$x=92$$

She needs a 92 on the final lab.

31. x = number of cricket chirps

F = Fahrenheit temperature

a. $F-40=\dfrac{x}{4}$

b. $90-40=\dfrac{x}{4}$

$$50=\dfrac{x}{4}$$

$$x=200$$

200 chirps per minute should be recorded.

c. $F-40=\dfrac{148}{4}$

$$F-40=37$$

$$F=77$$

The temperature would be 77°F.

Cumulative Review

32. $5x(2x^2-6x-3)=5x(2x^2)-5x(6x)-5x(3)$

$$=10x^3-30x^2-15x$$

33. $-2a(ab-3b+5a)$

$$=-2a(ab)-2a(-3b)-2a(5a)$$

$$=-2a^2b+6ab-10a^2$$

34. $7x-3y-12x-8y+5y=(7-12)x+(-3-8+5)y$

$$=-5x-6y$$

35. $5x^2y-7xy^2-8xy-9x^2y$

$$=(5-9)x^2y-7xy^2-8xy$$

$$=-4x^2y-7xy^2-8xy$$

Quick Quiz 2.6

1. x = the number

$$3x-15=36$$

$$3x-15+15=36+15$$

$$3x=51$$

$$\frac{3x}{3}=\frac{51}{3}$$

$$x=17$$

The number is 17.

2. Let x = the measure of the two equal angles.

$$x+x+70=180$$

$$2x+70=180$$

$$2x+70-70=180-70$$

$$2x=110$$

$$\frac{2x}{2}=\frac{110}{2}$$

$$x=55$$

The angles each measure 55°.

3. $x = $ score on next test
$$\frac{84 + 89 + 73 + 80 + x}{5} = 80$$
$$\frac{326 + x}{5} = 80$$
$$5\left(\frac{326 + x}{5}\right) = 5(80)$$
$$326 + x = 400$$
$$326 + x - 326 = 400 - 326$$
$$x = 74$$
James must score 74.

4. Answers may vary. Possible solution:
Since we want to know how many pairs of socks, let $x = $ the number of pairs of socks. Set up the equation to represent the total amount spent.
$2(\$23) + \$0.75x = \$60.25$

2.7 Exercises

1. $x = $ number of coffee products
$$0.50x + 11.50(30) = 364$$
$$0.50x + 345 = 364$$
$$0.50x = 19$$
$$\frac{0.50x}{0.50} = \frac{19}{0.50}$$
$$x = 38$$
Clyde sold 38 coffee products.

3. $x = $ overtime hours
$$32(40) + 48x = 1520$$
$$1280 + 48x = 1520$$
$$48x = 240$$
$$\frac{48x}{48} = \frac{240}{48}$$
$$x = 5$$
Eli needs 5 hours of overtime per week.

5. $x = $ time to pay off the uniforms
$$600 + 105x = 1817.75$$
$$105x = 1217.75$$
$$\frac{105x}{105} = \frac{1217.75}{105}$$
$$x \approx 11.6$$
It will take about 12 weeks.

7. $x = $ original price
$$0.28x = 100.80$$
$$\frac{0.28x}{0.28} = \frac{100.80}{0.28}$$
$$x = 360$$
The original price was $360.

9. $x = $ last year's salary
$0.03x = $ raise
$$x + 0.03x = 22,660$$
$$1.03x = 22,660$$
$$\frac{1.03x}{1.03} = \frac{22,660}{1.03}$$
$$x = 22,000$$
Last year's salary was $22,000.

11. $x = $ investment
$0.06x = $ interest
$$x + 0.06x = 12,720$$
$$1.06x = 12,720$$
$$\frac{1.06x}{1.06} = \frac{12,720}{1.06}$$
$$x = 12,000$$
She invested $12,000.

13. $x = $ amount earning 7%
$5000 - x = $ amount earning 5%
$$0.07x + 0.05(5000 - x) = 310$$
$$0.07x + 250 - 0.05x = 310$$
$$0.02x + 250 = 310$$
$$0.02x = 60$$
$$\frac{0.02x}{0.02} = \frac{60}{0.02}$$
$$x = 3000$$
$5000 - x = 2000$
They invested $3000 at 7% and $2000 at 5%.

15. $x = $ amount invested at 12% (growth fund)
$400,000 - x = $ amount invested at 8% (conservative fund)
$$0.12x + 0.08(400,000 - x) = 38,000$$
$$0.12x + 32,000 - 0.08x = 38,000$$
$$0.04x + 32,000 = 38,000$$
$$0.04x = 6000$$
$$\frac{0.04x}{0.04} = \frac{6000}{0.04}$$
$$x = 150,000$$
$400,000 - x = 250,000$
They invested $150,000 in the growth fund and $250,000 in the conservative fund.

17. $x = $ amount invested
$$\frac{x}{2} = \text{amount invested at 5\%}$$
$$\frac{x}{3} = \text{amount invested at 4\%}$$

$$x - \frac{x}{2} - \frac{x}{3} = \frac{6x - 3x - 2x}{6}$$
$$= \frac{x}{6}$$
$$= \text{amount invested at } 3.5\%$$

$$0.05\left(\frac{x}{2}\right) + 0.04\left(\frac{x}{3}\right) + (0.035)\left(\frac{x}{6}\right) = 530$$
$$\left(\frac{0.05}{2} + \frac{0.04}{3} + \frac{0.035}{6}\right)x = 530$$
$$\frac{53}{1200}x = 530$$
$$x = 12,000$$

He invested $12,000.

19. x = number of quarters
$x - 4$ = number of nickels
$$0.25x + 0.05(x - 4) = 3.70$$
$$25x + 5(x - 4) = 370$$
$$25x + 5x - 20 = 370$$
$$30x = 390$$
$$\frac{30x}{30} = \frac{390}{30}$$
$$x = 13$$

$x - 4 = 9$
She has 13 quarters and 9 nickels.

21. x = number of dimes
$x + 3$ = number of quarters
$2(x + 3) = 2x + 6$ = number of nickels
$$0.25(x + 3) + 0.10x + 0.05(2x + 6) = 3.75$$
$$25(x + 3) + 10x + 5(2x + 6) = 375$$
$$25x + 75 + 10x + 10x + 30 = 375$$
$$45x + 105 = 375$$
$$45x = 270$$
$$\frac{45x}{45} = \frac{270}{45}$$
$$x = 6$$

$x + 3 = 6 + 3 = 9$
$2x + 6 = 12 + 6 = 18$
He has 9 quarters, 6 dimes, and 18 nickels.

23. x = number of boxes of beige paper
$2x$ = number of boxes of white paper
$$3.50(2x) + 4.25x = 112.50$$
$$7x + 4.25x = 112.50$$
$$11.25x = 112.50$$
$$\frac{11.25x}{11.25} = \frac{112.50}{11.25}$$
$$x = 10$$

$2x = 20$
Huy should order 10 boxes of beige paper and 20 boxes of white paper.

25. x = number of $50 bills
$x + 16$ = number of $10 bills
$3x + 1$ = number of $20 bills
$$50x + 10(x + 16) + 20(3x + 1) = 1380$$
$$50x + 10x + 160 + 60x + 20 = 1380$$
$$120x + 180 = 1380$$
$$120x = 1200$$
$$\frac{120x}{120} = \frac{1200}{120}$$
$$x = 10$$

$x + 16 = 10 + 16 = 26$
$3x + 1 = 3(10) + 1 = 31$
They had ten $50 bills, twenty-six $10 bills, and thirty-one $20 bills.

27. x = amount of sales
$$18,000 + 0.04x = 55,000$$
$$0.04x = 37,000$$
$$\frac{0.04x}{0.04} = \frac{37,000}{0.04}$$
$$x = 925,000$$
She must sell $925,000 worth of furniture.

Cumulative Review

29. $5(3) + 6 \div (-2) = 15 + (-3) = 12$

30. $5(-3) - 2(12 - 15)^2 \div 9 = 5(-3) - 2(-3)^2 \div 9$
$$= 5(-3) - 2(9) \div 9$$
$$= -15 - 18 \div 9$$
$$= -15 - 2$$
$$= -17$$

31. If $a = -1$ and $b = 4$, then
$$a^2 - 2ab + b^2 = (-1)^2 - 2(-1)(4) + (4)^2$$
$$= 1 + 8 + 16$$
$$= 25$$

32. If $a = -1$ and $b = 4$, then
$$a^3 + ab^2 - b - 5 = (-1)^3 + (-1)(4)^2 - 4 - 5$$
$$= -1 + (-1)(16) - 4 - 5$$
$$= -1 - 16 - 4 - 5$$
$$= -26$$

Quick Quiz 2.7

1. x = number of months
$$224 + 114x = 1250$$
$$114x = 1026$$
$$\frac{114x}{114} = \frac{1026}{114}$$
$$x = 9$$
He can rent the machine for 9 months.

2. $x = $ last year's cost
    $0.7x = $ increase
    $x + 0.7x = 12,412$
    $1.07x = 12,412$
    $$\frac{1.07x}{1.07} = \frac{12,412}{1.07}$$
    $x = 11,600$
    It cost \$11,600 before the increase.

3. $x = $ amount invested at 4%
    $5000 - x = $ amount invested at 5%
    $0.04x + 0.05(5000 - x) = 228$
    $0.04x + 250 - 0.05x = 228$
    $250 - 0.01x = 228$
    $-0.01x = -22$
    $$\frac{-0.01x}{-0.01} = \frac{-22}{-0.01}$$
    $x = 2200$
    $5000 - x = 5000 - 2200 = 2800$
    They invested \$2200 at 4% and \$2800 at 5%.

4. Answers may vary. Possible solution:
    Since each amount is given in terms of quarters,
    let $x = $ the number of quarters. Then
    $2x = $ number of dimes, and
    $x + 1 = $ number of nickels. Now set up an
    equation, and solve, given the value of each coin
    and the total.
    $0.25x + 0.10(2x) + 0.05(x + 1) = 2.55$

2.8 Exercises

1. $5 > -6$ is equivalent to $-6 < 5$. Both statements
    imply that 5 is to the right of -6 on a number
    line.

3. $8 ? -6$
    Use $>$, since 8 is to the right of -6 on a number
    line.
    $8 > -6$

5. $0 ? -8$
    Use $>$, since 0 is to the right of -8 on a number
    line.
    $0 > -8$

7. $-4 ? -2$
    Use $<$, since -4 is to the left of -2 on a number
    line.
    $-4 < -2$

9. **a.** $-7 ? 2$
    Use $<$, since -7 is to the left of 2 on a
    number line.
    $-7 < 2$

    b. $2 ? -7$
    From part a, $2 > -7$ since $-7 < 2$ is
    equivalent to $2 > -7$.

11. **a.** $15 ? -15$
    Use $>$, since 15 is to the right of -15 on a
    number line.
    $15 > -15$

    b. $-15 ? 15$
    From part a, $-15 < 15$ since $15 > -15$ is
    equivalent to $-15 < 15$.

13. $\dfrac{1}{3} ? \dfrac{9}{10}$
    $\dfrac{10}{30} ? \dfrac{27}{30}$
    Use $<$, since $10 < 27$.
    $\dfrac{1}{3} < \dfrac{9}{10}$

15. $\dfrac{7}{8} ? \dfrac{25}{31}$
    $\dfrac{217}{248} ? \dfrac{200}{248}$
    Use $>$, since 217 is to the right of 200 on a
    number line.
    $\dfrac{7}{8} > \dfrac{25}{31}$

17. $-6.6 ? -8.9$
    Use $>$, since -6.6 is to the right of -8.9 on a
    number line.
    $-6.6 > -8.9$

19. $-4.2 ? 3.5$
    Use $<$, since -4.2 is to the left of 3.5 on a
    number line.
    $-4.2 < 3.5$

21. $-\dfrac{10}{3} ? -3$
    $-3\dfrac{1}{3} ? -3$
    Use $<$, since $-3\dfrac{1}{3}$ is to the left of -3 on a
    number line.
    $-\dfrac{10}{3} < -3$

23. $-\dfrac{5}{8} ? -\dfrac{3}{5}$

$-0.625 \ ? \ -0.6$

Use <, since -0.625 is to the left of -0.6 on a number line.

$-\dfrac{5}{8} < -\dfrac{3}{5}$

25. $x > 7$

x is greater than 7. All of the points to the right of 7 are shaded.

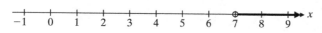

27. $x \geq -6$

x is greater than or equal to -6. All of the points to the right of -6 are shaded. The closed circle indicates that we do include the point for -6.

29. $x < -\dfrac{1}{4}$

x is less than $-\dfrac{1}{4}$. All of the points to the left of $-\dfrac{1}{4}$ are shaded.

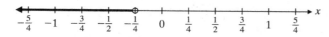

31. $x \leq -5.3$

x is less than or equal to -5.3. All of the points to the left of -5.3 are shaded. The closed circle indicates that we do include the point for -5.3.

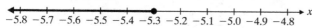

33. $25 < x$

25 is less than x is equivalent to x is greater than 25. All of the points to the right of 25 are shaded.

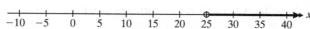

35. x is greater than or equal to $-\dfrac{2}{3}$.

$x \geq -\dfrac{2}{3}$

37. x is less than -20.

$x < -20$

39. x is less than or equal to 3.7.

$x \leq 3.7$

41. Since the number of hours must not be less than 12, then the number of hours must be greater than or equal to 12. Thus we write $c \geq 12$.

43. Since the height must be at least 48 inches, the height must be greater than or equal to 48 inches. Thus we write $h \geq 48$.

45. $x \leq 2, x > -3, \ x < \dfrac{5}{2}, x \geq -\dfrac{5}{2}$

x is less than or equal to 2.
x is greater than -3.

x is less than $\dfrac{5}{2}$.

x is greater than or equal to $-\dfrac{5}{2}$.

Since $-\dfrac{5}{2} = -2.5$ is greater than -3, x must be greater than or equal to $-\dfrac{5}{2}$.

Since 2 is less than $\dfrac{5}{2} = 2.5$, x must be less than or equal to 2.

$-\dfrac{5}{2} \leq x \leq 2$

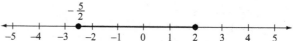

47. $x + 7 \leq 4$

$x + 7 - 7 \leq 4 - 7$

$x \leq -3$

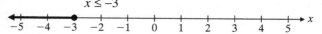

49. $5x \leq 25$

$\dfrac{5x}{5} \leq \dfrac{25}{5}$

$x \leq 5$

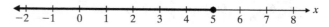

51. $-2x < 18$

$\dfrac{-2x}{-2} > \dfrac{18}{-2}$

$x > -9$

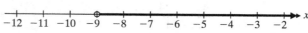

53. $\dfrac{1}{2}x \ge 4$

$2\left(\dfrac{1}{2}x\right) \ge 2(4)$

$x \ge 8$

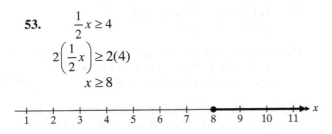

55. $-\dfrac{1}{4}x > 3$

$-4\left(-\dfrac{1}{4}x\right) < -4(3)$

$x < -12$

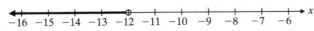

57. $8 - 5x > 13$

$8 - 5x - 8 > 13 - 8$

$-5x > 5$

$\dfrac{-5x}{-5} < \dfrac{5}{-5}$

$x < -1$

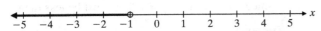

59. $-4 + 5x < -3x + 8$

$-4x + 5x + 3x < -3x + 3x + 8$

$-4 + 8x < 8$

$-4 + 4 + 8x < 8 + 4$

$8x < 12$

$\dfrac{8x}{8} < \dfrac{12}{8}$

$x < \dfrac{3}{2}$

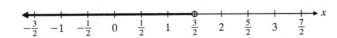

61. $\dfrac{5x}{6} - 5 > \dfrac{x}{6} - 9$

$6\left(\dfrac{5x}{6}\right) - 6(5) > 6\left(\dfrac{x}{6}\right) - 6(9)$

$5x - 30 > x - 54$

$5x - 30 - x > x - 54 - x$

$4x - 30 > -54$

$4x - 30 + 30 > -54 + 30$

$4x > -24$

$\dfrac{4x}{4} > \dfrac{-24}{4}$

$x > -6$

63. $2(3x + 4) > 3(x + 3)$

$6x + 8 > 3x + 9$

$6x + 8 - 3x > 3x + 9 - 3x$

$3x + 8 > 9$

$3x + 8 - 8 > 9 - 8$

$3x > 1$

$\dfrac{3x}{3} > \dfrac{1}{3}$

$x > \dfrac{1}{3}$

65. $5 > 3$

$5 + (-2) > 3 + (-2)$

$3 > 1$

Adding any number to both sides of an inequality does not reverse the direction.

67. $5x + 2 > 8x - 7$

$5x - 8x + 2 > 8x - 8x - 7$

$-3x + 2 > -7$

$-3x + 2 - 2 > -7 - 2$

$-3x > -9$

$\dfrac{-3x}{-3} < \dfrac{-9}{-3}$

$x < 3$

69. $6x - 2 \ge 4x + 6$

$6x - 4x - 2 \ge 4x - 4x + 6$

$2x - 2 \ge 6$

$2x - 2 + 2 \ge 6 + 2$

$2x \ge 8$

$\dfrac{2x}{2} \ge \dfrac{8}{2}$

$x \ge 4$

71. $0.3(x - 1) < 0.1x - 0.5$

$0.3x - 0.3 < 0.1x - 0.5$

$10(0.3x) - 10(0.3) < 10(0.1x) - 10(0.5)$

$3x - 3 < x - 5$

$3x - 3 - x < x - 5 - x$

$2x - 3 < -5$

$2x - 3 + 3 < -5 + 3$

$2x < -2$

$\dfrac{2x}{2} < \dfrac{-2}{2}$

$x < -1$

73.
$$3 + 5(2 - x) \geq -3(x + 5)$$
$$3 + 10 - 5x \geq -3x - 15$$
$$13 - 5x \geq -3x - 15$$
$$13 - 5x + 3x \geq -3x - 15 + 3x$$
$$-2x + 13 \geq -15$$
$$-2x + 13 - 13 \geq -15 - 13$$
$$-2x \geq -28$$
$$\frac{-2x}{-2} \geq \frac{-28}{-2}$$
$$x \leq 14$$

75.
$$\frac{x + 6}{7} - \frac{3}{7} > \frac{x + 3}{2}$$
$$14\left(\frac{x + 6}{7}\right) - 14\left(\frac{3}{7}\right) > 14\left(\frac{x + 3}{2}\right)$$
$$2(x + 6) - 6 > 7(x + 3)$$
$$2x + 12 - 6 > 7x + 21$$
$$2x + 6 > 7x + 21$$
$$2x + 6 - 7x > 7x + 21 - 7x$$
$$-5x + 6 > 21$$
$$-5x + 6 - 6 > 21 - 6$$
$$-5x > 15$$
$$\frac{-5x}{-5} < \frac{15}{-5}$$
$$x < -3$$

77.
$$\frac{75 + 83 + 86 + x}{4} \geq 80$$
$$\frac{244 + x}{4} \geq 80$$
$$4\left(\frac{244 + x}{4}\right) \geq 4(80)$$
$$244 + x \geq 320$$
$$x \geq 76$$
The student must get a 76 or higher.

79.
$$268 + 4x \geq 300$$
$$268 + 4x - 268 \geq 300 - 268$$
$$4x \geq 32$$
$$\frac{4x}{4} \geq \frac{32}{4}$$
$$x \geq 8$$
It will take 8 days or more.

Cumulative Review

81. $16\% \text{ of } 38 = 0.16 \times 38 = 6.08$

82. 18 is what percent of 120?
$$\frac{18}{120} = \frac{3}{20} = 0.15 = 15\%$$

83. 16 is what percent of 800?
$$\frac{16}{800} = 0.02 = 2\%$$
2% are accepted.

84. $\dfrac{3}{8} = 0.375 = 37.5\%$

Quick Quiz 2.8

1. $x \leq -3.5$
   x is less than or equal to -3.5. All of the points to the left of -3.5 are shaded. The closed circle indicates that we do include the point for -3.5.

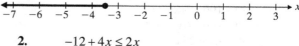

2.
$$-12 + 4x \leq 2x$$
$$-12 + 4x - 4x \leq 2x - 4x$$
$$-12 \leq -2x$$
$$\frac{-12}{-2} \geq \frac{-2x}{-2}$$
$$6 \geq x$$

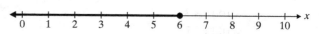

3.
$$\frac{x}{2} - 1 < \frac{3}{2}x + 4$$
$$2\left(\frac{x}{2}\right) - 2(1) < 2\left(\frac{3}{2}x\right) + 2(4)$$
$$x - 2 < 3x + 8$$
$$x - 2 - x < 3x + 8 - x$$
$$-2 < 2x + 8$$
$$-2 - 8 < 2x + 8 - 8$$
$$-10 < 2x$$
$$\frac{-10}{2} < \frac{2x}{2}$$
$$-5 < x$$

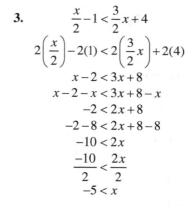

4. Answers may vary. Possible solution: $12 < x$ is the same as $x > 12$, but written in different ways. The graphs will be exactly the same.

Career Exploration Problems

1. Solve for W
    $$BMR = 10W + 6.25H - 5A + 5$$
    $$BMR - 6.25H + 5A - 5 = 10W$$
    $$\frac{BMR - 6.25H + 5A - 5}{10} = W$$
    Solve for H.
    $$BMR = 10W + 6.25H - 5A + 5$$
    $$BMR - 10W + 5A - 5 = 6.25H$$
    $$\frac{BMR - 10W + 5A - 5}{6.25} = H$$

2. $IBW = 106$ lb $+ 6$ (the number of inches
    over 5 feet tall)
    $$190 = 106 + 6(x - 60)$$
    $$190 = 106 + 6x - 360$$
    $$190 = 6x - 254$$
    $$444 = 6x$$
    $$74 = x$$
    The male's height is 74 inches or 6 feet 2 inches.

3. $IBW = 100$ lb $+ 5$ (the number of inches
    over 5 feet tall)
    $$= 100 + 5(5)$$
    $$= 100 + 25$$
    $$= 125$$
    Her ideal weight is 125 pounds.

You Try It

1. $$-8x - 1 + x = 13 - 6x - 2$$
    $$-7x - 1 = -6x + 11$$
    $$-7x - 1 + 6x = -6x + 11 + 6x$$
    $$-x - 1 = 11$$
    $$-x - 1 + 1 = 11 + 1$$
    $$-x = 12$$
    $$\frac{-x}{-1} = \frac{12}{-1}$$
    $$x = -12$$

2. $$\frac{1}{3}(y + 5) = \frac{1}{4}(5y - 8)$$
    $$\frac{1}{3}y + \frac{5}{3} = \frac{5}{4}y - 2$$
    $$12\left(\frac{1}{3}y\right) + 12\left(\frac{5}{3}\right) = 12\left(\frac{5}{4}y\right) - 12(2)$$
    $$4y + 20 = 15y - 24$$
    $$4y + 20 - 15y = 15y - 24 - 15y$$
    $$-11y + 20 = -24$$
    $$-11y + 20 - 20 = -24 - 20$$
    $$-11y = -44$$
    $$\frac{-11y}{-11} = \frac{-44}{-11}$$
    $$y = 4$$

3. $x = $ larger number
    The smaller number is one-fifth the larger.
    $$\frac{1}{5}x = \text{smaller number}$$
    The sum of the numbers is 60.
    $$x + \frac{1}{5}x = 60$$
    $$5(x) + 5\left(\frac{1}{5}x\right) = 5(60)$$
    $$5x + x = 300$$
    $$6x = 300$$
    $$x = 50$$
    $$\frac{1}{5}x = \frac{1}{5}(50) = 10$$
    The numbers are 10 and 50.

4. Let $x = $ the amount invested at 2%. Then
    $3600 - x = $ the amount invested at 4%.
    $$0.02x + 0.04(3600 - x) = 132$$
    $$0.02x + 144 - 0.04x = 132$$
    $$-0.02x + 144 = 132$$
    $$-0.02x = -12$$
    $$\frac{-0.02x}{-0.02} = \frac{-12}{-0.02}$$
    $$x = 600$$
    $3600 - x = 3600 - 600 = 3000$
    He invested \$600 at 2% and \$3000 at 4%.

5.
$$4 + 3x - 5 \geq \frac{1}{3}(10x + 1)$$
$$3x - 1 \geq \frac{10}{3}x + \frac{1}{3}$$
$$3(3x) - 3(1) \geq 3\left(\frac{10}{3}x\right) + 3\left(\frac{1}{3}\right)$$
$$9x - 3 \geq 10x + 1$$
$$9x - 3 - 10x \geq 10x + 1 - 10x$$
$$-x - 3 \geq 1$$
$$-x - 3 + 3 \geq 1 + 3$$
$$-x \geq 4$$
$$\frac{-x}{-1} \leq \frac{4}{-1}$$
$$x \leq -4$$

Chapter 2 Review Problems

1.
$$3x + 2x = -35$$
$$5x = -35$$
$$\frac{5x}{5} = \frac{-35}{5}$$
$$x = -7$$

2.
$$x - 19 = -29 + 7$$
$$x - 19 = -22$$
$$x - 19 + 19 = -22 + 19$$
$$x = -3$$

3.
$$18 - 10x = 63 + 5x$$
$$18 - 10x + 10x = 63 + 5x + 10x$$
$$18 = 63 + 15x$$
$$18 - 63 = 63 - 63 + 15x$$
$$-45 = 15x$$
$$\frac{-45}{15} = \frac{15x}{15}$$
$$-3 = x$$

4.
$$x - (0.5x + 2.6) = 17.6$$
$$x - 0.5x - 2.6 = 17.6$$
$$0.5x - 2.6 = 17.6$$
$$10(0.5x) - 10(2.6) = 10(17.6)$$
$$5x - 26 = 176$$
$$5x - 26 + 26 = 176 + 26$$
$$5x = 202$$
$$\frac{5x}{5} = \frac{202}{5}$$
$$x = 40.4 \text{ or } 40\frac{2}{5}$$

5.
$$3(x - 2) = -4(5 + x)$$
$$3x - 6 = -20 - 4x$$
$$3x + 4x - 6 = -20 - 4x + 4x$$
$$7x - 6 = -20$$
$$7x - 6 + 6 = -20 + 6$$
$$7x = -14$$
$$\frac{7x}{7} = \frac{-14}{7}$$
$$x = -2$$

6.
$$12 - 5x = -7x - 2$$
$$12 - 5x + 7x = -7x + 7x - 2$$
$$12 + 2x = -2$$
$$12 - 12 + 2x = -2 - 12$$
$$2x = -14$$
$$\frac{2x}{2} = \frac{-14}{2}$$
$$x = -7$$

7.
$$2(3 - x) = 1 - (x - 2)$$
$$6 - 2x = 1 - x + 2$$
$$6 - 2x + x = 3 - x + x$$
$$6 - x = 3$$
$$6 + (-6) - x = 3 + (-6)$$
$$-x = -3$$
$$\frac{-x}{-1} = \frac{-3}{-1}$$
$$x = 3$$

8.
$$4(x + 5) - 7 = 2(x + 3)$$
$$4x + 20 - 7 = 2x + 6$$
$$4x + 13 = 2x + 6$$
$$4x + 13 - 13 = 2x + 6 - 13$$
$$4x = 2x - 7$$
$$-2x + 4x = -2x + 2x - 7$$
$$2x = -7$$
$$\frac{2x}{2} = \frac{-7}{2}$$
$$x = -\frac{7}{2} \text{ or } -3\frac{1}{2} \text{ or } -3.5$$

9.
$$3 = 2x + 5 - 3(x - 1)$$
$$3 = 2x + 5 - 3x + 3$$
$$3 = -x + 8$$
$$3 + (-8) = -x + 8 + (-8)$$
$$-5 = -x$$
$$\frac{-5}{-1} = \frac{-x}{-1}$$
$$5 = x$$

10. $2(5x-1)-7 = 3(x-1)+5-4x$
$10x-2-7 = 3x-3+5-4x$
$10x-9 = -x+2$
$10x+x-9 = -x+x+2$
$11x-9 = 2$
$11x-9+9 = 2+9$
$11x = 11$
$\dfrac{11x}{11} = \dfrac{11}{11}$
$x = 1$

11. $\dfrac{3}{4}x-3 = \dfrac{1}{2}x+2$
$4\left(\dfrac{3}{4}x\right)-4(3) = 4\left(\dfrac{1}{2}x\right)+4(2)$
$3x-12 = 2x+8$
$3x-12+12 = 2x+8+12$
$3x = 2x+20$
$-2x+3x = -2x+2x+20$
$x = 20$

12. $1 = \dfrac{5x}{6}+\dfrac{2x}{3}$
$6(1) = 6\left(\dfrac{5x}{6}\right)+6\left(\dfrac{2x}{3}\right)$
$6 = 5x+4x$
$6 = 9x$
$\dfrac{6}{9} = \dfrac{9x}{9}$
$\dfrac{2}{3} = x$

13. $\dfrac{7x}{5} = 5+\dfrac{2x}{5}$
$5\left(\dfrac{7x}{5}\right) = 5(5)+5\left(+\dfrac{2x}{5}\right)$
$7x = 25+2x$
$7x-2x = 25+2x-2x$
$5x = 25$
$\dfrac{5x}{5} = \dfrac{25}{5}$
$x = 5$

14. $\dfrac{7x-3}{2}-4 = \dfrac{5x+1}{3}$
$6\left(\dfrac{7x-3}{2}\right)-6(4) = 6\left(\dfrac{5x+1}{3}\right)$
$3(7x-3)-24 = 2(5x+1)$
$21x-9-24 = 10x+2$
$21x-33 = 10x+2$
$21x+(-10x)-33 = 10x+(-10x)+2$
$11x-33 = 2$
$11x-33+33 = 2+33$
$11x = 35$
$\dfrac{11x}{11} = \dfrac{35}{11}$
$x = \dfrac{35}{11} \text{ or } 3\dfrac{2}{11}$

15. $\dfrac{3x-2}{2}+\dfrac{x}{4} = 2+x$
$4\left(\dfrac{3x-2}{2}\right)+4\left(\dfrac{x}{4}\right) = 4(2)+4(x)$
$2(3x-2)+x = 8+4x$
$6x-4+x = 8+4x$
$7x-4 = 4x+8$
$7x-4+4 = 4x+8+4$
$7x = 4x+12$
$-4x+7x = -4x+4x+12$
$3x = 12$
$\dfrac{3x}{3} = \dfrac{12}{3}$
$x = 4$

16. $\dfrac{-3}{2}(x+5) = 1-x$
$-\dfrac{3}{2}x-\dfrac{15}{2} = 1-x$
$2\left(-\dfrac{3}{2}x\right)-2\left(\dfrac{15}{2}\right) = 2(1)-2(x)$
$-3x-15 = 2-2x$
$-3x+3x-15 = 2-2x+3x$
$-15 = 2+x$
$-15+(-2) = 2+(-2)+x$
$-17 = x$

17.
$$-0.2(x+1) = 0.3(x+11)$$
$$10[-0.2(x+1)] = 10[0.3(x+11)]$$
$$-2(x+1) = 3(x+11)$$
$$-2x-2 = 3x+33$$
$$-2x-2-33 = 3x+33-33$$
$$-2x-35 = 3x$$
$$2x-2x-35 = 2x+3x$$
$$-35 = 5x$$
$$\frac{-35}{5} = \frac{5x}{5}$$
$$-7 = x$$

18.
$$1.2x-0.8 = 0.8x+0.4$$
$$1.2x-0.8-0.8x = 0.8x+0.4-0.8x$$
$$0.4x-0.8 = 0.4$$
$$0.4x-0.8+0.8 = 0.4+0.8$$
$$0.4x = 1.2$$
$$\frac{0.4x}{0.4} = \frac{1.2}{0.4}$$
$$x = 3$$

19.
$$3.2-0.6x = 0.4(x-2)$$
$$3.2-0.6x = 0.4x-0.8$$
$$3.2-0.6x+0.6x = 0.4x-0.8+0.6x$$
$$3.2 = x-0.8$$
$$3.2+0.8 = x-0.8+0.8$$
$$4 = x$$

20.
$$\frac{1}{3}(x-2) = \frac{x}{4}+2$$
$$\frac{1}{3}x-\frac{2}{3} = \frac{x}{4}+2$$
$$12\left(\frac{1}{3}x\right)-12\left(\frac{2}{3}\right) = 12\left(\frac{x}{4}\right)+12(2)$$
$$4x-8 = 3x+24$$
$$4x+(-3x)-8 = 3x+(-3x)+24$$
$$x-8 = 24$$
$$x-8+8 = 24+8$$
$$x = 32$$

21.
$$\frac{3}{4}-\frac{2}{3}x = \frac{1}{3}x+\frac{3}{4}$$
$$12\left(\frac{3}{4}\right)-12\left(\frac{2}{3}x\right) = 12\left(\frac{1}{3}x\right)+12\left(\frac{3}{4}\right)$$
$$9-8x = 4x+9$$
$$9-8x+8x = 4x+9+8x$$
$$9 = 12x+9$$
$$9-9 = 12x+9-9$$
$$0 = 12x$$
$$\frac{0}{12} = \frac{12x}{12}$$
$$0 = x$$

22.
$$-\frac{8}{3}x-8+2x-5 = -\frac{5}{3}$$
$$-\frac{8}{3}x-13+2x = -\frac{5}{3}$$
$$3\left(-\frac{8}{3}x\right)-3(13)+3(2x) = 3\left(-\frac{5}{3}\right)$$
$$-8x-39+6x = -5$$
$$-2x-39 = -5$$
$$-2x-39+39 = -5+39$$
$$-2x = 34$$
$$\frac{-2x}{-2} = \frac{34}{-2}$$
$$x = -17$$

23.
$$\frac{1}{6}+\frac{1}{3}(x-3) = \frac{1}{2}(x+9)$$
$$\frac{1}{6}+\frac{1}{3}x-1 = \frac{1}{2}x+\frac{9}{2}$$
$$\frac{1}{3}x-\frac{5}{6} = \frac{1}{2}x+\frac{9}{2}$$
$$6\left(\frac{1}{3}x\right)-6\left(\frac{5}{6}\right) = 6\left(\frac{1}{2}x\right)+6\left(\frac{9}{2}\right)$$
$$2x-5 = 3x+27$$
$$2x-2x-5 = 3x-2x+27$$
$$-5 = x+27$$
$$-5-27 = x+27-27$$
$$-32 = x$$

24.
$$\frac{1}{7}(x+5)-\frac{3}{7}=\frac{1}{2}(x+3)$$
$$\frac{1}{7}x+\frac{5}{7}-\frac{3}{7}=\frac{1}{2}x+\frac{3}{2}$$
$$\frac{1}{7}x+\frac{2}{7}=\frac{1}{2}x+\frac{3}{2}$$
$$14\left(\frac{1}{7}x\right)+14\left(\frac{2}{7}\right)=14\left(\frac{1}{2}x\right)+14\left(\frac{3}{2}\right)$$
$$2x+4=7x+21$$
$$2x-2x+4=7x-2x+21$$
$$4=5x+21$$
$$4-21=5x+21-21$$
$$-17=5x$$
$$\frac{-17}{5}=\frac{5x}{5}$$
$$-\frac{17}{5}=x \text{ or } x=-3.4$$

25. 19 more than a number: $x+19$

26. two-thirds of a number: $\frac{2}{3}x$

27. half a number: $\frac{1}{2}x$ or $\frac{x}{2}$

28. 18 less than a number: $x-18$

29. triple the sum of a number and 4: $3(x+4)$

30. twice a number decreased by 3: $2x-3$

31. Since the numbers of working people and unemployed people are being compared to the number of retired people, we let the variable represent the number of retired people.
r = number of retired people
The number of working people is four times the number of retired people.
$4r$ = number of working people
The number of unemployed people is one-half the number of retired people.
$$\frac{1}{2}r=0.5r = \text{number of unemployed people}$$

32. Since the length of the rectangle is being compared to the width, we let the variable represent the width of the rectangle.
w = width of the rectangle
The length is 5 meters more than triple the width.
$3w+5$ = length of the rectangle

33. Since the number of degrees in angles A and C are being compared to the number of degrees in angle B, we let the variable represent the number of degrees in angle B.
b = the number of degrees in angle B
The number of degrees in angle A is double the number of degrees in angle B.
$2b$ = number of degrees in angle A
The number of degrees in angle C is 17 less than the number of degrees in angle B.
$b-17$ = number of degrees in angle C

34. Since the numbers of students in biology and geology are being compared to the number of students in algebra, we let the variable represent the number of students in algebra.
a = number of students in algebra
There are 29 more students in biology than in algebra.
$a+29$ = number of students in biology
There are one-half as many students in geology as in algebra.
$$\frac{1}{2}a=0.5a = \text{number of students in geology}$$

35. x = the number
$$3x-14=-5$$
$$3x-14+14=-5+14$$
$$3x=9$$
$$\frac{3x}{3}=\frac{9}{3}$$
$$x=3$$
The number is 3.

36. x = the number
$$2x-7=-21$$
$$2x-7+7=-21+7$$
$$2x=-14$$
$$\frac{2x}{2}=\frac{-14}{2}$$
$$x=-7$$
The number is -7.

37. x = David's age
$2x$ = Jon's age
$$2x=32$$
$$\frac{2x}{2}=\frac{32}{2}$$
$$x=16$$
David is 16 years old.

38. x = score on last test

$$\frac{83+86+91+77+x}{5}=85$$

$$\frac{337+x}{5}=85$$

$$337+x=425$$

$$x=88$$

Zach needs a grade of 88.

39. t_1 = time for first car

t_2 = time for other car

$$800=60t_1$$

$$\frac{800}{60}=\frac{60t_1}{60}$$

$$13.3\approx t_1$$

$$800=65t_2$$

$$\frac{800}{65}=\frac{65t_2}{65}$$

$$12.3\approx t_2$$

The first car took 13.3 hours. The other car took 12.3 hours.

40. measure of 1st angle: x

measure of 2nd angle: $3x$

measure of 3rd angle: $2x-12$

$$x+3x+2x-12=180$$

$$6x-12=180$$

$$6x=192$$

$$x=32$$

$3x=3(32)=96$

$2x-12=2(32)-12=52$

The angles measure 32°, 96°, and 52°.

41. x = number of kilowatt-hours

$$25+0.15x=71.50$$

$$0.15x=46.50$$

$$x=310$$

310 kilowatt-hours were used.

42. x = number of miles driven

$$0.25x+39(3)=187$$

$$0.25x+117=187$$

$$0.25x=70$$

$$x=280$$

He drove 280 miles.

43. x = amount withdrawn

$$0.055(7400-x)=242$$

$$407-0.055x=242$$

$$-0.055x=-165$$

$$x=3000$$

They withdrew $3000.

44. x = original price

$$0.18x=36$$

$$x=200$$

The original price was $200.

45. x = amount invested at 12%

$9000-x$ = amount at 8%

$$0.12x+0.08(9000-x)=1000$$

$$12x+8(9000-x)=100,000$$

$$12x+72,000-8x=100,000$$

$$4x=28,000$$

$$x=7000$$

$9000-x=2000$

They invested $7000 at 12% and $2000 at 8%.

46. x = amount at 4.5%

$5000-x$ = amount at 6%

$$0.045x+0.06(5000-x)=270$$

$$45x+60(5000-x)=270,000$$

$$45x+300,000-60x=270,000$$

$$-15x=-30,000$$

$$x=2000$$

$5000-x=5000-2000=3000$

He invested $2000 at 4.5% and $3000 at 6%.

47. x = number of dimes

$x+3$ = number of quarters

$2(x+3)=2x+6$ = number of nickels

$$0.05(2x+6)+0.10x+0.25(x+3)=3.75$$

$$5(2x+6)+10x+25(x+3)=375$$

$$10x+30+10x+25x+75=375$$

$$45x=270$$

$$x=6$$

$x+3=6+3=9$

$2x+6=2(6)+6=18$

She has 18 nickels, 6 dimes, and 9 quarters.

48. n = number of nickels

$n+2$ = number of quarters

$n-3$ = number of dimes

$$0.05n+0.25(n+2)+0.10(n-3)=9.80$$

$$0.05n+0.25n+0.50+0.10n-0.30=9.80$$

$$0.4n+0.20=9.80$$

$$0.4n=9.60$$

$$n=24$$

$n+2=24+2=26$

$n-3=24-3=21$

There were 24 nickels, 21 dimes, and 26 quarters.

49.
$$9 + 2x \le 6 - x$$
$$9 + 2x + x \le 6 - x + x$$
$$9 + 3x \le 6$$
$$9 + 3x - 9 \le 6 - 9$$
$$3x \le -3$$
$$\frac{3x}{3} \le \frac{-3}{3}$$
$$x \le -1$$

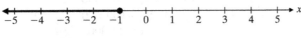

50.
$$2x - 3 + x > 5(x + 1)$$
$$3x - 3 > 5x + 5$$
$$3x - 3 - 5x > 5x + 5 - 5x$$
$$-2x - 3 > 5$$
$$-2x - 3 + 3 > 5 + 3$$
$$-2x > 8$$
$$\frac{-2x}{-2} < \frac{8}{-2}$$
$$x < -4$$

51.
$$-x + 4 < 3x + 16$$
$$-x + 4 - 4 < 3x + 16 - 4$$
$$-x < 3x + 12$$
$$-3x - x < -3x + 3x + 12$$
$$-4x < 12$$
$$\frac{-4x}{-4} > \frac{12}{-4}$$
$$x > -3$$

52.
$$8 - \frac{1}{3}x \le x$$
$$3(8) - 3\left(\frac{1}{3}x\right) \le 3x$$
$$24 - x \le 3x$$
$$24 - x + (-3x) \le 3x + (-3x)$$
$$24 - 4x \le 0$$
$$24 + (-24) - 4x \le 0 + (-24)$$
$$-4x \le -24$$
$$\frac{-4x}{-4} \ge \frac{-24}{-4}$$
$$x \ge 6$$

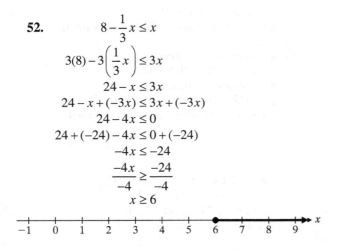

53.
$$7 - \frac{3}{5}x > 4$$
$$5(7) - 5\left(\frac{3}{5}x\right) > 5(4)$$
$$35 - 3x > 20$$
$$35 + (-35) - 3x > 20 + (-35)$$
$$-3x > -15$$
$$\frac{-3x}{-3} < \frac{-15}{-3}$$
$$x < 5$$

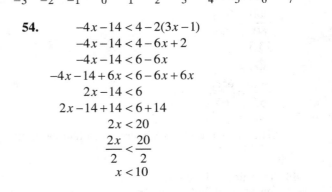

54.
$$-4x - 14 < 4 - 2(3x - 1)$$
$$-4x - 14 < 4 - 6x + 2$$
$$-4x - 14 < 6 - 6x$$
$$-4x - 14 + 6x < 6 - 6x + 6x$$
$$2x - 14 < 6$$
$$2x - 14 + 14 < 6 + 14$$
$$2x < 20$$
$$\frac{2x}{2} < \frac{20}{2}$$
$$x < 10$$

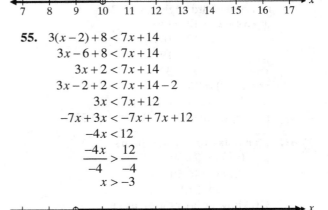

55.
$$3(x - 2) + 8 < 7x + 14$$
$$3x - 6 + 8 < 7x + 14$$
$$3x + 2 < 7x + 14$$
$$3x - 2 + 2 < 7x + 14 - 2$$
$$3x < 7x + 12$$
$$-7x + 3x < -7x + 7x + 12$$
$$-4x < 12$$
$$\frac{-4x}{-4} > \frac{12}{-4}$$
$$x > -3$$

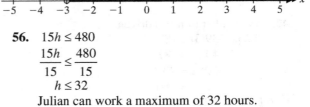

56.
$$15h \le 480$$
$$\frac{15h}{15} \le \frac{480}{15}$$
$$h \le 32$$
Julian can work a maximum of 32 hours.

57.
$$110n \le 2420$$
$$\frac{110n}{110} \le \frac{2420}{110}$$
$$n \le 22$$
A substitute teacher can be hired a maximum of 22 times.

58. $10(2x+4)-13=8(x+7)-3$
$20x+40-13=8x+56-3$
$20x+27=8x+53$
$20x+27-8x=8x+53-8x$
$12x+27=53$
$12x+27-27=53-27$
$12x=26$
$\dfrac{12x}{12}=\dfrac{26}{12}$
$x=\dfrac{13}{6}$

59. $-9x+15-2x=4-3x$
$-11x+15=4-3x$
$-11x+15+3x=4-3x+3x$
$-8x+15=4$
$-8x+15-15=4-15$
$-8x=-11$
$\dfrac{-8x}{-8}=\dfrac{-11}{-8}$
$x=\dfrac{11}{8}$

60. $-2(x-3)=-4x+3(3x+2)$
$-2x+6=-4x+9x+6$
$-2x+6=5x+6$
$-2x+6-6=5x+6-6$
$-2x=5x$
$2x-2x=2x+5x$
$0=7x$
$\dfrac{0}{7}=\dfrac{7x}{7}$
$0=x$

61. $\dfrac{1}{2}+\dfrac{5}{4}x=\dfrac{2}{5}x-\dfrac{1}{10}+4$
$20\left(\dfrac{1}{2}\right)+20\left(\dfrac{5}{4}x\right)=20\left(\dfrac{2}{5}x\right)-20\left(\dfrac{1}{10}\right)+20(4)$
$10+25x=8x-2+80$
$10+25x=8x+78$
$10+25x+(-8x)=8x+(-8x)+78$
$10+17x=78$
$10+(-10)+17x=78+(-10)$
$17x=68$
$\dfrac{17x}{17}=\dfrac{68}{17}$
$x=4$

62. $5-\dfrac{1}{2}x>4$
$2(5)-2\left(\dfrac{1}{2}x\right)>2(4)$
$10-x>8$
$-10+10-x>-10+8$
$-x>-2$
$\dfrac{-x}{-1}<\dfrac{-2}{-1}$
$x<2$

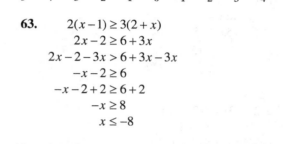

63. $2(x-1)\ge 3(2+x)$
$2x-2\ge 6+3x$
$2x-2-3x>6+3x-3x$
$-x-2\ge 6$
$-x-2+2\ge 6+2$
$-x\ge 8$
$x\le -8$

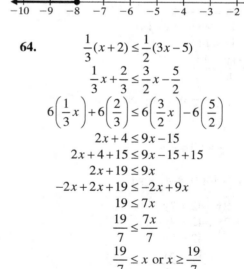

64. $\dfrac{1}{3}(x+2)\le\dfrac{1}{2}(3x-5)$
$\dfrac{1}{3}x+\dfrac{2}{3}\le\dfrac{3}{2}x-\dfrac{5}{2}$
$6\left(\dfrac{1}{3}x\right)+6\left(\dfrac{2}{3}\right)\le 6\left(\dfrac{3}{2}x\right)-6\left(\dfrac{5}{2}\right)$
$2x+4\le 9x-15$
$2x+4+15\le 9x-15+15$
$2x+19\le 9x$
$-2x+2x+19\le -2x+9x$
$19\le 7x$
$\dfrac{19}{7}\le\dfrac{7x}{7}$
$\dfrac{19}{7}\le x$ or $x\ge\dfrac{19}{7}$

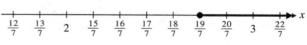

65. $4(2-x)-(-5x+1)\ge -8$
$8-4x+5x-1\ge -8$
$x+7\ge -8$
$x+7-7\ge -8-7$
$x\ge -15$

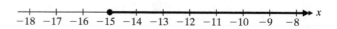

How Am I Doing? Chapter 2 Test

1.
$$3x + 5.6 = 11.6$$
$$3x + 5.6 - 5.6 = 11.6 - 5.6$$
$$3x = 6$$
$$\frac{3x}{3} = \frac{6}{3}$$
$$x = 2$$

2.
$$9x - 8 = -6x - 3$$
$$9x + 6x - 8 = -6x + 6x - 3$$
$$15x - 8 = -3$$
$$15x - 8 + 8 = -3 + 8$$
$$15x = 5$$
$$\frac{15x}{15} = \frac{5}{15}$$
$$x = \frac{1}{3}$$

3.
$$2(2y - 3) = 4(2y + 2)$$
$$4y - 6 = 8y + 8$$
$$4y - 6 + 6 = 8y + 8 + 6$$
$$4y = 8y + 14$$
$$-8y + 4y = -8y + 8y + 14$$
$$-4y = 14$$
$$\frac{-4y}{-4} = \frac{14}{-4}$$
$$y = -\frac{7}{2} \text{ or } -3\frac{1}{2} \text{ or } -3.5$$

4.
$$\frac{1}{7}y + 3 = \frac{1}{2}y$$
$$14\left(\frac{1}{7}y\right) + 14(3) = 14\left(\frac{1}{2}y\right)$$
$$2y + 42 = 7y$$
$$2y - 2y + 42 = 7y - 2y$$
$$42 = 5y$$
$$\frac{42}{5} = \frac{5y}{5}$$
$$y = \frac{42}{5} \text{ or } y = 8\frac{2}{5} \text{ or } y = 8.4$$

5.
$$4(7 - 4x) = 3(6 - 2x)$$
$$28 - 16x = 18 - 6x$$
$$28 - 16x + 6x = 18 - 6x + 6x$$
$$28 - 10x = 18$$
$$28 + (-28) - 10x = 18 + (-28)$$
$$-10x = -10$$
$$\frac{-10x}{-10} = \frac{-10}{-10}$$
$$x = 1$$

6.
$$0.8x + 0.18 - 0.4x = 0.3(x + 0.2)$$
$$0.4x + 0.18 = 0.3x + 0.06$$
$$100(0.4x) + 100(0.18) = 100(0.3x) + 100(0.06)$$
$$40x + 18 = 30x + 6$$
$$40x + 18 - 18 = 30x + 6 - 18$$
$$40x = 30x - 12$$
$$-30x + 40x = -30x + 30x - 12$$
$$10x = -12$$
$$\frac{10x}{10} = \frac{-12}{10}$$
$$x = -\frac{6}{5} \text{ or } -1.2$$

7.
$$\frac{2y}{3} + \frac{1}{5} - \frac{3y}{5} + \frac{1}{3} = 1$$
$$15\left(\frac{2y}{3}\right) + 15\left(\frac{1}{5}\right) - 15\left(\frac{3y}{5}\right) + 15\left(\frac{1}{3}\right) = 15(1)$$
$$10y + 3 - 9y + 5 = 15$$
$$y + 8 = 15$$
$$y + 8 - 8 = 15 - 8$$
$$y = 7$$

8.
$$3 - 2y = 2(3y - 2) - 5y$$
$$3 - 2y = 6y - 4 - 5y$$
$$3 - 2y = y - 4$$
$$3 - 2y + 2y = y + 2y - 4$$
$$3 = 3y - 4$$
$$3 + 4 = 3y - 4 + 4$$
$$7 = 3y$$
$$\frac{7}{3} = \frac{3y}{3}$$
$$\frac{7}{3} = y \text{ or } y = 2\frac{1}{3}$$

9.
$$5(20 - x) + 10x = 165$$
$$100 - 5x + 10x = 165$$
$$100 + 5x = 165$$
$$-100 + 100 + 5x = -100 + 165$$
$$5x = 65$$
$$\frac{5x}{5} = \frac{65}{5}$$
$$x = 13$$

10.
$$5(x+40)-6x=9x$$
$$5x+200-6x=9x$$
$$200-x=9x$$
$$200-x+x=9x+x$$
$$200=10x$$
$$\frac{200}{10}=\frac{10x}{10}$$
$$20=x$$

11.
$$-2(2-3x)=76-2x$$
$$-4+6x=76-2x$$
$$-76-4+6x=-76+76-2x$$
$$-80+6x=-2x$$
$$-80+6x-6x=-2x-6x$$
$$-80=-8x$$
$$\frac{-80}{-8}=\frac{-8x}{-8}$$
$$10=x$$

12.
$$20-(2x+6)=5(2-x)+2x$$
$$20-2x-6=10-5x+2x$$
$$-2x+14=-3x+10$$
$$3x-2x+14=3x-3x+10$$
$$x+14=10$$
$$x+14-14=10-14$$
$$x=-4$$

13.
$$2x-3=12-6x+3(2x+3)$$
$$2x-3=12-6x+6x+9$$
$$2x-3=21$$
$$2x-3+3=21+3$$
$$2x=24$$
$$\frac{2x}{2}=\frac{24}{2}$$
$$x=12$$

14.
$$\frac{1}{3}x-\frac{3}{4}x=\frac{1}{12}$$
$$12\left(\frac{1}{3}x\right)-12\left(\frac{3}{4}x\right)=12\left(\frac{1}{12}\right)$$
$$4x-9x=1$$
$$-5x=1$$
$$\frac{-5x}{-5}=\frac{1}{-5}$$
$$x=-\frac{1}{5}\text{ or }-0.2$$

15.
$$\frac{3}{5}x+\frac{7}{10}=\frac{1}{3}x+\frac{3}{2}$$
$$30\left(\frac{3}{5}x\right)+30\left(\frac{7}{10}\right)=30\left(\frac{1}{3}x\right)+30\left(\frac{3}{2}\right)$$
$$18x+21=10x+45$$
$$18x+21-21=10x+45-21$$
$$18x=10x+24$$
$$-10x+18x=-10x+10x+24$$
$$8x=24$$
$$\frac{8x}{8}=\frac{24}{8}$$
$$x=3$$

16.
$$\frac{15x-2}{28}=\frac{5x-3}{7}$$
$$28\left(\frac{15x-2}{28}\right)=28\left(\frac{5x-3}{7}\right)$$
$$15x-2=4(5x-3)$$
$$15x-2=20x-12$$
$$15x-2+12=20x-12+12$$
$$15x+10=20x$$
$$-15x+15x+10=-15x+20x$$
$$10=5x$$
$$\frac{10}{5}=\frac{5x}{5}$$
$$2=x$$

17.
$$\frac{2}{3}(x+8)+\frac{3}{5}=\frac{1}{5}(11-6x)$$
$$\frac{2}{3}x+\frac{16}{3}+\frac{3}{5}=\frac{11}{5}-\frac{6}{5}x$$
$$\frac{2}{3}x+\frac{89}{15}=\frac{11}{5}-\frac{6}{5}x$$
$$15\left(\frac{2}{3}x\right)+15\left(\frac{89}{15}\right)=15\left(\frac{11}{5}\right)-15\left(\frac{6}{5}x\right)$$
$$10x+89=33-18x$$
$$10x+18x+89=33-18x+18x$$
$$28x+89=33$$
$$28x+89+(-89)=33+(-89)$$
$$28x=-56$$
$$\frac{28x}{28}=\frac{-56}{28}$$
$$x=-2$$

18.
$$3(x-2) \geq 5x$$
$$3x-6 \geq 5x$$
$$3x+(-5x)-6 \geq 5x+(-5x)$$
$$-2x-6 \geq 0$$
$$-2x-6+6 \geq 0+6$$
$$-2x \geq 6$$
$$\frac{-2x}{-2} \leq \frac{6}{-2}$$
$$x \leq -3$$

19.
$$2-7(x+1)-5(x+2) < 0$$
$$2-7x-7-5x-10 < 0$$
$$-12x-15 < 0$$
$$-12x-15+15 < 0+15$$
$$-12x < 15$$
$$\frac{-12x}{-12} > \frac{15}{-12}$$
$$x > -\frac{5}{4}$$

20.
$$5+8x-4 < 2x+13$$
$$8x+1 < 2x+13$$
$$8x+1-1 < 2x+13-1$$
$$8x < 2x+12$$
$$-2x+8x < -2x+2x+12$$
$$6x < 12$$
$$\frac{6x}{6} = \frac{12}{6}$$
$$x < 2$$

21.
$$\frac{1}{4}x+\frac{1}{16} \leq \frac{1}{8}(7x-2)$$
$$\frac{1}{4}x+\frac{1}{16} \leq \frac{7}{8}x-\frac{1}{4}$$
$$16\left(\frac{1}{4}x\right)+16\left(\frac{1}{16}\right) \leq 16\left(\frac{7}{8}x\right)-16\left(\frac{1}{4}\right)$$
$$4x+1 \leq 14x-4$$
$$4x+1+4 \leq 14x-4+4$$
$$4x+5 \leq 14x$$
$$-4x+4x+5 \leq -4x+14x$$
$$5 \leq 10x$$
$$\frac{5}{10} \leq \frac{10x}{10}$$
$$\frac{1}{2} \leq x$$

22. x = number
$$2x-11 = 59$$
$$2x = 70$$
$$x = 35$$
The number is 35.

23. x = number
$$\frac{1}{2}x+\frac{1}{9}x+\frac{1}{12}x = 25$$
$$36\left(\frac{1}{2}x+\frac{1}{9}x+\frac{1}{12}x\right) = 36(25)$$
$$18x+4x+3x = 900$$
$$25x = 900$$
$$x = 36$$
The number is 36.

24. x = second number
The first number is six less than three times the second number.
$3x - 6$ = first number
The sum of the numbers is twenty-two.
$$x+(3x-6) = 22$$
$$4x-6 = 22$$
$$4x-6+6 = 22+6$$
$$4x = 28$$
$$\frac{4x}{4} = \frac{28}{4}$$
$$x = 7$$
$3x - 6 = 3(7) - 6 = 15$
The numbers are 7 and 15.

25. $t = \dfrac{d}{r}$

Jerome's time: $t = \dfrac{300}{50} = 6$ hours

Steven's time: $t = \dfrac{300}{60} = 5$ hours

$6 - 5 = 1$
Steven arrived 1 hour before Jerome.

26. $x =$ second angle
$3x =$ first angle
$x + 10 =$ third angle
$$3x + x + (x + 10) = 180$$
$$5x + 10 = 180$$
$$5x = 170$$
$$x = 34$$
$3x = 3(34) = 102$
$x + 10 = 34 + 10 = 44$
First angle = 102°
Second angle = 34°
Third angle = 44°

27. $x =$ number of months
$$116x + 200 = 1940$$
$$116x = 1740$$
$$x = 15$$
Raymond will be able to rent the computer for 15 months.

28. $x =$ last year's tuition
$$x + 0.08x = 34{,}560$$
$$1.08x = 34{,}560$$
$$x = 32{,}000$$
Last year's tuition was $32,000.

29. $x =$ amount at 14%
$4000 - x =$ amount at 11%
$$0.14x + 0.11(4000 - x) = 482$$
$$100[0.14x + 0.11(4000 - x)] = 100(482)$$
$$14x + 11(4000 - x) = 48{,}200$$
$$14x + 44{,}000 - 11x = 48{,}200$$
$$3x + 44{,}000 = 48{,}200$$
$$3x = 4200$$
$$x = 1400$$
$4000 - x = 4000 - 1400 = 2600$
He invested $1400 at 14% and $2600 at 11%.

30. $2x =$ number of nickels
$x - 1 =$ number of dimes
$x =$ number of quarters
$$0.05(2x) + 0.10(x - 1) + 0.25(x) = 3.50$$
$$5(2x) + 10(x - 1) + 25x = 350$$
$$10x + 10x - 10 + 25x = 350$$
$$45x - 10 = 350$$
$$45x = 360$$
$$x = 8$$
$2x = 2(8) = 16$
$x - 1 = 8 - 1 = 7$
She has: 16 nickels; 7 dimes; 8 quarters.

Chapter 3

3.1 Exercises

1. The *x*-coordinate of the origin is 0.

3. (5, 1) is an ordered pair because the order is important. The graphs of (5, 1) and (1, 5) are different.

5. They are not the same because the *x* and *y* coordinates are different. To plot (2, 7) we move 2 units to the right on the *x*-axis, but for the ordered pair (7, 2) we move 7 units to the right on the *x*-axis. Then to plot (2, 7) we move 7 units up on a line parallel to the *y*-axis, but for the ordered pair (7, 2) we move 2 units up.

7.

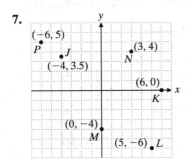

9. R: $(-3, -5)$

S: $\left(-4\frac{1}{2}, 0\right)$

X: $(3, -5)$

Y: $\left(2\frac{1}{2}, 6\right)$

11. $(-4, -1)$, $(-3, -2)$, $(-2, -3)$, $(-1, -5)$, $(0, -3)$, $(2, -1)$

13. Locate the grid for Lynbrook, NY: B5

15. Locate the grid for Athol, Mass: E1

17. Look for the grid for Hartford, CT: D3

19. a. (0, 803), (1, 746), (2, 767), (3, 705), (4, 615), (5, 511), (6, 368), (7, 293)

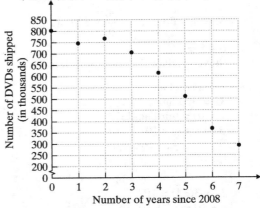

b. The number of DVDs shipped decreased overall between 2008 and 2015 with a slight increase in 2010.

21. a. (1, 3.2), (2, 3.6), (3, 3.9), (4, 4.2)

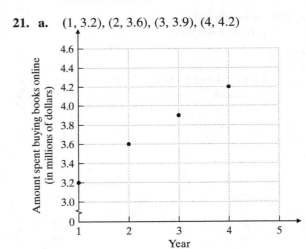

b. An estimated \$4.5 billion will be spent buying books online in year 5.

23. $2x - y = 6$; $(1, 0)$

$2(1) - 0 \stackrel{?}{=} 6$

$2 - 0 \stackrel{?}{=} 6$

$2 = 6$ False

No, $(1, 0)$ is not a solution.

25. $2x - y = 6$; $(2, -2)$

$2(2) - (-2) \stackrel{?}{=} 6$

$4 + 2 \stackrel{?}{=} 6$

$6 = 6$ ✓

Yes, $(2, -2)$ is a solution.

27. Jon is right because for the ordered pair (5, 3), $x = 5$ and $y = 3$. When we substitute these values in the equation, we get $2(5) - 2(3) = 4$. For the ordered pair (3, 5), when we replace $x = 3$ and $y = 5$ in the equation we get $2(3) - 2(5)$, which equals -4, not 4.

29. a. $y - 4 = -\dfrac{2}{3}x$

$$y = -\dfrac{2}{3}x + 4$$

     b.   (6,):

$$y = -\dfrac{2}{3}(6) + 4$$
$$y = -4 + 4$$
$$y = 0$$
$$(6, 0)$$

31. a. $4x + y = 11$

$$y = -4x + 11$$

     b.   (2,):

$$y = -4(2) + 11$$
$$y = -8 + 11$$
$$y = 3$$
$$(2, 3)$$

33. a. $2x + 3y = 6$

$$3y = -2x + 6$$
$$\dfrac{3y}{3} = \dfrac{-2x + 6}{3}$$
$$y = -\dfrac{2}{3}x + 2$$

     b.   (3,):

$$y = -\dfrac{2}{3}(3) + 2$$
$$y = -2 + 2$$
$$y = 0$$
$$(3, 0)$$

35. $y = 4x + 7$

     a.   (0,):

$$y = 4(0) + 7$$
$$y = 0 + 7$$
$$y = 7$$
$$(0, 7)$$

     b.   (2,), y

$$y = 4(2) + 7$$
$$y = 8 + 7$$
$$y = 15$$
$$(2, 15)$$

37. $y + 6x = 5$

     a.   (−1,):

$$y + 6(-1) = 5$$
$$y - 6 = 5$$
$$y = 11$$
$$(-1, 11)$$

     b.   (3,):

$$y + 6(3) = 5$$
$$y + 18 = 5$$
$$y = -13$$
$$(3, -13)$$

39. $3x - 4y = 11$

     a.   (−3,):

$$3(-3) - 4y = 11$$
$$-9 - 4y = 11$$
$$-4y = 20$$
$$y = -5$$
$$(-3, -5)$$

     b.   (, 1):

$$3x - 4(1) = 11$$
$$3x - 4 = 11$$
$$3x = 15$$
$$x = 5$$
$$(5, 1)$$

41. $3x + 2y = -6$

     a.   (−2,):

$$3(-2) + 2y = -6$$
$$-6 + 2y = -6$$
$$2y = 0$$
$$y = 0$$
$$(-2, 0)$$

     b.   (, 3):

$$3x + 2(3) = -6$$
$$3x + 6 = -6$$
$$3x = -12$$
$$x = -4$$
$$(-4, 3)$$

43. $y - 1 = \dfrac{2}{7}x$

   a. (7,):

$$y - 1 = \dfrac{2}{7}(7)$$
$$y - 1 = 2$$
$$y = 3$$
$$(7, 3)$$

   b. $\left(\ , \dfrac{5}{7} \right)$:

$$\dfrac{5}{7} - 1 = \dfrac{2}{7}x$$
$$\dfrac{5}{7} - \dfrac{7}{7} = \dfrac{2}{7}x$$
$$-\dfrac{2}{7} = \dfrac{2}{7}x$$
$$-1 = x$$
$$\left(-1, \dfrac{5}{7} \right)$$

45. $3x + \dfrac{1}{2}y = 7$

   a. (, 2):

$$3x + \dfrac{1}{2}(2) = 7$$
$$3x + 1 = 7$$
$$3x = 6$$
$$x = 2$$
$$(2, 2)$$

   b. $\left(\dfrac{3}{2}, \ \right)$:

$$3\left(\dfrac{3}{2} \right) + \dfrac{1}{2}y = 7$$
$$\dfrac{9}{2} + \dfrac{1}{2}y = 7$$
$$\dfrac{1}{2}y = \dfrac{5}{2}$$
$$y = 5$$
$$\left(\dfrac{3}{2}, 5 \right)$$

Cumulative Review

47. $A = \pi r^2;\ r = 19$ yards

$$A \approx 3.14(19)^2 = 3.14(361) = 1133.54$$
The area is 1133.54 square yards.

48. Let $x =$ the number.
$$2x - 3 = 21$$
$$2x - 3 + 3 = 21 + 3$$
$$2x = 24$$
$$\dfrac{2x}{2} = \dfrac{24}{2}$$
$$x = 12$$
The number is 12.

Quick Quiz 3.1

1.

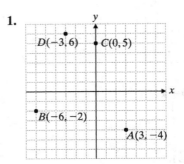

2. $y = -5x - 7$

   a. $(-2, \ ),\ y = -5(-2) - 7 = 10 - 7 = 3$
      $(-2, 3)$

   b. $(3, \ ),\ y = -5(3) - 7 = -15 - 7 = -22$
      $(3, -22)$

   c. $(0, \ ),\ y = -5(0) - 7 = 0 - 7 = -7$
      $(0, -7)$

3. $4x - 3y = -12$
$$-3y = -12 - 4x$$
$$y = \dfrac{-12 - 4x}{-3}$$
or
$$4x = -12 + 3y$$
$$x = \dfrac{-12 + 3y}{4}$$

   a. $(3, \ ),\ y = \dfrac{-12 - 4(3)}{-3} = \dfrac{-24}{-3} = 8$
      $(3, 8)$

b. $(\ \ , -8)$, $x = \dfrac{-12 + 3(-8)}{4} = \dfrac{-36}{4} = -9$

$(-9, -8)$

c. $(\ \ , 10)$, $x = \dfrac{-12 + 3(10)}{4} = \dfrac{18}{4} = \dfrac{9}{2} = 4.5$

$(4.5, 10)$

4. Answers may vary. Possible solution:
Isolate x on the left side of the equation.
Substitute the given value for y and solve for x.

3.2 Exercises

1. No, replacing x by -2 and y by 5 in $2x + 5y = 0$ does not result in a true statement so $(-2, 5)$ is not a solution.

3. The x-intercept of a line is the point where the line crosses the <u>x-axis</u>.

5. $y = x - 4$

x	$y = x - 4$	y
0	$y = 0 - 4$	-4
2	$y = 2 - 4$	-2
4	$y = 4 - 4$	0

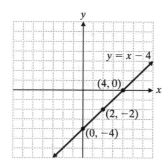

7. $y = -2x + 1$

x	$y = -2x + 1$	y
0	$-2(0) + 1 = 0 + 1$	1
-2	$-2(-2) + 1 = 4 + 1$	5
1	$-2(1) + 1 = -2 + 1$	-1

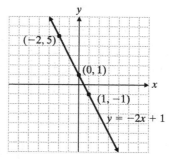

9. $y = 3x - 1$

x	$y = 3x - 1$	y
0	$y = 3(0) - 1 = 0 - 1$	-1
2	$y = 3(2) - 1 = 6 - 1$	5
-1	$y = -3(-1) - 1 = -3 - 1$	-4

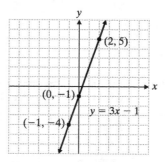

11. $y = 2x - 5$

x	$y = 2x - 5$	y
0	$y = 2(0) - 5 = 0 - 5$	-5
2	$y = 2(2) - 5 = 4 - 5$	-1
4	$y = 2(4) - 5 = 8 - 5$	3

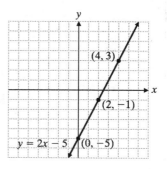

13. $y = -x + 3$

x	$y = -x + 3$	y
-1	$y = -(-1) + 3$	4
0	$y = -(0) + 3$	3
2	$y = -(2) + 3$	1

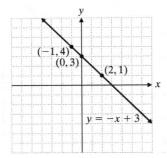

15. $3x - 2y = 0$
$$2y = 3x$$
$$y = \frac{3x}{2}$$

x	$y = \frac{3x}{2}$	y
-2	$y = \frac{3(-2)}{2} = \frac{-6}{2}$	-3
0	$y = \frac{3(0)}{2} = \frac{0}{2}$	0
2	$y = \frac{3(2)}{2} = \frac{6}{2}$	3

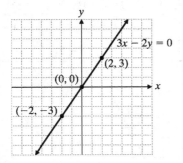

17. $y = -\frac{3}{4}x + 3$

x	$y = -\frac{3}{4}x + 3$	y
-4	$y = -\frac{3}{4}(-4) + 3 = 3 + 3$	6
0	$y = -\frac{3}{4}(0) + 3 = 0 + 3$	3
4	$y = -\frac{3}{4}(4) + 3 = -3 + 3$	0

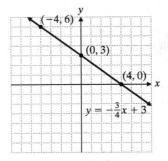

19. $4x + 6 + 3y = 18$
$$4x + 3y = 12$$
$$3y = -4x + 12$$
$$y = -\frac{4}{3}x + 4$$

x	$y = -\frac{4}{3}x + 4$	y
0	$y = -\frac{4}{3}(0) + 4 = 0 + 4$	4
3	$y = -\frac{4}{3}(3) + 4 = -4 + 4$	0
6	$y = -\frac{4}{3}(6) + 4 = -8 + 4$	-4

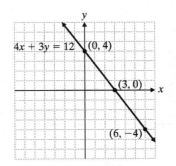

21. $y = 6 - 2x$

   a. Let $y = 0$.
$$0 = 6 - 2x$$
$$2x = 6$$
$$x = 3$$
      x-intercept: $(3, 0)$
      Let $x = 0$.
$$y = 6 - 2(0) = 6 - 0 = 6$$
      y-intercept: $(0, 6)$

   b. Let $x = 1$.
$$y = 6 - 2(1) = 6 - 2 = 4$$

x	y
0	6
1	4
3	0

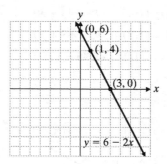

23. $x + 3 = 6y$

   a. Let $y = 0$.
$$x + 3 = 6(0)$$
$$x = -3$$
      x-intercept: $(3, 0)$
      Let $x = 0$.
$$0 + 3 = 6y$$
$$\frac{1}{2} = y$$
      y-intercept: $\left(0, \frac{1}{2}\right)$

   b. Let $x = 3$.
$$3 + 3 = 6y$$
$$1 = y$$

x	y
-3	0
0	$\frac{1}{2}$
3	1

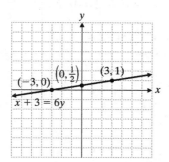

25. $x = 4$
This is a vertical line 4 units to the right of the y-axis.

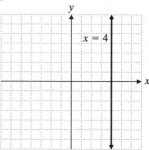

27. $y - 2 = 3y$
$$-2 = 2y$$
$$-1 = y$$

This is a horizontal line 1 unit below the x-axis.

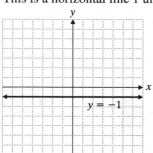

29. $2x + 5y - 2 = -12$
$$2x + 5y = -10$$
$$5y = -2x - 10$$
$$y = -\frac{2}{5}x - 2$$

x	$y = -\frac{2}{5}x - 2$	y
-5	$y = -\frac{2}{5}(-5) - 2 = 2 - 2$	0
0	$y = -\frac{2}{5}(0) - 2 = 0 - 2$	-2
5	$y = -\frac{2}{5}(5) - 2 = -2 - 2$	-4

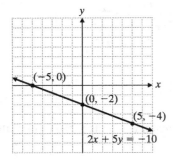

31. $2x + 9 = 5x$
$$-3x + 9 = 0$$
$$-3x = -9$$
$$x = 3$$

This is a vertical line 3 units to the right of the *y*-axis.

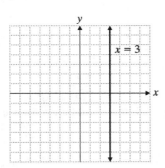

33. $C = 8m$

m	0	15	30	45	60	75
C	0	120	240	360	480	600

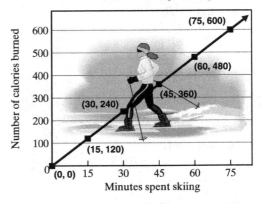

Calories Burned While Cross-Country Skiing

35. $S = 11t + 395$

t	$S = 11t + 395$	S
0	$S = 11(0) + 395 = 0 + 395$	395
4	$S = 11(4) + 395 = 44 + 395$	439
8	$S = 11(8) + 395 = 88 + 395$	483
16	$S = 11(16) + 395 = 176 + 395$	571

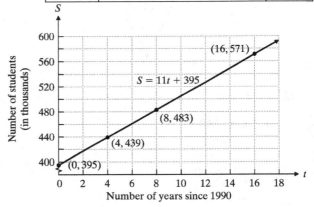

Cumulative Review

37. $2(x + 3) + 5x = 3x - 2$
$$2x + 6 + 5x = 3x - 2$$
$$7x + 6 = 3x - 2$$
$$7x + 6 - 3x = 3x - 2 - 3x$$
$$4x + 6 = -2$$
$$4x + 6 - 6 = -2 - 6$$
$$4x = -8$$
$$\frac{4x}{4} = \frac{-8}{4}$$
$$x = -2$$

38. $4 - 3x \le 18$

$-3x \le 14$

$x \ge -\dfrac{14}{3}$

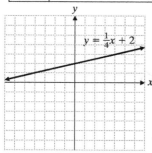

Quick Quiz 3.2

1. $4y + 1 = x + 9$

$x = 4y - 8$ or $y = \dfrac{1}{4}x + 2$

x	$x = 4y - 8$	y
-4	$x = 4(1) - 8 = 4 - 8$	1
0	$x = 4(2) - 8 = 8 - 8$	2
4	$x = 4(3) - 8 = 12 - 8$	3

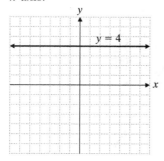

2.

$3y = 2y + 4$

$3y - 2y = 2y + 4 - 2y$

$y = 4$

This is a horizontal line four units above the x-axis.

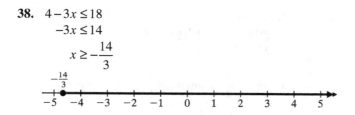

3. $y = -2x + 4$

a. Let $y = 0$.

$0 = -2x + 4$

$2x = 4$

$x = 2$

x-intercept: $(2, 0)$

Let $x = 0$.

$y = -2(0) + 4$

$y = 0 + 4$

$y = 4$

y-intercept: $(0, 4)$

b. Find a third point.

Let $x = 4$.

$y = -2(4) + 4$

$y = -8 + 4$

$y = -4$

$(4, -4)$

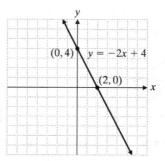

4. Answers may vary. Possible solution:
The most important ordered pair gives the y-intercept. With the y-intercept and the slope (covered in the next section), the line may be graphed.

3.3 Exercises

1. You cannot find the slope of the line passing through $(5, -12)$ and $(5, -6)$ because division by zero is impossible so the slope is undefined.

Use $m = \dfrac{y_2 - y_1}{x_2 - x_1}$ in Exercises 3–15.

3. $(2, 3)$ and $(5, 9)$

$m = \dfrac{9 - 3}{5 - 2} = \dfrac{6}{3} = 2$

5. $(5, 10)$ and $(6, 5)$

$m = \dfrac{5 - 10}{6 - 5} = \dfrac{-5}{1} = -5$

7. $(-2, 1)$ and $(3, 4)$

$m = \dfrac{4 - 1}{3 - (-2)} = \dfrac{3}{5}$

9. $(-6, -5)$ and $(2, -7)$

$$m = \frac{-7-(-5)}{2-(-6)} = \frac{-2}{8} = -\frac{1}{4}$$

11. $(-3, 0)$ and $(0, -4)$

$$m = \frac{-4-0}{0-(-3)} = \frac{-4}{3} = -\frac{4}{3}$$

13. $(5, -1)$ and $(-7, -1)$

$$m = \frac{(-1)-(-1)}{-7-5} = \frac{0}{-12} = 0$$

15. $\left(\frac{3}{4}, -4\right)$ and $(2, -8)$

$$m = \frac{-8-(-4)}{2-\frac{3}{4}} = \frac{-4}{\frac{5}{4}} = -\frac{16}{5}$$

Use $y = mx + b$ in Exercises 17–35.

17. $y = 8x + 9$; $m = 8$, $b = 9$, y-intercept $(0, 9)$

19. $3x + y - 4 = 0$, $y = -3x - 4$, $m = -3$, $b = 4$, y-intercept $(0, 4)$

21. $y = -\frac{8}{7}x + \frac{3}{4}$; $m = -\frac{8}{7}$, $b = \frac{3}{4}$,

y-intercept $\left(0, \frac{3}{4}\right)$

23. $y = -6x$; $m = -6$, $b = 0$, y-intercept $(0, 0)$

25. $y = -2$; $m = 0$, $b = -2$
y-intercept $(0, -2)$

27. $7x - 3y = 4$

$$y = \frac{7}{3}x - \frac{4}{3}; \ m = \frac{7}{3}, \ b = -\frac{4}{3},$$

y-intercept $\left(0, -\frac{4}{3}\right)$

29. $m = \frac{3}{5}$, y-intercept $(0, 3)$, $b = 3$

$$y = \frac{3}{5}x + 3$$

31. $m = 4$, y-intercept $(0, -5)$, $b = -5$
$y = 4x - 5$

33. $m = -1$,
y-intercept $(0, 0)$, $b = 0$
$y = -x$

35. $m = -\frac{5}{4}$, y-intercept $\left(0, -\frac{3}{4}\right)$, $b = -\frac{3}{4}$

$$y = -\frac{5}{4}x - \frac{3}{4}$$

37. $m = \frac{3}{4}$, $b = -4$

$$y = \frac{3}{4}x - 4$$

Graph $(0, -4)$. Find another point using the slope. Begin at $(0, -4)$. Go up 3 units and right 4 units: $(4, -1)$.

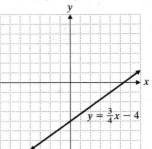

39. $m = -\frac{5}{3}$, $b = 2$

$$y = -\frac{5}{3}x + 2$$

Graph $(0, 2)$. Find another point using the slope. Begin at $(0, 2)$. Go down 5 units and right 3 units: $(3, -3)$

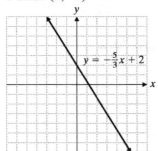

41. $y = \dfrac{2}{3}x + 2$

$m = \dfrac{2}{3}$, $b = 2$, y-intercept $(0, 2)$

Find another point using the slope: $(3, 4)$.

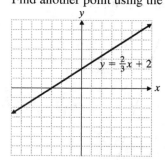

43. $y + 2x = 3$

$\qquad y = -2x + 3$

$m = -2$ or $\dfrac{-2}{1}$, $b = 3$, y-intercept $(0, 3)$

Find another point using the slope: $(1, 1)$

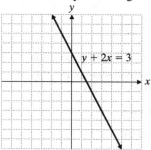

45. $y = 2x$

$m = 2$, $b = 0$, y-intercept $(0, 0)$

Find another point using the slope: $(1, 2)$

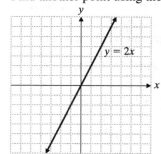

47. a. Parallel lines have the same slopes, so the slope is $\dfrac{5}{6}$.

    b. Perpendicular lines have slopes whose product is -1.

$$m_1 m_2 = -1$$
$$\frac{5}{6}m_2 = -1$$
$$\frac{6}{5}\left(\frac{5}{6}m_2\right) = -1\left(\frac{6}{5}\right)$$
$$m_2 = -\frac{6}{5}$$

The slope is $-\dfrac{6}{5}$.

49. a. Parallel lines have the same slopes, so the slope is -8.

    b. Perpendicular lines have slopes whose product is -1.

$$m_1 m_2 = -1$$
$$-8m_2 = -1$$
$$-\frac{1}{8}(-8m_2) = -1\left(-\frac{1}{8}\right)$$
$$m_2 = \frac{1}{8}$$

The slope is $\dfrac{1}{8}$.

51. The slope of the given line is $m_1 = \dfrac{2}{3}$.

    a. Parallel lines have the same slopes, so the slope is $\dfrac{2}{3}$.

    b. Perpendicular lines have slopes whose product is -1.

$$m_1 m_2 = -1$$
$$\frac{2}{3}m_2 = -1$$
$$\frac{3}{2}\left(\frac{2}{3}m_2\right) = -1\left(\frac{3}{2}\right)$$
$$m_2 = -\frac{3}{2}$$

The slope is $-\dfrac{3}{2}$.

53. Yes; after graphing the points, one can see they possibly all lie on the same line. Notice the slope between each set of points is equal.

$$m_1 = \frac{6-(-4)}{18-3} = \frac{10}{15} = \frac{2}{3}, \quad m_2 = \frac{0-6}{9-18} = \frac{-6}{-9} = \frac{2}{3},$$

$$m_3 = \frac{0-(-4)}{9-3} = \frac{4}{6} = \frac{2}{3}$$

From the graph, $b = -6$.

$$y = \frac{2}{3}x - 6 \text{ or } 2x - 3y = 18$$

55. $y = 5(7x + 125)$

a. $y = 5 \cdot 7x + 5 \cdot 125$
$y = 35x + 625$

b. $m = 35$, $b = 625$, y-intercept $(0, 625)$

c. The slope is the amount of increase in thousands in the number of cell phone accessories in the United States per year from 2010 to 2020.

Cumulative Review

57.
$$\frac{1}{4}x + 3 > \frac{2}{3}x + 2$$
$$12\left(\frac{1}{4}x + 3\right) > 12\left(\frac{2}{3}x + 2\right)$$
$$3x + 12 \cdot 3 > 4 \cdot 2x + 12 \cdot 2$$
$$3x + 36 > 8x + 24$$
$$-5x > -12$$
$$x < \frac{12}{5}$$

$$\xleftarrow{\quad} \underset{\frac{11}{5}}{\mid} \underset{\frac{12}{5}}{\overset{\oplus}{\mid}} \underset{\frac{13}{5}}{\mid} \underset{\frac{14}{5}}{\mid} \underset{3}{\mid} \xrightarrow{\quad}$$

58.
$$\frac{1}{2}(x + 2) \le \frac{1}{3}x + 5$$
$$6\left[\frac{1}{2}(x + 2)\right] \le 6\left(\frac{1}{3}x + 5\right)$$
$$3(x + 2) \le 2x + 30$$
$$3x + 6 \le 2x + 30$$
$$x \le 24$$

$$\xleftarrow{\quad} \underset{22}{\mid} \underset{23}{\mid} \underset{24}{\overset{\bullet}{\mid}} \underset{25}{\mid} \underset{26}{\mid} \xrightarrow{\quad}$$

Quick Quiz 3.3

1. $(-2, 5)$, $(-6, 3)$
$$m = \frac{3-5}{-6-(-2)} = \frac{-2}{-4} = \frac{1}{2}$$

2. $6x + 2y - 4 = 0$

a. $2y = -6x + 4$
$y = -3x + 2$
$m = -3$
$b = 2$, y-intercept $(0, 2)$

b.

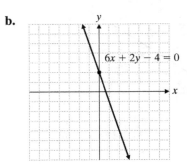

3. $m = -\frac{5}{7}$
$b = -5$
$y = mx + b$
$y = -\frac{5}{7}x - 5$

4. Answers may vary. Possible solution: Slope measures the vertical change per one unit of horizontal change.

Use Math to Save Money

1. Divide his annual salary by 12.
$$\frac{\$42{,}000}{12} = \$3500$$
Louis earns $3500 per month.

2. Find 5% of $3500.
$0.05 \times \$3500 = \$175$
Louis will contribute $175 each month.

3. Replace *PMT* with 175, *I* with 0.0067, and *N* with 480.
$$FV = PMT \times \frac{(1+I)^N - 1}{I}$$
$$= 175 \times \frac{(1+0.0067)^{480} - 1}{0.0067}$$
$$\approx \$618{,}044$$
The future value is approximately $618,044.

4. Find 4% of $618,044.
$0.04 \times \$618{,}044 \approx \$24{,}722$
Louis can withdraw $24,722 the first year.

5. Divide the annual withdrawal by 12.

$$\frac{\$24,722}{12} \approx \$2060$$

His monthly income will be about $2060.

6. Subtract to find the difference.
$3500 - 2060 = 1440$
The retirement income will be $1440 less per month than his current income.

7. Yes, he needs to increase the amount.

How Am I Doing? Sections 3.1–3.3
(Available online through MyMathLab.)

1.

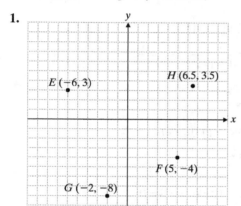

2. $A(3, 9)$
$B(7, -7)$
$C(-7, -8)$
$D(-4, 3)$

3. $y = -7x + 3$
$y = -7(4) + 3 = -25,\ (4, -25)$
$y = -7(0) + 3 = 3,\ (0, 3)$
$y = -7(-2) + 3 = 17,\ (-2, 17)$

4. $4x + y = -3$
$\quad\ y = -4x - 3$

x	$y = -4x - 3$	y
-2	$y = -4(-2) - 3 = 8 - 3$	5
-1	$y = -4(-1) - 3 = 4 - 3$	1
0	$y = -4(0) - 3 = 0 - 3$	-3

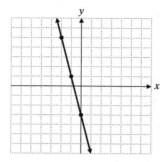

5. $y = \dfrac{3}{4}x - 1$

x	$y = \frac{3}{4}x - 1$	y
-4	$y = \frac{3}{4}(-4) - 1 = -3 - 1$	-4
0	$y = \frac{3}{4}(0) - 1 = 0 - 1$	-1
4	$y = \frac{3}{4}(4) - 1 = 3 - 1$	2

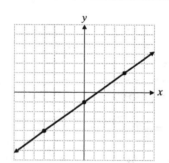

6. $6x - 5y = 30$

  a. Let $y = 0$.
    $6x - 5(0) = 30$
       $6x = 30$
        $x = 5$
    x-intercept $(5, 0)$
    Let $x = 0$.
    $6(0) - 5y = 30$
       $-5y = 30$
        $y = -6$
    y-intercept: $(0, -6)$

b. Find a third point.
Let $y = -3$.
$$6x - 5(-3) = 30$$
$$6x + 15 = 30$$
$$6x = 15$$
$$x = \frac{15}{6} = \frac{5}{2} \text{ or } 2\frac{1}{2}$$
$$\left(\frac{5}{2}, -3\right)$$

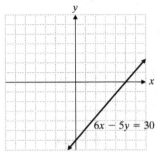

7. $3x - 5y = 0$
$$5y = 3x$$
$$y = \frac{3}{5}x$$

x	$y = \frac{3}{5}x$	y
-5	$y = \frac{3}{5}(-5)$	-3
0	$y = \frac{3}{5}(0)$	0
5	$y = \frac{3}{5}(5)$	3

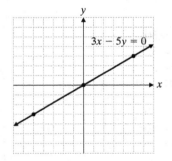

8. $-5x + 2 = -13$
$$-5x = -15$$
$$x = 3$$
This is a vertical line with *x*-intercept (3, 0).

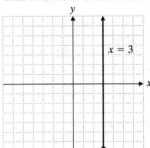

9. $(-2, 5), (0, 1)$
$$m = \frac{1-5}{0-(-2)} = \frac{-4}{2} = -2$$

10. $m = \frac{2}{7}$
y-intercept $(0, -2)$, $b = -2$
$$y = \frac{2}{7}x - 2$$

11. a. $-12x + 4y + 8 = 0$
$$4y = 12x - 8$$
$$y = \frac{12x - 8}{4}$$
$$y = 3x - 2$$
$m = 3$
$b = -2$, *y*-intercept $(0, -2)$

b.

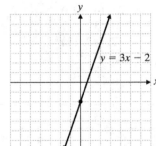

12. Perpendicular lines have slopes whose product is
-1.

$$m_1 m_2 = -1$$

$$\frac{3}{4} m_2 = -1$$

$$\frac{4}{3}\left(\frac{3}{4} m_2\right) = -1\left(\frac{4}{3}\right)$$

$$m_2 = -\frac{4}{3}$$

The slope is $-\dfrac{4}{3}$.

13. $3y - 5x + 10 = 0$

$$3y = 5x - 10$$

$$y = \frac{5}{3}x - \frac{10}{3}$$

The given line has slope $\dfrac{5}{3}$. The slope of a

parallel line is also $\dfrac{5}{3}$.

3.4 Exercises

Use $m = \dfrac{y_2 - y_1}{x_2 - x_1}$ and $y = mx + b$ as needed in

Exercises 1–37.

1. $m = 3,\ (-2, 0)$
$y = mx + b$
$0 = 3(-2) + b$
$b = 6$
$y = 3x + 6$

3. $m = -2,\ (3, 5)$
$y = mx + b$
$5 = (-2)3 + b$
$b = 11$
$y = -2x + 11$

5. $m = -3,\ \left(\dfrac{1}{2}, 2\right)$

$y = mx + b$

$2 = -3\left(\dfrac{1}{2}\right) + b$

$b = \dfrac{7}{2}$

$y = -3x + \dfrac{7}{2}$

7. $m = \dfrac{1}{4},\ (4, 5)$

$y = mx + b$

$5 = \dfrac{1}{4}(4) + b$

$b = 4$

$y = \dfrac{1}{4}x + 4$

9. $(3, -12)$ and $(-4, 2)$

$m = \dfrac{2 - (-12)}{-4 - 3} = \dfrac{14}{-7} = -2$

$y = mx + b$

$2 = -2(-4) + b \Rightarrow b = -6$

$y = -2x - 6$

11. $(2, -6)$ and $(-1, 6)$

$m = \dfrac{6 - (-6)}{-1 - 2} = \dfrac{12}{-3} = -4$

$y = mx + b$

$6 = -4(-1) + b \Rightarrow b = 2$

$y = -4x + 2$

13. $(3, 5)$ and $(-1, -15)$

$m = \dfrac{-15 - 5}{-1 - 3} = \dfrac{-20}{-4} = 5$

$y = mx + b$

$5 = 5(3) + b \Rightarrow b = -10$

$y = 5x - 10$

15. $\left(1, \dfrac{5}{6}\right)$ and $\left(3, \dfrac{3}{2}\right)$

$m = \dfrac{\frac{3}{2} - \frac{5}{6}}{3 - 1} = \dfrac{\frac{4}{6}}{2} = \dfrac{1}{3}$

$y = mx + b$

$\dfrac{5}{6} = \dfrac{1}{3}(1) + b \Rightarrow b = \dfrac{1}{2}$

$y = \dfrac{1}{3}x + \dfrac{1}{2}$

17. $m = -3,\ (-1, 3)$
$y = mx + b$
$3 = -3(-1) + b$
$b = 0$
$y = -3x$

19. $(2, -3)$ and $(-1, 6)$

$m = \dfrac{6 - (-3)}{-1 - 2} = \dfrac{9}{-3} = -3$

$y = mx + b$

$3 = -3(2) + b \Rightarrow b = 3$

$y = -3x + 3$

21. $b = 1, \; m = -\dfrac{2}{3}$

$y = -\dfrac{2}{3}x + 1$

23. $b = -4, \; m = \dfrac{2}{3}$

$y = \dfrac{2}{3}x - 4$

25. $b = 0, \; m = -\dfrac{2}{3}$

$y = -\dfrac{2}{3}x$

27. $b = -2, \; m = 0$
$y = -2$

29. $m = 0, \; (7, -5)$
The line is horizontal so $y = -5$.

31. $(4, -6)$ perpendicular to x-axis so m is undefined.
$x = 4$ for all values of y.
$x = 4$

33. $(0, 5)$ parallel to $y = \dfrac{1}{3}x + 4$.

$5 = \dfrac{1}{3}(0) + b$
$b = 5$
$y = \dfrac{1}{3}x + 5$

35. Perpendicular to $y = 2x - 9 \Rightarrow m_1 = 2$

$m_2 = -\dfrac{1}{m_1} = -\dfrac{1}{2}$

$y = mx + b, \; (2, 3)$

$3 = -\dfrac{1}{2}(2) + b \Rightarrow b = 4$

$y = -\dfrac{1}{2}x + 4$

37. $(0, 227)$ and $(10, 251)$

$m = \dfrac{251 - 227}{10 - 0} = \dfrac{24}{10} = 2.4, \; b = 227$

$y = mx + b$
$y = 2.4x + 227$

Cumulative Review

39.
$$10 - 3x > 14 - 2x$$
$$10 - 3x + 2x > 14 - 2x + 2x$$
$$10 - x > 14$$
$$10 - x - 10 > 14 - 10$$
$$-x > 4$$
$$\dfrac{-x}{-1} < \dfrac{4}{-1}$$
$$x < -4$$

40.
$$2x - 3 \geq 7x - 18$$
$$2x - 3 - 7x \geq 7x - 18 - 7x$$
$$-5x - 3 \geq -18$$
$$-5x - 3 + 3 \geq -18 + 3$$
$$-5x \geq -15$$
$$\dfrac{-5x}{-5} \leq \dfrac{-15}{-5}$$
$$x \leq 3$$

41. 2nd week: Price $= 80 - 0.15(80) = \$68$
3rd week: Price $= 68 - 0.1(68) = \$61.20$

42. $x =$ minutes beyond 200
$0.21x + 50 = 68.90$
$0.21x = 18.90$
$x = 90$
Total time $= 200 + 90 = 290$ minutes

Quick Quiz 3.4

1. $(3, -5), \; m = \dfrac{2}{3}$

$y = mx + b$

$-5 = \dfrac{2}{3}(3) + b$

$-7 = b$

$y = \dfrac{2}{3}x - 7$

2. $(-2, 7), \; (-4, -5)$

$m = \dfrac{-5 - 7}{-4 - (-2)} = \dfrac{-12}{-2} = 6$

$m = 6$
$y = mx + b$
$-5 = 6(-4) + b$
$19 = b$
$y = 6x + 19$

3. $(4, 5), (4, -2)$

$m = \dfrac{-2-5}{4-4} = \dfrac{-7}{0} =$ undefined

This is a vertical line $x = 4$.

4. Answers may vary. Possible solution:
Zero slope indicates a horizontal line. Since the line is horizontal and it passes through $(-2, -3)$, the equation must be $y = -3$.

3.5 Exercises

1. No. All points in one region will be solutions to the inequality, while points in the other region will not be solutions. Thus testing any point will give the same result, as long as the test point is not on the boundary line.

3. $y \ge 4x$
Graph $y = 4x$ with a solid line.
Test point: $(1, 0)$
$0 \ge 4(1)$
$0 \ge 4$ False
Shade the region not containing $(1, 0)$.

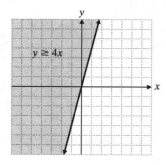

5. $2x - 3y < 6$
Graph $2x - 3y = 6$ using a dashed line.
Test point: $(0, 0)$
$2(0) - 3(0) < 6$
$\qquad 0 < 6$ True
Shade the region containing $(0, 0)$.

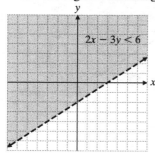

7. $2x - y \ge 3$
Graph $2x - y = 3$ with a solid line.
Test point: $(0, 0)$
$2(0) - (0) \ge 3$
$\qquad 0 \ge 3$ False
Shade the region not containing $(0, 0)$.

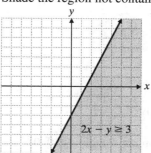

9. $y < 2x - 4$
Graph $y = 2x - 4$ with a dashed line.
Test point: $(2, -2)$
$-2 < 2(2) - 4$
$-2 < 0$ True
Shade the region containing $(2, -2)$.

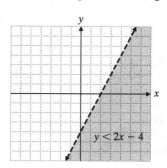

11. $y < -\dfrac{1}{2}x$

Graph $y = -\dfrac{1}{2}x$ with a dashed line.

Test point: $(1, 0)$

$0 < -\dfrac{1}{2}(1)$

$0 < -\dfrac{1}{2}$ False

Shade the region not containing $(1, 0)$.

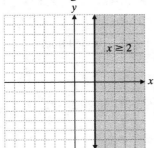

13. $x \geq 2$

Graph $x = 2$ with a solid line.

Test point: $(0, 0)$

$0 \geq 2$ False

Shade the region not containing $(0, 0)$.

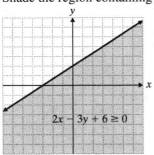

15. $2x - 3y + 6 \geq 0$

Graph $2x - 3y + 6 = 0$ or $2x - 3y = -6$ with a solid line.

Test point: $(0, 0)$

$2(0) - 3(0) + 6 \geq 0$

$6 \geq 0$ True

Shade the region containing $(0, 0)$.

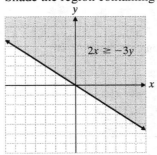

17. $x > -2y$

Graph $x = -2y$ with a dashed line.

Test point: $(1, 0)$

$1 > -2(0)$

$1 > 0$ True

Shade the region containing $(1, 0)$.

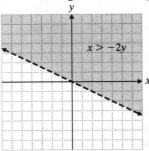

19. $2x > 3 - y$

Graph $2x = 3 - y$ or $y = -2x + 3$ with a dashed line.

Test point: $(0, 0)$

$2(0) > 3 - (0)$

$0 > 3$ False

Shade the region not containing $(0, 0)$.

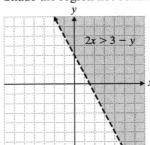

21. $2x \geq -3y$

Graph $2x = -3y$ with a dashed line.

Test point: $(1, 1)$

$2(1) \geq -3(1)$

$2 \geq -3$ True

Shade the region containing $(1, 1)$.

Cumulative Review

23. $6(2) + 10 \div (-2) = 12 + 10 \div (-2) = 12 + (-5) = 7$

24. $3(-3) + 2(12 - 15)^2 \div 9 = 3(-3) + 2(-3)^2 \div 9$
$$= 3(-3) + 2(9) \div 9$$
$$= -9 + 2(9) \div 9$$
$$= -9 + 18 \div 9$$
$$= -9 + 2$$
$$= -7$$

25. When $x = -2$ and $y = 3$,
$$2x^2 + 3xy - 2y^2 = 2(-2)^2 + 3(-2)(3) - 2(3)^2$$
$$= 2(4) + 3(-2)(3) - 2(9)$$
$$= 8 - 18 - 18$$
$$= -28$$

26. When $x = -2$ and $y = 3$,
$$x^3 - 5x^2 + 3y - 1 = (-2)^3 - 5(-2)^2 + 3(3) - 1$$
$$= -8 - 5(4) + 3(3) - 1$$
$$= -8 - 20 + 9 - 1$$
$$= -20$$

27. $\dfrac{22,400}{200} = 112$

The average cost is originally $112 per part.
15% of $112 = 0.15 \times 112 = 16.8$
The discount is $16.8 per part.
$112 - 16.8 = 95.2$
The average cost at the discounted price is $95.20 per part.

28. 15% of $22,400 = 0.15 \times 22,400 = 3360$
The total discount is $3360.
$22,400 - 3360 = 19,040$
The total bill at the discounted price is $19,040.

Quick Quiz 3.5

1. Use a dashed line. If the inequality has a > or a < symbol, the points on the line itself are not included. This is indicated by a dashed line.

2. $3y \leq -7x$
Graph $3y = -7x$ with a solid line.
Test point: (1, 1)
$3(1) \leq -7(1)$
$3 \leq -7$ False
Shade the region not containing (1, 1).

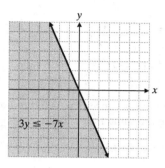

3. $-5x + 2y > -3$
Graph $-5x + 2y = -3$ with a dashed line.
Test point: (0, 0)
$-5(0) + 2(0) > -3$
$0 > -3$ True
Shade the region containing (0, 0

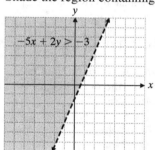

4. Answers may vary. Possible solution:
The inequality is first graphed without shading. The test point coordinates are then substituted into the inequality. If the result of the substitution results in a true statement, the area where the test point lies is shaded. If false, the opposite area is shaded.

3.6 Exercises

1. You can describe a function using a table of values, an algebraic equation, or a graph.

3. The domain of a function is the set of <u>possible values</u> of the <u>independent</u> variable.

5. If a vertical line can intersect the graph more than once, the relation is not a function. If no such line exists, then the relation is a function.

7. $\left\{ \left(\dfrac{3}{7}, 4 \right), \left(3, \dfrac{3}{7} \right), \left(-3, \dfrac{3}{7} \right), \left(\dfrac{3}{7}, -1 \right) \right\}$

   a. The domain consists of all the first coordinates and the range consists of all the second coordinates.

$\text{Domain} = \left\{-3, \dfrac{3}{7}, 3\right\}$

$\text{Range} = \left\{-1, \dfrac{3}{7}, 4\right\}$

b. Not a function because two different ordered pairs have the same first coordinate.

9. {(6, 2.5), (3, 1.5), (0, 0.5)}

   a. The domain consists of all the first coordinates and the range consists of all the second coordinates.
Domain = {0, 3, 6}
Range = {0.5, 1.5, 2.5}

   b. Function, because no two ordered pairs have the same first coordinate.

11. {(12, 1), (14, 3), (1, 12), (9, 12)}

   a. The domain consists of all the first coordinates and the range consists of all the second coordinates.
Domain = {1, 9, 12, 14}
Range = {1, 3, 12}

   b. Function, because no two ordered pairs have the same first coordinate.

13. {(3, 75), (5, 95), (3, 85), (7, 100)}

   a. The domain consists of all the first coordinates and the range consists of all the second coordinates.
Domain = {3, 5, 7}
Range = {75, 85, 95, 100}

   b. Not a function because two different ordered pairs have the same first coordinate.

15. $y = x^2 + 3$

x	$y = x^2 + 3$	y
-2	$y = (-2)^2 + 3 = 4 + 3$	7
-1	$y = (-1)^2 + 3 = 1 + 3$	4
0	$y = 0^2 + 3 = 0 + 3$	3
1	$y = 1^2 + 3 = 1 + 3$	4
2	$y = 2^2 + 3 = 4 + 3$	7

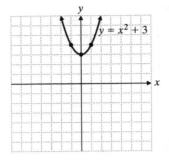

17. $y = 2x^2$

x	$y = 2x^2$	y
-2	$y = 2(-2)^2 = 2(4)$	8
-1	$y = 2(-1)^2 = 2(1)$	2
0	$y = 2(0)^2 = 2(0)$	0
1	$y = 2(1)^2 = 2(1)$	2
2	$y = 2(2)^2 = 2(4)$	8

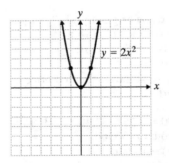

19. $x = -2y^2$

y	$x = -2y^2$	x
-2	$x = -2(-2)^2 = -2(4)$	-8
-1	$x = -2(-1)^2 = -2(1)$	-2
0	$x = -2(0)^2 = -2(0)$	0
1	$x = -2(1)^2 = -2(1)$	-2
2	$x = -2(2)^2 = -2(4)$	-8

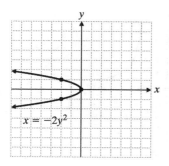

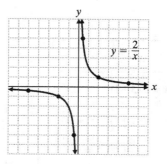

21. $x = y^2 - 4$

y	$x = y^2 - 4$	x
-2	$x = (-2)^2 - 4 = 4 - 4$	0
-1	$x = (-1)^2 - 4 = 1 - 4$	-3
0	$x = (0)^2 - 4 = 0 - 4$	-4
1	$x = 1^2 - 4 = 1 - 4$	-3
2	$x = 2^2 - 4 = 4 - 4$	0

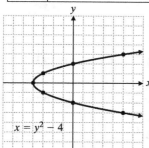

23. $y = \dfrac{2}{x}$

x	$y = \dfrac{2}{x}$	y
-4	$y = \dfrac{2}{-4}$	$-\dfrac{1}{2}$
-2	$y = \dfrac{2}{-2}$	-1
-1	$y = \dfrac{2}{-1}$	-2
1	$y = \dfrac{2}{1}$	2
2	$y = \dfrac{2}{1}$	1
4	$y = \dfrac{2}{4}$	$\dfrac{1}{2}$

25. $y = \dfrac{4}{x^2}$

x	$y = \dfrac{4}{x^2}$	y
-4	$y = \dfrac{4}{(-4)^2} = \dfrac{4}{16}$	$\dfrac{1}{4}$
-2	$y = \dfrac{4}{(-2)^2} = \dfrac{4}{4}$	1
-1	$y = \dfrac{4}{(-1)^2} = \dfrac{4}{1}$	4
1	$y = \dfrac{4}{1^2} = \dfrac{4}{1}$	4
2	$y = \dfrac{4}{2^2} = \dfrac{4}{4}$	1
4	$y = \dfrac{4}{4^2} = \dfrac{4}{16}$	$\dfrac{1}{4}$

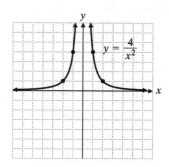

27. $x = (y+1)^2$

y	$x = (y+1)^2$	x
-3	$x = (-3+1)^2 = (-2)^2$	4
-2	$x = (-2+1)^2 = (-1)^2$	1
-1	$x = (-1+1)^2 = 0^2$	0
0	$x = (0+1)^2 = 1^2$	1
1	$x = (1+1)^2 = 2^2$	4

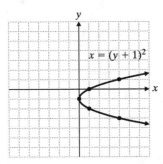

29. $y = \dfrac{4}{x-2}$

x	$y = \dfrac{4}{x-2}$	y
0	$y = \dfrac{4}{0-2} = \dfrac{4}{-2}$	-2
1	$y = \dfrac{4}{1-2} = \dfrac{4}{-1}$	-4
$\dfrac{3}{2}$	$y = \dfrac{4}{\frac{3}{2}-2} = \dfrac{4}{-\frac{1}{2}}$	-8
$\dfrac{5}{2}$	$y = \dfrac{4}{\frac{5}{2}-2} = \dfrac{4}{\frac{1}{2}}$	8
3	$y = \dfrac{4}{3-2} = \dfrac{4}{1}$	4
4	$y = \dfrac{4}{4-2} = \dfrac{4}{2}$	2

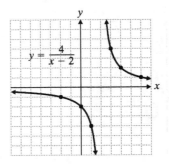

31. Function, passes the vertical line test.

33. Not a function, fails the vertical line test.

35. Function, passes the vertical line test.

37. Not a function, fails the vertical line test.

39. $f(x) = 2 - 3x$

    a. $f(-8) = 2 - 3(-8) = 2 + 24 = 26$

    b. $f(0) = 2 - 3(0) = 2 - 0 = 2$

    c. $f(2) = 2 - 3(2) = 2 - 6 = -4$

41. $f(x) = 2x^2 - x + 3$

    a. $f(0) = 2(0)^2 - 0 + 3 = 0 - 0 + 3 = 3$

    b. $f(-3) = 2(-3)^2 - (-3) + 3 = 18 + 3 + 3 = 24$

    c. $f(2) = 2(2)^2 - (2) + 3 = 8 - 2 + 3 = 9$

43. $f(x) = 0.02x^2 + 0.08x + 31.6$

$f(0) = 0.02(0)^2 + 0.08(0) + 31.6 = 31.6$

$f(4) = 0.02(4)^2 + 0.08(4) + 31.6 = 32.24$

$f(10) = 0.02(10)^2 + 0.08(10) + 31.6 = 34.4$

The curve slopes more steeply for larger values of x. The increase in pet ownership is increasing as x gets larger.

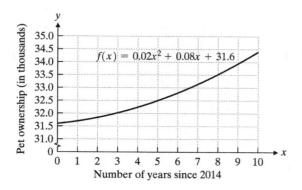

Number of years since 2014

Cumulative Review

45. $-4x(2x^2 - 3x + 8) = -4x(2x^2) - (-4x)(3x) + (-4x)(8)$
$$= -8x^3 + 12x^2 - 32x$$

46. $5a(ab + 6b - 2a) = 5a(ab) + 5a(6b) - 5a(2a)$
$$= 5a^2b + 30ab - 10a^2$$

47. $-7x + 10y - 12x - 8y - 2 = -7x - 12x + 10y - 8y - 2$
$$= -19x + 2y - 2$$

48. $3x^2y - 6xy^2 + 7xy + 6x^2y = 3x^2y + 6x^2y - 6xy^2 + 7xy$
$$= 9x^2y - 6xy^2 + 7xy$$

Quick Quiz 3.6

1. No; two different ordered pairs have the same first coordinate.

2. $f(x) = 3x^2 - 4x + 2$

    a. $f(-3) = 3(-3)^2 - 4(-3) + 2$
$$= 27 + 12 + 2$$
$$= 41$$

    b. $f(4) = 3(4)^2 - 4(4) + 2 = 48 - 16 + 2 = 34$

3. $g(x) = \dfrac{7}{x - 3}$

    a. $g(2) = \dfrac{7}{2 - 3} = \dfrac{7}{-1} = -7$

    b. $g(-5) = \dfrac{7}{-5 - 3} = \dfrac{7}{-8} = -\dfrac{7}{8}$

4. Answers may vary. Possible solution: Because duplicate elements are recorded only once in the domain and in the range.

Career Exploration Problems

1.

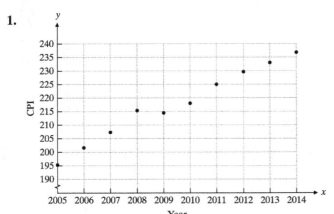

Source: U.S. Bureau of Labor Statistics

2. 2005–2008: $m = \dfrac{y_2 - y_1}{x_2 - x_1} = \dfrac{215.30 - 195.30}{2008 - 2005} = \dfrac{20}{3} \approx 6.67$

Average rate of change is \$6.67.

2008–2009: $m = \dfrac{y_2 - y_1}{x_2 - x_1} = \dfrac{214.54 - 215.30}{2009 - 2008} = \dfrac{-0.76}{1} = -0.76$

Average rate of change is –\$0.76.

2010–2014: $m = \dfrac{y_2 - y_1}{x_2 - x_1} = \dfrac{236.74 - 218.06}{2014 - 2010} = \dfrac{18.68}{4} = 4.67$

Average rate of change is \$4.67.

3. $m = \dfrac{y_2 - y_1}{x_2 - x_1} = \dfrac{236.74 - 195.30}{2014 - 2005} = \dfrac{41.44}{9} \approx 4.60$

$y = mx + b$
$(x, y) = (2014, 236.74)$
$\quad 236.74 = 4.60(2014) + b$
$\quad 236.74 = 9264.4 + b$
$-9027.66 = b$
The equation of the line is $y = 4.60x - 9027.66$.

4. 2016: $y = 4.60(2016) - 9027.66$
$\qquad = 9273.6 - 9027.66$
$\qquad = 245.94$
2020: $y = 4.60(2020) - 9027.66$
$\qquad = 9292 - 9027.66$
$\qquad = 264.34$
The set of goods and services is predicted to cost \$245.94 in 2016 and \$264.34 in 2020.

You Try It

1. $2x - y = 6$
$\quad 2x - 6 = y$
$\qquad y = 2x - 6$

x	$y = 2x - 6$	y
0	$y = 2(0) - 6$	-6
1	$y = 2(1) - 6$	-4
2	$y = 2(2) - 6$	-2

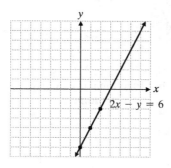

2. $-2x + 4y = -8$

  a. Let $y = 0$.

$$-2x + 4(0) = -8$$
$$-2x = -8$$
$$x = 4$$

    x-intercept $(4, 0)$
    Let $x = 0$.

$$-2(0) + 4y = -8$$
$$4y = -8$$
$$y = -2$$

    y-intercept $(0, -2)$

  b. Find a third point.
    Let $x = 2$.

$$-2(2) + 4y = -8$$
$$-4 + 4y = -8$$
$$4y = -4$$
$$y = -1$$

    $(2, -1)$

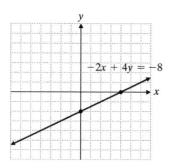

3. a. $x = -2$

This is a vertical line two units to the left of the y-axis.

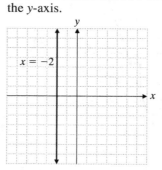

  b. $y = 5$

This is a horizontal line five units above the x-axis.

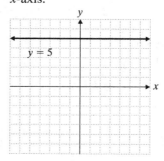

4. $m = \dfrac{1 - 5}{0 - (-2)} = \dfrac{-4}{2} = -2$

5. $5x + 3y = 9$

$$3y = -5x + 9$$
$$y = \frac{-5x + 9}{3}$$
$$y = -\frac{5}{3}x + 3$$
$$m = -\frac{5}{3}$$

$b = 3$, y-intercept $(0, 3)$

6. $m = 2$

y-intercept $(0, -3)$, $b = -3$

$y = 2x - 3$

7. $y = 3x - 4$

$m = 3$ or $\dfrac{3}{1}$, $b = -4$,

y-intercept $(0, -4)$

Find another point using the slope: $(1, -1)$

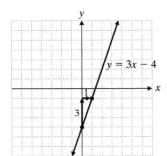

8. $m_1 = 3$

   a. $m_2 = m_1 = 3$

   b. $m_2 = -\dfrac{1}{m_1} = -\dfrac{1}{3}$

9. $y = mx + b$

$$3 = -\frac{1}{2}(1) + b$$

$$3 = -\frac{1}{2} + b$$

$$3 + \frac{1}{2} = b$$

$$b = \frac{7}{2}$$

An equation is $y = -\dfrac{1}{2}x + \dfrac{7}{2}$.

10. $m = \dfrac{0-2}{-1-3} = \dfrac{-2}{-4} = \dfrac{1}{2}$

$$y = mx + b$$

$$2 = \frac{1}{2}(3) + b$$

$$2 = \frac{3}{2} + b$$

$$2 - \frac{3}{2} = b$$

$$\frac{1}{2} = b$$

An equation is $y = \dfrac{1}{2}x + \dfrac{1}{2}$.

11. $y \le 2x - 4$
Graph the line $y = 2x - 4$. Use a solid line.
Test $(0, 0)$.
$0 \le 2(0) - 4$
$0 \le -4$ False
Shade the region that does not contain $(0, 0)$.

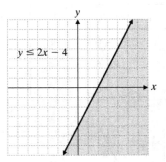

12. $\{(2, -1), (3, 0), (-1, 4), (0, 4)\}$
This relation is a function since no two ordered pairs have the same *x*-coordinate.

13. $y = (x+2)^2$

x	$y = (x+2)^2$	y
-4	$y = (-4+2)^2 = (-2)^2$	4
-3	$y = (-3+2)^2 = (-1)^2$	1
-2	$y = (-2+2)^2 = 0^2$	0
-1	$y = (-1+2)^2 = 1^2$	1
0	$y = (0+2)^2 = 2^2$	4

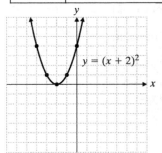

14. Not a function, fails the vertical line test.

15. $f(x) = x^2 - x$

   a. $f(5) = (5)^2 - (5) = 25 - 5 = 20$

   b. $f(-2) = (-2)^2 - (-2) = 4 + 2 = 6$

Chapter 3 Review Problems

1. A: (2, −3), B: (−1, 0), C: (3, 2), D: (−2, −3)

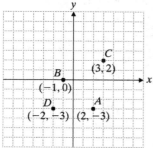

2. E: (4, 4), F: (0, 3), G: (1, −4), H: (−4, −1)

3. $y = 7 - 3x$

   a.   (0,   ): $y = 7 - 3(0)$
                      $y = 7 - 0$
                      $y = 7$: (0, 7)

   b.   (, 10): $10 = 7 - 3x$
                    $3x = -3$
                    $x = -1$: (−1, 10)

4. $2x + 5y = 12$

   a.   (1,   ): $2(1) + 5y = 12$
                     $5y = 10$
                      $y = 2$: (1, 2)

   b.   (, 4): $2x + 5(4) = 12$
                     $2x = -8$
                      $x = -4$: (−4, 4)

5. $x = 6$

   a.   (, −1): $x = 6$: (6, −1)

   b.   (, 3): $x = 6$: (6, 3)

6. $5y + x = -15$
       $x = -5y - 15$

y	$x = -5y - 15$	x
−2	$x = -5(-2) - 15$ $x = 10 - 15$	−5
−3	$x = -5(-3) - 15$ $x = 15 - 15$	0
−4	$x = -5(-4) - 15$ $x = 20 - 15$	5

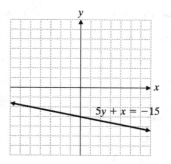

7. $2y + 4x = -8 + 2y$
         $4x = -8$
          $x = -2$

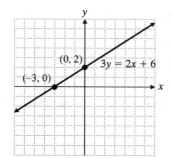

8. $3y = 2x + 6$

x	$3y = 2x + 6$	y
−3	$3y = 2(-3) + 6$ $3y = 0$	0
0	$3y = 2(0) + 6$ $3y = 6$	2
3	$3y = 2(3) + 6$ $3y = 12$	4

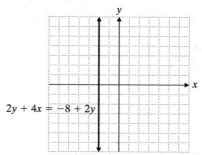

9. (5, −3) and $\left(2, -\dfrac{1}{2}\right)$

$$m = \frac{y_2 - y_1}{x_2 - x_1} = \frac{-\frac{1}{2} - (-3)}{2 - 5} = \frac{\frac{5}{2}}{-3} = -\frac{5}{6}$$

10. Perpendicular lines have slopes whose product is
-1, so the slope is $m = -\dfrac{5}{3}$.

11. $9x - 11y + 15 = 0$
$$9x + 15 = 11y$$
$$\frac{9}{11}x + \frac{15}{11} = y$$
$$y = \frac{9}{11}x + \frac{15}{11}$$
$m = \dfrac{9}{11}$, $b = \dfrac{15}{11}$, y-intercept $\left(0, \dfrac{15}{11}\right)$

12. $m = -\dfrac{1}{2}$, $b = 3$
$$y = mx + b$$
$$y = -\frac{1}{2}x + 3$$

13. $y = -\dfrac{1}{2}x + 3$

$m = -\dfrac{1}{2}$, $b = 3$, y-intercept $(0, 3)$

Find another point using the slope $(2, 2)$.

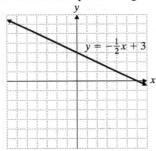

14. $2x - 3y = -12$
$$-3y = -2x - 12$$
$$y = \frac{2}{3}x + 4$$
$m = \dfrac{2}{3}$, $b = 4$, y-intercept $(0, 4)$

Find another point using the slope: $(3, 6)$.

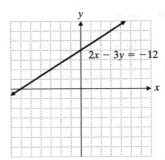

15. $y = -2x$
$m = -2$, $b = 0$,
y-intercept $(0, 0)$
Find another point using the slope: $(1, -2)$

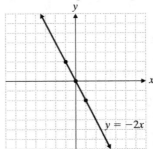

16. $(3, -4)$, $m = -6$
$$y = mx + b$$
$$-4 = -6(3) + b$$
$$-4 = -18 + b$$
$$b = 14$$
$$y = -6x + 14$$

17. $m = -\dfrac{1}{3}$, $(-1, 4)$
$$y = mx + b$$
$$4 = -\frac{1}{3}(-1) + b$$
$$4 = \frac{1}{3} + b$$
$$\frac{11}{3} = b$$
$$y = -\frac{1}{3}x + \frac{11}{3}$$

18. $m = 1$, $(2, 5)$
$$y = mx + b$$
$$5 = 1(2) + b$$
$$5 = 2 + b$$
$$3 = b$$
$$y = x + 3$$

19. (3, 7) and (−6, 7)

$$m = \frac{y_2 - y_1}{x_2 - x_1} = \frac{7 - 7}{-6 - 3} = 0$$

Horizontal line with $y = 7$

$y = 7$

20. (0, −3) and (3, −1)

$$m = \frac{-3 - (-1)}{0 - 3} = \frac{2}{3}, \; b = -3$$

$$y = \frac{2}{3}x - 3$$

21. (0, 1) and (1, −2)

$$m = \frac{-2 - 1}{1 - 0} = -3, \; b = 1$$

$$y = -3x + 1$$

22. $x = 5$

23. $y < \frac{1}{3}x + 2$

Graph $y = \frac{1}{3}x + 2$ with a dashed line.

Test point: (0, 0)

$$0 < \frac{1}{3}(0) + 2$$

$0 < 2$ True

Shade the region containing (0, 0).

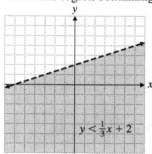

24. $3y + 2x \geq 12$
Graph $3y + 2x = 12$ with a solid line.
Test point: (0, 0)
$3(0) + 2(0) \geq 12$
$0 \geq 12$ False
Shade the region not containing (0, 0).

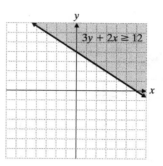

25. $x \leq 2$
Graph $x = 2$ with a solid line.
Test point: (0, 0)
$0 \leq 2$ True
Shade the region containing (0, 0).

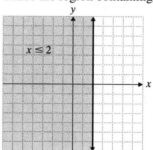

26. The domain consists of all the first coordinates and the range consists of all the second coordinates.
{(5, −6), (−6, 5), (−5, 5), (−6, −6)}
Domain: {−6, −5, 5}
Range: {−6, 5}
Not a function, because two different ordered pairs have the same first coordinate.

27. The domain consists of all the first coordinates and the range consists of all the second coordinates.
{(2, −3), (5, −3), (6, 4), (−2, 4)}
Domain: {−2, 2, 5, 6}
Range: {−3, 4}
Function, because no two ordered pairs have the same first coordinate.

28. Function, passes the vertical line test.

29. Not a function, fails the vertical line test.

30. Function, passes the vertical line test.

31. $y = x^2 - 5$

x	$y = x^2 - 5$	y
-2	$y = (-2)^2 - 5 = 4 - 5$	-1
-1	$y = (-1)^2 - 5 = 1 - 5$	-4
0	$y = 0^2 - 5 = 0 - 5$	-5
1	$y = 1^2 - 5 = 1 - 5$	-4
2	$y = 2^2 - 5 = 4 - 5$	-1

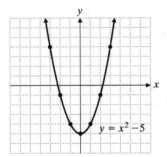

33. $y = (x - 3)^2$

x	$y = (x - 3)^2$	y
1	$y = (1 - 3)^2 = (-2)^2$	4
2	$y = (2 - 3)^2 = (-1)^2$	1
3	$y = (3 - 3)^2 = 0^2$	0
4	$y = (4 - 3)^2 = 1^2$	1
5	$y = (5 - 3)^2 = 2^2$	4

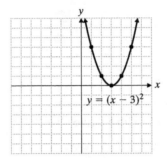

32. $x = y^2 + 3$

y	$x = y^2 + 3$	x
-2	$x = (-2)^2 + 3 = 4 + 3$	7
-1	$x = (-1)^2 + 3 = 1 + 3$	4
0	$x = 0^2 + 3 = 0 + 3$	3
1	$x = 1^2 + 3 = 1 + 3$	4
2	$x = 2^2 + 3 = 4 + 3$	7

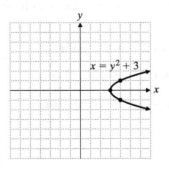

34. $f(x) = 7 - 6x$

   a. $f(0) = 7 - 6(0) = 7 - 0 = 7$

   b. $f(-4) = 7 - 6(-4) = 7 + 24 = 31$

35. $g(x) = -2x^2 + 3x + 4$

   a. $g(-1) = -2(-1)^2 + 3(-1) + 4 = -2 - 3 + 4 = -1$

   b. $g(3) = -2(3)^2 + 3(3) + 4 = -18 + 9 + 4 = -5$

36. $f(x) = \dfrac{2}{x + 4}$

   a. $f(-2) = \dfrac{2}{-2 + 4} = \dfrac{2}{2} = 1$

   b. $f(6) = \dfrac{2}{6 + 4} = \dfrac{2}{10} = \dfrac{1}{5}$

Copyright © 2017 Pearson Education, Inc.

37. $f(x) = x^2 - 2x + \dfrac{3}{x}$

  a. $f(-1) = (-1)^2 - 2(-1) + \dfrac{3}{(-1)} = 1 + 2 - 3 = 0$

  b. $f(3) = (3)^2 - 2(3) + \dfrac{3}{3} = 9 - 6 + 1 = 4$

38. $5x + 3y = -15$
$\qquad 3y = -5x - 15$
$\qquad y = -\dfrac{5}{3}x - 5$

x	$y = -\dfrac{5}{3}x - 5$	y
0	$y = -\dfrac{5}{3}(0) - 5$	-5
-3	$y = -\dfrac{5}{3}(-3) - 5$	0

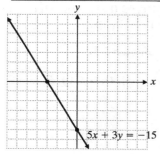

39. $y = \dfrac{3}{4}x - 3$

x	$y = \dfrac{3}{4}x - 3$	y
0	$y = \dfrac{3}{4}(0) - 3$	-3
4	$y = \dfrac{3}{4}(4) - 3$	0

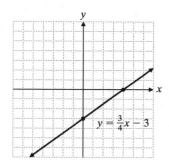

40. $y < -2x + 1$
Graph $y = -2x + 1$ with a dashed line.
Test point: (0, 0)
$0 < -2(0) + 1$
$0 < 1$ True
Shade the region containing (0, 0).

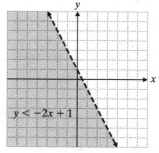

41. (2, −7) and (−3, −5)
$$y = \frac{-5 - (-7)}{-3 - 2} = \frac{2}{-5} = -\frac{2}{5}$$

42. $7x + 6y - 10 = 0$
$\qquad 6y - 10 = -7x$
$\qquad 6y = -7x + 10$
$\qquad y = -\dfrac{7}{6}x + \dfrac{5}{3}$
$\qquad m = -\dfrac{7}{6}, \ b = \dfrac{5}{3}, \ y\text{-intercept} \left(0, \dfrac{5}{3}\right)$

43. $m = \dfrac{2}{3}, \ (3, -5)$
$\qquad y = mx + b$
$\qquad -5 = \dfrac{2}{3}(3) + b \Rightarrow b = -7$
$\qquad y = \dfrac{2}{3}x - 7$

44. (−1, 4) and (2, 1)
$$m = \frac{y_2 - y_1}{x_2 - x_1} = \frac{1 - 4}{2 - (-1)} = \frac{-3}{3} = -1$$
$\qquad y = mx + b$
$\qquad 1 = -1(2) + b$
$\qquad 1 = -2 + b$
$\qquad 3 = b$
An equation of the line is $y = -x + 3$.

Use $y = 30 + 0.09x$ in Exercises 45–50.

45. $y = 30 + 0.09(2000) = 210$
Their monthly bill would be $210.

46. $y = 30 + 0.09(1600) = 174$
Their monthly bill would be $174.

47. $y = 0.09x + 30$
$y = mx + b$, $b = 30$, y-intercept (0, 30)
It tells us that if no electricity is used, the cost is a minimum of $30.

48. $y = 0.09x + 30$
$y = mx + b$, $m = 0.09$
It tells us that the bill increases $0.09 for each kilowatt-hour of use.

49. $147 = 0.09x + 30$
$117 = 0.09x$
$x = 1300$
They used 1300 kilowatt-hours.

50. $246 = 0.09x + 30$
$216 = 0.09x$
$x = 2400$
They used 2400 kilowatt-hours.

51. $x = 1994 - 1994 = 0$
$y = -269(0) + 17,020$
$y = 17,020$
17,020,000 people in 1994

$x = 2008 - 1994 = 14$
$y = -269(14) + 17,020$
$y = 13,254$
13,254,000 people in 2008

$x = 2014 - 1994 = 20$
$y = -269(20) + 17,020$
$y = 11,640$
11,640,000 people in 2014

52. Plot the points and connect (0, 17,020), (14, 13,254), (20, 11,640).

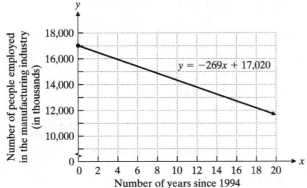

53. The slope is −269. The slope tells us that the number of people employed in manufacturing decreases each year by 269 thousand. In other words, employment in manufacturing goes down 269,000 people each year.

54. The y-intercept is (0, 17,020). This tells us that in the year of 1994, the number of manufacturing jobs was 17,020 thousand which is 17,020,000.

55. $y = -269x + 17,020$
$11,102 = -269x + 17,020$
$-5918 = -269x$
$22 = x$
There will be 11,102,000 manufacturing jobs 22 years after 1994, in 2016.

56. $y = -269x + 17,020$
$10,295 = -269x + 17,020$
$-6725 = -269x$
$25 = x$
There will be 10,295,000 manufacturing jobs 25 years after 1994, in 2019.

How Am I Doing? Chapter 3 Test

1. B: (6, 1), C: (−4, −3), D: (−3, 0), E: (5, −2)

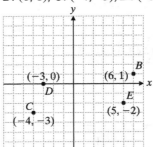

2. $6x - 3 = 5x - 2y$
$x + 2y = 3$
Find the x- and y-intercepts.
Let $x = 0$.
$0 + 2y = 3$
$y = \dfrac{3}{2}$
$\left(0, \dfrac{3}{2}\right)$
Let $y = 0$.
$x + 2(0) = 3$
$x = 3$
(3, 0)

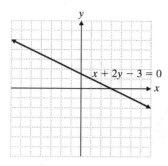

3. $-3x + 9 = 6x$

$\quad\quad 9 = 9x$

$\quad\quad 1 = x$

This is a vertical line one unit to the right of the
y-axis.

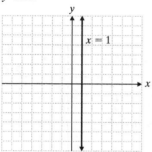

4. $y = \dfrac{2}{3}x - 4$

x	$y = \frac{2}{3}x - 4$	y
0	$y = \frac{2}{3}(0) - 4$	-4
3	$y = \frac{2}{3}(3) - 4$	-2
6	$y = \frac{2}{3}(6) - 4$	0

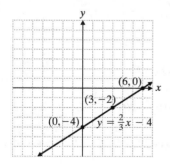

5. $4x + 2y = -8$

   a. Let $y = 0$.

$\quad 4x + 2(0) = -8$

$\quad\quad\quad 4x = -8$

$\quad\quad\quad\quad x = -2$

$\quad$ *x*-intercept $(-2, 0)$

$\quad$ Let $x = 0$.

$\quad 4(0) + 2y = -8$

$\quad\quad\quad 2y = -8$

$\quad\quad\quad\quad y = -4$

$\quad$ *y*-intercept $(0, -4)$

   b. Find a third point.

$\quad$ Let $x = -1$.

$\quad 4(-1) + 2y = -8$

$\quad\quad -4 + 2y = -8$

$\quad\quad\quad\quad 2y = -4$

$\quad\quad\quad\quad\quad y = -2$

$\quad (-1, -2)$

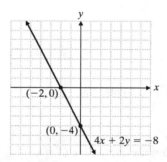

6. $(8, 6)$ and $(-3, -5)$

$$m = \frac{y_2 - y_1}{x_2 - x_1} = \frac{-5 - 6}{-3 - 8} = \frac{-11}{-11} = 1$$

7. $3x + 2y - 5 = 0$

$\quad\quad\quad 2y = -3x + 5$

$\quad\quad\quad\quad y = -\dfrac{3}{2}x + \dfrac{5}{2}$

$\quad m = -\dfrac{3}{2},\ b = \dfrac{5}{2},\ $ *y*-intercept $\left(0, \dfrac{5}{2}\right)$

8. $m = \dfrac{3}{4}$

$\quad$ *y*-intercept $(0, -6),\ b = -6$

$\quad\quad y = \dfrac{3}{4}x - 6$

9. a. $(4, -2)$, $m = \dfrac{1}{2}$

$$y = mx + b$$

$$-2 = \frac{1}{2}(4) + b \Rightarrow b = -4$$

$$y = \frac{1}{2}x - 4 \ \text{ or } x - 2y = 8$$

b. $m_2 = -\dfrac{1}{m_1} = -2$

10. $(5, -4)$ and $(-3, 8)$

$$m = \frac{y_2 - y_1}{x_2 - x_1} = \frac{8 - (-4)}{-3 - 5} = \frac{12}{-8} = -\frac{3}{2}$$

$$y = mx + b$$

$$-4 = \left(-\frac{3}{2}\right)(5) + b$$

$$-8 = -15 + 2b$$

$$\frac{7}{2} = b$$

$$y = -\frac{3}{2}x + \frac{7}{2}$$

11. $4y \le 3x$

Graph $4y = 3x$ with a solid line.
Test point: $(1, 1)$
$4(1) \le 3(1)$
$\quad 4 \le 3$ False
Shade the region not containing $(1, 1)$.

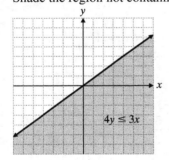

12. $-3x - 2y > 10$

Graph $-3x - 2y = 10$ with a dashed line.
Test point: $(0, 0)$
$-3(0) - 2(0) > 10$
$\qquad\qquad 0 > 10$ False
Shade the region not containing $(0, 0)$.

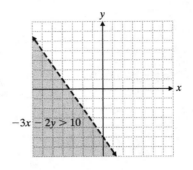

13. $\{(2, -8), (3, -7), (2, 5)\}$
No; two different ordered pairs have the same first coordinate.

14. Function, passes the vertical line test.

15. $y = 2x^2 - 3$

x	$y = 2x^2 - 3$	y
-2	$y = 2(-2)^2 - 3$	5
-1	$y = 2(-1)^2 - 3$	-1
0	$y = 2(0)^2 - 3$	-3
1	$y = 2(1)^2 - 3$	-1
2	$y = 2(2)^2 - 3$	5

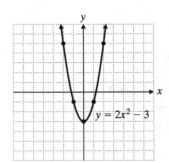

16. $f(x) = -x^2 - 2x - 3$

a. $f(0) = -(0)^2 - 2(0) - 3 = 0 - 0 - 3 = -3$

b. $f(-2) = -(-2)^2 - 2(-2) - 3 = -4 + 4 - 3 = -3$

Cumulative Test for Chapters 0–3

1.
$$2.4 \wedge \overline{)8.8 \wedge 56} \quad \frac{3.69}{}$$
$$\underline{7\ 2}$$
$$1\ 6\quad 5$$
$$\underline{1\ 4\quad 4}$$
$$2\quad 16$$
$$\underline{2\quad 16}$$

2. $\dfrac{3}{8} + \dfrac{5}{12} + \dfrac{1}{2} = \dfrac{9}{24} + \dfrac{10}{24} + \dfrac{12}{24} = \dfrac{9+10+12}{24} = \dfrac{31}{24}$ or

$1\dfrac{7}{24}$

3. $1.386 + (-2.9) = -1.514$

4.
$$\begin{array}{r} 9.00 \\ -\ 0.48 \\ \hline 8.52 \end{array}$$

5.
$$\begin{array}{r} 12.04 \\ \times\ 0.72 \\ \hline 2208 \\ 8\ 428 \\ \hline 8.6488 \end{array}$$
$-12.04 \times 0.72 = -8.6488$

6. $-4(4x - y + 5) - 3(6x - 2y)$
$= -16x + 4y - 20 - 18x + 6y$
$= -34x + 10y - 20$

7. $12 - 3(2 - 4) + 12 \div 4 = 12 - 3(-2) + 12 \div 4$
$= 12 + 6 + 3$
$= 21$

8. $(-3)(-5)(-1)(2)(-1) = (3)(5)(1)(2)(1) = 30$

9. $3st - 8s^2 t + 12st^2 + s^2 t - 5st + 2st^2$
$= (3 - 5)st + (-8 + 1)s^2 t + (12 + 2)st^2$
$= -2st - 7s^2 t + 14st^2$

10. $(5x)^2 = 5^2 (x)^2 = 25x^2$

11. $2\{3x - 4[5 - 3y(2 - x)]\}$
$= 2[3x - 4(5 - 6y + 3xy)]$
$= 2(3x - 20 + 24y - 12xy)$
$= 6x - 40 + 48y - 24xy$

12.
$$-2(x + 5) = 4x - 15$$
$$-2x - 10 = 4x - 15$$
$$-2x - 10 + 2x = 4x - 15 + 2x$$
$$-10 = 6x - 15$$
$$-10 + 15 = 6x - 15 + 15$$
$$5 = 6x$$
$$\frac{5}{6} = \frac{6x}{6}$$
$$\frac{5}{6} = x$$

13.
$$\frac{1}{3}(x + 5) = 2x - 5$$
$$\frac{1}{3}x + \frac{5}{3} = 2x - 5$$
$$3\left(\frac{1}{3}x\right) + 3\left(\frac{5}{3}\right) = 3(2x) - 3(5)$$
$$x + 5 = 6x - 15$$
$$x + 5 + 15 = 6x - 15 + 15$$
$$x + 20 = 6x$$
$$-x + x + 20 = -x + 6x$$
$$20 = 5x$$
$$\frac{20}{5} = \frac{5x}{5}$$
$$4 = x$$

14.
$$\frac{2y}{3} - \frac{1}{4} = \frac{1}{6} + \frac{y}{4}$$
$$12\left(\frac{2y}{3}\right) - 12\left(\frac{1}{4}\right) = 12\left(\frac{1}{6}\right) + 12\left(\frac{y}{4}\right)$$
$$8y - 3 = 2 + 3y$$
$$8y - 3 - 3y = 2 + 3y - 3y$$
$$5y - 3 = 2$$
$$5y - 3 + 3 = 2 + 3$$
$$5y = 5$$
$$\frac{5y}{5} = \frac{5}{5}$$
$$y = 1$$

15.
$$\frac{1}{2}(x-5) \geq x-4$$
$$\frac{1}{2}x - \frac{5}{2} \geq x-4$$
$$2\left(\frac{1}{2}x\right) - 2\left(\frac{5}{2}\right) \geq 2(x) - 2(4)$$
$$x - 5 \geq 2x - 8$$
$$x - 5 - 2x \geq 2x - 8 - 2x$$
$$-x - 5 \geq -8$$
$$-x - 5 + 5 \geq -8 + 5$$
$$-x \geq -3$$
$$\frac{-x}{-1} \leq \frac{-3}{-1}$$
$$x \leq 3$$

16. Area $= \pi r^2 = 3.14(3)^2 = 28.26$ in.^{2}

17. $b = 25$, $a = 13$
$$A = \frac{1}{2}ab = \frac{1}{2}(13)(25) = 162.5 \text{ m}^2$$
Cost $= 4.50(162.5) = 731.25$
The total cost is $731.25.

18. $x =$ the number
$$2x + 15 = 1$$
$$2x = -14$$
$$x = -7$$
The number is −7.

19. Let $x =$ number of office staff Ian can invite.
$$250 + 30x \leq 2000$$
$$30x \leq 1750$$
$$\frac{30x}{30} \leq \frac{1750}{30}$$
$$x \leq 58\frac{1}{3}$$
Ian can invite at most 58 office staff and stay within his budget.

20.
$$\frac{0 + 82 + 89 + 87 + x}{5} \geq 70$$
$$\frac{258 + x}{5} \geq 70$$
$$5\left(\frac{258 + x}{5}\right) \geq 5(70)$$
$$258 + x \geq 350$$
$$258 + x - 258 \geq 350 - 258$$
$$x \geq 92$$
He must receive a minimum score of 92 in order for him to pass the course and play football.

21. $(-8, -3)$ and $(11, -3)$
$$m = \frac{y_2 - y_1}{x_2 - x_1} = \frac{-3 - (-3)}{11 - (-8)} = \frac{0}{19} = 0$$

22. $(3, -5)$, $m = -\frac{2}{3}$
$$y = mx + b$$
$$-5 = \left(-\frac{2}{3}\right)(3) + b$$
$$-5 = -2 + b$$
$$-3 = b$$
$$y = -\frac{2}{3}x - 3$$

23. $y = \frac{2}{3}x - 4$

x	$y = \frac{2}{3}x - 4$	y
−3	$y = \frac{2}{3}(-3) - 4$	−6
0	$y = \frac{2}{3}(0) - 4$	−4
6	$y = \frac{2}{3}(6) - 4$	0

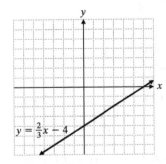

24. $3x + 8 = 5x$
$$8 = 2x$$
$$4 = x$$

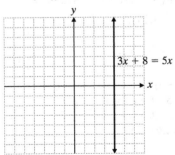

25. $2x + 5y \le -10$

Graph $2x + 5y = -10$ with a solid line.
Test point: (0, 0)
$2(0) + 5(0) \le -10$
$\quad\quad 0 \le -10$ False
Shade the region not containing (0, 0).

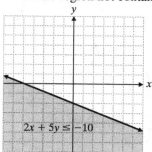

26. $f(x) = 3x^2 - 2x + \dfrac{4}{x}$

a. $f(-1) = 3(-1)^2 - 2(-1) + \dfrac{4}{-1} = 3 + 2 - 4 = 1$

b. $f(2) = 3(2)^2 - 2(2) + \dfrac{4}{2} = 12 - 4 + 2 = 10$

Chapter 4

4.1 Exercises

1. There is no solution. There is no point (x, y) that satisfies both equations. The graph of such a system yields two parallel lines.

3. A system of two linear equations in two unknowns can have an infinite number of solutions, one solution, or no solutions.

5.
$$2x - 8 = y - 5$$
$$2\left(\frac{5}{2}\right) - 8 \stackrel{?}{=} 2 - 5$$
$$5 - 8 \stackrel{?}{=} -3$$
$$-3 = -3$$

$$4x - 3y = 4$$
$$4\left(\frac{5}{2}\right) - 3(2) \stackrel{?}{=} 4$$
$$10 - 6 \stackrel{?}{=} 4$$
$$4 = 4$$

$\left(\frac{5}{2}, 2\right)$ is a solution.

7. $3x + y = 2$
$2x - y = 3$

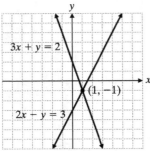

$(1, -1)$ is the solution.
Check:
(1) $3(1) + (-1) \stackrel{?}{=} 2, 2 = 2$
(2) $2(1) - (-1) \stackrel{?}{=} 3, 3 = 3$

9. $3x - 2y = 6$
$4x + y = -3$

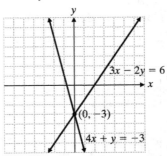

$(0, -3)$ is the solution.
Check:
(1) $3(0) - 2(-3) \stackrel{?}{=} 6, 6 = 6$
(2) $4(0) + (-3) \stackrel{?}{=} -3, -3 = -3$

11.
$$y = -x + 3$$
$$x + y = -\frac{2}{3}$$

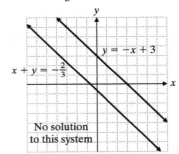

The system has no solution.
Check:
$$y = -x + 3$$
$$x + y = -\frac{2}{3} \Rightarrow y = -x - \frac{2}{3}$$
The lines have the same slope and different y-intercepts.

13.
$$y = -2x + 5$$
$$3y + 6x = 15$$

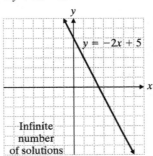

There is an infinite number of solutions.
Check:
$y = -2x + 5$
$3y + 6x = 15 \Rightarrow 3y = -6x + 15 \Rightarrow y = -2x + 5$
The lines are the same.

15. $3x + 4y = 14$ **(1)**
   $x + 2y = -2$ **(2)**

Solve **(2)** for x and substitute into **(1)**.
$$x = -2 - 2y$$
$$3(-2 - 2y) + 4y = 14$$
$$-6 - 6y + 4y = 14$$
$$-2y = 20$$
$$y = -10$$

Substitute $y = -10$ into **(2)** and solve for x.
$$x + 2(-10) = -2$$
$$x - 20 = -2$$
$$x = 18$$
$(18, -10)$ is the solution.
Check:
(1) $3(18) + 4(-10) \stackrel{?}{=} 14, \quad 14 = 14$
(2) $18 + 2(-10) \stackrel{?}{=} -2, \quad -2 = -2$

17. $-x + 3y = -8$ **(1)**
   $2x - y = 6$ **(2)**

Solve **(2)** for y and substitute into **(1)**.
$$y = 2x - 6$$
$$-x + 3(2x - 6) = -8$$
$$-x + 6x - 18 = -8$$
$$5x = 10$$
$$x = 2$$
Substitute $x = 2$ into **(1)**.
$$-2 + 3y = -8$$
$$3y = -6$$
$$y = -2$$

$(2, -2)$ is the solution.
Check:
(1) $-2 + 3(-2) \stackrel{?}{=} -8, -8 = -8$
(2) $2(2) - (-2) \stackrel{?}{=} 6, 6 = 6$

19. $2x - \dfrac{1}{2}y = -3$ **(1)**

   $\dfrac{x}{5} + 2y = \dfrac{19}{5}$ **(2)**

Solve **(2)** for y and substitute into **(1)**.
$$\frac{x}{5} + 2y = \frac{19}{5}$$
$$x + 10y = 19$$
$$y = \frac{19 - x}{10}$$

$$2x - \frac{1}{2}\left(\frac{19 - x}{10}\right) = -3$$
$$40x - 19 + x = -60$$
$$41x = -41$$
$$x = -1$$
Substitute $x = -1$ into **(1)** and solve for y.
$$2(-1) - \frac{1}{2}y = -3$$
$$-4 - y = -6$$
$$-y = -2$$
$$y = 2$$
$(-1, 2)$ is the solution.

21. $\dfrac{1}{2}x - \dfrac{1}{8}y = 3$ **(1)**

   $\dfrac{2}{3}x + \dfrac{3}{4}y = 4$ **(2)**

Solve **(2)** for x and substitute into **(1)**.
$$12\left(\frac{2}{3}x + \frac{3}{4}y\right) = 12(4)$$
$$8x + 9y = 48$$
$$8x = 48 - 9y$$
$$x = 6 - \frac{9}{8}y$$

$$\frac{1}{2}\left(6 - \frac{9}{8}y\right) - \frac{1}{8}y = 3$$
$$3 - \frac{9}{16}y - \frac{1}{8}y = 3$$
$$-\frac{11}{16}y = 0$$
$$y = 0$$
Substitute $y = 0$ into **(2)** and solve for x.
$$\frac{2}{3}x + \frac{3}{4}(0) = 4$$
$$x = 4\left(\frac{3}{2}\right)$$
$$x = 6$$
$(6, 0)$ is the solution.

23. $9x + 2y = 2$ **(1)**
   $3x + 5y = 5$ **(2)**

Multiply **(2)** by -3 and add to **(1)**.
$$9x + 2y = 2$$
$$\underline{-9x - 15y = -15}$$
$$-13y = -13$$
$$y = 1$$
Substitute $y = 1$ into **(1)**.

$$9x + 2(1) = 2$$
$$9x + 2 = 2$$
$$9x = 0$$
$$x = 0$$

(0, 1) is the solution.

Check:

(1) $9(0) + 2(1) \stackrel{?}{=} 2$, $2 = 2$

(2) $3(0) + 5(1) \stackrel{?}{=} 5$, $5 = 5$

25. $3s + 3t = 10$ **(1)**
    $4s - 9t = -4$ **(2)**

Multiply **(1)** by 3 and add to **(2)**.
$$9s + 9t = 30$$
$$+ \ 4s - 9t = -4$$
$$\overline{\qquad 13s = 26}$$
$$s = 2$$

Substitute $s = 2$ into **(1)** and solve for t.
$$3(2) + 3t = 10$$
$$6 + 3t = 10$$
$$3t = 4$$
$$t = \frac{4}{3}$$

$\left(2, \dfrac{4}{3}\right)$ is the solution.

Check:

(1) $3(2) + 3\left(\dfrac{4}{3}\right) \stackrel{?}{=} 10$, $10 = 10$

(2) $4(2) - 9\left(\dfrac{4}{3}\right) \stackrel{?}{=} -4$, $-4 = -4$

27. $\dfrac{7}{2}x + \dfrac{5}{2}y = -4$ **(1)**

    $3x + \dfrac{2}{3}y = 1$ **(2)**

Clear fractions.
$$7x + 5y = -8 \quad \textbf{(1)}$$
$$9x + 2y = 3 \quad \textbf{(2)}$$

Multiply **(1)** by 2 and **(2)** by −5 and add.
$$14x + 10y = -16$$
$$-45x - 10y = -15$$
$$\overline{\qquad -31x = -31}$$
$$x = 1$$

Substitute $x = 1$ into **(2)**.
$$9(1) + 2y = 3$$
$$2y = -6$$
$$y = -3$$

(1, −3) is the solution.

29. $1.6x + 1.5y = 1.8 \rightarrow 16x + 15y = 18$ **(1)**
    $0.4x + 0.3y = 0.6 \rightarrow 4x + 3y = 6$ **(2)**

Multiply **(2)** by −5 and add to **(1)**.
$$16x + 15y = 18$$
$$-20x - 15y = -30$$
$$\overline{\qquad -4x = -12}$$
$$x = 3$$

Substitute $x = 3$ into **(1)**.
$$16(3) + 15y = 18$$
$$48 + 15y = 18$$
$$15y = -30$$
$$y = -2$$

(3, −2) is the solution.

31. $7x - y = 6$ **(1)**
    $3x + 2y = 22$ **(2)**

Solve **(1)** for y and substitute into **(2)**.
$$7x - y = 6$$
$$y = 7x - 6$$
$$3x + 2(7x - 6) = 22$$
$$3x + 14x - 12 = 22$$
$$17x = 34$$
$$x = 2$$

Substitute $x = 2$ into **(1)**.
$$7(2) - y = 6$$
$$y = 8$$

(2, 8) is the solution.

33. $3x + 4y = 8$ **(1)**
    $5x + 6y = 10$ **(2)**

Multiply **(1)** by 5 and **(2)** by −3 and add.
$$15x + 20y = 40$$
$$-15x - 18y = -30$$
$$\overline{\qquad 2y = 10}$$
$$y = 5$$

Substitute $y = 5$ into **(1)**.
$$3x + 4(5) = 8$$
$$3x + 20 = 8$$
$$3x = -12$$
$$x = -4$$

(−4, 5) is the solution.

35. $2x + y = 4$ **(1)**

    $\dfrac{2}{3}x + \dfrac{1}{4}y = 2 \xrightarrow{\times 12} 8x + 3y = 24$ **(2)**

Solve **(1)** for y and substitute into **(2)**.
$$y = 4 - 2x$$

$$8x + 3(4 - 2x) = 24$$
$$8x + 12 - 6x = 24$$
$$2x = 12$$
$$x = 6$$

Substitute $x = 6$ into **(1)**.
$$2(6) + y = 4$$
$$12 + y = 4$$
$$y = -8$$

$(6, -8)$ is the solution.

37. $0.2x = 0.1y - 1.2$ **(1)**
$2x - y = 6$ **(2)**

Solve **(2)** for y and substitute into **(1)**.
$$y = 2x - 6$$
$$0.2x = 0.1(2x - 6) - 1.2$$
$$0.2x = 0.2x - 0.6 - 1.2$$
$$0 = -1.8$$

This is an inconsistent system of equations and has no solution.

39. $5x - 7y = 12$ **(1)**
$-10x + 14y = -24$ **(2)**

Multiply **(1)** by 2 and add to **(2)**.
$$10x - 14y = 24$$
$$\underline{-10x + 14y = -24}$$
$$0 = 0$$

This is a dependent system of equations and has an infinite number of solutions.

41. $0.4x + 0.1y = -0.9$
$-0.1x + 0.2y = 1.8$

Multiply both equations by 10 to clear decimals.
$4x + y = -9$ **(1)**
$-x + 2y = 18$ **(2)**

Solve **(1)** for y and substitute into **(2)**.
$$y = -4x - 9$$
$$-x + 2(-4x - 9) = 18$$
$$-x - 8x - 18 = 18$$
$$-9x = 36$$
$$x = -4$$

Substitute $x = -4$ into **(2)**.
$$-(-4) + 2y = 18$$
$$2y = 14$$
$$y = 7$$

$(-4, 7)$ is the solution.

43. $\dfrac{5}{3}b = \dfrac{1}{3} + a$ **(1)**
$9a - 15b = 2$ **(2)**

Multiply **(1)** by 9, rearrange and add to **(2)**.

$$-9a + 15b = 3$$
$$\underline{9a - 15b = 2}$$
$$0 = 5$$

There is no solution; this is an inconsistent system of equations.

45. $\dfrac{2}{3}x - y = 4$ **(1)**

$2x - \dfrac{3}{4}y = 21$ **(2)**

Solve **(1)** for y and substituting into **(2)**,

$$y = \dfrac{2}{3}x - 4$$

$$2x - \dfrac{3}{4}\left(\dfrac{2}{3}x - 4\right) = 21$$

$$2x - \dfrac{6}{12}x + 3 = 21$$

$$2x - \dfrac{1}{2}x = 18$$

$$\dfrac{3}{2}x = 18$$

$$x = 12$$

Substituting $x = 12$ in **(1)** and solving for y yields:

$$\dfrac{2}{3}(12) - y = 4$$

$$8 - y = 4$$

$$-y = -4$$

$$y = 4$$

$(12, 4)$ is the solution.

47. $3.2x - 1.5y = -3 \Rightarrow 32x - 15y = -30$ **(1)**
$0.7x + y = 2 \Rightarrow 7x + 10y = 20$ **(2)**

Multiply **(1)** by 2 and **(2)** by 3 and add.
$$64x - 30y = -60$$
$$\underline{21x + 30y = 60}$$
$$85x = 0$$
$$x = 0$$

Substitute $x = 0$ into **(2)**.
$$7(0) + 10y = 20$$
$$10y = 20$$
$$y = 2$$

$(0, 2)$ is the solution.

49. $3 - (2x + 1) = y + 6$ **(1)**
$x + y + 5 = 1 - x$ **(2)**

Solve **(1)** for y and substitute in **(2)**.
$$3 - (2x + 1) = y + 6$$
$$y = 3 - 2x - 1 - 6$$
$$y = -2x - 4$$

$$x - 2x - 4 + 5 = 1 - x$$
$$1 = 1$$

This is a dependent system and has an infinite number of solutions.

51. a. $y = 200 + 50x$ for Old World Tile
$y = 300 + 30x$ for Modern Bath

b.

x	$y = 300 + 30x$	x	$y = 200 + 50x$
0	300	0	200
4	420	4	400
8	540	8	600

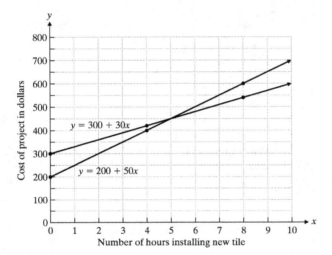

c. From the graph, the cost will be the same for 5 hours of installing new tile.

d. From the graph, Modern Bathroom Headquarters will cost less to remove old tile and install new tile if the time needed to install the new tile is 6 hours.

53. $y_1 = -1.7x + 3.8$
$y_2 = 0.7x - 2.1$

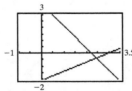

To the nearest hundredth, the point of intersection is (2.46, −0.38).

55. $0.5x + 1.1y = 5.5 \Rightarrow y_1 = \dfrac{(5.5 - 0.5x)}{1.1}$

$-3.1x + 0.9y = 13.1 \Rightarrow y_2 = \dfrac{13.1 + 3.1x}{0.9}$

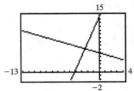

The point of intersection is (−2.45, 6.11).

Cumulative Review

56. $\dfrac{200,000,000}{9,000,000 \text{ tons}} \left(\dfrac{1 \text{ ton}}{2000 \text{ pounds}} \right)$
≈ 0.01

This is approximately $0.01 per pound.

57. $\dfrac{4}{5}x = 273,511$

$x = 341,888.75$

During rush hour about 341,889 cars enter the city.

Quick Quiz 4.1

1. $5x - 3y = 14$ **(1)**
$2x - y = 6$ **(2)**

Solve (2) for y and substitute into **(1)**.
$$y = 2x - 6$$
$$5x - 3(2x - 6) = 14$$
$$5x - 6x + 18 = 14$$
$$-x = -4$$
$$x = 4$$

Substitute $x = 4$ into **(2)** and solve for y.
$$2(4) - y = 6$$
$$8 - y = 6$$
$$-y = -2$$
$$y = 2$$

(4, 2) is the solution.

2. $6x + 7y = 26$ **(1)**
$5x - 2y = 6$ **(2)**

Multiply **(1)** by 2 and **(2)** by 7 then add equations.
$$12x + 14y = 52$$
$$\underline{35x - 14y = 42}$$
$$47x = 94$$
$$x = 2$$

Substitute $x = 2$ into **(2)** and solve for y.

$$5(2) - 2y = 6$$
$$10 - 2y = 6$$
$$-2y = 6 - 10$$
$$y = 2$$

$(2, 2)$ is the solution.

3. $\dfrac{2}{3}x + \dfrac{3}{5}y = -17$ **(1)**

$\dfrac{1}{2}x - \dfrac{1}{3}y = -1$ **(2)**

Solve **(2)** for x then substitute into **(1)**.

$$\dfrac{1}{2}x - \dfrac{1}{3}y = -1$$
$$\dfrac{1}{2}x = \dfrac{1}{3}y - 1$$
$$x = \dfrac{2}{3}y - 2$$

$$\dfrac{2}{3}\left(\dfrac{2}{3}y - 2\right) + \dfrac{3}{5}y = -17$$
$$\dfrac{4}{9}y - \dfrac{4}{3} + \dfrac{3}{5}y = -17$$
$$\dfrac{20}{45}y + \dfrac{27}{45}y = -17 + \dfrac{4}{3}$$
$$\dfrac{47}{45}y = -\dfrac{47}{3}$$
$$y = -15$$

Substitute $y = -15$ into **(2)** then solve for x.

$$\dfrac{1}{2}x - \dfrac{1}{3}(-15) = -1$$
$$\dfrac{1}{2}x + 5 = -1$$
$$\dfrac{1}{2}x = -6$$
$$x = -12$$

$(-12, -15)$ is the solution.

4. Answers may vary. Possible solution:
The equation results in an identity $(0 = 0)$.
Therefore the system is dependent and there is
an infinite number of solutions.

4.2 Exercises

1. $2x - 3y + 2z = -7$ **(1)**
    $x + 4y - z = 10$ **(2)**
    $3x + 2y + z = 4$ **(3)**

Check:

(1) $2(2) - 3(1) + 2(-4) \overset{?}{=} -7, \; -7 = -7$

(2) $2 + 4(1) - (-4) \overset{?}{=} 10, \; 10 = 10$

(3) $3(2) + 2(1) + (-4) \overset{?}{=} 4, \; 4 = 4$

$(2, 1, -4)$ is a solution.

3. $3x + 2y - z = 6$ **(1)**
    $x - y - 2z = -8$ **(2)**
    $4x + y + 2z = 5$ **(3)**

Check:

(1) $3(-1) + 2(5) - 1 \overset{?}{=} 6, \; 6 = 6$

(2) $-1 - 5 - 2(1) \overset{?}{=} -8, \; -8 = -8$

(3) $4(-1) + 5 + 2(1) \overset{?}{=} 5, \; 3 \ne 5$

$(-1, 5, 1)$ is not a solution to the system.

5. $x + y + 2z = 0$ **(1)**
    $2x - y - z = 1$ **(2)**
    $x + 2y + 3z = 1$ **(3)**

Add **(1)** and **(2)**.
$3x + z = 1$ **(4)**
Add $2 \cdot$ **(2)** and **(3)**.
$5x + z = 3$ **(5)**
Subtract **(5)** from **(4)**.
$$-2x = -2$$
$$x = 1$$
Substitute $x = 1$ into **(5)**.
$$5(1) + z = 3$$
$$z = -2$$
Substitute $x = 1$, $z = -2$ into **(1)**.
$$1 + y + 2(-2) = 0$$
$$1 + y - 4 = 0$$
$$y = 3$$

$(1, 3, -2)$ is the solution.

7. $x + 3y - z = -5$ **(1)**
   $-2x - y + z = 8$ **(2)**
    $x - y + 3z = 3$ **(3)**

Add **(1)** and **(2)**.
$-x + 2y = 3$ **(4)**
Add $3 \cdot$ **(1)** and **(3)**.
$4x + 8y = -12$ **(5)**
Add $4 \cdot$ **(4)** and **(5)**.
$$16y = 0$$
$$y = 0$$
Substitute $y = 0$ into **(4)** and solve for x.
$$-x + 2(0) = 3$$
$$-x = 3$$
$$x = -3$$
Substitute $x = -3$, $y = 0$ into **(2)**, then solve for z.
$$-2(-3) - 0 + z = 8$$
$$6 + z = 8$$
$$z = 2$$

$(-3, 0, 2)$ is the solution.

9. $8x - 5y + z = 15$ **(1)**
  $3x + y - z = -7$ **(2)**
  $x + 4y + z = -3$ **(3)**

Add **(1)** and **(2)**.
$8x - 5y + z = 15$
$\underline{3x + y - z = -7}$
$11x - 4y = 8$ **(4)**

Add **(2)** and **(3)**.
$3x + y - z = -7$
$\underline{x + 4y + z = -3}$
$4x + 5y\quad = -10$ **(5)**

Add 5 times **(4)** and 4 times **(5)**.
$55x - 20y = 40$
$\underline{16x + 20y = -40}$
$76x\qquad = 0$
  $x = 0$

Substitute $x = 0$ into **(5)**.
$4(0) + 5y = -10$
  $5y = -10$
  $y = -2$

Substitute $x = 0$, $y = -2$ into **(1)**.
$8(0) - 5(-2) + z = 15$
  $10 + z = 15$
  $z = 5$

$(0, -2, 5)$ is the solution.

11. $x + 4y - z = -5$ **(1)**
  $-2x - 3y + 2z = 5$ **(2)**
  $x - \dfrac{2}{3}y + z = \dfrac{11}{3}$ **(3)**

Add **(1)** and **(3)**.
$2x + \dfrac{10}{3}y = -\dfrac{4}{3}$ **(4)**

Add $2 \cdot$ **(1)** and **(2)**.
$5y = -5$
$y = -1$

Substitute $y = -1$ into **(4)**.
$2x + \dfrac{10}{3}(-1) = -\dfrac{4}{3}$
$6x - 10 = -4$
  $6x = 6$
  $x = 1$

Substitute $x = 1$ and $y = -1$ into **(3)**.
$1 - \dfrac{2}{3}(-1) + z = \dfrac{11}{3}$
$3 + 2 + 3z = 11$
  $3z = 6$
  $z = 2$

$(1, -1, 2)$ is the solution.

13. $2x + 2z = -7 + 3y$
  $\dfrac{3}{2}x + y + \dfrac{1}{2}z = 2$
    $x + 4y = 10 + z$

Rearrange terms in the first and third equations. Multiply the second equation by 2.
$2x - 3y + 2z = -7$ **(1)**
  $3x + 2y + z = 4$ **(2)**
  $x + 4y - z = 10$ **(3)**

Add **(1)** and 2 times **(3)**.
$4x + 5y = 13$ **(4)**
Add **(3)** and **(2)**.
$4x + 6y = 14$ **(5)**
Subtract **(5)** from **(4)**.
$-y = -1$
  $y = 1$

Substitute $y = 1$ into **(5)**.
$4x + 6(1) = 14$
  $4x = 8$
  $x = 2$

Substitute $x = 2$, $y = 1$ into **(3)**.
$2 + 4(1) - z = 10$
  $6 - z = 10$
  $-z = 4$
  $z = -4$

$(2, 1, -4)$ is the solution.

15. $0.2a + 0.1b + 0.2c = 0.1$
  $0.3a + 0.2b + 0.4c = -0.1$
  $0.6a + 1.1b + 0.2c = 0.3$

Multiply all three equations by 10 to clear decimals.
  $2a + b + 2c = 1$ **(1)**
  $3a + 2b + 4c = -1$ **(2)**
  $6a + 11b + 2c = 3$ **(3)**

Subtract **(3)** from **(1)**.
$-4a - 10b = -2$
  $2a + 5b = 1$ **(4)**

Add $-2 \cdot$ **(1)** and **(2)**.
$-a = -3$
  $a = 3$

Substitute $a = 3$ into **(4)**.
$2(3) + 5b = 1$
  $6 + 5b = 1$
  $5b = -5$
  $b = -1$

Substitute $a = 3$, $b = -1$ into **(1)**.

$2(3) + (-1) + 2c = 1$
$6 - 1 + 2c = 1$
$2c = -4$
$c = -2$
$(3, -1, -2)$ is the solution.

17. When a calculator is used it is convenient to keep all three equations together as the operations are performed.
$x - 4y + 4z = -3.72186$
$-x + 3y - z = 5.98115$
$2x - y + 5z = 7.93645$

Now perform two operations on the system; first, add the first equation to the second and add -2 times the first to the third. Note that this *does not* change the first equation, only the second and third.
$x - 4y + 4z = -3.72186$
$-y + 3z = 2.25929$
$7y - 3z = 15.38017$

Add the second equation to the third.
$x - 4y + 4z = -3.72186$
$-y + 3z = 2.25929$
$6y = 17.63946$

From the third equation $y = 2.93991$ which may be substituted into the second equation to give $z = 1.73307$. Substituting these values for x and y into the first equation gives $x = 1.10551$.
$(1.10551, 2.93991, 1.73307)$ is the solution.

19.
$x - y = 5$ **(1)**
$2y - z = 1$ **(2)**
$3x + 3y + z = 6$ **(3)**

Add **(2)** and **(3)**.
$3x + 5y = 7$ **(4)**
Solve **(1)** for x and substitute into **(4)**.
$x = 5 + y$
$3(5 + y) + 5y = 7$
$15 + 3y + 5y = 7$
$8y = -8$
$y = -1$
Substitute $y = -1$ into **(1)**.
$x + 1 = 5$
$x = 4$
Substitute $y = -1$ into **(2)**.
$2(-1) - z = 1$
$-2 - z = 1$
$-z = 3$
$z = -3$
$(4, -1, -3)$ is the solution.

21. $-y + 2z = 1$ **(1)**
$x + y + z = 2$ **(2)**
$-x + 3z = 2$ **(3)**
Add **(2)** and **(3)**.
$y + 4z = 4$ **(4)**
Add **(1)** and **(4)**.
$6z = 5$

Substitute $z = \dfrac{5}{6}$ into **(4)**.

$y + 4\left(\dfrac{5}{6}\right) = 4$
$6y + 20 = 24$
$6y = 4$
$y = \dfrac{2}{3}$

Substitute $z = \dfrac{5}{6}$, $y = \dfrac{2}{3}$ into **(2)**.

$x + \dfrac{2}{3} + \dfrac{5}{6} = 2$
$x = \dfrac{1}{2}$

$\left(\dfrac{1}{2}, \dfrac{2}{3}, \dfrac{5}{6}\right)$ is the solution.

23. $x - 2y + z = 0$ **(1)**
$-3x - y = -6$ **(2)**
$y - 2z = -7$ **(3)**
Multiply **(1)** by 2 and add to **(3)**.
$2x - 3y = -7$ **(4)**
Multiply **(2)** by -3 and add to **(4)**.
$11x = 11$
$x = 1$
Substitute $x = 1$ into **(2)**.
$-3(1) - y = -6$
$y = 3$
Substitute $x = 1$, $y = 3$ into **(1)**.
$1 - 6 + z = 0$
$z = 5$
$(1, 3, 5)$ is the solution.

25. $a - \dfrac{b}{2} - 2c = -3$ **(1)**
$3a - b = -12$ **(2)**
$2a + \dfrac{3b}{2} + 2c = 3$ **(3)**
Add **(1)** and **(3)**.
$3a + b = 0$ **(4)**
Add **(2)** and **(4)**.

$$6a = -12$$
$$a = -2$$

Substitute $a = -2$ into (**2**).
$$3(-2) - b = -12$$
$$-b = -6$$
$$b = 6$$

Substitute $a = -2$, $b = 6$ into (**1**).
$$-2 - \frac{6}{2} - 2c = -3$$
$$-2 - 3 - 2c = -3$$
$$-2c = 2$$
$$c = -1$$

$(-2, 6, -1)$ is the solution.

27. $$2x + y = -3 \quad (\mathbf{1})$$
$$2y + 16z = -18 \quad (\mathbf{2})$$
$$-7x - 3y + 4z = 6 \quad (\mathbf{3})$$

Add (**2**) and $-4 \cdot$ (**3**).
$$2y + 16z = -18$$
$$28x + 12y - 16z = -24$$
$$\overline{28x + 14y = -42} \text{ and dividing by 14}$$

$2x + y = -3$ which is (**1**).
The system of equations is a dependent system and has an infinite number of solutions.

29. $$3x + 3y - 3z = -1 \quad (\mathbf{1})$$
$$4x + y - 2z = 1 \quad (\mathbf{2})$$
$$-2x + 4y - 2z = -8 \quad (\mathbf{3})$$

Subtract (**3**) from (**2**).
$$6x - 3y = 9 \quad (\mathbf{4})$$
Multiply (**1**) by -2 and add to 3 times (**2**).
$$6x - 3y = 5 \quad (\mathbf{5})$$
Comparing (**4**) and (**5**) gives $5 = 9$ which is false. This is an inconsistent system of equations and has no solution.

31. $$a = 8 + 3b - 2c$$
$$4a + 2b - 3c = 10$$
$$c = 10 + b - 2a$$
Rearrange terms in the first and third equations.
$$a - 3b + 2c = 8 \quad (\mathbf{1})$$
$$4a + 2b - 3c = 10 \quad (\mathbf{2})$$
$$2a - b + c = 10 \quad (\mathbf{3})$$
Add $3 \cdot$ (**1**) and $2 \cdot$ (**2**).
$$3a - 9b + 6c = 24$$
$$8a + 4b - 6c = 20$$
$$\overline{11a - 5b = 44} \quad (\mathbf{4})$$
Add (**2**) and $3 \cdot$ (**3**).

$$4a + 2b - 3c = 10$$
$$6a - 3b + 3c = 30$$
$$\overline{10a - b = 40} \quad (\mathbf{5})$$
Add (**4**) and -5 times (**5**).
$$11a - 5b = 44$$
$$-50a + 5b = -200$$
$$\overline{-39a = -156}$$
$$a = 4$$
Substitute $a = 4$ into (**5**).
$$10(4) - b = 40$$
$$40 - b = 40$$
$$-b = 0$$
$$b = 0$$
Substitute $a = 4$, $b = 0$ into (**3**).
$$2(4) - 0 + c = 10$$
$$c = 2$$
$(4, 0, 2)$ is the solution.

Cumulative Review

33. $$-2(3x - 4) + 12 = 6x - 4$$
$$-6x + 8 + 12 = 6x - 4$$
$$-6x + 20 = 6x - 4$$
$$-12x + 20 = -4$$
$$-12x = -24$$
$$x = 2$$

34. $$m = \frac{y_2 - y_1}{x_2 - x_1} = \frac{-3 - 6}{4 - (-1)} = \frac{-9}{5} = -\frac{9}{5}$$

35. $$m = \frac{y_2 - y_1}{x_2 - x_1} = \frac{4 - 3}{1 - (-2)} = \frac{1}{3}$$
$$y - y_1 = m(x - x_1)$$
$$y - 4 = \frac{1}{3}(x - 1)$$
$$y - 4 = \frac{1}{3}x - \frac{1}{3}$$
$$y = \frac{1}{3}x + \frac{11}{3}$$

36. $$y = -\frac{2}{3}x + 4 \Rightarrow m = -\frac{2}{3}, \; m_\perp = \frac{3}{2}$$

Quick Quiz 4.2

1. $$4x - y + 2z = 0 \quad (\mathbf{1})$$
$$2x + y + z = 3 \quad (\mathbf{2})$$
$$3x - y + z = -2 \quad (\mathbf{3})$$
Add (**1**) and (**2**).

$6x + 3z = 3$ **(4)**

Add **(2)** and **(3)**.

$5x + 2z = 1$ **(5)**

Solve **(4)** for x then substitute in **(5)**.

$$6x + 3z = 3$$
$$6x = 3 - 3z$$
$$x = \frac{3}{6} - \frac{3}{6}z$$
$$x = \frac{1}{2} - \frac{1}{2}z$$
$$5\left(\frac{1}{2} - \frac{1}{2}z\right) + 2z = 1$$
$$\frac{5}{2} - \frac{5}{2}z + 2z = 1$$
$$-\frac{5}{2}z + 2z = 1 - \frac{5}{2}$$
$$-\frac{1}{2}z = -\frac{3}{2}$$
$$z = \frac{3}{2} \cdot \frac{2}{1}$$
$$z = 3$$

Substitute $z = 3$ into **(4)**.

$$5x + 2(3) = 1$$
$$5x + 6 = 1$$
$$5x = -5$$
$$x = -1$$

Substitute $x = -1$, $z = 3$ into **(3)**.

$$3(-1) - y + 3 = -2$$
$$-3 - y + 3 = -2$$
$$-y = -2$$
$$y = 2$$

$(-1, 2, 3)$ is the solution.

2. $x + 2y + 2z = -1$ **(1)**
    $2x - y + z = 1$ **(2)**
    $x + 3y - 6z = 7$ **(3)**

Add **(1)** and $-$**(3)** then solve for y.

$$x + 2y + 2z = -1$$
$$\underline{-x - 3y + 6z = -7}$$
$$-y + 8z = -8$$
$$-y = -8 - 8z$$
$$y = 8 + 8z \quad \textbf{(4)}$$

Add -2**(1)** and **(2)**, substitute **(4)** into **(5)**.

$$-2x - 4y - 4z = 2$$
$$\underline{2x - y + z = 1}$$
$$-5y - 3z = 3 \quad \textbf{(5)}$$

$$-5(8 + 8z) - 3z = 3$$
$$-40 - 40z - 3z = 3$$
$$-43z = 43$$
$$z = -1$$

Substitute $z = -1$ into **(5)**.

$$-5y - 3(-1) = 3$$
$$-5y + 3 = 3$$
$$-5y = 0$$
$$y = 0$$

Substitute $y = 0$, $z = -1$ into **(1)**.

$$x + 2(0) + 2(-1) = -1$$
$$x + 0 - 2 = -1$$
$$x = 1$$

$(1, 0, -1)$ is the solution.

3. $4x - 2y + 6z = 0$ **(1)**
    $6y + 3z = 3$ **(2)**
    $x + 2y - z = 5$ **(3)**

Add **(1)** and -4**(3)** then solve for y.

$$4x - 2y + 6z = 0$$
$$\underline{-4x - 8y + 4z = -20}$$
$$-10y + 10z = -20$$
$$-10y = -20 - 10z$$
$$y = 2 + z \quad \textbf{(4)}$$

Substitute **(4)** into **(2)**.

$$6(2 + z) + 3z = 3$$
$$12 + 6z + 3z = 3$$
$$9z = -9$$
$$z = -1$$

Substitute $z = -1$ into **(2)**.

$$6y + 3(-1) = 3$$
$$6y - 3 = 3$$
$$6y = 6$$
$$y = 1$$

Substitute $y = 1$, $z = -1$ into **(3)**.

$$x + 2(1) - (-1) = 5$$
$$x + 2 + 1 = 5$$
$$x + 3 = 5$$
$$x = 2$$

$(2, 1, -1)$ is the solution.

4. Answers may vary. Possible solution:
    To eliminate z and obtain two equations with only variables x and y one could do the following.
    Add $5 \cdot$ **(1)** to $2 \cdot$ **(2)** and add $-3 \cdot$ **(2)** to $5 \cdot$ **(3)**

Use Math to Save Money

1. Job 1: Divide the annual salary by 52 since there are 52 weeks in one year.
   $50,310 ÷ 52 = $967.50
   The weekly gross income is $967.50.
   Job 2: We multiply to find the gross weekly earnings.
   $6 × 5 × $20.50 = $615
   The gross pay per week is $615.

2. Job 1:
   15% of $967.50 = 0.15 × $967.50 ≈ $145.13
   The weekly payroll deductions are $145.13.
   Job 2: 12% of $615 = 0.12 × $615 = $73.80
   The weekly payroll deductions are $73.80.

3. Job 1:
   $175 + $75 + $25 + $50 + ($0.35 × 45 × 5)
   = $175 + $75 + $25 + $50 + $78.75
   = $403.75
   Job-related expenses will be $403.75.
   Job 2:
   $150 + $50 + $25 + $25 + ($0.35 × 5 × 5)
   = $150 + $50 + $25 + $25 + $8.75
   = $258.75
   Job-related expenses will be $258.75.

4. Job 1:
   $967.50 − ($145.13 + $403.75)
   = $967.50 − $548.88
   = $418.62
   Sharon will have $418.62 left to spend.
   Job 2:
   $615 − ($73.80 + $258.75) = $615 − $332.55
   $\qquad\qquad\qquad\qquad\qquad = $282.45
   Sharon will have $282.45 left to spend.

5. Job 1:
   Total time: 45 + 5 = 50 hours
   $\dfrac{\$418.62}{50 \text{ hr}} \approx \$8.37$ per hour
   The real hourly wage is $8.37.
   Job 2:
   Total time: 30 + 1 = 31 hours
   $\dfrac{\$282.45}{31 \text{ hr}} \approx \$9.11$
   The real hourly wage is $9.11.

6. She should take job 2, since it has the greater real hourly wage.

7. She should take Job 1, since she will have more left to spend each week.

How Am I Doing? Sections 4.1–4.2
(Available online through MyMathLab.)

1. $4x - y = -1$ **(1)**
   $3x + 2y = 13$ **(2)**
   Solve **(1)** for y and substitute into **(2)**.
   $4x - y = -1$
   $-y = -1 - 4x$
   $y = 1 + 4x$
   $3x + 2(1 + 4x) = 13$
   $3x + 2 + 8x = 13$
   $11x = 11$
   $x = 1$
   Substitute $x = 1$ into **(2)**.
   $3(1) + 2y = 13$
   $2y = 10$
   $y = 5$
   (1, 5) is the solution.

2. $5x + 2y = 0$ **(1)**
   $-3x - 4y = 14$ **(2)**
   Add 2**(1)** to **(2)**.
   $10x + 4y = 0$
   $\underline{-3x - 4y = 14}$
   $7x \qquad = 14$
   $x = 2$
   Substitute $x = 2$ into **(1)**.
   $5(2) + 2y = 0$
   $2y = -10$
   $y = -5$
   (2, −5) is the solution.

3. $5x - 2y = 27$ **(1)**
   $3x - 5y = -18$ **(2)**
   Multiply **(1)** by 5, and **(2)** by −2 and add.
   $25x - 10y = 135$
   $\underline{-6x + 10y = 36}$
   $19x \qquad = 171$
   $x = 9$
   Substitute $x = 9$ into **(1)**.
   $5(9) - 2y = 27$
   $45 - 2y = 27$
   $2y = 18$
   $y = 9$
   The solution is (9, 9).

4. $7x + 3y = 15$ **(1)**
   $\dfrac{1}{3}x - \dfrac{1}{2}y = 2$ **(2)**
   Multiply **(2)** by 6 to clear fractions.

$2x - 3y = 12$ **(3)**
Add **(1)** and **(3)**.

$$\begin{array}{r} 7x + 3y = 15 \\ 2x - 3y = 12 \\ \hline 9x \quad\;\; = 27 \\ x = 3 \end{array}$$

Substitute $x = 3$ into **(1)**.

$$\begin{array}{r} 7(3) + 3y = 15 \\ 3y = -6 \\ y = -2 \end{array}$$

The solution is $(3, -2)$.

5. $2x = 3 + y$ **(1)**
    $3y = 6x - 9$ **(2)**

Solve **(1)** for y.
$y = 2x - 3$ and substitute into **(2)**.

$$\begin{array}{r} 3(2x - 3) = 6x - 9 \\ 6x - 9 = 6x - 9 \\ 0 = 0 \end{array}$$

The equations are dependent and the system has an infinite number of solutions.

6. $0.2x + 0.7y = -1$
    $0.5x + 0.6y = -0.2$

Multiply both equations by 10 to clear decimals.
$2x + 7y = -10$ **(1)**
$5x + 6y = -2$ **(2)**

Multiply **(1)** by 5 and **(2)** by -2 and add.

$$\begin{array}{r} 10x + 35y = -50 \\ -10x - 12y = 4 \\ \hline 23y = -46 \\ y = -2 \end{array}$$

Substitute $y = -2$ into **(1)**.

$$\begin{array}{r} 2x + 7(-2) = -10 \\ 2x - 14 = -10 \\ 2x = 4 \\ x = 2 \end{array}$$

The solution is $(2, -2)$.

7. $6x - 9y = 15$ **(1)**
    $-4x + 6y = 8$ **(2)**

Add 2 times **(1)** to 3 times **(2)**.

$$\begin{array}{r} 12x - 18y = 30 \\ -12x + 18y = 54 \\ \hline 0 = 84 \end{array}$$

This is an inconsistent system with no solution. The lines are parallel.

8. $3x + y - 2z = 5$
    $4x + 3y - z = 12$
    $-2x - y + 3z = -2$

First equation
$3(3) + (-4) - 2(0) \overset{?}{=} 5, \; 5 = 5$
Second equation
$4(3) + 3(-4) - 0 \overset{?}{=} 12, \; 0 \neq 12$
Third equation
$-2(3) - (-4) + 3(0) \overset{?}{=} -2, \; -2 = -2$
$(3, -4, 0)$ does not satisfy all three equations, so it is not a solution.

9. $5x - 2y + z = -1$ **(1)**
    $3x + y - 2z = 6$ **(2)**
    $-2x + 3y - 5z = 7$ **(3)**

Multiply **(1)** by 2 and add to **(2)**.

$$\begin{array}{r} 10x - 4y + 2z = -2 \\ 3x + y - 2z = 6 \\ \hline 13x - 3y = 4 \quad \textbf{(4)} \end{array}$$

Multiply **(1)** by 5 and add to **(3)**.

$$\begin{array}{r} 25x - 10y + 5z = -5 \\ -2x + 3y - 5z = 7 \\ \hline 23x - 7y = 2 \quad \textbf{(5)} \end{array}$$

Multiply **(4)** by 7 and **(5)** by -3 and add.

$$\begin{array}{r} 91x - 21y = 28 \\ -69x + 21y = -6 \\ \hline 22x \quad\;\; = 22 \\ x = 1 \end{array}$$

Substitute $x = 1$ into **(5)**.

$$\begin{array}{r} 23(1) - 7y = 2 \\ 7y = 21 \\ y = 3 \end{array}$$

Substitute $x = 1$, $y = 3$ into **(1)**.

$$\begin{array}{r} 5(1) - 2(3) + z = -1 \\ 5 - 6 + z = -1 \\ z = 0 \end{array}$$

The solution is $(1, 3, 0)$.

10. $2x - y + 3z = -1$ **(1)**
    $5x + y + 6z = 0$ **(2)**
    $2x - 2y + 3z = -2$ **(3)**

Add **(1)** and **(2)**.

$$\begin{array}{r} 2x - y + 3z = -1 \\ 5x + y + 6z = 0 \\ \hline 7x \quad\;\; + 9z = -1 \quad \textbf{(4)} \end{array}$$

Add 2 times **(2)** to **(3)**.

$$\begin{array}{r} 10x + 2y + 12z = 0 \\ 2x - 2y + 3z = -2 \\ \hline 12x \quad\quad\;\; + 15z = -2 \quad \textbf{(5)} \end{array}$$

Add 5 times (**4**) to −3 times (**5**).

$$35x + 45z = -5$$
$$\underline{-36x - 45z = 6}$$
$$-x \qquad = 1$$
$$x = -1$$

Substitute $x = -1$ into (**4**).

$$7(-1) + 9z = -1$$
$$9z = 6$$
$$z = \frac{2}{3}$$

Substitute $x = -1$, $z = \frac{2}{3}$ into (**2**).

$$5(-1) + y + 6\left(\frac{2}{3}\right) = 0$$
$$-5 + y + 4 = 0$$
$$y - 1 = 0$$
$$y = 1$$

$\left(-1, 1, \dfrac{2}{3}\right)$ is the solution.

11.
$$x + y + 2z = 9 \quad (\mathbf{1})$$
$$3x + 2y + 4z = 16 \quad (\mathbf{2})$$
$$2y + z = 10 \quad (\mathbf{3})$$

Multiply (**1**) by −3 and add to (**2**).

$$-3x - 3y - 6z = -27$$
$$\underline{3x + 2y + 4z = 16}$$
$$-y - 2z = -11 \quad (\mathbf{4})$$

Multiply (**4**) by 2 and add to (**3**).

$$2y + z = 10$$
$$\underline{-2y - 4z = -22}$$
$$-3z = -12$$
$$z = 4$$

Substitute $z = 4$ into (**3**).

$$2y + 4 = 10$$
$$2y = 6$$
$$y = 3$$

Substitute $y = 3$, $z = 4$ into (**1**).

$$x + 3 + 2(4) = 9$$
$$x + 3 + 8 = 9$$
$$x = -2$$

The solution is (−2, 3, 4).

12.
$$x - 2z = -5 \quad (\mathbf{1})$$
$$y - 3z = -3 \quad (\mathbf{2})$$
$$2x - z = -4 \quad (\mathbf{3})$$

Solve (**3**) for z and substitute into (**1**).

$$z = 2x + 4$$

$$x - 2(2x + 4) = -5$$
$$x - 4x - 8 = -5$$
$$-3x = 3$$
$$x = -1$$

Substitute $x = -1$ into (**3**).

$$2(-1) - z = -4$$
$$-z = -2$$
$$z = 2$$

Substitute $z = 2$ into (**2**).

$$y - 3(2) = -3$$
$$y - 6 = -3$$
$$y = 3$$

The solution is (−1, 3, 2).

4.3 Exercises

1. Let $x =$ the smaller number and
$y =$ the larger number.

$$x + y = 87 \quad (\mathbf{1})$$
$$y - 2x = 12 \quad (\mathbf{2})$$

Solve (**2**) for y and substitute into (**1**).

$$y - 2x = 12$$
$$y = 2x + 12$$

$$x + 2x + 12 = 87$$
$$3x = 75$$
$$x = 25$$

Substitute $x = 25$ into (**1**).

$$25 + y = 87$$
$$y = 62$$

The smaller number is 25 and the larger number
is 62.

3. Let $x =$ number of heavy equipment operators
and $y =$ number of general laborers.

$$x + y = 35 \quad (\mathbf{1})$$
$$140x + 90y = 3950 \quad (\mathbf{2})$$

Solve (**1**) for y and substitute into (**2**).

$$x + y = 35$$
$$y = 35 - x$$

$$140x + 90(35 - x) = 3950$$
$$140x + 3150 - 90x = 3950$$
$$50x = 800$$
$$x = 16$$

Substitute $x = 16$ into (**1**).

$$16 + y = 35$$
$$y = 19$$

16 heavy equipment operators and 19 general
laborers were employed.

5. x = number of tickets for regular coach seats
y = number of tickets for sleeper car seats
$$x + y = 98 \qquad \textbf{(1)}$$
$$120x + 290y = 19,750 \quad \textbf{(2)}$$
Solve **(1)** for y and substitute into **(2)**.
$$y = 98 - x$$
$$120x + 290(98 - x) = 19,750$$
$$120x + 28,420 - 290x = 19,750$$
$$-170x = -8670$$
$$x = 51$$
Substitute 51 for x in **(1)**.
$$51 + y = 98$$
$$y = 47$$
Number of coach tickets = 51
Number of sleeper tickets = 47

7. Let x = no. of experienced employees
y = no. of new employees
$$3x + 4y = 115 \quad \textbf{(1)}$$
$$4x + 7y = 170 \quad \textbf{(2)}$$
Add $-4 \cdot$ **(1)** and $3 \cdot$ **(2)**.
$$-12x - 16y = -460$$
$$\underline{12x + 21y = 510}$$
$$5y = 50$$
$$y = 10$$
Substitute $y = 10$ into **(1)**.
$$3x + 4(10) = 115$$
$$3x + 40 = 115$$
$$3x = 75$$
$$x = 25$$
They can send 25 experienced and 10 new employees.

9. x = number of packages of old fertilizer
y = number of packages of new fertilizer
$$50x + 65y = 3125 \quad \textbf{(1)}$$
$$60x + 45y = 2925 \quad \textbf{(2)}$$
Add $6 \cdot$ **(1)** and $-5 \cdot$ **(2)**.
$$300x + 390y = 18,750$$
$$\underline{-300x - 225y = 14,625}$$
$$165y = 4125$$
$$y = 25$$
Substitute $y = 25$ into **(1)**.
$$50x + 65(25) = 3125$$
$$50x + 1625 = 3125$$
$$50x = 1500$$
$$x = 30$$
He should use 30 old packages and 25 new packages.

11. x = cost of one scone
y = cost of one large coffee
$$5x + 6y = 25.15 \quad \textbf{(1)}$$
$$4x + 7y = 24.30 \quad \textbf{(2)}$$
Multiply **(1)** by 4 and add to $-5 \cdot$ **(2)**.
$$20x + 24y = 100.6$$
$$\underline{-20x - 35y = -121.5}$$
$$-11y = -20.9$$
$$y = 1.9$$
Substitute 1.9 for y in **(1)** and solve for x.
$$5x + 6(1.9) = 25.15$$
$$5x = 13.75$$
$$x = 2.75$$
The cost of one scone is \$2.75. The cost of one large coffee is \$1.90.

13. x = speed of plane in still air
y = speed of wind
$$(x - y)\left(\frac{7}{6}\right) = 210$$
$$(x + y)\left(\frac{5}{6}\right) = 210$$
$$x - y = 180$$
$$\underline{x + y = 252}$$
$$2x \quad = 432$$
$$x = 216$$
$$x + y = 252$$
$$216 + y = 252$$
$$y = 36$$
The speed of the wind was 36 mph; the speed of the plane in still air was 216 mph.

15. x = speed in still water
y = speed of current
$$\frac{2}{3}(x - y) = 8$$
$$\frac{1}{2}(x + y) = 8$$
$$x - y = 12$$
$$\underline{x + y = 16}$$
$$2x \quad = 28$$
$$x = 14$$
$$x + y = 16$$
$$14 + y = 16$$
$$y = 2$$
The speed of the boat in still water is 14 mph; the speed of the current is 2 mph.

17. Let x = no. of free throws
y = no. of 2-point shots
$$x + 2y = 100 \quad (1)$$
$$x + y = 64 \quad (2)$$
Solve **(2)** for x then substitute into **(1)**.
$$x = 64 - y$$
$$64 - y + 2y = 100$$
$$y = 36$$
Substitute $y = 36$ into **(2)**.
$$x + 36 = 64$$
$$x = 28$$
He made 28 free throws and 36 regular shots.

19. Let x = number of regular text messages, and
y = number of multimedia text messages.
$$x + y = 315 \quad (1)$$
$$0.2x + 0.5y = 87 \quad (2)$$
Add $-2 \cdot$ **(1)** and $10 \cdot$ **(2)**
$$-2x - 2y = -630$$
$$\underline{2x + 5y = 870}$$
$$3y = 240$$
$$y = 80$$
Substitute $y = 80$ in **(1)**.
$$x + 80 = 315$$
$$x = 235$$
He sent 235 regular text messages and
80 multimedia text messages.

21. x = cost of car, y = cost of truck
$$256x + 183y = 5{,}791{,}948 \quad (1)$$
$$64x + 107y = 2{,}507{,}612 \quad (2)$$

Add $-4 \cdot$ **(2)** and **(1)**.
$$-256x - 428y = -10{,}030{,}448$$
$$\underline{256x + 183y = 5{,}791{,}948}$$
$$-245y = -4{,}238{,}500$$
$$y = 17{,}300$$

Substitute $y = 17{,}300$ into **(2)**.
$$64x + 107(17{,}300) = 2{,}507{,}612$$
$$64x + 1{,}851{,}100 = 2{,}507{,}612$$
$$64x = 656{,}512$$
$$x = 10{,}258$$
The department pays \$10,258 for a car and
\$17,300 for a truck.

23. Let x = no. of pens
y = no. of notebooks
z = no. of highlighters
$$x + y + z = 15 \quad (1)$$
$$0.50x + 3y + 1.50z = 23 \quad (2)$$
$$y = z + 2 \quad (3)$$
Add **(1)** and $-2 \cdot$ **(2)**.

$$x + y + z = 15$$
$$\underline{-x - 6y - 3z = -46}$$
$$-5y - 2z = -31 \quad (4)$$
Substitute $y = z + 2$ into **(4)**.
$$-5(z + 2) - 2z = -31$$
$$-5z - 10 - 2z = -31$$
$$-7z = -21$$
$$z = 3$$
Substitute $z = 3$ into **(3)**.
$$y = 3 + 2 = 5$$
Substitute $z = 3$, $y = 5$ into **(1)**.
$$x + 5 + 3 = 15$$
$$x + 8 = 15$$
$$x = 7$$
She bought 7 pens, 5 notebooks, and
3 highlighters.

25. x = number of adults
y = number of high school students
z = number of children
$$12x + 9y + 5z = 2460 \quad (1)$$
$$15x + 10y + 6z = 2920 \quad (2)$$
$$x + y + z = 300 \quad (3)$$
Multiply -15 times **(3)** and add to **(2)**.
$$-5y - 9z = -1580 \quad (4)$$
Multiply -12 times **(3)** and add to **(1)**.
$$-3y - 7z = -1140 \quad (5)$$
Add $3 \cdot$ **(4)** and $-5 \cdot$ **(5)**.
$$8z = 960$$
$$z = 120$$
Substitute $z = 120$ into **(5)**.
$$-3y - 7(120) = -1140$$
$$-3y = -300$$
$$y = 100$$
Substitute $y = 100$, $z = 120$ into **(3)**.
$$x + 100 + 120 = 300$$
$$x = 80$$
The number of adults attending was 80, the
number of high school students was 100, and the
number of children was 120.

27. Let x = number of junior and high school
students, y = number of adults, and
z = number of senior citizens.
$$x + y + z = 12{,}000 \quad (1)$$
$$1.05x + 2.10y + 1.05z = 23{,}100 \quad (2)$$
$$1.25x + 2.40y + 1.05z = 26{,}340 \quad (3)$$
Add $-1.05 \cdot$ **(1)** to **(2)**.

$$-1.05x - 1.05y - 1.05z = -12,600$$
$$\underline{1.05x + 2.10y + 1.05z = 23,100}$$
$$1.05y \qquad\qquad = 10,500$$
$$y = 10,000$$

Add $-1.05 \cdot$ (**1**) to (**3**).
$$-1.05x - 1.05y - 1.05z = -12,600$$
$$\underline{1.25x + 2.40y + 1.05z = 26,340}$$
$$0.2z + 1.35y \qquad = 13,740 \quad (\mathbf{4})$$

Substitute $y = 10,000$ into (**4**).
$$0.2x + 1.35(10,000) = 13,740$$
$$0.2x + 13,500 = 13,740$$
$$0.2x = 240$$
$$x = 1200$$

Substitute $x = 1200$, $y = 10,000$ into (**1**).
$$1200 + 10,000 + z = 12,000$$
$$z = 800$$

1200 junior and high school students, 10,000 adults, and 800 senior citizens normally ride.

29. x = number of small pizzas
y = number of medium pizzas
z = number of large pizzas
$$x \ \ + y \ \ + z = 20 \qquad\qquad (\mathbf{1})$$
$$8x + 11y + 15z = 233 \qquad (\mathbf{2})$$
$$3x \ + 4y \ \ + 5z = 5(16) + 2 = 82 \quad (\mathbf{3})$$

Add $-8 \cdot$ (**1**) and (**2**).
$$-8x - 8y - 8z = -160$$
$$\underline{8x + 11y + 15z = 233}$$
$$3y + 7z = 73 \qquad (\mathbf{4})$$

Add $-3 \cdot$ (**1**) and (**3**).
$$-3x - 3y - 3z = -60$$
$$\underline{3x + 4y + 5z = 82}$$
$$y + 2z = 22 \qquad (\mathbf{5})$$

Add (**4**) and $-3 \cdot$ (**5**).
$$3y + 7z = 73$$
$$\underline{-3y - 6z = -66}$$
$$z = 7$$

Substitute $z = 7$ into (**5**).
$$y + 2(7) = 22$$
$$y = 8$$

Substitute $y = 8$, $z = 7$ into (**1**).
$$x + 8 + 7 = 20$$
$$x = 5$$

There were 5 small, 8 medium, and 7 large pizzas delivered.

31. Let x = no. of A boxes
y = no. of B boxes
z = no. of C boxes

$$12x + 10y + 5z = 91 \quad (\mathbf{1})$$
$$5x + 6y + 8z = 63 \quad (\mathbf{2})$$
$$3x + 4y + 5z = 40 \quad (\mathbf{3})$$

Add (**1**) and $-$(**3**).
$$12x + 10y + 5z = 91$$
$$\underline{-3x - 4y - 5z = -40}$$
$$9x + 6y = 51 \quad (\mathbf{4})$$

Add $-5 \cdot$ (**2**) and $8 \cdot$ (**3**) then solve for x.
$$-25x - 30y - 40z = -315$$
$$\underline{24x + 32y + 40z = 320}$$
$$-x + 2y = 5$$
$$-x = 5 - 2y$$
$$x = 2y - 5 \quad (\mathbf{5})$$

Substitute (**5**) into (**4**).
$$9(2y - 5) + 6y = 51$$
$$18y - 45 + 6y = 51$$
$$24y = 96$$
$$y = 4$$

Substitute $y = 4$ into (**5**).
$$x = 2(4) - 5$$
$$x = 8 - 5$$
$$x = 3$$

Substitute $x = 3$, $y = 4$ into (**3**).
$$3(3) + 4(4) + 5z = 40$$
$$9 + 16 + 5z = 40$$
$$5z = 15$$
$$z = 3$$

She can prepare 3 A boxes, 4 B boxes, and 3 C boxes.

Cumulative Review

32. $$\frac{1}{3}(4 - 2x) = \frac{1}{2}x - 3$$
$$2(4 - 2x) = 3x - 18$$
$$8 - 4x = 3x - 18$$
$$7x = 26$$
$$x = \frac{26}{7} \text{ or } 3\frac{5}{7}$$

33. $$0.06x + 0.15(0.5 - x) = 0.04$$
$$0.06x + 0.075 - 0.15x = 0.04$$
$$-0.09x = -0.035$$
$$-90x = -35$$
$$x = \frac{35}{90}$$
$$x = \frac{7(5)}{18(5)} = \frac{7}{18}$$

34. $2(y-3)-(2y+4)=-6y$
$2y-6-2y-4=-6y$
$6y=10$
$$y=\frac{5}{3} \text{ or } 1\frac{2}{3}$$

35. $4(3x+1)=-6+x$
$12x+4=-6+x$
$11x+4=-6$
$11x=-10$
$$x=\frac{-10}{11}$$
$$x=-\frac{10}{11}$$

Quick Quiz 4.3

1. Let x = speed of plane in still air
y = speed of wind
$2.5x+2.5y=1200$ **(1)**
$3x-3y=1200$ **(2)**
Solve **(2)** for x and substitute into **(1)**.
$3x=1200+3y$
$x=400+y$
$2.5(x)+2.5y=1200$
$2.5(400+y)+2.5y=1200$
$1000+2.5y+2.5y=1200$
$5y=200$
$y=40$
Substitute y = 40 into **(2)**.
$3x-3(40)=1200$
$3x-120=1200$
$3x=1320$
$x=440$
speed of plane in still air = 440 mph
speed of wind = 40 mph

2. Let x = mileage fee
y = daily fee
$300x+8y=355$ **(1)**
$260x+9y=380$ **(2)**
Add $9 \cdot$ **(1)** and $-8 \cdot$ **(2)**.
$2700x+72y=3195$
$\underline{-2080x-72y=-3040}$
$620x\qquad\quad=155$
$$x=\frac{1}{4}$$
Substitute $x=\frac{1}{4}$ into **(1)**.

$$300\left(\frac{1}{4}\right)+8y=355$$
$75+8y=355$
$8y=280$
$y=35$
The mileage fee is $0.25 per mile.
The daily fee is $35 per day.

3. Let x = price of drawing
y price of carved elephant
z = price of drum set
$3x+2y+z=55$ **(1)**
$2x+3y+z=65$ **(2)**
$4x+3y+2z=85$ **(3)**
Add **(1)** and $-$**(2)**.
$3x+2y+z=55$
$\underline{-2x-3y-z=-65}$
$x-y\qquad=-10$
$x=y-10$ **(4)**
Add -2**(2)** and **(3)**.
$-4x-6y-2z=-130$
$\underline{4x+3y+2z=85}$
$-3y\qquad=-45$
$y=15$
Substitute y = 15 into **(4)**.
$x=15-10$
$x=5$
Substitute x = 5, y = 15 into **(1)**.
$3(5)+2(15)+z=55$
$15+30+z=55$
$z=10$
Drawing price is $5
Carved elephant price is $15
Drum Set price is $10

4. Answers may vary. Possible solution:
The equations would be set up the same except the right sides would be set to 1500 rather than 1200.

4.4 Exercises

1. In the graph of the system $y > 3x + 1$,
$y < -2x + 5$ the boundary lines should be dashed because they are not included in the solution. The system contains only < or > symbols.

3. Test the point in both regions.
$y < -2x + 3$, $-4 < -2(3) + 3 = -3$, True
$y > 5x - 3$, $-4 > 5(3) - 3 = 12$, False
The point (3, −4) does not lie in the solution region.

5. $y \geq 2x - 2$
Test point: $(0, 0)$
$0 \geq 2(0) - 2$
$0 \geq -2$ True
$x + y \leq 4$
Test point: $(0, 0)$
$0 + 0 \leq 4$
$\quad 0 \leq 4$ True

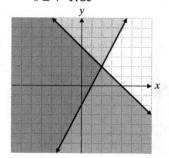

7. $y \geq -2x$
$y \geq 3x + 5$
Test point: $(-1, 4)$
$y \geq -2x$
$4 \geq -2(-1)$
$4 \geq -3$ True
$y \geq 3x + 5$
$4 \geq 3(-1) + 5$
$4 \geq -3 + 5$
$4 \geq 2$ True

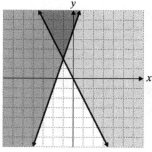

9. $y \geq 2x - 3$
Test point: $(0, -1)$
$-1 \geq 2(0) - 3$
$-1 \geq -3$ True
$y \leq \dfrac{2}{3} x$
$-1 \leq \dfrac{2}{3}(0)$
$-1 \leq 0$ True

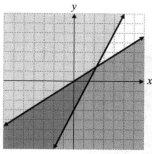

11. $x - y \geq -1$
Test point: $(0, 0) \Rightarrow 0 - 0 \geq -1$
$\qquad\qquad\qquad 0 \geq -1$ True
$-3x - y \leq 4$, Test point: $(0, 0)$
$-3(0) - 0 \leq 4$
$\qquad 0 \leq 4$ True

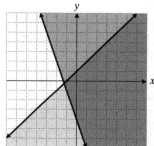

13. $x + 2y < 6$
Test point: $(0, 0)$
$0 + 2(0) < 6$
$\qquad 0 < 6$ True
$y < 3$
Test point: $(0, 0)$
$0 < 3$ True

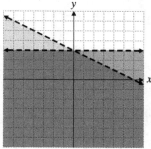

15. $y < 4, x > -2$

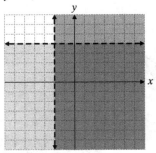

17. $x - 4y \geq -4$

Test point: $(0, 0)$

$0 - 4(0) \geq -4$

$\quad 0 \geq -4$ True

$3x + y \leq 3$

Test point: $(0, 0)$

$3(0) + 0 \leq 3$

$\quad 0 \leq 3$ True

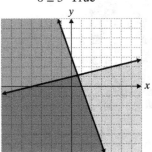

19. $3x + 2y < 6$

Test point: $(0, 0)$

$3(0) + 2(0) < 6$

$\quad 0 < 6$ True

$3x + 2y > -6$

Test point: $(0, 0)$

$3(0) + 2(0) > -6$

$\quad 0 > -6$ True

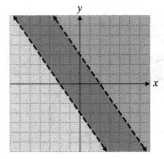

21. $x + y \leq 5$

Test point: $(0, 0)$

$0 + 0 \leq 5$

$\quad 0 \leq 5$ True

$2x - y \geq 1$

Test point: $(0, 0)$

$2(0) - 0 \geq 1$

$\quad 0 \geq 1$ False

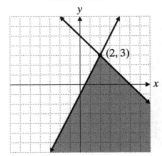

23. $x + 4y < -20$

Test point: $(0, 0)$

$0 + 4(0) < -20$

$\quad 0 < -20$ False

$y \leq x$

Test point: $(2, 0)$

$0 \leq 2$ True

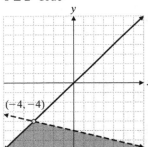

25. $y \leq x$ **(1)**

$x + y \geq 1$ **(2)**

$\quad x \leq 5$ **(3)**

Test point: $(2, 1)$

$y \leq x$ **(1)**

$1 \leq 2$ True

$x + y \geq 1$ **(2)**

$1 + 2 \geq 1$

$\quad 3 \geq 1$ True

$x \leq 5$ **(3)**

$2 \leq 5$ True

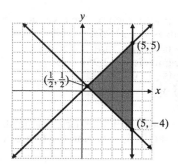

27. $y \leq 3x + 6$
Test point: (0, 0)
$0 \leq 3(0) + 6$
$0 \leq 6$ True
$4y + 3x \leq 3$
Test point: (0, 0)
$4(0) + 3(0) \leq 3$
$\qquad 0 \leq 3$ True
$x \geq -2,\ y \geq -3$

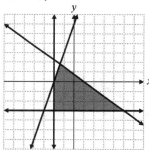

29. a. $N \leq 2D$, Test point: (2, 2)
$2 \leq 2(2)$
$2 \leq 4$ True
$4N + 3D \leq 20$, Test point: (2, 2)
$4(2) + 3(2) \leq 20$
$\qquad 8 + 6 \leq 20$
$\qquad\quad 14 \leq 20$ True
$N \geq 0,\ D \geq 0$

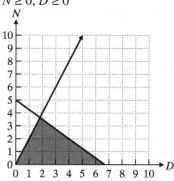

b. Yes, (3, 2) is in the shaded region.

c. No, (1, 4) is not in the shaded region.

Cumulative Review

31. $-3x^2 y - x^2 + 5y^2 = -3(2)^2(-1) - (2)^2 + 5(-1)^2$
$\qquad\qquad\qquad\qquad = 12 - 4 + 5$
$\qquad\qquad\qquad\qquad = 13$

32. $2x - 2[y + 3(x - y)] = 2x - 2[y + 3x - 3y]$
$\qquad\qquad\qquad\qquad\quad = 2x - 2[3x - 2y]$
$\qquad\qquad\qquad\qquad\quad = 2x - 6x + 4y$
$\qquad\qquad\qquad\qquad\quad = -4x + 4y$

33. x = average revenue on sunny day
y = average revenue on rainy day
$\quad 6x + y = 6050$ **(1)**
$\quad 3x + 4y = 7400$ **(2)**
Solve **(1)** for y, substitute into **(2)**.
$y = 6050 - 6x$
$\quad 3x + 4(6050 - 6x) = 7400$
$\quad 3x + 24,200 - 24x = 7400$
$\qquad\qquad\qquad -21x = -16,800$
$\qquad\qquad\qquad\qquad x = 800$
Substitute $x = 800$ into **(2)**.
$\quad 3(800) + 4y = 7400$
$\quad 2400 + 4y = 7400$
$\qquad\qquad 4y = 5000$
$\qquad\qquad\ y = 1250$
Average revenue on a sunny day is $800.
Average revenue on a rainy day is $1250.

34. Let x = cost of a chicken sandwich,
y = cost of a side salad, and
z = cost of a soda.
$\quad 3x + 2y + 3z = 27.75$ **(1)**
$\quad 3x + 4y + 4z = 36.75$ **(2)**
$\quad 4x + 3y + 5z = 39.75$ **(3)**
Subtract **(1)** from **(2)**.
$2y + z = 9$ **(4)**
Add $-4 \cdot$ **(1)** and $3 \cdot$ **(3)**.
$-12x - 8y - 12z = -111$
$\underline{\quad 12x + 9y + 15z = 119.25\quad}$
$\qquad\qquad y + 3z = 8.25$ **(5)**
Add $-1 \cdot$ **(4)** and $2 \cdot$ **(5)**.
$-2y - z = -9$
$\underline{\quad 2y + 6z = 16.5\quad}$
$\qquad\quad 5z = 7.5$
$\qquad\qquad z = 1.5$
Substitute $z = 1.5$ in **(4)**.

$$2y + 1.5 = 9$$
$$2y = 7.5$$
$$y = 3.75$$

Substitute $y = 3.75$, $z = 1.5$ into (**1**).
$$3x + 2(3.75) + 3(1.5) = 27.75$$
$$3x + 7.5 + 4.5 = 27.75$$
$$3x = 15.75$$
$$x = 5.25$$

A chicken sandwich cost $5.25, a side salad cost $3.75, and a soda cost $1.50.

Quick Quiz 4.4

1. Below the line since (0, 0) is in the solution region.

2. Solid lines since the inequality symbols used are $\geq$ and $\geq$.

3. $3x + 2y > 6$
   $x - 2y < 2$

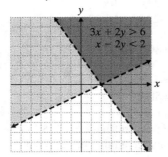

4. Answers may vary. Possible solution:
   $y > x + 2$ is graphed using a dashed line and shaded above the line.
   $x < 3$ is graphed using a dashed line and shaded to the left of the line.
   The region that satisfies both inequalities is the overlapping shaded regions.

Career Exploration Problems

1. The revenue y for selling x units is $y = 19.95x$.
   The cost y of producing x units is
   $y = 7.50x + 19,920$.
   The break-even point is when cost equals revenue.
   $$19.95x = 7.50x + 19,920$$
   $$12.45x = 19,920$$
   $$x = 1600$$
   RealTek must produce and sell 1600 units monthly to break even.

2. For the new price structure, the revenue y for selling x units is $y = 22.95x$.
   For the new price structure, the cost y of producing x units is $y = 2.95x + 19,920$.
   $$22.95x = 2.95x + 19,920$$
   $$20x = 19,920$$
   $$x = 996$$
   With the new price structure, RealTek needs to produce and sell 996 units monthly to break even.

3. If RealTek moves to a smaller location, the revenue equation is $y = 19.95x$, but the cost equation is $y = 7.50x + 11,205$.
   $$19.95x = 7.50x + 11,205$$
   $$12.45x = 11,205$$
   $$x = 900$$
   In the smaller location, RealTek needs to produce and sell 900 units monthly to break even.

You Try It

1. $-x + 3y = 6$
   $2x - 3y = -9$

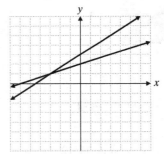

The solution is (−3, 1).

2. $-3x + 4y = -10$ (**1**)
   $x - 5y = 7$ (**2**)
   Solve (**2**) for x and substitute into (**1**).
   $$x = 5y + 7$$
   $$-3(5y + 7) + 4y = -10$$
   $$-15y - 21 + 4y = -10$$
   $$-11y = 11$$
   $$y = -1$$
   Substitute $y = -1$ into (**2**) and solve for x.
   $$x - 5(-1) = 7$$
   $$x = 2$$
   The solution is (2, −1).

3. $4x - 5y = 5$ **(1)**

$-3x + 7y = 19$ **(2)**

Add $3 \cdot$ **(1)** and $4 \cdot$ **(2)**.

$$12x - 15y = 15$$
$$\underline{-12x + 28y = 76}$$
$$13y = 91$$
$$y = 7$$

Substitute $y = 7$ into **(1)** and solve for x.

$$4x - 5(7) = 5$$
$$4x = 40$$
$$x = 10$$

The solution is $(10, 7)$.

4. $2x + 12y = 3$ **(1)**

$x + 6y = 8$ **(2)**

Add **(1)** and $-2 \cdot$ **(2)**.

$$2x + 12y = 3$$
$$\underline{-2x - 12y = -16}$$
$$0 = -13$$

This is an inconsistent system of equations and has no solution.

5. $x - y = 2$ **(1)**

$-4x + 4y = -8$ **(2)**

Add $4 \cdot$ **(1)** and **(2)**.

$$4x - 4y = 8$$
$$\underline{-4x + 4y = -8}$$
$$0 = 0$$

This is a dependent system of equations and has an infinite number of solutions.

6. $x + 2y - z = -13$ **(1)**

$-2x + 3y - 2z = -8$ **(2)**

$x - y + z = 3$ **(3)**

Add $-2 \cdot$ **(1)** and **(2)**.

$-4x - y = 18$ **(4)**

Add **(1)** and **(3)**.

$2x + y = -10$ **(5)**

Add **(4)** and **(5)**.

$$-2x = 8$$
$$x = -4$$

Substitute $x = -4$ into **(5)**.

$$2(-4) + y = -10$$
$$y = -2$$

Substitute $x = -4$, $y = -2$ into **(3)**.

$$-4 - (-2) + z = 3$$
$$z = 5$$

The solution is $(-4, -2, 5)$.

7. $x - 3y < 6$ $2x + y > 5$

Test point: $(0, 0)$ Test point: $(0, 0)$

$0 - 3(0) < 6$ $2(0) + 0 > 5$

$\quad\quad 0 < 6$ True $0 > 5$ False

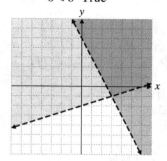

Chapter 4 Review Problems

1. $x + 2y = 8$

$x - y = 2$

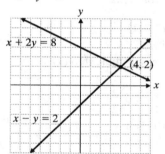

The solution is $(4, 2)$.

2. $2x + y = 6$

$3x + 4y = 4$

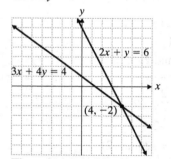

The solution is $(4, -2)$.

3. $3x - 2y = -9$ **(1)**

$2x + y = 1$ **(2)**

Solve **(2)** for y, then substitute in **(1)**.

$$y = -2x + 1$$
$$3x - 2(-2x + 1) = -9$$
$$3x + 4x - 2 = -9$$
$$7x = -7$$
$$x = -1$$

Substitute $x = -1$ into (2).
$$2(-1) + y = 1$$
$$y = 3$$
The solution is $(-1, 3)$.

4. $4x + 5y = 2$ (1)
   $3x - y = 11$ (2)
Solve (2) for y, then substitute into (1).
$$y = 3x - 11$$
$$4x + 5(3x - 11) = 2$$
$$4x + 15x - 55 = 2$$
$$19x = 57$$
$$x = 3$$
Substitute $x = 3$ into (2).
$$3(3) - y = 11$$
$$-y = 2$$
$$y = -2$$
$(3, -2)$ is the solution.

5. $3x - 4y = -12$ (1)
   $x + 2y = -4$ (2)
Solve (2) for x, substitute into (1).
$$x = -4 - 2y$$
$$3(-4 - 2y) - 4y = -12$$
$$-12 - 6y - 4y = -12$$
$$-10y = 0$$
$$y = 0$$
Substitute $y = 0$ into (2).
$$x + 0 = -4$$
$$x = -4$$
$(-4, 0)$ is the solution.

6. $-2x + 5y = -12$ (1)
   $3x + y = 1$ (2)
Add $3 \cdot$ (1) and $2 \cdot$ (2).
$$-6x + 15y = -36$$
$$\underline{6x + 2y = 2}$$
$$17y = -34$$
$$y = -2$$
Substitute $y = -2$ into (2).
$$3x + (-2) = 1$$
$$3x = 3$$
$$x = 1$$
The solution is $(1, -2)$.

7. $7x - 4y = 2$ (1)
   $6x - 5y = -3$ (2)
Add $-5 \cdot$ (1) and $4 \cdot$ (2).

$$-35x + 20y = -10$$
$$\underline{24x - 20y = -12}$$
$$-11x = -22$$
$$x = 2$$
Substitute $x = 2$ into (2).
$$6(2) - 5y = -3$$
$$-5y = -15$$
$$y = 3$$
The solution is $(2, 3)$.

8. $5x + 2y = 3$ (1)
   $7x + 5y = -20$ (2)
Add $7 \cdot$ (1) and $-5 \cdot$ (2).
$$35x + 14y = 21$$
$$\underline{-35x - 25y = 100}$$
$$-11y = 121$$
$$y = -11$$
Substitute $y = -11$ into (1).
$$5x + 2(-11) = 3$$
$$5x = 25$$
$$x = 5$$
The solution is $(5, -11)$.

9. $x = 3 - 2y$ (1)
   $3x + 6y = 8$ (2)
Substitute (1) into (2).
$$3(3 - 2y) + 6y = 8$$
$$9 - 6y + 6y = 8$$
$$9 = 8 \text{ False}$$
This is an inconsistent system. The system has no solution.

10. $x + 5y = 10$ (1)
    $y = 2 - \dfrac{1}{5}x$ (2)
Substitute (2) into (1).
$$x + 5\left(2 - \frac{1}{5}x\right) = 10$$
$$x + 10 - x = 10$$
$$10 = 10 \text{ True}$$
The equations are dependent. The system has an infinite number of solutions.

11. $5x - 2y = 15$ (1)
   $3x + y = -2$ (2)
Solve (2) for y, substitute into (1).
$$y = -2 - 3x$$
$$5x - 2(-2 - 3x) = 15$$
$$5x + 4 + 6x = 15$$
$$11x = 11$$
$$x = 1$$

Substitute $x = 1$ into **(2)**.
$$3(1) + y = -2$$
$$3 + y = -2$$
$$y = -5$$
$(1, -5)$ is the solution.

12. $x + \dfrac{1}{3}y = 1$

$$\dfrac{1}{4}x - \dfrac{3}{4}y = -\dfrac{9}{4}$$

Multiply the first equation by 3 and the second equation by 4 to clear fractions.
$$3x + y = 3 \quad \textbf{(1)}$$
$$x - 3y = -9 \quad \textbf{(2)}$$
Add $3 \cdot$ **(1)** and **(2)**.
$$9x + 3y = 9$$
$$\underline{x - 3y = -9}$$
$$10x \quad\quad = 0$$
$$x = 0$$
Substitute $x = 0$ into **(1)**.
$$3(0) + y = 3$$
$$y = 3$$
The solution is $(0, 3)$.

13. $3a + 8b = 0 \quad \textbf{(1)}$
$$9a + 2b = 11 \quad \textbf{(2)}$$
Add **(1)** and $-4 \cdot$ **(2)**.
$$3a + 8b = 0$$
$$\underline{-36a - 8b = -44}$$
$$-33a \quad\quad = -44$$
$$a = \dfrac{4}{3}$$
Substitute $a = \dfrac{4}{3}$ into **(1)**.

$$3 \cdot \dfrac{4}{3} + 8b = 0$$
$$8b = -4$$
$$b = -\dfrac{1}{2}$$
The solution is $\left(\dfrac{4}{3}, -\dfrac{1}{2} \right)$.

14. $x + 3 = 3y + 1$
$$1 - 2(x - 2) = 6y + 1$$
Solve the first equation for x.
$$x = 3y - 2 \quad \textbf{(1)}$$
Expand the second equation and combine like terms.

$$1 - 2x + 4 = 6y + 1$$
$$-2x + 4 = 6y \quad \textbf{(2)}$$
Substitute **(1)** into **(2)**.
$$-2(3y - 2) + 4 = 6y$$
$$-6y + 4 + 4 = 6y$$
$$8 = 12y$$
$$\dfrac{2}{3} = y$$
Substitute $y = \dfrac{2}{3}$ into **(1)**.

$$x = 3\left(\dfrac{2}{3} \right) - 2 = 2 - 2 = 0$$

The solution is $\left(0, \dfrac{2}{3} \right)$.

15. $10(x + 1) - 13 = -8y$
$$4(2 - y) = 5(x + 1)$$
Expand the equations and combine like terms.
First equation: $10x + 10 - 13 = -8y$
$$10x + 8y = 3 \quad \textbf{(1)}$$
Second equation: $8 - 4y = 5x + 5$
$$-5x - 4y = -3 \quad\quad \textbf{(2)}$$
Add **(1)** and $2 \cdot$ **(2)**.
$$10x + 8y = 3$$
$$\underline{-10x - 8y = -6}$$
$$0 = -3 \quad \text{False}$$
The system is inconsistent. The system has no solution.

16. $0.2x - 0.1y = 0.8$
$$0.1x + 0.3y = 1.1$$
Multiply both equations by 10 to clear decimals.
$$2x - y = 8 \quad \textbf{(1)}$$
$$x + 3y = 11 \quad \textbf{(1)}$$
Add $3 \cdot$ **(1)** and **(2)**.
$$6x - 3y = 24$$
$$\underline{x + 3y = 11}$$
$$7x \quad\quad = 35$$
$$x = 5$$
Substitute $x = 5$ into **(1)**.
$$2(5) - y = 8$$
$$-y = -2$$
$$y = 2$$
The solution is $(5, 2)$.

17. $3x - 2y - z = 3 \quad \textbf{(1)}$
$$2x + y + z = 1 \quad \textbf{(2)}$$
$$-x - y + z = -4 \quad \textbf{(3)}$$
Add **(1)** and **(2)**.

$5x - y = 4$ **(4)**

Add **(1)** and **(3)**.

$2x - 3y = -1$ **(5)**

$-15x + 3y = -12$

$\underline{2x - 3y = -1}$

$-13x = -13$

$ x = 1$

Substitute $x = 1$ into **(4)**.

$5(1) - y = 4$

$ -y = -1$

$ y = 1$

Substitute $x = 1$, $y = 1$ into **(2)**.

$2(1) + 1 + z = 1$

$ z = -2$

The solution is $(1, 1, -2)$.

18. $-2x + y - z = -7$ **(1)**

$ x - 2y - z = 2$ **(2)**

$ 6x + 4y + 2z = 4$ **(3)**

Add 2 times **(2)** to **(3)**.

$2x - 4y - 2z = 4$

$\underline{6x + 4y + 2z = 4}$

$8x = 8$

$ x = 1$

Substitute $x = 1$ into **(1)** and **(2)**.

$-2 + y - z = -7 \Rightarrow y - z = -5$ **(4)**

$1 - 2y - z = 2 \Rightarrow -2y - z = 1$ **(5)**

Add -1 times **(5)** to **(4)**.

$ y - z = -5$

$\underline{ 2y + z = -1}$

$3y = -6$

$ y = -2$

Substitute $x = 1$, $y = -2$ into **(3)**.

$6(1) + 4(-2) + 2z = 4$

$ 2z = 6$

$ z = 3$

The solution is $(1, -2, 3)$.

19. $2x + 5y + z = 3$ **(1)**

$ x + y + 5z = 42$ **(2)**

$ 2x + y = 7$ **(3)**

Solve **(1)** for z and substitute into **(2)**.

$z = 3 - 2x - 5y$

$x + y + 5(3 - 2x - 5y) = 42$

$x + y + 15 - 10x - 25y = 42$

$ -9x - 24y = 27$ **(4)**

Solve **(3)** for y, and substitute into **(4)**.

$y = 7 - 2x$

$-9x - 24(7 - 2x) = 27$

$-9x - 168 + 48x = 27$

$ 39x = 195$

$ x = 5$

Substitute $x = 5$ into $y = 7 - 2x$.

$y = 7 - 2(5) = 7 - 10 = -3$

$y = -3$

Substitute $x = 5$, $y = -3$ into

$z = 3 - 2x - 5y$

$z = 3 - 2(5) - 5(-3) = 8$

$z = 8$

The solution is $(5, -3, 8)$.

20. $ x + 2y + z = 5$ **(1)**

$ 3x - 8y = 17$ **(2)**

$ 2y + z = -2$ **(3)**

Add $-3 \cdot$ **(1)** to **(2)**.

$-14y - 3z = 2$ **(4)**

Add $3 \cdot$ **(3)** and **(4)**.

$-8y = -4$

$ y = \dfrac{1}{2}$

Substitute $y = \dfrac{1}{2}$ into **(3)**.

$2\left(\dfrac{1}{2}\right) + z = -2$

$ 1 + z = -2$

$ z = -3$

Substitute $y = \dfrac{1}{2}$ into **(2)**.

$3x - 8\left(\dfrac{1}{2}\right) = 17$

$ 3x - 4 = 17$

$ 3x = 21$

$ x = 7$

The solution is $\left(7, \dfrac{1}{2}, -3 \right)$.

21. $-3x - 4y + z = -4$ **(1)**

$ x + 6y + 3z = -8$ **(2)**

$ 5x + 3y - z = 14$ **(3)**

Add $-3 \cdot$ **(1)** and **(2)**.

$10x + 18y = 4$ **(4)**

Add **(1)** and **(3)**.

$2x - y = 10$ **(5)**

Add **(4)** and $-5 \cdot$ **(5)**.

$23y = -46$

$ y = -2$

Substitute $y = -2$ into **(5)**.

$$2x - (-2) = 10$$
$$2x = 8$$
$$x = 4$$

Substitute $x = 4$, $y = -2$ into (1).
$$-3(4) - 4(-2) + z = -4$$
$$z = 0$$
$(4, -2, 0)$ is the solution.

22. $3x + 2y = 7$ (1)
     $2x + 7z = -26$ (2)
     $5y + z = 6$ (3)

Add $-2 \cdot$ (1) and $3 \cdot$ (2).
$-4y + 21z = -92$ (4)
Add $-21 \cdot$ (3) and (4).
$$-109y = -218$$
$$y = 2$$

Substitute $y = 2$ into (3).
$$5(2) + z = 6$$
$$10 + z = 6$$
$$z = -4$$

Substitute $y = 2$ into (1).
$$3x + 2(2) = 7$$
$$3x + 4 = 7$$
$$3x = 3$$
$$x = 1$$
The solution is $(1, 2, -4)$.

23. $v =$ speed of plane in still air
     $w =$ speed of wind
     $720 = (v - 2) \cdot 3$
     $720 = (v + w)(2.5)$

$$\begin{array}{r} v - w = 240 \\ v + w = 288 \\ \hline 2v \quad\;\; = 528 \\ v = 264 \end{array}$$

$w = 288 - v = 288 - 264 = 24$
The speed of the plane in still air is 264 mph and the wind speed is 24 mph.

24. Let $x =$ no. of touchdowns (6 points)
     $y =$ no. of field goals (3 points)
     $6x + 3y = 54$ (1)
     $x + y = 10$ (2)

Solve (2) for x and substitute into (1).
$$x = 10 - y$$
$$6(10 - y) + 3y = 54$$
$$60 - 6y + 3y = 54$$
$$-3y = -6$$
$$y = 2$$

Substitute $y = 2$ into (2).

$$x + 2 = 10$$
$$x = 8$$
8 touchdowns and 2 field goals were scored.

25. $x =$ number of laborers
     $y =$ number of mechanics
     $70x + 90y = 1950$
     $80x + 100y = 2200$

Divide both equations by 10.
     $7x + 9y = 195$ (1)
     $8x + 10y = 220$ (2)

Add -8 times (1) to 7 times (2).
$$\begin{array}{r} -56x - 72y = -1560 \\ 56x + 70y = \;\;1540 \\ \hline -2y = -20 \\ y = 10 \end{array}$$

Substitute $y = 10$ into (2).
$$8x + 10(10) = 220$$
$$8x = 120$$
$$x = 15$$
The circus hired 15 laborers and 10 mechanics.

26. Let $x =$ number of children's tickets
     $y =$ number of adult tickets
     $x + y = 330$ (1)
     $8x + 13y = 3215$ (2)

Solve (1) for x, substitute into (2).
$$x = 330 - y$$
$$8(330 - y) + 13y = 3215$$
$$2640 - 8y + 13y = 3215$$
$$5y = 575$$
$$y = 115$$

Substitute $y = 115$ into (1).
$$x + 115 = 330$$
$$x = 215$$
115 adult tickets sold.
215 children's tickets sold.

27. $x =$ cost of hat
     $y =$ cost of shirt
     $z =$ cost of pants
     $2x + 5y + 4z = 177$ (1)
     $x + y + 2z = 66$ (2)
     $2x + 3y + z = 81$ (3)

Add -2 times (2) and (1).
$$3y = 45$$
$$y = 15$$

Substitute $y = 15$ into (2) and solve for x.
$$x + 15 + 2z = 66$$
$$x = 51 - 2z$$

Substitute $x = 51 - 2z$ and $y = 15$ into (3).

$$2(51-2z)+3(15)+z=81$$
$$102-4z+45+z=81$$
$$-3z=-66$$
$$z=22$$

Substitute $y=15$, $z=22$ into (2).
$$x+15+2(22)=66$$
$$x+15+44=66$$
$$x=7$$

The hats cost $7, shirts $15, and pants $22.

28. $J=$ Jess' score
$N=$ Nick's score
$C=$ Chris' score

$$J+C+N=249 \qquad (1)$$
$$J=C+20 \qquad (2)$$
$$2N=J+C+6 \qquad (3)$$

Substitute J from (2) into (1) and (3).
$$C+20+C+N=249 \Rightarrow 2C=229-N$$
$$2N=C+20+C+6 \Rightarrow 2C=2N-26$$

from which $229-N=2N-26$
$$3N=255$$
$$N=85$$

and $2C=229-85=144$
$$C=72$$

and $J=C+20=72+20$
$$J=92$$

Jess scored 92 points, Nick scored 85 points, and Chris scored 72 points.

29. $x=$ cost of jelly
$y=$ cost of peanut butter
$z=$ cost of honey

$$4x+3y+5z=32.50 \quad (1)$$
$$2x+2y+z=14.80 \quad (2)$$
$$3x+4y+2z=27.00 \quad (3)$$

Add (1) and -5 times (2).
$$-6x-7y=-41.5 \quad (4)$$

Add -2 times (2) to (3).
$$-x=-2.6$$
$$x=2.6$$

Substitute $x=2.6$ into (4).
$$-6(2.6)-7y=-41.5$$
$$-7y=-25.9$$
$$y=3.7$$

Substitute $x=2.6$, $y=3.7$ into (2).
$$2(2.6)+2(3.7)+z=14.80$$
$$z=2.20$$

The cost of a jar of jelly is $2.60, the cost of a jar of peanut butter is $3.70, and the cost of a jar of honey is $2.20.

30. $x=$ number of buses
$y=$ number of station wagons
$z=$ number of sedans

$$x+y+z=9 \qquad (1)$$
$$40x+8y+5z=127 \quad (2)$$
$$8(3y)+5(2z)=126$$
$$24y+10z=126 \quad (3)$$

Add -40 times (1) to (2).
$$-32y-35z=-233 \quad (4)$$

Add 32 times (3) to 24 times (4).
$$-520z=-1560$$
$$z=3$$

Substitute $z=3$ into (3).
$$24y+10(3)=126$$
$$24y=96$$
$$y=3$$

Substitute $y=4$, $z=3$ into (1).
$$x+4+3=9$$
$$x=2$$

They are planning to use 2 buses, 4 station wagons, and 3 sedans.

31.
$$x-y=1$$
$$5x+y=7$$

Adding gives
$$6x=8$$
$$x=\frac{4}{3} \text{ and}$$
$$5 \cdot \frac{4}{3}+y=7$$
$$y=\frac{1}{3}$$

The solution is $\left(\frac{4}{3},\frac{1}{3}\right)$.

32.
$$2x+5y=4 \qquad (1)$$
$$5x-7y=-29 \quad (2)$$

Add $7 \cdot (1)$ and $5 \cdot (2)$.
$$14x+35y=28$$
$$\underline{25x-35y=-145}$$
$$39x \qquad =-117$$
$$x=-3$$

Substitute $x=-3$ into (1).
$$2(-3)+5y=4$$
$$-6+5y=4$$
$$5y=10$$
$$y=2$$

The solution is $(-3, 2)$.

33. $\dfrac{x}{2} - 3y = -6$

$\dfrac{4}{3}x + 2y = 4$

Multiply the first equation by 2 and the second equation by 3 to clear fractions.

$x - 6y = -12$ **(1)**

$4x + 6y = 12$ **(2)**

Add **(1)** and **(2)**.

$5x = 0$

$x = 0$

Substitute $x = 0$ into **(1)**.

$0 - 6y = -12$

$\quad\quad y = 2$

The solution is (0, 2).

34. $\dfrac{x}{2} - y = -12$

$x + \dfrac{3}{4}y = 9$

Multiply the first equation by 2 and the second equation by 4 to clear fractions.

$x - 2y = -24$ **(1)**

$4x + 3y = 36$ **(2)**

Add $-4 \cdot$ **(1)** and **(2)**.

$-4x + 8y = 96$

$\underline{4x + 3y = 36}$

$\quad\quad 11y = 132$

$\quad\quad\quad y = 12$

Substitute $y = 12$ into **(1)**.

$x - 2(12) = -24$

$x - 24 = -24$

$\quad\quad x = 0$

The solution is (0, 12).

35. $7(x + 3) = 2y + 25$

$7x + 21 = 2y + 25$

$7x - 2y = 4$ **(1)**

$3(x - 6) = -2(y + 1)$

$3x - 18 = -2y - 2$

$3x + 2y = 16$ **(2)**

Add **(1)** and **(2)**.

$10x = 20$

$\quad x = 2$

Substitute $x = 2$ into **(2)**.

$3(2) + 2y = 16$

$\quad\quad 2y = 10$

$\quad\quady = 5$

The solution is (2, 5).

36. $0.3x - 0.4y = 0.9$

$0.2x - 0.3y = 0.4$

Multiply both equations by 10 to clear decimals.

$3x - 4y = 9$ **(1)**

$2x - 3y = 4$ **(2)**

Add $2 \cdot$ **(1)** and $-3 \cdot$ **(2)**.

$6x - 8y = 18$

$\underline{-6x + 9y = -12}$

$\quad\quad\quad y = 6$

Substitute $y = 6$ into **(2)**.

$2x - 3(6) = 4$

$\quad\quad 2x = 22$

$\quad\quadx = 11$

The solution is (11, 6).

37. $1.2x - y = 1.6$

$x + 1.5y = 6$

Solve the first equation for y and substitute into the second equation.

$y = 1.2x - 1.6$

$x + 1.5(1.2x - 1.6) = 6$

$x + 1.8x - 2.4 = 6$

$\quad\quad 2.8x = 8.4$

$\quad\quad\quad x = 3$

$y = 1.2(3) - 1.6$

$y = 2$

The solution is (3, 2).

38. $x - \dfrac{y}{2} + \dfrac{1}{2}z = -1$

$2x + \dfrac{5}{2}z = -1$

$\dfrac{3}{2}y + 2z = 1$

Multiply each equation by 2 to clear fractions.

$2x - y + z = -2$ **(1)**

$4x + 5z = -2$ **(2)**

$3y + 4z = 2$ **(3)**

Add $3 \cdot$ **(1)** and **(3)**.

$6x + 7z = -4$ **(4)**

Add $-3 \cdot$ **(2)** and $2 \cdot$ **(4)**.

$-12x - 15z = 6$

$\underline{12x + 14z = -8}$

$\quad\quad\quad -z = -2$

$\quad\quad\quadz = 2$

Substitute $z = 2$ into **(2)**.

$4x + 5(2) = -2$

$\quad\quad 4x = -12$

$\quad\quadx = -3$

Substitute $z = 2$ into **(3)**.

$$3y + 4(2) = 2$$
$$3y = -6$$
$$y = -2$$

The solution is $(-3, -2, 2)$.

39. $x - 4y + 4z = -1$ **(1)**
$\quad 2x - y + 5z = -3$ **(2)**
$\quad\quad x - 3y + z = 4$ **(3)**

Add **(1)** and -4 times **(3)**.
$-3x + 8y = -17$ **(4)**
Add **(2)** and -5 times **(3)**.
$-3x + 14y = -23$ **(5)**
Subtract **(5)** from **(4)**.
$-6y = 6$
$\quad y = -1$

Substitute $y = -1$ into **(4)**.
$-3x + 8(-1) = -17$
$\quad\quad -3x = -9$
$\quad\quad\quad x = 3$

Substitute $x = 3$, $y = -1$ into **(3)**.
$3 - 3(-1) + z = 4$
$\quad\quad\quad z = -2$

The solution is $(3, -1, -2)$.

40. $x - 2y + z = -5$ **(1)**
$\quad 2x \quad\quad + z = -10$ **(2)**
$\quad\quad\quad y - z = 15$ **(3)**

Add **(1)** and **(3)**.
$x - y = 10$ **(4)**
Add **(2)** and **(3)**.
$2x + y = 5$ **(5)**
Add **(4)** and **(5)**.
$3x = 15$
$\quad x = 5$

Substitute $x = 5$ into **(4)**.
$5 - y = 10$
$\quad -y = 5$
$\quad\quad y = -5$

Substitute $x = 5$, $y = -5$ into **(1)**.
$5 - 2(-5) + z = -5$
$\quad\quad\quad\quad z = -20$

The solution is $(5, -5, -20)$.

41. $\quad y \geq -\dfrac{1}{2}x - 1$ **(1)**
$\quad -x + y \leq 5$ **(2)**

Test point: $(0, 0)$

(1) $0 \geq -\dfrac{1}{2}(0) - 1$
$\quad 0 \geq -1$ True

(2) $0 + 0 \leq 5$ True

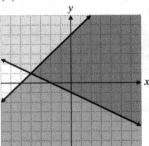

42. $-2x + 3y < 6$
Test point: $(0, 0)$
$-2(0) + 3(0) < 6$
$\quad\quad\quad 0 < 6$ True

$y > -2$

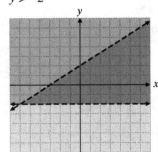

43. $x + y > 1$
Test point: $(0, 0)$
$0 + 0 > 1$
$\quad\quad 0 > 1$ False
$2x - y < 5$
Test point: $(0, 0)$
$2(0) - 0 < 5$
$\quad\quad 0 < 5$ True

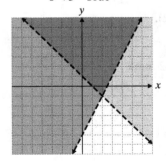

44. $x + y \geq 4$
Test point: (0, 0)
$0 + 0 \geq 4$
$\quad 0 \geq 4$ False
$y \leq x$
Test point: (2, 0)
$0 \leq 2$ True
$x \leq 6$

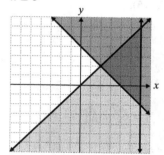

How Am I Doing? Chapter 4 Test

1. $x - y = 3$ **(1)**
$2x - 3y = -1$ **(2)**
Solve the first equation for *y* and substitute into the second equation.
$y = x - 3$
$2x - 3(x - 3) = -1$
$2x - 3x + 9 = -1$
$\qquad -x = -10$
$\qquad\quad x = 10$
$y = 10 - 3$
$y = 7$
The solution is (10, 7).

2. $3x + 2y = 1$ **(1)**
$5x + 3y = 3$ **(2)**
Multiply the first equation by 3 and the second equation by −2 and add.
$9x + 6y = 3$
$\underline{-10x - 6y = -6}$
$\quad -x \qquad = -3$
$\qquad\quad x = 3$
Substitute $x = 3$ into the first equation.
$3(3) + 2y = 1$
$\qquad 2y = -8$
$\qquad\ y = -4$
The solution is (3, −4).

3. $5x - 3y = 3$ **(1)**
$7x + y = 25$ **(2)**
Add 3 times **(2)** to **(1)**.
$5x - 3y = 3$
$\underline{21x + 3y = 75}$
$\quad 26x = 78$
$\qquad x = 3$
$7x + y = 25$
$7(3) + y = 25$
$\qquad\ y = 4$
The solution is (3, 4).

4. $\dfrac{1}{4}a - \dfrac{3}{4}b = -1$ **(1)**
$\dfrac{1}{3}a + b = \dfrac{5}{3}$ **(2)**
Multiply **(1)** by 4 and **(2)** by 3 to clear fractions.
$a - 3b = -4$ **(3)**
$a + 3b = 5$ **(4)**
Add **(3)** and **(4)**.
$2a = 1$
$a = \dfrac{1}{2}$

Substitute $a = \dfrac{1}{2}$ into **(3)**.

$\dfrac{1}{2} - 3b = -4$

$\qquad -3b = -\dfrac{9}{2}$

$\qquad\quad b = \dfrac{3}{2}$

The solution is $\left(\dfrac{1}{2}, \dfrac{3}{2}\right)$.

5. $\dfrac{1}{3}x + \dfrac{5}{6}y = 2$ **(1)**
$\dfrac{3}{5}x - y = -\dfrac{7}{5}$ **(2)**
Multiply **(1)** by 6 and **(2)** by 5 to clear fractions.
$2x + 5y = 12$ **(3)**
$3x - 5y = -7$ **(4)**
Add **(3)** and **(4)**.
$5x = 5$
$\ x = 1$
Substitute $x = 1$ into **(3)**.
$2(1) + 5y = 12$
$\qquad 5y = 10$
$\qquad\ y = 2$
The solution is (1, 2).

6. $8x - 3y = 5$
$-16x + 6y = 8$
Multiply the first equation by 2 and add to the second equation.
$16x - 6y = 10$
$\underline{-16x + 6y = 8}$
$0 = 18$

Inconsistent system; no solution

7. $3x + 5y - 2z = -5$ **(1)**
$2x + 3y - z = -2$ **(2)**
$2x + 4y + 6z = 18$ **(3)**

Add $-2 \cdot$ **(2)** to **(1)**.
$-x - y = -1$ **(4)**
Add $6 \cdot$ **(2)** to **(3)**.
$14x + 22y = 6$ **(5)**
Add $14 \cdot$ **(4)** to **(5)**.
$8y = -8$
$y = -1$

Substitute $y = -1$ into **(4)**.
$-x - (-1) = -1$
$-x = -2$
$x = 2$
Substitute $x = 2$, $y = -1$ into **(2)**.
$2(2) + 3(-1) - z = -2$
$-z = -3$
$z = 3$
The solution is $(2, -1, 3)$.

8. $3x + 2y = 0$ **(1)**
$2x - y + 3z = 8$ **(2)**
$5x + 3y + z = 4$ **(3)**

Add **(2)** and $-3 \cdot$ **(3)**.
$-13x - 10y = -4$ **(4)**
Add $5 \cdot$ **(1)** and **(4)**.
$2x = -4$
$x = -2$
Substitute $x = -2$ into **(1)**.
$3(-2) + 2y = 0$
$2y = 6$
$y = 3$
Substitute $x = -2$, $y = 3$ into **(3)**.
$5(-2) + 3(3) + z = 4$
$z = 5$
The solution is $(-2, 3, 5)$.

9. $x + 5y + 4z = -3$ **(1)**
$x - y - 2z = -3$ **(2)**
$x + 2y + 3z = -5$ **(3)**

Add $-1 \cdot$ **(1)** and **(2)**.
$-6y - 6z = 0$ **(4)**
Add $-1 \cdot$ **(1)** and **(3)**.
$-3y - z = -2$ **(5)**
Add **(4)** and $-2 \cdot$ **(5)**.
$-4z = 4$
$z = -1$
Substitute $z = -1$ into **(5)**.
$-3y - (-1) = -2$
$-3y = -3$
$y = 1$
Substitute $y = 1$, $z = -1$ into **(1)**.
$x + 5(1) + 4(-1) = -3$
$x = -4$
The solution is $(-4, 1, -1)$.

10. $v =$ speed of plane in still air
$w =$ speed of wind
$1000 = (v + w)(2) \Rightarrow v + w = 500$
$1000 = (v - w)(2.5) \Rightarrow \underline{ v - w = 400}$
$ 2v = 900$
$ v = 450$

$450 + w = 500$
$w = 50$
The speed of the plane in still air is 450 mph.
The speed of the wind is 50 mph.

11. $p =$ price of a pen
$m =$ price of a mug
$s =$ price of a T-shirt
$4p + m + s = 20$ **(1)**
$2p + 2m = 11$ **(2)**
$6p + m + 2s = 33$ **(3)**

Add -1 times **(1)** to **(3)**.
$2p + s = 13$ **(4)**
Add -2 times **(3)** to **(2)**.
$-10p - 4s = -55$ **(5)**
Add 5 times **(4)** to **(5)**.
$s = 10$
Substitute $s = 10$ into **(4)**.
$2p + 10 = 13$
$2p = 3$
$p = 1.50$
Substitute $s = 10$, $p = 1.5$ into **(1)**.
$4(1.5) + m + 10 = 20$
$m = 4$
Each pen cost $1.50, each mug cost $4.00, and each T-shirt cost $10.00.

12. $x =$ daily charge
    $y =$ mileage charge
    $5x + 150y = 180$ **(1)**
    $7x + 320y = 274$ **(2)**
    Add $-7 \cdot$ **(1)** and $5 \cdot$ **(2)**.
    $-35x - 1050y = -1260$
    $\underline{35x + 1600y = 1370}$
             $550y = 110$
                $y = 0.2$
    Substitute $y = 0.2$ into **(1)**.
    $5x + 150(0.2) = 180$
           $5x = 150$
            $x = 30$
    They charge \$30 per day and \$0.20 per mile.

13. $x + 2y \le 6$
    Test point: (0, 0)
    $0 + 2(0) \le 6$
         $0 \le 6$ True
    $-2x + y \ge -2$
    Test point: (0, 0)
    $-2(0) + 0 \ge -2$
         $0 \ge -2$ True

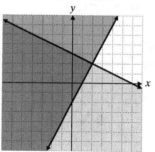

14. $3x + y > 8$
    Test point: (0, 0)
    $3(0) + 0 > 8$
         $0 > 8$ False
    $x - 2y > 5$
    Test point: (0, 0)
    $0 - 2(0) > 5$
         $0 > 5$ False

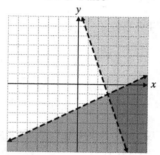

Chapter 5

5.1 Exercises

1. When you multiply exponential expressions with the same base, keep the base the same and add the exponents.

3. A sample example is: $\dfrac{2^2}{2^3} \stackrel{?}{=} \dfrac{1}{2^{3-2}}$

$$\dfrac{4}{8} \stackrel{?}{=} \dfrac{1}{2}$$

$$\dfrac{1}{2} = \dfrac{1}{2}$$

5. $6x^{11}y$: Numerical coefficient is 6, bases are x, y and exponents are 11, 1.

7. $2 \cdot 2 \cdot a \cdot a \cdot a \cdot b = 2^2 a^3 b$

9. $(-5)(x)(y)(z)(y)(x)(x)(z)$
$\quad = -5(x \cdot x \cdot x)(y \cdot y)(z \cdot z)$
$\quad = -5x^3 y^2 z^2$

11. $(7^4)(7^6) = 7^{4+6} = 7^{10}$

13. $(8^9)(8^{12}) = 8^{9+12} = 8^{21}$

15. $x^4 \cdot x^8 = x^{4+8} = x^{12}$

17. $t^{15} \cdot t = t^{15} \cdot t^1 = t^{15+1} = t^{16}$

19. $-5x^4(4x^2) = (-5 \cdot 4)(x^4 \cdot x^2)$
$\quad\quad\quad\quad = -20x^{4+2}$
$\quad\quad\quad\quad = -20x^6$

21. $(5x)(10x^2) = (5 \cdot 10)(x \cdot x^2) = 50x^{1+2} = 50x^3$

23. $(2xy^3)(9x^2y^5) = (2 \cdot 9)(x \cdot x^2)(y^3 \cdot y^5)$
$\quad\quad\quad\quad\quad = 18x^{1+2}y^{3+5}$
$\quad\quad\quad\quad\quad = 18x^3 y^8$

25. $\left(\dfrac{2}{5}xy^3\right)\left(\dfrac{1}{3}x^2y^2\right) = \left(\dfrac{2}{5} \cdot \dfrac{1}{3}\right)(x \cdot x^2)(y^3 \cdot y^2)$
$\quad\quad\quad\quad\quad\quad\quad\quad = \dfrac{2}{15}x^{1+2}y^{3+2}$
$\quad\quad\quad\quad\quad\quad\quad\quad = \dfrac{2}{15}x^3 y^5$

27. $(1.1x^2z)(-2.5xy) = (1.1)(-2.5)(x^2 \cdot x)yz$
$\quad\quad\quad\quad\quad\quad = -2.75x^{2+1}yz$
$\quad\quad\quad\quad\quad\quad = -2.75x^3 yz$

29. $(8a)(2a^3b)(0) = 0$

31. $(-16x^2y^4)(-5xy^3) = (-16)(-5)(x^2 \cdot x)(y^4 \cdot y^3)$
$\quad\quad\quad\quad\quad\quad\quad = 80x^{2+1}y^{4+3}$
$\quad\quad\quad\quad\quad\quad\quad = 80x^3 y^7$

33. $(-8x^3y^2)(3xy^5) = (-8)(3)(x^3 \cdot x)(y^2 \cdot y^5)$
$\quad\quad\quad\quad\quad\quad\quad = -24x^{3+1}y^{2+5}$
$\quad\quad\quad\quad\quad\quad\quad = -24x^4 y^7$

35. $(-2x^3y^2)(0)(-3x^4y) = 0$

37. $(8a^4b^3)(-3x^2y^5) = (8)(-3)a^4b^3x^2y^5$
$\quad\quad\quad\quad\quad\quad\quad = -24a^4b^3x^2y^5$

39. $(2x^2y)(-3y^3z^2)(5xz^4)$
$\quad = (2)(-3)(5)(x^2 \cdot x)(y \cdot y^3)(z^2 \cdot z^4)$
$\quad = -30x^{2+1}y^{1+3}z^{2+4}$
$\quad = -30x^3 y^4 z^6$

41. $\dfrac{y^{12}}{y^5} = y^{12-5} = y^7$

43. $\dfrac{y^5}{y^8} = \dfrac{1}{y^{8-5}} = \dfrac{1}{y^3}$

45. $\dfrac{11^{18}}{11^{30}} = \dfrac{1}{11^{30-18}} = \dfrac{1}{11^{12}}$

47. $\dfrac{2^{17}}{2^{10}} = 2^{17-10} = 2^7$

49. $\dfrac{a^{13}}{4a^5} = \dfrac{a^{13-5}}{4} = \dfrac{a^8}{4}$

51. $\dfrac{x^7}{y^9} = \dfrac{x^7}{y^9}$

53. $\dfrac{48x^5y^3}{24xy^3} = 2x^{5-1}y^{3-3} = 2x^4y^0 = 2x^4$

55. $\dfrac{16x^5y}{-32x^2y^3} = \dfrac{x^{5-2}}{-2y^{3-1}} = -\dfrac{x^3}{2y^2}$

57. $\dfrac{1.8f^4g^3}{54f^2g^8} = \dfrac{f^{4-2}}{30g^{8-3}} = \dfrac{f^2}{30g^5}$

59. $\dfrac{(-17x^5y^4)(5y^6)}{-5xy^7} = \dfrac{-85x^5y^{10}}{-5xy^7}$
$$= 17x^{5-1}y^{10-7}$$
$$= 17x^4y^3$$

61. $\dfrac{8^0x^2y^3}{16x^5y} = \dfrac{y^{3-1}}{16x^{5-2}} = \dfrac{y^2}{16x^3}$

63. $\dfrac{18a^6b^3c^0}{24a^5b^3} = \dfrac{3}{4}a^{6-5}b^{3-3}c^0 = \dfrac{3}{4}ab^0 = \dfrac{3}{4}a$

65. $(x^2)^6 = x^{2\cdot6} = x^{12}$

67. $(x^3y)^5 = (x^3)^5 \cdot y^5 = x^{3\cdot5}y^5 = x^{15}y^5$

69. $(rs^2)^6 = r^6(s^2)^6 = r^6s^{2\cdot6} = r^6s^{12}$

71. $(3a^3b^2c)^3 = 3^3(a^3)^3(b^2)^3c^3$
$$= 3^3a^{3\cdot3}b^{2\cdot3}c^3$$
$$= 27a^9b^6c^3$$

73. $(-3a^4)^2 = (-3)^2(a^4)^2 = 9a^{4\cdot2} = 9a^8$

75. $\left(\dfrac{x}{2m^4}\right)^7 = \dfrac{x^7}{2^7m^{4\cdot7}} = \dfrac{x^7}{128m^{28}}$

77. $\left(\dfrac{5x}{7y^2}\right)^2 = \dfrac{5^2x^2}{7^2y^{2\cdot2}} = \dfrac{25x^2}{49y^4}$

79. $(-3a^2b^3c^0)^4 = (-3a^2b^3)^4$
$$= (-3)^4(a^2)^4(b^3)^4$$
$$= 81a^8b^{12}$$

81. $(-2x^3y^0z)^3 = (-2x^3z)^3$
$$= (-2)^3(x^3)^3z^3$$
$$= -8x^9z^3$$

83. $\dfrac{(3x)^5}{(3x^2)^3} = \dfrac{3^5x^5}{3^3x^{2\cdot3}} = \dfrac{3^5x^5}{3^3x^6} = \dfrac{3^{5-3}}{x^{6-5}} = \dfrac{9}{x}$

85. $(-5a^2b^3)^2(ab) = (-5)^2(a^2)^2(b^3)^2(ab)$
$$= 25a^4b^6(ab)$$
$$= 25a^{4+1}b^{6+1}$$
$$= 25a^5b^7$$

87. $\left(\dfrac{7}{a^5}\right)^2 = \dfrac{7^2}{a^{5\cdot2}} = \dfrac{49}{a^{10}}$

89. $\left(\dfrac{2x}{y^3}\right)^4 = \dfrac{2^4x^4}{y^{3\cdot4}} = \dfrac{16x^4}{y^{12}}$

91. $\dfrac{(10ac^3)(7a)}{40b} = \dfrac{70a^2c^3}{40b} = \dfrac{7a^2c^3}{4b}$

93. $\dfrac{11x^7y^2}{33x^8y^3} = \dfrac{1}{3x^{8-7}y^{3-2}} = \dfrac{1}{3xy}$

Cumulative Review

94. $-3 - 8 = -3 + (-8) = -11$

95. $-17 + (-32) + (-24) + 27 = -49 + (-24) + 27$
$$= -73 + 27$$
$$= -46$$

96. $\left(-\dfrac{3}{5}\right)\left(-\dfrac{2}{15}\right) = \dfrac{3}{5}\cdot\dfrac{2}{15} = \dfrac{\cancel{3}\cdot2}{5\cdot\cancel{3}\cdot5} = \dfrac{2}{25}$

97. $-\dfrac{5}{4} \div \dfrac{5}{16} = -\dfrac{5}{4}\cdot\dfrac{16}{5} = -\dfrac{\cancel{5}\cdot\cancel{4}\cdot4}{\cancel{4}\cdot\cancel{5}} = -4$

98. $\dfrac{4,966,400}{7,760,000} = 0.64 = 64\%$

64% of the Amazon rainforest lay in Brazil in 2012.

99. $\dfrac{\text{sum of losses in each year}}{\text{area in Brazil}} = \dfrac{7500 + 4(5250)}{4,966,400}$

$= \dfrac{28,500}{4,966,400}$

≈ 0.006

$= 0.6\%$

About 0.6% of the Amazon rainforest in Brazil will be lost from 2013 through 2017.

Quick Quiz 5.1

1. $(2x^2 y^3)(-5xy^4) = (2)(-5)(x^2 \cdot x)(y^3 \cdot y^4)$

$= -10 x^{2+1} y^{3+4}$

$= -10 x^3 y^7$

2. $\dfrac{-28 x^6 y^6}{35 x^3 y^8} = -\dfrac{4 x^{6-3}}{5 y^{8-6}} = -\dfrac{4 x^3}{5 y^2}$

3. $(-3 x^3 y^5)^4 = (-3)^4 (x^3)^4 (y^5)^4$

$= 81 x^{3 \cdot 4} y^{5 \cdot 4}$

$= 81 x^{12} y^{20}$

4. Answers may vary. Possible solution:

$\dfrac{(4x^3)^2}{(2x^4)^3}$

In the numerator and the denominator, raise each factor inside the parentheses to the power.

$\dfrac{4^2 (x^3)^2}{2^3 (x^4)^3}$

Evaluate the constants to their respective powers, and multiply exponents on the variable expressions.

$\dfrac{16 x^6}{8 x^{12}}$

Divide the numbers and subtract exponents on the variable expressions. Because the larger exponent is in the denominator, the variable expression will be in the denominator.

$\dfrac{2}{x^6}$

5.2 Exercises

1. $x^{-4} = \dfrac{1}{x^4}$

3. $3^{-4} = \dfrac{1}{3^4} = \dfrac{1}{81}$

5. $\dfrac{1}{y^{-8}} = y^8$

7. $\dfrac{x^{-4} y^{-5}}{z^{-6}} = \dfrac{z^6}{x^4 y^5}$

9. $a^3 b^{-2} = \dfrac{a^3}{b^2}$

11. $(2 x^{-3})^{-3} = 2^{-3} x^9 = \dfrac{x^9}{2^3} = \dfrac{x^9}{8}$

13. $3 x^{-2} = \dfrac{3}{x^2}$

15. $(3 x y^2)^{-2} = 3^{-2} x^{-2} y^{-4} = \dfrac{1}{3^2 x^2 y^4} = \dfrac{1}{9 x^2 y^4}$

17. $\dfrac{3 x y^{-2}}{z^{-3}} = \dfrac{3 x z^3}{y^2}$

19. $\dfrac{(4xy)^{-1}}{(4xy)^{-2}} = \dfrac{(4xy)^2}{(4xy)^1} = (4xy)^{2-1} = 4xy$

21. $a^{-1} b^3 c^{-4} d = \dfrac{b^3 d}{a c^4}$

23. $(8^{-2})(2^3) = \dfrac{1}{8^2} \cdot 2^3 = \dfrac{1}{64} \cdot 8 = \dfrac{1}{8}$

25. $\left(\dfrac{3 x^0 y^2}{z^4} \right)^{-2} = \left(\dfrac{3 y^2}{z^4} \right)^{-2}$

$= \dfrac{3^{-2} y^{-4}}{z^{-8}}$

$= \dfrac{z^8}{3^2 y^4}$

$= \dfrac{z^8}{9 y^4}$

27. $\dfrac{x^{-2}y^{-3}}{x^4 y^{-2}} = \dfrac{y^2}{x^4 x^2 y^3} = \dfrac{1}{x^{4+2}y^{3-2}} = \dfrac{1}{x^6 y}$

29. $123,780 = 1.2378 \cdot 10,000 = 1.2378 \times 10^5$

31. $0.063 = 6.3 \times 0.01 = 6.3 \times 10^{-2}$

33. Move the decimal point 11 places to the left.
$889,610,000,000 = 8.8961 \times 10^{11}$

35. Move the decimal point 6 places to the right.
$0.00000342 = 3.42 \times 10^{-6}$

37. $3.02 \times 10^5 = 3.02 \times 100,000 = 302,000$

39. $4.7 \times 10^{-4} = 4.7 \times \dfrac{1}{10,000} = 0.00047$

41. $9.83 \times 10^5 = 9.83 \times 100,000 = 983,000$

43. $0.0000237 = 2.37 \times 10^{-5}$ miles per hour

45. $1.496 \times 10^8 = 149,600,000$ km

47. $(42,000,000)(150,000,000)$
$= (4.2 \times 10^7)(1.5 \times 10^8)$
$= 6.3 \times 10^{15}$

49. $\dfrac{(5,000,000)(16,000)}{8,000,000,000} = \dfrac{(5 \times 10^6)(1.6 \times 10^4)}{8 \times 10^9}$
$= \dfrac{8 \times 10^{10}}{8 \times 10^9}$
$= 1.0 \times 10^1$

51. $(0.003)^4 = (3 \times 10^{-3})^4$
$= 3^4 \times 10^{-3(4)}$
$= 81 \times 10^{-12}$
$= 8.1 \times 10^{-11}$

53. $(150,000,000)(0.00005)(0.002)(30,000)$
$= (1.5 \times 10^8)(5 \times 10^{-5})(2 \times 10^{-3})(3 \times 10^4)$
$= 45 \times 10^4$
$= 4.5 \times 10^5$

55. $\dfrac{1.8 \times 10^{13}}{3.21 \times 10^8} = \dfrac{1.8}{3.21} \times 10^{13-8}$
$\approx 0.561 \times 10^5$
$= 5.61 \times 10^4$
Each individual would be assigned approximately $\$5.61 \times 10^4$, or $\$56,100$.

57. $d = rt$
$d = (0.00000275)(24)$
$= (2.75 \times 10^{-6})(24)$
$= 66 \times 10^{-6}$
$= 6.6 \times 10^{-5}$
It traveled 6.6×10^{-5} mile in a day.

59. $r = \dfrac{d}{t} = \dfrac{3.5 \times 10^9}{9.5}$
$= \dfrac{3.5}{9.5} \times 10^9$
$\approx 0.368 \times 10^9$
$= 3.68 \times 10^8$
New Horizons traveled approximately 3.68×10^8 miles per year.

61. $\dfrac{6.29 \times 10^{10} - 5.46 \times 10^{10}}{6.29 \times 10^{10}} = \dfrac{0.83 \times 10^{10}}{6.29 \times 10^{10}}$
≈ 0.132
$= 13.2\%$

Cumulative Review

63. $-2.7 - (-1.9) = -2.7 + 1.9 = -0.8$

64. $(-1)^{33} = -1$

65. $-\dfrac{3}{4} + \dfrac{5}{7} = \dfrac{-21}{28} + \dfrac{20}{28} = \dfrac{-21 + 20}{28} = -\dfrac{1}{28}$

Quick Quiz 5.2

1. $3x^{-3}y^2 z^{-4} = \dfrac{3y^2}{x^3 z^4}$

2. $\dfrac{4a^3 b^{-4}}{8a^{-5}b^{-3}} = \dfrac{a^3 a^5 b^3}{2b^4} = \dfrac{a^{3+5}}{2b^{4-3}} = \dfrac{a^8}{2b}$

3. Move the decimal point 3 places to the right.
$0.00876 = 8.76 \times 10^{-3}$

4. Answers may vary. Possible solution:

$(4x^{-3}y^4)^{-3}$

Raise each factor inside the parentheses to the power.

$4^{-3}(x^{-3})^{-3}(y^4)^{-3}$

Multiply exponents on the variable expressions.

$4^{-3}x^9y^{-12}$

Rewrite as a fraction.

$\dfrac{x^9}{4^3 y^{12}}$

Evaluate 4^3.

$\dfrac{x^9}{64y^{12}}$

5.3 Exercises

1. A polynomial in x is the sum of a finite number of terms of the form ax^n, where a is any real number and n is a whole number. An example is $3x^2 - 5x - 9$.

3. The degree of a polynomial in x is the largest exponent of x in any of the terms of the polynomial.

5. $6x^3 y$

   The sum of the exponents is $3 + 1 = 4$. Therefore this is a polynomial of degree 4. It has one term, so it is a monomial.

7. $20x^5 + 6x^3 - 7x$

   The greatest degree of any term is 5, so this polynomial is of degree 5. It has three terms, so it is a trinomial.

9. $4x^2 y^3 - 7x^3 y^3$

   The sum of the exponents on the first term is $2 + 3 = 5$, and the sum of the exponents on the second term is $3 + 3 = 6$. The greater of these is 6, so the degree of the polynomial is 6. The polynomial has two terms, so it is a binomial.

11. $(6x - 11) + (-9x - 4) = [6x + (-9x)] + [-11 + (-4)]$
    $= [(6 - 9)x] + [-11 + (-4)]$
    $= -3x - 15$

13. $(6x^2 + 5x - 6) + (-8x^2 - 3x + 5)$
    $= [6x^2 + (-8x^2)] + [5x + (-3x)] + [-6 + 5]$
    $= [(6 - 8)x^2] + [(5 - 3)x] + [-6 + 5]$
    $= (-2x^2) + 2x + (-1)$
    $= -2x^2 + 2x - 1$

15. $\left(\dfrac{1}{2}x^2 + \dfrac{1}{3}x - 4\right) + \left(\dfrac{1}{3}x^2 + \dfrac{1}{6}x - 5\right)$
    $= \left[\dfrac{1}{2}x^2 + \dfrac{1}{3}x^2\right] + \left[\dfrac{1}{3}x + \dfrac{1}{6}x\right] + [(-4) + (-5)]$
    $= \left[\left(\dfrac{1}{2} + \dfrac{1}{3}\right)x^2\right] + \left[\left(\dfrac{1}{3} + \dfrac{1}{6}\right)x\right] + [(-4) + (-5)]$
    $= \left[\left(\dfrac{3}{6} + \dfrac{2}{6}\right)x^2\right] + \left[\left(\dfrac{2}{6} + \dfrac{1}{6}\right)x\right] + (-9)$
    $= \dfrac{5}{6}x^2 + \dfrac{3}{6}x + (-9)$
    $= \dfrac{5}{6}x^2 + \dfrac{1}{2}x - 9$

17. $(3.4x^3 - 7.1x + 3.4) + (2.2x^2 - 6.1x - 8.8)$
    $= 3.4x^3 + 2.2x^2 + (-7.1 - 6.1)x + (3.4 - 8.8)$
    $= 3.4x^3 + 2.2x^2 - 13.2x - 5.4$

19. $(2x - 19) - (-3x + 5) = (2x - 19) + (3x - 5)$
    $= (2 + 3)x + (-19 - 5)$
    $= 5x - 24$

21. $\left(\dfrac{2}{5}x^2 - \dfrac{1}{2}x + 5\right) - \left(\dfrac{1}{3}x^2 - \dfrac{3}{7}x - 6\right)$
    $= \left(\dfrac{2}{5}x^2 - \dfrac{1}{2}x + 5\right) + \left(-\dfrac{1}{3}x^2 + \dfrac{3}{7}x + 6\right)$
    $= \left(\dfrac{2}{5} - \dfrac{1}{3}\right)x^2 + \left(-\dfrac{1}{2} + \dfrac{3}{7}\right)x + (5 + 6)$
    $= \left(\dfrac{6}{15} - \dfrac{5}{15}\right)x^2 + \left(-\dfrac{7}{14} + \dfrac{6}{14}\right)x + (5 + 6)$
    $= \dfrac{1}{15}x^2 - \dfrac{1}{14}x + 11$

23. $(4x^3 + 3x) - (x^3 + x^2 - 5x)$
    $= (4x^3 + 3x) + (-x^3 - x^2 + 5x)$
    $= (5 - 1)x^3 - x^2 + (3 + 5)x$
    $= 3x^3 - x^2 + 8x$

25. $(0.5x^4 - 0.7x^2 + 8.3) - (5.2x^4 + 1.6x + 7.9)$
$= (0.5x^4 - 0.7x^2 + 8.3) + (-5.2x^4 - 1.6x - 7.9)$
$= (0.5 - 5.2)x^4 - 0.7x^2 - 1.6x + (8.3 - 7.9)$
$= -4.7x^4 - 0.7x^2 - 1.6x + 0.4$

27. $(8x + 2) + (x - 7) - (3x + 1)$
$= (8x + 2) + (x - 7) + (-3x - 1)$
$= (8 + 1 - 3)x + (2 - 7 - 1)$
$= 6x - 6$

29. $(-4x^2 y^2 + 9xy - 3) + (8x^2 y^2 - 5xy - 7)$
$= (-4 + 8)x^2 y^2 + (9 - 5)xy + (-3 - 7)$
$= 4x^2 y^2 + 4xy - 10$

31. $(3x^4 - 4x^2 - 18) - (2x^4 + 3x^3 + 6)$
$= (3x^4 - 4x^2 - 18) + (-2x^4 - 3x^3 - 6)$
$= (3 - 2)x^4 - 3x^3 - 4x^2 + (-18 - 6)$
$= x^4 - 3x^3 - 4x^2 - 24$

33. $x = 1990 - 1990 = 0$
$-2.06(0)^2 + 77.82(0) + 743 = 0 + 0 + 743$
$ = 743$
There were 743 thousand, or 743,000, prisoners in 1990.

35. Find the population for each year.
2002: $x = 2002 - 1990 = 12$
$-2.06(12)^2 + 77.82(12) + 743$
$= -2.06(144) + 77.82(12) + 743$
$= -296.64 + 933.84 + 743$
$= 1380.2$ thousand
2007: $x = 2007 - 1990 = 17$
$-2.06(17)^2 + 77.82(17) + 743$
$= -2.06(289) + 77.82(17) + 743$
$= -595.34 + 1322.94 + 743$
$= 1470.6$ thousand
Subtract to find the increase.
$1470.6 - 1380.2 = 90.4$
The prison population increased by 90.4 thousand, or 90,400.

37. $(x)^2 + (12)(x) + (2x)(x) = x^2 + 12x + 2x^2$
$ = (1 + 2)x^2 + 12x$
$ = 3x^2 + 12x$

Cumulative Review

39. $3y - 8x = 2$
$3y = 8x + 2$
$y = \dfrac{8x + 2}{3}$
$y = \dfrac{8}{3}x + \dfrac{2}{3}$
$m = \dfrac{8}{3}$; *y*-intercept: $\left(0, \dfrac{2}{3}\right)$

40. $\dfrac{5x}{7} - 4 > \dfrac{2x}{7} - 1$
$7\left(\dfrac{5x}{7} - 4\right) > 7\left(\dfrac{2x}{7} - 1\right)$
$5x - 28 > 2x - 7$
$3x - 28 > -7$
$3x > 21$
$x > 7$

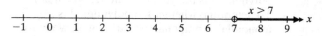

41. $-2(x - 5) + 6 = 2^2 - 9 + x$
$-2(x - 5) + 6 = 4 - 9 + x$
$-2x + 10 + 6 = 4 - 9 + x$
$-2x + 16 = -5 + x$
$-2x + 16 - x = -5 + x - x$
$-3x + 16 = -5$
$-3x + 16 - 16 = -5 - 16$
$-3x = -21$
$\dfrac{-3x}{-3} = \dfrac{-21}{-3}$
$x = 7$

42. $\dfrac{x}{6} + \dfrac{x}{2} = \dfrac{4}{3}$
$6\left(\dfrac{x}{6}\right) + 6\left(\dfrac{x}{2}\right) = 6\left(\dfrac{4}{3}\right)$
$x + 3x = 8$
$4x = 8$
$\dfrac{4x}{4} = \dfrac{8}{4}$
$x = 2$

Quick Quiz 5.3

1. $(3x^2 - 5x + 8) + (-7x^2 - 6x - 3)$
$= (3 - 7)x^2 + (-5 - 6)x + (8 - 3)$
$= -4x^2 - 11x + 5$

2. $(2x^2 - 3x - 7) - (-4x^2 + 6x + 9)$

$= (2x^2 - 3x - 7) + (4x^2 - 6x - 9)$

$= (2 + 4)x^2 + (-3 - 6)x + (-7 - 9)$

$= 6x^2 - 9x - 16$

3. $(5x - 3) - (2x - 4) + (-6x + 7)$

$= (5x - 3) + (-2x + 4) + (-6x + 7)$

$= (5 - 2 - 6)x + (-3 + 4 + 7)$

$= -3x + 8$

4. Answers may vary. Possible solution:

$2xy^2 - 5x^3y^4$

To determine the degree of the polynomial, first determine the degree of each term by finding the sum of the exponents on the variables in each term. The degree of the first term, $2xy^2$, is $1 + 2 = 3$, and the degree of the second term, $-5x^3y^4$, is $3 + 4 = 7$. The degree of the polynomial is the greater of these, which is 7. To determine whether the polynomial is a monomial, a binomial, or a trinomial, we must count the number of terms in the polynomial. The polynomial has two terms, $2xy^2$ and $-5x^3y^4$, so it is a binomial.

Use Math to Save Money

1. $\dfrac{\$450 + \$425 + \$460}{3} = \dfrac{\$1335}{3} = \$445$

Jenny's monthly average for grocery expenses is $445.

2. 6% of \$445 = $0.06 \times \$445 = \$26.70$

$\$445 - \$26.70 - \$30 = \$388.30$

In the fourth month, Jenny's grocery bill was $388.30.

3. 20% of \$388.30 = $0.20 \times \$388.30 = \$77.66$

$\$388.30 - \$77.66 = \$310.64$

In the fifth month, her grocery bill was $310.64.

4. $\$445 - \$310.64 = \$134.36$

$\dfrac{\$134.36}{\$445} \approx 0.302 = 30.2\%$

Jenny's savings in the fifth month were \$134.36 or 30.2%.

5. $\$134.36 \times 12 = \$1612.32$

They will have saved \$1612.32 at the end of one year.

How Am I Doing? Sections 5.1–5.3
(Available online through MyMathLab.)

1. $(8x^2y^3)(-3xy^2) = (8)(-3)(x^2 \cdot x)(y^3 \cdot y^2)$

$= -24x^{2+1}y^{3+2}$

$= -24x^3y^5$

2. $(-6a^3)(a^7)\left(\dfrac{1}{2}a\right) = -6 \cdot \dfrac{1}{2}a^{3+7+1} = -3a^{11}$

3. $-\dfrac{35xy^6}{25x^8y^3} = -\dfrac{7y^{6-3}}{5x^{8-1}} = -\dfrac{7y^3}{5x^7}$

4. $\dfrac{60x^7y^0}{15x^2y^9} = \dfrac{60x^7}{15x^2y^9} = \dfrac{4x^{7-2}}{y^9} = \dfrac{4x^5}{y^9}$

5. $(-3x^5y)^4 = (-3)^4(x^5)^4y^4 = 81x^{5 \cdot 4}y^4 = 81x^{20}y^4$

6. $\left(\dfrac{3x^2}{y}\right)^3 = \dfrac{3^3(x^2)^3}{y^3} = \dfrac{27x^6}{y^3}$

7. $(4x^{-3}y^4)^{-2} = (4)^{-2}(x^{-3})^{-2}(y^4)^{-2}$

$= (4)^{-2}x^6y^{-8}$

$= \dfrac{x^6}{(4)^2y^8}$

$= \dfrac{x^6}{16y^8}$

8. $\dfrac{4x^4y^{-3}}{12x^{-1}y^2} = \dfrac{x^4 \cdot x}{3y^2 \cdot y^3} = \dfrac{x^{4+1}}{3y^{2+3}} = \dfrac{x^5}{3y^5}$

9. Move the decimal point four places to the left.

$58,740 = 5.874 \times 10^4$

10. Move the decimal point five places to the right.

$0.00009362 = 9.362 \times 10^{-5}$

11. $(42,000,000)(1,500,000,000)$

$= (4.2 \times 10^7)(1.5 \times 10^9)$

$= (4.2 \times 1.5) \times 10^{7+9}$

$= 6.3 \times 10^{16}$

12. $(2x^2 + 0.5x - 2) + (0.3x^2 - 0.9x - 3.4)$
$= (2 + 0.3)x^2 + (0.5 - 0.9)x + (-2 - 3.4)$
$= 2.3x^2 - 0.4x - 5.4$

13. $(3x^2 + 7x - 10) - (-x^2 + 5x - 2)$
$= (3x^2 + 7x - 10) + (x^2 - 5x + 2)$
$= (3 + 1)x^2 + (7 - 5)x + (-10 + 2)$
$= 4x^2 + 2x - 8$

14. $\left(\dfrac{1}{2}x^3 + \dfrac{1}{4}x^2 - 2x\right) - \left(\dfrac{1}{3}x^3 - \dfrac{1}{8}x^2 - 5x\right)$
$= \left(\dfrac{1}{2}x^3 + \dfrac{1}{4}x^2 - 2x\right) + \left(-\dfrac{1}{3}x^3 + \dfrac{1}{8}x^2 + 5x\right)$
$= \left(\dfrac{1}{2} - \dfrac{1}{3}\right)x^3 + \left(\dfrac{1}{4} + \dfrac{1}{8}\right)x^2 + (-2 + 5)x$
$= \left(\dfrac{3}{6} - \dfrac{2}{6}\right)x^3 + \left(\dfrac{2}{8} + \dfrac{1}{8}\right)x^2 + 3x$
$= \dfrac{1}{6}x^3 + \dfrac{3}{8}x^2 + 3x$

15. $\left(\dfrac{1}{16}x^2 + \dfrac{1}{8}\right) + \left(\dfrac{1}{4}x^2 - \dfrac{3}{10}x - \dfrac{1}{2}\right)$
$= \left(\dfrac{1}{16} + \dfrac{1}{4}\right)x^2 - \dfrac{3}{10}x + \left(\dfrac{1}{8} - \dfrac{1}{2}\right)$
$= \left(\dfrac{1}{16} + \dfrac{4}{16}\right)x^2 - \dfrac{3}{10}x + \left(\dfrac{1}{8} - \dfrac{4}{8}\right)$
$= \dfrac{5}{16}x^2 - \dfrac{3}{10}x - \dfrac{3}{8}$

5.4 Exercises

1. $-2x(6x^3 - x) = -2x(6x^3) - 2x(-x)$
$= -12x^4 + 2x^2$

3. $4x^2(6x - 1) = 4x^2(6x) + 4x^2(-1) = 24x^3 - 4x^2$

5. $2x^3(-2x^3 + 5x - 1)$
$= 2x^3(-2x^3) + 2x^3(5x) + 2x^3(-1)$
$= -4x^6 + 10x^4 - 2x^3$

7. $\dfrac{1}{2}(2x + 3x^2 + 5x^3) = \dfrac{1}{2}(2x) + \dfrac{1}{2}(3x^2) + \dfrac{1}{2}(5x^3)$
$= x + \dfrac{3}{2}x^2 + \dfrac{5}{2}x^3$

9. $(2x^3 - 4x^2 + 5x)(-x^2y) = -2x^5y + 4x^4y - 5x^3y$

11. $(3x^3 + x^2 - 8x)(3xy) = 9x^4y + 3x^3y - 24x^2y$

13. $(x^3 - 3x^2 + 5x - 2)(3x) = 3x^4 - 9x^3 + 15x^2 - 6x$

15. $(x^2y^2 - 6xy + 8)(-2xy)$
$= -2x^3y^3 + 12x^2y^2 - 16xy$

17. $(-7x^3 + 3x^2 + 2x - 1)(4x^2y)$
$= -28x^5y + 12x^4y + 8x^3y - 4x^2y$

19. $(3d^4 - 4d^2 + 6)(-2c^2d)$
$= -6c^2d^5 + 8c^2d^3 - 12c^2d$

21. $6x^3(2x^4 - x^2 + 3x + 9)$
$= 12x^7 - 6x^5 + 18x^4 + 54x^3$

23. $-2x^3(8x^3 - 5x^2 + 6x) = -16x^6 + 10x^5 - 12x^4$

25. $(x + 5)(x + 7) = x^2 + 7x + 5x + 35 = x^2 + 12x + 35$

27. $(x + 6)(x + 2) = x^2 + 2x + 6x + 12 = x^2 + 8x + 12$

29. $(x - 8)(x + 2) = x^2 + 2x - 8x - 16 = x^2 - 6x - 16$

31. $(x - 5)(x - 4) = x^2 - 4x - 5x + 20 = x^2 - 9x + 20$

33. $(5x - 2)(-4x - 3) = -20x^2 - 15x + 8x + 6$
$= -20x^2 - 7x + 6$

35. $(2x - 5)(x + 3y) = 2x^2 + 6xy - 5x - 15y$

37. $(5x + 2)(3x - y) = 15x^2 - 5xy + 6x - 2y$

39. $(4y + 1)(5y - 3) = 20y^2 - 12y + 5y - 3$
$= 20y^2 - 7y - 3$

41. $(5x^2 + 4y^3)(2x^2 + 3y^3)$
$= 10x^4 + 15x^2y^3 + 8x^2y^3 + 12y^6$
$= 10x^4 + 23x^2y^3 + 12y^6$

43. The signs are incorrect.
The result should be: $(x - 2)(-3) = -3x + 6$

45. $(5x + 2)(5x + 2) = 25x^2 + \underline{20x} + 4$

47. $(4x-3y)(5x-2y) = 20x^2 - 8xy - 15xy + 6y^2$
$$= 20x^2 - 23xy + 6y^2$$

49. $(7x-2)^2 = (7x-2)(7x-2)$
$$= 49x^2 - 14x - 14x + 4$$
$$= 49x^2 - 28x + 4$$

51. $(4a+2b)^2 = (4a+2b)(4a+2b)$
$$= 16a^2 + 8ab + 8ab + 4b^2$$
$$= 16a^2 + 16ab + 4b^2$$

53. $(0.2x+3)(4x-0.3) = 0.8x^2 - 0.06x + 12x - 0.9$
$$= 0.8x^2 + 11.94x - 0.9$$

55. $\left(\dfrac{1}{2}x+\dfrac{1}{3}\right)\left(\dfrac{1}{2}x-\dfrac{1}{4}\right) = \dfrac{1}{4}x^2 - \dfrac{1}{8}x + \dfrac{1}{6}x - \dfrac{1}{12}$
$$= \dfrac{1}{4}x^2 - \dfrac{3}{24}x + \dfrac{4}{24}x - \dfrac{1}{12}$$
$$= \dfrac{1}{4}x^2 + \dfrac{1}{24}x - \dfrac{1}{12}$$

57. $(2x^2+4y^3)(3x^2+2y^3)$
$$= 6x^4 + 4x^2y^3 + 12x^2y^3 + 8y^6$$
$$= 6x^4 + 16x^2y^3 + 8y^6$$

59. $(2x-3)(5x+2) = 10x^2 + 4x - 15x - 6$
$$= 10x^2 - 11x - 6$$
The area is $(10x^2 - 11x - 6)$ square units.

Cumulative Review

61. $3(x-6) = -2(x+4) + 6x$
$3x - 18 = -2x - 8 + 6x$
$3x - 18 = 4x - 8$
$-18 = x - 8$
$-10 = x$

62. $3(w-7) - (4-w) = 11w$
$3w - 21 - 4 + w = 11w$
$4w - 25 = 11w$
$-25 = 7w$
$-\dfrac{25}{7} = w$ or $w = -3\dfrac{4}{7}$

63. x = number of \$10 bills
$x + 1$ = number of \$20 bills
$3x - 1$ = number of \$5 bills
$10x + 20(x+1) + 5(3x-1) = 375$
$10x + 20x + 20 + 15x - 5 = 375$
$45x + 15 = 375$
$45x = 360$
$x = 8$

$x + 1 = 8 + 1 = 9$
$3x - 1 = 3(8) - 1 = 23$
She had eight \$10's, nine \$20's, and twenty-three \$5's.

64. $x = 2000 - 2000 = 0$
$2.23x + 25 = 2.23(0) + 25 = 0 + 25 = 25$
The estimate is \$25 billion sent in 2000.

65. $x = 2005 - 2000 = 5$
$2.23x + 25 = 2.23(5) + 25 = 11.15 + 25 \approx 36.2$
The estimate in \$36.2 billion sent in 2005.

66. $x = 2015 - 2000 = 15$
$2.23x + 25 = 2.23(15) + 25 = 33.45 + 25 \approx 58.5$
The prediction is \$58.5 billion sent in 2015.

67. $x = 2019 - 2000 = 19$
$2.23x + 25 = 2.23(19) + 25 = 42.37 + 25 \approx 67.4$
The prediction is \$67.4 billion sent in 2019.

Quick Quiz 5.4

1. $(2x^2y^2 - 3xy + 4)(4xy^2)$
$$= 8x^3y^4 - 12x^2y^3 + 16xy^2$$

2. $(2x+3)(3x-5) = 6x^2 - 10x + 9x - 15$
$$= 6x^2 - x - 15$$

3. $(6a-4b)(2a-3b) = 12a^2 - 18ab - 8ab + 12b^2$
$$= 12a^2 - 26ab + 12b^2$$

4. Answers may vary. Possible solution:
$(7x-3)^2$
First write the square of the binomial as the product of the binomial and itself.
$(7x-3)(7x-3)$
Then use FOIL and collect like terms.
$49x^2 - 21x - 21x + 9 = 49x^2 - 42x + 9$

5.5 Exercises

1. In the special case of $(a + b)(a - b)$, a binomial times a binomial is a <u>binomial</u>.

3. $(4x - 7)^2 = 16x^2 - 56x + 49$

   The student left out the middle term which comes from the product of the two outer terms and the product of the two inner terms.

5. $(y - 7)(y + 7) = y^2 - 7^2 = y^2 - 49$

7. $(x - 9)(x + 9) = x^2 - 9^2 = x^2 - 81$

9. $(6x - 5)(6x + 5) = (6x)^2 - 5^2 = 36x^2 - 25$

11. $(2x - 7)(2x + 7) = (2x)^2 - (7)^2 = 4x^2 - 49$

13. $(5x - 3y)(5x + 3y) = (5x)^2 - (3y)^2 = 25x^2 - 9y^2$

15. $(0.6x + 3)(0.6x - 3) = (0.6x)^2 - 3^2 = 0.36x^2 - 9$

17. $(2y + 5)^2 = (2y)^2 + (2)(2y)(5) + (5)^2$
    $= 4y^2 + 20y + 25$

19. $(5x - 4)^2 = (5x)^2 - 2(5x)(4) + 4^2$
    $= 25x^2 - 40x + 16$

21. $(7x + 3)^2 = (7x)^2 + 2(7x)(3) + 3^2$
    $= 49x^2 + 42x + 9$

23. $(3x - 7)^2 = (3x)^2 - 2(3x)(7) + 7^2$
    $= 9x^2 - 42x + 49$

25. $\left(\dfrac{2}{3}x + \dfrac{1}{4}\right)^2 = \left(\dfrac{2}{3}x\right)^2 + 2\left(\dfrac{2}{3}x\right)\left(\dfrac{1}{4}\right) + \left(\dfrac{1}{4}\right)^2$
    $= \dfrac{4}{9}x^2 + \dfrac{1}{3}x + \dfrac{1}{16}$

27. $(9xy + 4z)^2 = (9xy)^2 + 2(9xy)(4z) + (4z)^2$
    $= 81x^2y^2 + 72xyz + 16z^2$

29. $(7x + 3y)(7x - 3y) = (7x)^2 - (3y)^2 = 49x^2 - 9y^2$

31. $(3c - 5d)^2 = (3c)^2 - (2)(3c)(5d) + (5d)^2$
    $= 9c^2 - 30cd + 25d^2$

33. $(9a-10b)(9a+10b) = (9a)^2 - (10b)^2$
$$= 81a^2 - 100b^2$$

35. $(5x+9y)^2 = (5x)^2 + 2(5x)(9y) + (9y)^2$
$$= 25x^2 + 90xy + 81y^2$$

37. $(x^2-x+5)(x-3) = (x^2-x+5)x + (x^2-x+5)(-3)$
$$= x^3 - x^2 + 5x - 3x^2 + 3x - 15$$
$$= x^3 - 4x^2 + 8x - 15$$

39. $(2x+1)(x^3+3x^2-x+4) = 2x(x^3+3x^2-x+4) + 1(x^3+3x^2-x+4)$
$$= 2x(x^3) + 2x(3x^2) + 2x(-x) + 2x(4) + x^3 + 3x^2 - x + 4$$
$$= 2x^4 + 6x^3 - 2x^2 + 8x + x^3 + 3x^2 - x + 4$$
$$= 2x^4 + 7x^3 + x^2 + 7x + 4$$

41. $(a^2-3a+2)(a^2+4a-3) = a^2(a^2+4a-3) - 3a(a^2+4a-3) + 2(a^2+4a-3)$
$$= a^4 + 4a^3 - 3a^2 - 3a^3 - 12a^2 + 9a + 2a^2 + 8a - 6$$
$$= a^4 + a^3 - 13a^2 + 17a - 6$$

43. $(x+3)(x-1)(3x-8) = (x^2-x+3x-3)(3x-8)$
$$= (x^2+2x-3)(3x) + (x^2+2x-3)(-8)$$
$$= 3x^3 + 6x^2 - 9x - 8x^2 - 16x + 24$$
$$= 3x^3 - 2x^2 - 25x + 24$$

45. $(2x-5)(x-1)(x+3) = (2x^2-2x-5x+5)(x+3)$
$$= (2x^2-7x+5)(x+3)$$
$$= (2x^2-7x+5)x + (2x^2-7x+5)3$$
$$= 2x^3 - 7x^2 + 5x + 6x^2 - 21x + 15$$
$$= 2x^3 - x^2 - 16x + 15$$

47. $(a-5)(2a+3)(a+5) = (2a^2+3a-10a-15)(a+5)$
$$= (2a^2-7a-15)(a+5)$$
$$= (2a^2-7a-15)a + (2a^2-7a-15)5$$
$$= 2a^3 - 7a^2 - 15a + 10a^2 - 35a - 75$$
$$= 2a^3 + 3a^2 - 50a - 75$$

49. $V = (2x+1)(3x-2)(4x+3)$
$$= (6x^2 - 4x + 3x - 2)(4x+3)$$
$$= (6x^2 - x - 2)(4x+3)$$
$$= (6x^2 - x - 2)(4x) + (6x^2 - x - 2)(3)$$
$$= 24x^3 - 4x^2 - 8x + 18x^2 - 3x - 6$$
$$= 24x^3 + 14x^2 - 11x - 6$$

Cumulative Review

51. Let $x =$ the first number, then
$2x + 3 =$ the second number.
$$x + 2x + 3 = 60$$
$$3x + 3 = 60$$
$$3x = 57$$
$$x = 19$$
$2x + 3 = 2(19) + 3 = 38 + 3 = 41$
The numbers are 19 and 41.

52. Let $x =$ length.
Then width $= 2 + \dfrac{1}{2}x$.
$$2x + 2\left(2 + \frac{1}{2}x\right) = 34$$
$$2x + 4 + x = 34$$
$$3x + 4 = 34$$
$$3x = 30$$
$$x = 10$$
$$2 + \frac{1}{2}x = 2 + \frac{1}{2}(10) = 7$$
The dimensions of the room are: width = 7 m, length = 10 m.

Quick Quiz 5.5

1. $(7x - 12y)(7x + 12y) = (7x)^2 - (12y)^2$
$$= 49x^2 - 144y^2$$

2. $(2x + 3)(x - 2)(3x + 1) = (2x^2 - 4x + 3x - 6)(3x + 1)$
$$= (2x^2 - x - 6)(3x + 1)$$
$$= (2x^2 - x - 6)(3x) + (2x^2 - x - 6)(1)$$
$$= 6x^3 - 3x^2 - 18x + 2x^2 - x - 6$$
$$= 6x^3 - x^2 - 19x - 6$$

3. $(3x - 2)(5x^3 - 2x^2 - 4x + 3) = (3x)(5x^3 - 2x^2 - 4x + 3) + (-2)(5x^3 - 2x^2 - 4x + 3)$
$$= 15x^4 - 6x^3 - 12x^2 + 9x - 10x^3 + 4x^2 + 8x - 6$$
$$= 15x^4 - 16x^3 - 8x^2 + 17x - 6$$

4. Answers may vary. Possible solution:
To use the formula $(a + b)^2 = a^2 + 2ab + b^2$ to multiply $(6x - 9y)^2$, first identify a and b: $a = 6x$ and $b = -9y$.
Then substitute these values for a and b in the formula and simplify.
$$(6x - 9y)^2 = (6x)^2 + 2(6x)(-9y) + (-9y)^2$$
$$= 36x^2 - 108xy + 81y^2$$

5.6 Exercises

1. $\dfrac{25x^4 - 15x^2 + 20x}{5x} = \dfrac{25x^4}{5x} - \dfrac{15x^2}{5x} + \dfrac{20x}{5x}$
$$= 5x^3 - 3x + 4$$

3. $\dfrac{3y^5 + 21y^3 - 9y^2}{3y^2} = \dfrac{3y^5}{3y^2} + \dfrac{21y^3}{3y^2} - \dfrac{9y^2}{3y^2}$
$$= y^3 + 7y - 3$$

5. $\dfrac{81x^7 - 36x^5 - 63x^3}{9x^3} = \dfrac{81x^7}{9x^3} - \dfrac{36x^5}{9x^3} - \dfrac{63x^3}{9x^3}$
$$= 9x^4 - 4x^2 - 7$$

7. $(48x^7 - 54x^4 + 36x^3) \div 6x^3$
$$= \dfrac{48x^7}{6x^3} - \dfrac{54x^4}{6x^3} + \dfrac{36x^3}{6x^3}$$
$$= 8x^4 - 9x + 6$$

9.
$$\begin{array}{r} 3x+5 \\ 2x+1\overline{\smash{\big)}\,6x^2+13x+5} \\ \underline{6x^2+\ 3x} \\ 10x+5 \\ \underline{10x+5} \\ 0 \end{array}$$

$\dfrac{6x^2 + 13x + 5}{2x+1} = 3x + 5$

Check: $(3x+5)(2x+1) = 6x^2 + 3x + 10x + 5$
$$= 6x^2 + 13x + 5$$

11.
$$\begin{array}{r} x-3 \\ x-5\overline{\smash{\big)}\,x^2-8x-17} \\ \underline{x^2-5x} \\ -3x-17 \\ \underline{-3x+15} \\ -32 \end{array}$$

$\dfrac{x^2 - 8x - 17}{x-5} = x - 3 - \dfrac{32}{x-5}$

Check:

$(x-5)\left(x - 3 - \dfrac{32}{x-5}\right) = x^2 - 3x - 5x + 15 - 32$
$$= x^2 - 8x - 17$$

13.
$$\begin{array}{r} 3x^2-4x+8 \\ x+1\overline{\smash{\big)}\,3x^3-\ x^2+4x-2} \\ \underline{3x^3+3x^2} \\ -4x^2+4x \\ \underline{-4x^2-4x} \\ 8x-2 \\ \underline{8x+8} \\ -10 \end{array}$$

$\dfrac{3x^3 - x^2 + 4x - 2}{x+1} = 3x^2 - 4x + 8 - \dfrac{10}{x+1}$

Check: $(x+1)\left(3x^2 - 4x + 8 - \dfrac{10}{x+1}\right)$
$$= 3x^3 - x^2 + 4x + 8 - 10$$
$$= 3x^3 - x^2 + 4x - 2$$

15.
$$\begin{array}{r} 2x^2-3x-2 \\ 2x+5\overline{\smash{\big)}\,4x^3+\ 4x^2-19x-15} \\ \underline{4x^3+10x^2} \\ -6x^2-19x \\ \underline{-6x^2-15x} \\ -4x-15 \\ \underline{-4x-10} \\ -5 \end{array}$$

$\dfrac{4x^3 + 4x^2 - 19x - 15}{2x+5} = 2x^2 - 3x - 2 - \dfrac{5}{2x+5}$

Check: $(2x+5)\left(2x^2 - 3x - 2 - \dfrac{5}{2x+5}\right)$
$$= 4x^3 + 4x^2 - 19x - 10 - 5$$
$$= 4x^3 + 4x^2 - 19x - 15$$

17.
$$\begin{array}{r} 2x^2+3x-1 \\ 5x-2\overline{\smash{\big)}\,10x^3+11x^2-11x+2} \\ \underline{10x^3-4x^2} \\ 15x^2-11x \\ \underline{15x^2-6x} \\ -5x+2 \\ \underline{-5x+2} \\ 0 \end{array}$$

$\dfrac{10x^3 + 11x^2 - 11x + 2}{5x-2} = 2x^2 + 3x - 1$

19.

$$
2x-3 \overline{\smash{\big)}\, \begin{array}{r} 2x^2+3x+6 \\ 4x^3+0x^2+3x\ +5 \end{array}}
$$

$$
\underline{4x^3-6x^2}
$$
$$
6x^2+3x
$$
$$
\underline{6x^2-9x}
$$
$$
12x+\ 5
$$
$$
\underline{12x-18}
$$
$$
23
$$

$$
\frac{4x^3+3x+5}{2x-3}=2x^2+3x+6+\frac{23}{2x-3}
$$

21.

$$
y+3 \overline{\smash{\big)}\, \begin{array}{r} y^2-4y-1 \\ y^3\ -y^2-13y-12 \end{array}}
$$

$$
\underline{y^3+3y^2}
$$
$$
-4y^2-13y
$$
$$
\underline{-4y^2-12y}
$$
$$
-y-12
$$
$$
\underline{-y-\ 3}
$$
$$
-9
$$

$$
(y^3-y^2-13y-12)\div(y+3)=y^2-4y-1-\frac{9}{y+3}
$$

23.

$$
y-2 \overline{\smash{\big)}\, \begin{array}{r} y^3+2y^2-5y-10 \\ y^4+0y^3-9y^2+\ 0y-5 \end{array}}
$$

$$
\underline{y^4-2y^3}
$$
$$
2y^3-9y^2
$$
$$
\underline{2y^3-4y^2}
$$
$$
-5y^2+\ 0y
$$
$$
\underline{-5y^2+10y}
$$
$$
-10y-\ 5
$$
$$
\underline{-10y+20}
$$
$$
-25
$$

$$
(y^4-9y^2-5)\div(y-2)
$$
$$
=y^3+2y^2-5y-10-\frac{25}{y-2}
$$

Cumulative Review

25. Decrease of 16% = 100% − 16% = 84% or 0.84
2010: $3.87(0.84) ≈ $3.25
The average price of a gallon of whole milk was
$3.25 in January 2010.
Increase of 16% = 100% + 16% = 116% or 1.16
2015: $3.25(1.16) = $3.77

The average price of a gallon of whole milk was
$3.77 in January 2015.

26. x = first page number
$x + 1$ = second page number
$$x+(x+1)=341$$
$$2x+1=341$$
$$2x=340$$
$$x=170$$
$x + 1 = 170 + 1 = 171$
The page numbers are 170 and 171.

27. a. $\dfrac{5+9+11+5}{4}=\dfrac{30}{4}=7.5$ hurricanes per year

b. $\dfrac{6+8+3+12}{4}=\dfrac{29}{4}\approx 7.3$ hurricanes per year

c. $\dfrac{7+10+2+6}{4}=\dfrac{25}{4}\approx 6.3$ hurricanes per year

d. $\dfrac{7.5-7.3}{7.5}\approx 0.027$ or 2.7%
There was a 2.7% decrease.

e. $\dfrac{7.3-6.3}{7.3}=\dfrac{1.0}{7.3}\approx 0.137$ or 1.37%
There was a 1.37% decrease.

Quick Quiz 5.6

1. $\dfrac{20x^5-64x^4-8x^3}{4x^2}=\dfrac{20x^5}{4x^2}-\dfrac{64x^4}{4x^2}-\dfrac{8x^3}{4x^2}$
$$=5x^3-16x^2-2x$$

2.

$$
2x+3 \overline{\smash{\big)}\, \begin{array}{r} 4x^2-5x-2 \\ 8x^3+\ 2x^2-19x-6 \end{array}}
$$

$$
\underline{8x^3+12x^2}
$$
$$
-10x^2-19x
$$
$$
\underline{-10x^2-15x}
$$
$$
-4x-6
$$
$$
\underline{-4x-6}
$$
$$
0
$$

$$(8x^3+2x^2-19x-6)\div(2x+3)=4x^2-5x-2$$

3.
$$x - 2 \overline{\smash{\big)}\, x^3 + 0x^2 + 4x - 3} \quad \overset{\displaystyle x^2 + 2x + 8}{}$$

$$\underline{x^3 - 2x^2}$$
$$2x^2 + 4x$$
$$\underline{2x^2 - 4x}$$
$$8x - 3$$
$$\underline{8x - 16}$$
$$13$$

$$(x^3 + 4x - 3) \div (x - 2) = x^2 + 2x + 8 + \frac{13}{x - 2}$$

4. Answers may vary. Possible solution:
Multiply the quotient and the divisor. Then add the remainder. You should get the original dividend.

$$(x - 2)(x^2 + 2x + 8) + 13$$
$$= x^3 + 2x^2 + 8x - 2x^2 - 4x - 16 + 13$$
$$= x^3 + 4x - 3$$

Yes, the answer checks.

Career Exploration Problems

1. First set:

$$140 \text{ transformants} \div \left[\frac{2.5 \times 10^2}{2.5 \times 10^6} \right]$$

$$= (1.4 \times 10^2) \times \left(\frac{2.5 \times 10^6}{2.5 \times 10^2} \right)$$

$$= (1.4 \times 10^2) \times (1 \times 10^4)$$

$$= 1.4 \times 10^6 \text{ transformants}$$

Second set:

$$164 \text{ transformants} \times \left[\frac{1}{5 \times 10^{-3}} \right]$$

$$= (1.64 \times 10^2) \times \left[\frac{1}{5 \times 10^{-3}} \right]$$

$$= \frac{1.64 \times 10^2}{5 \times 10^{-3}}$$

$$= 3.28 \times 10^4 \text{ transformants}$$

2. a. Cell density: $5.01 \times 10^7 = 50,100,000$

b. Absorption: $3.02 \times 10^{-3} = 0.00302$

c. Number of bacteria:
$$237 \times 10 \times 10^5 = 237 \times 10^6 = 237,000,000$$

3.
$$(0.5X + 0.5y)^2$$
$$= (0.5X + 0.5y)(0.5X + 0.5y)$$
$$= 0.25X^2 + 0.25Xy + 0.25Xy + 0.25y^2$$
$$= 0.25X^2 + 0.5Xy + 0.25y^2$$

You Try It

1. a. $2^9 \cdot 2^{14} = 2^{9+14} = 2^{23}$

b. $(-8a^3)(-2a^5) = (-8)(-2)a^{3+5} = 16a^8$

c. $(-ab^2)(3a^4b^2) = -3a^{1+4}b^{2+2} = -3a^5b^4$

2. a. $\dfrac{21x^5}{3x^2} = 7x^{5-2} = 7x^3$

b. $\dfrac{-3x}{9x^2} = -\dfrac{1}{3x^{2-1}} = -\dfrac{1}{3x}$

c. $\dfrac{14ab^7}{28a^3b} = \dfrac{b^{7-1}}{2a^{3-1}} = \dfrac{b^6}{2a^2}$

3. a. $9^0 = 1$

b. $m^0 = 1$

c. $\dfrac{a^5}{a^5} = a^{5-5} = a^0 = 1$

d. $6ab^0 = 6a(1) = 6a$

4. a. $(a^4)^5 = a^{4 \cdot 5} = a^{20}$

b. $(2n^3)^2 = 2^2(n^3)^2 = 4n^{3 \cdot 2} = 4n^6$

c. $\left(\dfrac{3x^3}{y} \right)^3 = \dfrac{3^3(x^3)^3}{y^3} = \dfrac{27x^{3 \cdot 3}}{y^3} = \dfrac{27x^9}{y^3}$

d. $(-5s^2t^5)^2 = (-5)^2(s^2)^2(t^5)^2$
$$= 25s^{2 \cdot 2}t^{5 \cdot 2}$$
$$= 25s^4t^{10}$$

e. $(-a^2b)^5 = (-1)^5(a^2)^5b^5$
$$= -1a^{2 \cdot 5}b^5$$
$$= -a^{10}b^5$$

5. a. $a^{-3} = \dfrac{1}{a^3}$

b. $\dfrac{1}{x^{-1}} = \dfrac{x}{1} = x$

c. $\dfrac{m^{-9}}{n^{-6}} = \dfrac{n^6}{m^9}$

d. $3^{-2} = \dfrac{1}{3^2} = \dfrac{1}{9}$

6. a. $386{,}400 = 3.864 \times 10^5$

b. $0.000052 = 5.2 \times 10^{-5}$

7. a. $(3.1 \times 10^6)(2.5 \times 10^4) = 3.1 \times 2.5 \times 10^6 \times 10^4$
$= 7.75 \times 10^{10}$

b. $\dfrac{3.8 \times 10^9}{1.25 \times 10^5} = \dfrac{3.8}{1.25} \times \dfrac{10^9}{10^5}$
$= 3.04 \times 10^4$

8. $(x^4 - 5x^3 + 2x^2) + (-7x^4 + x^3 - x^2)$
$= (x^4 - 7x^4) + (-5x^3 + x^3) + (2x^2 - x^2)$
$= -6x^4 - 4x^3 + x^2$

9. $(8 - x^2) - (5 + 2x^2) = (8 - x^2) + (-5 - 2x^2)$
$\qquad\qquad = (8 - 5) + (-x^2 - 2x^2)$
$\qquad\qquad = 3 - 3x^2$

10. a. $-2a(3a^2 - 5a + 1)$
$= -2a(3a^2) + (-2a)(-5a) + (-2a)(1)$
$= -6a^3 + 10a^2 - 2a$

b. $(-x^2 + 3xy - 3y^2)(4xy)$
$= -x^2(4xy) + 3xy(4xy) - 3y^2(4xy)$
$= -4x^3y + 12x^2y^2 - 12xy^3$

11. a. $(2a + 5b)(2a - 5b) = (2a)^2 - (5b)^2$
$\qquad\qquad\qquad = 4a^2 - 25b^2$

b. $(2a + 5b)^2 = (2a)^2 + 2(2a)(5b) + (5b)^2$
$\qquad\qquad = 4a^2 + 20ab + 25b^2$

c. $(2a - 5b)^2 = (2a)^2 - 2(2a)(5b) + (5b)^2$
$\qquad\qquad = 4a^2 - 20ab + 25b^2$

d. $(2a - b)(3a + 5b) = 6a^2 + 10ab - 3ab - 5b^2$
$\qquad\qquad\qquad = 6a^2 + 7ab - 5b^2$

12. a.
$$
\begin{array}{r}
6x^2 - 5x + 3 \\
2x + 1 \\
\hline
6x^2 - 5x + 3 \\
12x^3 - 10x^2 + 6x \\
\hline
12x^3 - 4x^2 + x + 3
\end{array}
$$

b. $(x - 5)(3x^2 - 2x + 1)$
$= 3x^3 - 2x^2 + x - 15x^2 + 10x - 5$
$= 3x^3 - 17x^2 + 11x - 5$

13. $(x + 5)(x - 1)(3x + 2) = (x^2 + 4x - 5)(3x + 2)$
$\qquad\qquad\qquad\qquad = 3x^3 + 14x^2 - 7x - 10$

14. $(18a^3 - 9a^2 + 3a) \div (3a) = \dfrac{18a^3}{3a} - \dfrac{9a^2}{3a} + \dfrac{3a}{3a}$
$= 6a^2 - 3a + 1$

15.
$$
\begin{array}{r}
x^2 + 5x - 1 \\
3x - 2 \overline{)\, 3x^3 + 13x^2 - 13x + 2} \\
\underline{3x^3 - 2x^2} \\
15x^2 - 13x \\
\underline{15x^2 - 10x} \\
-3x + 2 \\
\underline{-3x + 2} \\
0
\end{array}
$$

$\dfrac{3x^3 + 13x^2 - 13x + 2}{3x - 2} = x^2 + 5x - 1$

Chapter 5 Review Problems

1. $(-6a^2)(3a^5) = (-6)(3)(a^2 \cdot a^5)$
$\qquad\qquad = -18a^{2+5}$
$\qquad\qquad = -18a^7$

2. $(5^{10})(5^{13}) = 5^{10+13} = 5^{23}$

3. $(3xy^2)(2x^3y^4) = (3 \cdot 2)(x \cdot x^3)(y^2 \cdot y^4)$
$$= 6x^{1+3}y^{2+4}$$
$$= 6x^4y^6$$

4. $(2x^3y^4)(-7xy^5) = (2)(-7)(x^3 \cdot x)(y^4 \cdot y^5)$
$$= -14x^{3+1}y^{4+5}$$
$$= -14x^4y^9$$

5. $\dfrac{7^{15}}{7^{27}} = \dfrac{1}{7^{27-15}} = \dfrac{1}{7^{12}}$

6. $\dfrac{x^{12}}{x^{17}} = \dfrac{1}{x^{17-12}} = \dfrac{1}{x^5}$

7. $\dfrac{y^{30}}{y^{16}} = y^{30-16} = y^{14}$

8. $\dfrac{9^{13}}{9^{24}} = \dfrac{1}{9^{24-13}} = \dfrac{1}{9^{11}}$

9. $\dfrac{-15xy^2}{25x^6y^6} = -\dfrac{3}{5x^{6-1}y^{6-2}} = -\dfrac{3}{5x^5y^4}$

10. $\dfrac{-12a^3b^6}{18a^2b^{12}} = -\dfrac{2a^{3-2}}{3b^{12-6}} = -\dfrac{2a}{3b^6}$

11. $(x^3)^8 = x^{3 \cdot 8} = x^{24}$

12. $\dfrac{(2b^2)^4}{(5b^3)^6} = \dfrac{2^4b^8}{5^6b^{18}} = \dfrac{2^4}{5^6b^{18-8}} = \dfrac{2^4}{5^6b^{10}}$

13. $(-3a^3b^2)^2 = (-3)^2(a^3)^2(b^2)^2$
$$= 9a^{3 \cdot 2}b^{2 \cdot 2}$$
$$= 9a^6b^4$$

14. $(3x^3y)^4 = 3^4(x^3)^4y^4 = 81x^{3 \cdot 4}y^4 = 81x^{12}y^4$

15. $\left(\dfrac{5ab^2}{c^3}\right)^2 = \dfrac{5^2a^2(b^2)^2}{(c^3)^2} = \dfrac{25a^2b^4}{c^6}$

16. $\left(\dfrac{x^0y^3}{4w^5z^2}\right)^3 = \dfrac{(y^3)^3}{4^3(w^5)^3(z^2)^3} = \dfrac{y^9}{64w^{15}z^6}$

17. $a^{-3}b^5 = \dfrac{b^5}{a^3}$

18. $m^8p^{-5} = \dfrac{m^8}{p^5}$

19. $\dfrac{2x^{-6}}{y^{-3}} = \dfrac{2y^3}{x^6}$

20. $(2x^{-5}y)^{-3} = 2^{-3}(x^{-5})^{-3}y^{-3}$
$$= 2^{-3}x^{15}y^{-3}$$
$$= \dfrac{x^{15}}{2^3y^3}$$
$$= \dfrac{x^{15}}{8y^3}$$

21. $(6a^4b^5)^{-2} = 6^{-2}(a^4)^{-2}(b^5)^{-2}$
$$= 6^{-2}a^{-8}b^{-10}$$
$$= \dfrac{1}{6^2a^8b^{10}}$$
$$= \dfrac{1}{36a^8b^{10}}$$

22. $\dfrac{3x^{-3}}{y^{-2}} = \dfrac{3y^2}{x^3}$

23. $\dfrac{4x^{-5}y^{-6}}{w^{-2}z^8} = \dfrac{4w^2}{x^5y^6z^8}$

24. $\dfrac{3^{-3}a^{-2}b^5}{c^{-3}d^{-4}} = \dfrac{b^5c^3d^4}{3^3a^2} = \dfrac{b^5c^3d^4}{27a^2}$

25. $156{,}340{,}200{,}000 = 1.563402 \times 10^{11}$

26. $179{,}632 = 1.79632 \times 10^5$

27. $0.00092 = 9.2 \times 10^{-4}$

28. $0.00000174 = 1.74 \times 10^{-6}$

29. $1.2 \times 10^5 = 120{,}000$

30. $6.034 \times 10^6 = 6{,}034{,}000$

31. $2.5 \times 10^{-1} = 0.25$

32. $4.32 \times 10^{-5} = 0.0000432$

33. $\dfrac{(28,0000,000)(5,000,000,000)}{7000}$

$= \dfrac{(2.8 \times 10^{7})(5 \times 10^{9})}{7 \times 10^{3}}$

$= \dfrac{14 \times 10^{16}}{7 \times 10^{3}}$

$= 2.0 \times 10^{13}$

34. $(3.12 \times 10^{5})(2.0 \times 10^{6})(1.5 \times 10^{8})$

$= 9.36 \times 10^{5+6+8}$

$= 9.36 \times 10^{19}$

35. $\dfrac{(0.00078)(0.000005)(0.00004)}{0.002}$

$= \dfrac{(7.8 \times 10^{-4})(5.0 \times 10^{-6})(4.0 \times 10^{-5})}{2.0 \times 10^{-3}}$

$= \dfrac{156 \times 10^{-15}}{2.0 \times 10^{-3}}$

$= 78 \times 10^{-12}$

$= 7.8 \times 10^{-11}$

36. $(3.5 \times 10^{9}) \times (0.20) = (3.5 \times 10^{9}) \times (2 \times 10^{-1})$

$= (3.5 \times 2) \times 10^{9+(-1)}$

$= 7 \times 10^{8}$

The total cost is $\$7 \times 10^{8}$.

37. Seconds in one day $= 60 \times 60 \times 24$

$= 86,400$

$= 8.64 \times 10^{4}$

Cycles of radiation in one second

$= 9,192,631,770$

$= 9.19 \times 10^{9}$

Cycles of radiation in one day

$= (8.64 \times 10^{4})(9.19 \times 10^{9})$

$= 79.4016 \times 10^{13}$

$= 7.94 \times 10^{14}$

In one day, there are 7.94×10^{14} cycles.

38. $\dfrac{60}{1 \times 10^{-11}} = 60 \times 10^{11} = 6 \times 10^{12}$

In one minute the computer can perform 6×10^{12} operations.

39. $(2.8x^{2} - 1.5x + 3.4) + (2.7x^{2} + 0.5x - 5.7)$

$= (2.8 + 2.7)x^{2} + (-1.5 + 0.5)x + (3.4 - 5.7)$

$= 5.5x^{2} - x - 2.3$

40. $(4x^{3} - x^{2} - x + 3) - (-3x^{3} + 2x^{2} + 5x - 1)$

$= (4x^{3} - x^{2} - x + 3) + (3x^{3} - 2x^{2} - 5x + 1)$

$= (4 + 3)x^{3} + (-1 - 2)x^{2} + (-1 - 5)x + (3 + 1)$

$= 7x^{3} - 3x^{2} - 6x + 4$

41. $\left(\dfrac{3}{5}x^{2}y - \dfrac{1}{3}x + \dfrac{3}{4} \right) - \left(\dfrac{1}{2}x^{2}y + \dfrac{2}{7}x + \dfrac{1}{3} \right)$

$= \left(\dfrac{3}{5}x^{2}y - \dfrac{1}{3}x + \dfrac{3}{4} \right) + \left(-\dfrac{1}{2}x^{2}y - \dfrac{2}{7}x - \dfrac{1}{3} \right)$

$= \left(\dfrac{3}{5} - \dfrac{1}{2} \right)x^{2}y + \left(-\dfrac{1}{3} - \dfrac{2}{7} \right)x + \left(\dfrac{3}{4} - \dfrac{1}{3} \right)$

$= \dfrac{1}{10}x^{2}y - \dfrac{13}{21}x + \dfrac{5}{12}$

42. $\dfrac{1}{2}x^{2} - \dfrac{3}{4}x + \dfrac{1}{5} - \left(\dfrac{1}{4}x^{2} - \dfrac{1}{2}x + \dfrac{1}{10} \right)$

$= \left(\dfrac{1}{2}x^{2} - \dfrac{3}{4}x + \dfrac{1}{5} \right) + \left(-\dfrac{1}{4}x^{2} + \dfrac{1}{2}x - \dfrac{1}{10} \right)$

$= \left(\dfrac{1}{2} - \dfrac{1}{4} \right)x^{2} + \left(-\dfrac{3}{4} + \dfrac{1}{2} \right)x + \left(\dfrac{1}{5} - \dfrac{1}{10} \right)$

$= \dfrac{1}{4}x^{2} - \dfrac{1}{4}x + \dfrac{1}{10}$

43. $(x^{2} - 9) - (4x^{2} + 5x) + (5x - 6)$

$= (x^{2} - 9) + (-4x^{2} - 5x) + (5x - 6)$

$= (1 - 4)x^{2} + (-5 + 5)x + (-9 - 6)$

$= -3x^{2} + 0x - 15$

$= -3x^{2} - 15$

44. $(3x + 1)(5x - 1) = 15x^{2} - 3x + 5x - 1$

$= 15x^{2} + 2x - 1$

45. $(7x - 2)(4x - 3) = 28x^{2} - 21x - 8x + 6$

$= 28x^{2} - 29x + 6$

46. $(2x+3)(10x+9) = 20x^2 + 18x + 30x + 27$
$$= 20x^2 + 48x + 27$$

47. $5x(2x^2 - 6x + 3) = 10x^3 - 30x^2 + 15x$

48. $(xy^2 + 5xy - 6)(-4xy^2)$
$$= xy^2(-4xy^2) + 5xy(-4xy^2) - 6(-4xy^2)$$
$$= -4x^2y^4 - 20x^2y^3 + 24xy^2$$

49. $(5a + 7b)(a - 3b) = 5a^2 - 15ab + 7ab - 21b^2$
$$= 5a^2 - 8ab - 21b^2$$

50. $(2x^2 - 3)(4x^2 - 5y) = 8x^4 - 10x^2y - 12x^2 + 15y$

51. $(4x + 3)^2 = (4x)^2 + (2)(4x)(3) + (3)^2$
$$= 16x^2 + 24x + 9$$

52. $(a + 5b)(a - 5b) = (a)^2 - (5b)^2 = a^2 - 25b^2$

53. $(7x + 6y)(7x - 6y) = (7x)^2 - (6y)^2$
$$= 49x^2 - 36y^2$$

54. $(5a - 2b)^2 = (5a)^2 - 2(5a)(2b) + (2b)^2$
$$= 25a^2 - 20ab + 4b^2$$

55. $(x^2 + 7x + 3)(4x - 1)$
$$= 4x^3 + 28x^2 + 12x - x^2 - 7x - 3$$
$$= 4x^3 + 27x^2 + 5x - 3$$

56. $(x - 6)(2x - 3)(x + 4)$
$$= (2x^2 - 15x + 18)(x + 4)$$
$$= 2x^3 - 15x^2 + 18x + 8x^2 - 60x + 72$$
$$= 2x^3 - 7x^2 - 42x + 72$$

57. $(12y^3 + 18y^2 + 24y) \div (6y) = \dfrac{12y^3}{6y} + \dfrac{18y^2}{6y} + \dfrac{24y}{6y}$
$$= 2y^2 + 3y + 4$$

58. $(30x^5 + 35x^4 - 90x^3) \div (5x^2)$
$$= \dfrac{30x^5}{5x^2} + \dfrac{35x^4}{5x^2} - \dfrac{90x^3}{5x^2}$$
$$= 6x^3 + 7x^2 - 18x$$

59. $(16x^3 - 24x^2 + 32x) \div (4x)$
$$= \dfrac{16x^3}{4x} - \dfrac{24x^2}{4x} + \dfrac{32x}{4x}$$
$$= 4x^2 - 6x + 8$$

60.
$$\begin{array}{r} 3x - 2 \\ 5x+7\overline{)15x^2 + 11x - 14} \\ \underline{15x^2 + 21x} \\ -10x - 14 \\ \underline{-10x - 14} \\ 0 \end{array}$$

$$\dfrac{15x^2 + 11x - 14}{5x + 7} = 3x - 2$$

61.
$$\begin{array}{r} 3x - 7 \\ 4x+9\overline{)12x^2 -\ x - 63} \\ \underline{12x^2 + 27x} \\ -28x - 63 \\ \underline{-28x - 63} \\ 0 \end{array}$$

$$\dfrac{12x^2 - x - 63}{4x + 9} = 3x - 7$$

62.
$$\begin{array}{r} 2x^2 - 5x + 13 \\ x+2\overline{)2x^3 -\ x^2 +\ 3x -\ 1} \\ \underline{2x^3 + 4x^2} \\ -5x^2 +\ 3x \\ \underline{-5x^2 - 10x} \\ 13x -\ 1 \\ \underline{13x + 26} \\ -27 \end{array}$$

$$\dfrac{2x^3 - x^2 + 3x - 1}{x + 2} = 2x^2 - 5x + 13 - \dfrac{27}{x + 2}$$

63.
$$\begin{array}{r} 3x - 4 \\ 2x+3\overline{)6x^2 +\ x\ -9} \\ \underline{6x^2 + 9x} \\ -8x -\ 9 \\ \underline{-8x - 12} \\ 3 \end{array}$$

$$(6x^2 + x - 9) \div (2x + 3) = 3x - 4 + \dfrac{3}{2x + 3}$$

64.

$$x-3\overline{\smash{\big)}\,x^3+0x^2-x-24}$$

with quotient x^2+3x+8

$$\underline{x^3-3x^2}$$
$$3x^2-x$$
$$\underline{3x^2-9x}$$
$$8x-24$$
$$\underline{8x-24}$$
$$0$$

$(x^3-x-24)\div(x-3)=x^2+3x+8$

65.

$$x-2\overline{\smash{\big)}\,2x^3+0x^2-3x+1}$$

with quotient $2x^2+4x+5$

$$\underline{2x^3-4x^2}$$
$$4x^2-3x$$
$$\underline{4x^2-8x}$$
$$5x+1$$
$$\underline{5x-10}$$
$$11$$

$(2x^3-3x+1)\div(x-2)=2x^2+4x+5+\dfrac{11}{x-2}$

66. $\dfrac{4.5\times10^9}{3.1\times10^8}=\dfrac{4.5}{3.1}\times10^{9-8}$

$$\approx1.452\times10^1$$
$$=14.52$$

The United States spent about \$14.52 per person.

67. $1.25\times10^9+5.97\times10^8=1.25\times10^9+0.597\times10^9$

$$=(1.25+0.597)\times10^9$$
$$=1.847\times10^9$$

The total population is projected to be 1.847×10^9 people.

68. $(9.11\times10^{-28})(30,000)=(9.11\times10^{-28})(3\times10^4)$

$$=27.33\times10^{-24}$$
$$=2.733\times10^{-23}$$

The mass is 2.733×10^{-23} gram.

69. $(3.4\times10^5)\times140=(3.4\times10^5)\times(1.4\times10^2)$

$$=(3.4\times1.4)\times10^{5+2}$$
$$=4.76\times10^7$$

A gray whale will consume a total of 4.76×10^7 pounds of food.

70. $A=2x(2y+1)-xy=4xy+2x-xy=3xy+2x$

71. $A=2x(x)-4y(y)=2x^2-4y^2$

How Am I Doing? Chapter 5 Test

1. $(3^{10})(3^{24})=3^{10+24}=3^{34}$

2. $\dfrac{25^{18}}{25^{34}}=\dfrac{1}{25^{34-18}}=\dfrac{1}{25^{16}}$

3. $(8^4)^6=8^{4\cdot6}=8^{24}$

4. $(-3xy^4)(-4x^3y^6)=(-3)(-4)(x\cdot x^3)(y^4\cdot y^6)$

$$=12x^{1+3}y^{4+6}$$
$$=12x^4y^{10}$$

5. $\dfrac{-35x^8y^{10}}{25x^5y^{10}}=-\dfrac{7x^{8-5}}{5y^{10-10}}=-\dfrac{7x^3}{5y^0}=-\dfrac{7x^3}{5}$

6. $(-5xy^6)^3=(-5)^3x^3(y^6)^3=-125x^3y^{18}$

7. $\left(\dfrac{7a^7b^2}{3c^0}\right)^2=\left(\dfrac{7a^7b^2}{3}\right)^2$

$$=\dfrac{7^2(a^7)^2(b^2)^2}{3^2}$$
$$=\dfrac{49a^{14}b^4}{9}$$

8. $\dfrac{(3x^2)^3}{(6x)^2}=\dfrac{3^3(x^2)^3}{6^2x^2}=\dfrac{27x^6}{36x^2}=\dfrac{3x^{6-2}}{4}=\dfrac{3x^4}{4}$

9. $4^{-3}=\dfrac{1}{4^3}=\dfrac{1}{64}$

10. $6a^{-4}b^{-3}c^5=\dfrac{6c^5}{a^4b^3}$

11. $\dfrac{3x^{-3}y^2}{x^{-4}y^{-5}}=\dfrac{3x^4y^2y^5}{x^3}=3x^{4-3}y^{2+5}=3xy^7$

12. $0.0005482=5.482\times10^{-4}$

13. $5.82\times10^8=582,000,000$

14. $(4.0 \times 10^{-3})(3.0 \times 10^{-8})(2.0 \times 10^4) = (4.0 \times 3.0 \times 2.0) \times 10^{-3-8+4}$
$$= 24.0 \times 10^{-7}$$
$$= 2.4 \times 10^{-6}$$

15. $(2x^2 - 3x - 6) + (-4x^2 + 8x + 6) = (2-4)x^2 + (-3+8)x + 6 - 6$
$$= -2x^2 + 5x$$

16. $(3x^3 - 4x^2 + 3) - (14x^3 - 7x + 11) = (3x^3 - 4x^2 + 3) + (-14x^3 + 7x - 11)$
$$= (3-14)x^3 - 4x^2 + 7x + (3-11)$$
$$= -11x^3 - 4x^2 + 7x - 8$$

17. $-7x^2(3x^3 - 4x^2 + 6x - 2) = -7x^2(3x^3) + (-7x^2)(-4x^2) + (-7x^2)(6x) + (-7x^2)(-2)$
$$= -21x^5 + 28x^4 - 42x^3 + 14x^2$$

18. $(5x^2y^2 - 6xy + 2)(3x^2y) = 5x^2y^2(3x^2y) - 6xy(3x^2y) + 2(3x^2y)$
$$= 15x^4y^3 - 18x^3y^2 + 6x^2y$$

19. $(5a - 4b)(2a + 3b) = 10a^2 + 15ab - 8ab - 12b^2$
$$= 10a^2 + 7ab - 12b^2$$

20. $(3x + 2)(2x + 1)(x - 3) = (6x^2 + 7x + 2)(x - 3)$
$$= 6x^3 + 7x^2 + 2x - 18x^2 - 21x - 6$$
$$= 6x^3 - 11x^2 - 19x - 6$$

21. $(7x^2 + 2y^2)^2 = (7x^2)^2 + 2(7x^2)(2y^2) + (2y^2)^2$
$$= 49x^4 + 28x^2y^2 + 4y^4$$

22. $(5s - 11t)(5s + 11t) = (5s)^2 - (11t)^2$
$$= 25s^2 - 121t^2$$

23. $(3x - 2)(4x^3 - 2x^2 + 7x - 5) = 3x(4x^3 - 2x^2 + 7x - 5) - 2(4x^3 - 2x^2 + 7x - 5)$
$$= 12x^4 - 6x^3 + 21x^2 - 15x - 8x^3 + 4x^2 - 14x + 10$$
$$= 12x^4 - 14x^3 + 25x^2 - 29x + 10$$

24. $(3x^2 - 5xy)(x^2 + 3xy) = 3x^4 + 9x^3y - 5x^3y - 15x^2y^2$
$$= 3x^4 + 4x^3y - 15x^2y^2$$

25. $\dfrac{15x^6 - 5x^4 + 25x^3}{5x^3} = \dfrac{15x^6}{5x^3} - \dfrac{5x^4}{5x^3} + \dfrac{25x^3}{5x^3}$
$$= 3x^3 - x + 5$$

26.

$$
\require{enclose}
\begin{array}{r}
2x^2 - 7x + 4 \\
4x+3 \enclose{longdiv}{8x^3 - 22x^2 - 5x + 12} \\
\underline{8x^3 + 6x^2 } \\
-28x^2 - 5x \\
\underline{-28x^2 - 21x} \\
16x + 12 \\
\underline{16x + 12} \\
0
\end{array}
$$

$$\frac{8x^3 - 22x^2 - 5x + 12}{4x+3} = 2x^2 - 7x + 4$$

27.

$$
\require{enclose}
\begin{array}{r}
2x^2 + 6x + 12 \\
x-3 \enclose{longdiv}{2x^3 + 0x^2 - 6x - 36} \\
\underline{2x^3 - 6x^2} \\
6x^2 - 6x \\
\underline{6x^2 - 18x} \\
12x - 36 \\
\underline{12x - 36} \\
0
\end{array}
$$

$$\frac{2x^3 - 6x - 36}{x-3} = 2x^2 + 6x + 12$$

28. $\dfrac{2.6 \times 10^{11}}{69} \approx 0.0377 \times 10^{11} = 3.77 \times 10^{9}$

They would pump 3.77×10^{9} barrels per year.

29. $d = rt$

$= 2.49 \times 10^{4}(24)(7)$

$= 418 \times 10^{4}$

$= 4.18 \times 10^{6}$

In one week, the space probe would travel 4.18×10^{6} miles.

Chapter 6

1. In the expression $3x^2 \cdot 5x^3$, $3x^2$ and $5x^3$ are called <u>factors</u>.

3. The factoring is not complete because $6a^3 + 3a^2 - 9a$ contains a common factor of $3a$.

5. $8a^2 + 8a = 8a(a+1)$
   Check: $8a(a+1) = 8a^2 + 8a$

7. $21ab - 14ab^2 = 7ab(3 - 2b)$
   Check: $7ab(3-2b) = 21ab - 14ab^2$

9. $2\pi rh + 2\pi r^2 = 2\pi r(h + r)$
   Check: $2\pi r(h+r) = 2\pi rh + 2\pi r^2$

11. $5x^3 + 25x^2 - 15x = 5x(x^2 + 5x - 3)$
    Check: $5x(x^2 + 5x - 3) = 5x^3 + 25x^2 - 15x$

13. $12ab - 28bc + 20ac = 4(3ab - 7bc + 5ac)$
    Check:
    $4(3ab - 7bc + 5ac) = 12ab - 28bc + 20ac$

15. $16x^5 + 24x^3 - 32x^2 = 8x^2(2x^3 + 3x - 4)$
    Check: $8x^2(2x^3 + 3x - 4) = 16x^5 + 24x^3 - 32x^2$

17. $14x^2y - 35xy - 63x = 7x(2xy - 5y - 9)$
    Check: $7x(2xy - 5y - 9) = 14x^2y - 35xy - 63x$

19. $54x^2 - 45xy + 18x = 9x(6x - 5y + 2)$
    Check: $9x(6x - 5y + 2) = 54x^2 - 45xy + 18x$

21. $3xy^2 - 2ay + 5xy - 2y = y(3xy - 2a + 5x - 2)$
    Check:
    $y(3xy - 2a + 5x - 2) = 3xy^2 - 2ay + 5xy - 2y$

23. $24x^2y - 40xy^2 = 8xy(3x - 5y)$
    Check: $8xy(3x - 5y) = 24x^2y - 40xy^2$

25. $7x^3y^2 + 21x^2y^2 = 7x^2y^2(x+3)$
    Check: $7x^2y^2(x+3) = 7x^3y^2 + 21x^2y^2$

27. $16x^4y^2 - 24x^2y^2 - 8x^2y$
    $= 8x^2y(2x^2y - 3y - 1)$
    Check: $8x^2y(2x^2y - 3y - 1)$
    $\quad = 16x^4y^2 - 24x^2y^2 - 8x^2y$

29. $7a(x + 2y) - b(x + 2y) = (x + 2y)(7a - b)$

31. $3x(x - 4) - 2(x - 4) = (x - 4)(3x - 2)$

33. $6b(2a - 3c) - 5d(2a - 3c) = (2a - 3c)(6b - 5d)$

35. $7c(b - a^2) - 5d(b - a^2) + 2f(b - a^2)$
    $= (b - a^2)(7c - 5d + 2f)$

37. $3a(ab - 4) - 5(ab - 4) - b(ab - 4)$
    $= (ab - 4)(3a - 5 - b)$

39. $4a^3(a - 3b) + (a - 3b) = (a - 3b)(4a^3 + 1)$

41. $(a + 2) - x(a + 2) = (a + 2)(1 - x)$

43. The circumferences are $2\pi x$, $2\pi y$, and $2\pi z$.
    $2\pi x + 2\pi y + 2\pi z = 2\pi(x + y + z)$

Cumulative Review

45. 26% of $6,400,000 = 0.26(6,400,000)$
    $\qquad\qquad\qquad = 1,664,000$
    1,664,000 metric tons were produced in Vietnam.

46. 43% of $6,400,000 = 0.43(6,400,000)$
    $\qquad\qquad\qquad = 2,752,000$
    2,752,000 metric tons were produced in Brazil.

47. $\left(\dfrac{1,664,000 \text{ metric tons}}{89,700,000 \text{ people}} \right) \left(\dfrac{2205 \text{ lb}}{1 \text{ metric ton}} \right) \approx 41$
    About 41 pounds per person were produced in Vietnam.

48. $\left(\dfrac{2,752,000 \text{ metric tons}}{200,400,000 \text{ people}} \right) \left(\dfrac{2205 \text{ lb}}{1 \text{ metric ton}} \right) \approx 30$
    About 30 pounds per person were produced in Brazil.

Quick Quiz 6.1

1. $3x - 4x^2 + 2xy = x(3 - 4x + 2y)$

2. $20x^3 - 25x^2 - 5x = 5x(4x^2 - 5x - 1)$

3. $8a(a + 3b) - 7b(a + 3b) = (a + 3b)(8a - 7b)$

4. Answers may vary. Possible solution:
   Determine that the largest integer that will divide into the coefficients of all terms is 36.
   Determine that the variables common to all terms are a^2 and b^2.
   Write the above common factors as the first part of the answer (the first factor).
   Remove common factors, and what remains is the second part of the answer (the second factor).

6.2 Exercises

1. We must remove a common factor of 5 from the last two terms. This will give us:
$$3x^2 - 6xy + 5x - 10y = 3x(x - 2y) + 5(x - 2y)$$
$$= (x - 2y)(3x + 5)$$

3. $ab - 4a + 6b - 24 = a(b - 4) + 6(b - 4)$
$$= (b - 4)(a + 6)$$
Check: $(b - 4)(a + 6) = ab - 4a + 6b - 24$

5. $x^3 - 4x^2 + 3x - 12 = x^2(x - 4) + 3(x - 4)$
$$= (x - 4)(x^2 + 3)$$
Check: $(x - 4)(x^2 + 3) = x^3 + 3x - 4x^2 - 12$
$$= x^3 - 4x^2 + 3x - 12$$

7. $2ax + 6bx - ay - 3by = 2x(a + 3b) - y(a + 3b)$
$$= (a + 3b)(2x - y)$$
Check: $(a + 3b)(2x - y) = 2ax - ay + 6bx - 3by$
$$= 2ax + 6bx - ay - 3by$$

9. $3ax + bx - 6a - 2b = x(3a + b) - 2(3a + b)$
$$= (3a + b)(x - 2)$$
Check: $(3a + b)(x - 2) = 3ax - 6a + bx - 2b$
$$= 3ax + bx - 6a - 2b$$

11. $5a + 12bc + 10b + 6ac = 5a + 10b + 6ac + 12bc$
$$= 5(a + 2b) + 6c(a + 2b)$$
$$= (a + 2b)(5 + 6c)$$
Check: $(a + 2b)(5 + 6c) = 5a + 6ac + 10b + 12bc$
$$= 6a + 12bc + 10b + 6ac$$

13. $6c - 12d + cx - 2dx = 6(c - 2d) + x(c - 2d)$
$$= (c - 2d)(6 + x)$$
Check: $(c - 2d)(6 + x) = 6c + cx - 12d - 2dx$
$$= 6c - 12d + cx - 2dx$$

15. $y^2 - 2y - 3y + 6 = y(y - 2) - 3(y - 2)$
$$= (y - 2)(y - 3)$$
Check: $(y - 2)(y - 3) = y^2 - 2y - 3y + 6$

17. $54 - 6y + 9y - y^2 = 6(9 - y) + y(9 - y)$
$$= (9 - y)(6 + y)$$
Check: $(9 - y)(6 + y) = 54 + 9y - 6y - y^2$
$$= 54 - 6y + 9y - y^2$$

19. $6ax - y + 2ay - 3x = 6ax - 3x + 2ay - y$
$$= 3x(2a - 1) + y(2a - 1)$$
$$= (2a - 1)(3x + y)$$
Check: $(2a - 1)(3x + y) = 6ax - y + 2ay - 3x$

21. $2x^2 + 8x - 3x - 12 = 2x(x + 4) - 3(x + 4)$
$$= (x + 4)(2x - 3)$$
Check: $(x + 4)(2x - 3) = 2x^2 + 8x - 3x - 12$

23. $t^3 - t^2 + t - 1 = t^2(t - 1) + (t - 1) = (t - 1)(t^2 + 1)$
Check: $(t - 1)(t^2 + 1) = t^3 - t^2 + t - 1$

25. $6x^2 + 15xy^2 + 8xw + 20y^2w$
$$= 3x(2x + 5y^2) + 4w(2x + 5y^2)$$
$$= (2x + 5y^2)(3x + 4w)$$
Check: $(2x + 5y^2)(3x + 4w)$
$$= 6x^2 + 15xy^2 + 8xw + 20y^2w$$

27. Rearrange the terms so that factoring gives the same expression in parentheses.
$$6a^2 - 12bd - 8ad + 9ab$$
$$= 6a^2 - 8ad + 9ab - 12bd$$
$$= 2a(3a - 4d) + 3b(3a - 4d)$$
$$= (3a - 4d)(2a + 3b)$$

Cumulative Review

29. $\dfrac{6}{7} \div \left(-\dfrac{2}{5}\right) = \dfrac{6}{7} \cdot \left(-\dfrac{5}{2}\right) = -\dfrac{6 \cdot 5}{7 \cdot 2} = -\dfrac{30}{14} = -\dfrac{15}{7}$

30. $-\dfrac{2}{3}+\dfrac{4}{5}=-\dfrac{2\cdot5}{3\cdot5}+\dfrac{4\cdot3}{5\cdot3}$

$\phantom{-\dfrac{2}{3}+\dfrac{4}{5}}=-\dfrac{10}{15}+\dfrac{12}{15}$

$\phantom{-\dfrac{2}{3}+\dfrac{4}{5}}=\dfrac{-10+12}{15}$

$\phantom{-\dfrac{2}{3}+\dfrac{4}{5}}=\dfrac{2}{15}$

31. $\dfrac{-5a^2b^8}{25ab^{10}}=-\dfrac{5}{25}a^{2-1}b^{8-10}=-\dfrac{1}{5}a^1b^{-2}=-\dfrac{a}{5b^2}$

32. $(2x-5)^2=(2x)^2-2(2x)(5)+5^2$

$=4x^2-20x+25$

33. x = salary for a pharmacy technician
$3x + 22,000$ = salary for a pharmacist
$x+3x+22,000=162,000$
$4x+22,000=162,000$
$4x=140,000$
$x=35,000$
$3x + 22,000 = 3(35,000) + 22,000 = 127,000$
The average salary of a pharmacy technician is
\$35,000 and the average salary of a pharmacist
is \$127,000.

34. 13% of 17,000 = 0.13(17,000) = 2210
17,000 + 2210 = 19,210
In 2013, in North America 19,210 thousand
barrels of oil were produced.
11% of 19,210 = 0.11(19,210) = 2113.1
19,210 + 2113.1 = 21,323.1
In 2015, in North America about
21,323.1 thousand barrels of oil were produced.

Quick Quiz 6.2

1. $7ax+12a-14x-24=7ax-14x+12a-24$

$=7x(a-2)+12(a-2)$

$=(7x+12)(a-2)$

2. $2xy^2-15+6x-5y^2=2xy^2+6x-15-5y^2$

$=2x(y^2+3)-5(3+y^2)$

$=(y^2+3)(2x-5)$

3. $10xy-3x+40by-12b=10xy+40by-3x-12b$

$=10y(x+4b)-3(x+4b)$

$=(10y-3)(x+4b)$

4. Answers may vary. Possible solution:
Start by grouping terms $10ax$ with $5ab$ and $2bx$
with b^2.
$(2x + b)$ can be factored out of both groups
leaving the second factor to be $(5a + b)$.

6.3 Exercises

1. To factor $x^2+5x+6,$ find two numbers whose
<u>product</u> is 6 and whose <u>sum</u> is 5.

3. $x^2+8x+16;$ product: 16, sum: 8, + signs

$x^2+8x+16=(x+4)(x+4)$

5. $x^2+12x+35;$ product: 35, sum: 12, + signs

$x^2+12x+35=(x+5)(x+7)$

7. $x^2-4x+3;$ product: 3, sum: $-4,$ $-$ signs

$x^2-4x+3=(x-1)(x-3)$

9. $x^2-11x+28;$ product: 28, sum: $-11,$ $-$ signs

$x^2-11x+28=(x-4)(x-7)$

11. $x^2+5x-24;$ product: $-24,$ sum: 5
opposite signs with larger absolute value +

$x^2+5x-24=(x+8)(x-3)$

13. $x^2-13x-14;$ product: $-14,$ sum: $-13,$
opposite signs with larger absolute value $-$

$x^2-13x-14=(x-14)(x+1)$

15. $x^2+2x-35;$ product: $-35,$ sum: 2
opposite signs with larger absolute value +

$x^2+2x-35=(x+7)(x-5)$

17. $x^2-2x-24;$ product: $-24,$ sum: -2
opposite signs with larger absolute value $-$

$x^2-2x-24=(x-6)(x+4)$

19. $x^2+15x+36;$ product: 36, sum: 15, + signs

$x^2+15x+36=(x+12)(x+3)$

21. $x^2-10x+24;$ product: 24, sum: $-10,$ $-$ signs

$x^2-10x+24=(x-6)(x-4)$

23. $x^2 + 13x + 30$; product: 30, sum: 13, + signs
$$x^2 + 13x + 30 = (x+3)(x+10)$$

25. $x^2 - 6x + 5$; product: 5, sum: -6, $-$ signs
$$x^2 - 6x + 5 = (x-1)(x-5)$$

27. $a^2 + 6a - 16 = (a+8)(a-2)$
Check:
$$(a+8)(a-2) = a^2 - 2a + 8a - 16 = a^2 + 6a - 16$$

29. $x^2 - 12x + 32 = (x-4)(x-8)$
Check:
$$(x-4)(x-8) = x^2 - 8x - 4x + 32 = x^2 - 12x + 32$$

31. $x^2 + 4x - 21 = (x+7)(x-3)$
Check:
$$(x+7)(x-3) = x^2 - 3x + 7x - 21 = x^2 + 4x - 21$$

33. $x^2 + 15x + 56 = (x+7)(x+8)$
Check:
$$(x+7)(x+8) = x^2 + 8x + 7x + 56 = x^2 + 15x + 56$$

35. $y^2 + 4y - 45 = (y+9)(y-5)$
Check:
$$(y+9)(y-5) = y^2 - 5y + 9y - 45 = y^2 + 4y - 45$$

37. $x^2 + 9x - 36 = (x+12)(x-3)$
Check: $(x+12)(x-3) = x^2 - 3x + 12x - 36$
$$= x^2 + 9x - 36$$

39. $x^2 - 2xy - 15y^2 = (x-5y)(x+3y)$
Check: $(x-5y)(x+3y) = x^2 + 3xy - 5xy - 15y^2$
$$= x^2 - 2xy - 15y^2$$

41. $x^2 - 16xy + 63y^2 = (x-9y)(x-7y)$
Check: $(x-9y)(x-7y) = x^2 - 7xy - 9xy + 63y^2$
$$= x^2 - 16xy + 63y^2$$

43. $4x^2 + 24x + 20 = 4(x^2 + 6x + 5) = 4(x+1)(x+5)$

45. $6x^2 + 18x + 12 = 6(x^2 + 3x + 2) = 6(x+2)(x+1)$

47. $5x^2 - 30x + 25 = 5(x^2 - 6x + 5) = 5(x-1)(x-5)$

49. $3x^2 - 6x - 72 = 3(x^2 - 2x - 24) = 3(x-6)(x+4)$

51. $7x^2 + 21x - 70 = 7(x^2 + 3x - 10)$
$$= 7(x+5)(x-2)$$

53. $5x^2 - 35x + 30 = 5(x^2 - 7x + 6) = 5(x-1)(x-6)$

55. A = large rectangle area $-$ small rectangle area
$A = 10(12) - x(x+2)$
$$= 120 - x^2 - 2x$$
$$= 120 - 2x - x^2$$
$$= (10-x)(12+x)$$

Cumulative Review

57. $(9ab^3)(2a^5b^6c^0) = (9 \cdot 2)a^{1+5}b^{3+6}c^0$
$$= 18a^6b^9(1)$$
$$= 18a^6b^9$$

58. $(-5y^6)^2 = (-5)^2(y^6)^2 = 25y^{6 \cdot 2} = 25y^{12}$

59. $\dfrac{x^4y^{-3}}{x^{-2}y^5} = \dfrac{x^{4-(-2)}}{y^{5-(-3)}} = \dfrac{x^6}{y^8}$

60. $(2x+3y)(4x-2y) = 8x^2 - 4xy + 12xy - 6y^2$
$$= 8x^2 + 8xy - 6y^2$$

61. c = car's speed
$c = \dfrac{d}{t} = \dfrac{d}{2}$
$c + 20$ = train's speed
$\dfrac{d}{2} + 20 = \dfrac{d}{1.5}$
$1.5d + 60 = 2d$
$60 = 0.5d$
$120 = d$
The distance is 120 miles.

62. 4% of $80,000 = 0.04 \times 80,000 = 3200$
$600 + 3200 = 3800$
She will earn $3800 for the month.

63. $M = 3$
$T = 19 + 2M$
$T = 19 + 2(3)$
$T = 25$
The average temperature during April is 25°C.

64. $T = 19 + 2M$, $T = 29$
$29 = 19 + 2M$
$10 = 2M$
$5 = M$
Five months after January will be June.

Quick Quiz 6.3

1. $x^2 + 17x + 70$; product: 70, sum: 17, + signs
$x^2 + 17x + 70 = (x + 7)(x + 10)$

2. $x^2 - 14x + 48$; product: 48, sum: −14, − signs
$x^2 - 14x + 48 = (x - 6)(x - 8)$

3. $2x^2 - 4x - 96 = 2(x^2 - 2x - 48)$
product: −48, sum: −2
opposite signs with larger absolute value −
$2(x^2 - 2x - 48) = 2(x + 6)(x - 8)$

4. Answers may vary. Possible solution:
The first step is to factor out the greatest
common factor of 4, leaving $4(x^2 - x - 30)$.
Next, write expression in factored form using
variables n and m, $4(x + m)(x + n)$.
Next determine that the product of m and n is
−30, and the sum is −1.
m and n equal 5 and −6.
Substitute values of m and n, then check.

6.4 Exercises

1. $4x^2 + 21x + 5$
Factorizations of 4: (2)(2) or (1)(4)
Factorization of 5: (1)(5)
Each factor will be positive. Find the factoring
combination that yields a middle term of $21x$.
$4x^2 + 21x + 5 = (4x + 1)(x + 5)$
Check: $(4x + 1)(x + 5) = 4x^2 + 20x + x + 5$
$= 4x^2 + 21x + 5$

3. $5x^2 + 7x + 2$
Factorization of 5: (1)(5)
Factorization of 2: (1)(2)
Each factor will be positive. Find the factoring
combination that yields a middle term of $7x$.
$5x^2 + 7x + 2 = (5x + 2)(x + 1)$
Check:
$(5x + 2)(x + 1) = 5x^2 + 5x + 2x + 2 = 5x^2 + 7x + 2$

5. $4x^2 + 5x - 6$
Factorizations of 4: (2)(2) or (1)(4)
Factorizations of 6: (1)(6) or (2)(3)
The constants in the factors will have opposite
signs. Find the factoring combination that yields
a middle term of $5x$.
$4x^2 + 5x - 6 = (4x - 3)(x + 2)$
Check:
$(4x - 3)(x + 2) = 4x^2 + 8x - 3x - 6 = 4x^2 + 5x - 6$

7. $2x^2 - 5x - 3$
Factorization of 2: (1)(2)
Factorization of 3: (1)(3)
The constants in the factors will have opposite
signs. Find the factoring combination that yields
a middle term of $-5x$.
$2x^2 - 5x - 3 = (x - 3)(2x + 1)$
Check:
$(x - 3)(2x + 1) = 2x^2 + x - 6x - 3 = 2x^2 - 5x - 3$

9. $9x^2 + 9x + 2$; grouping number: 18
$9x^2 + 9x + 2 = 9x^2 + 3x + 6x + 2$
$= 3x(3x + 1) + 2(3x + 1)$
$= (3x + 1)(3x + 2)$
Check: $(3x + 1)(3x + 2) = 9x^2 + 6x + 3x + 2$
$= 9x^2 + 9x + 2$

11. $15x^2 - 34x + 15$; grouping number: 225
$15x^2 - 34x + 15 = 15x^2 - 25x - 9x + 15$
$= 5x(3x - 5) - 3(3x - 5)$
$= (5x - 3)(3x - 5)$
Check: $(5x - 3)(3x - 5) = 15x^2 - 25x - 9x + 15$
$= 15x^2 - 34x + 15$

13. $2x^2 + 3x - 20$; grouping number: −40
$2x^2 + 3x - 20 = 2x^2 + 8x - 5x - 20$
$= 2x(x + 4) - 5(x + 4)$
$= (x + 4)(2x - 5)$
Check: $(x + 4)(2x - 5) = 2x^2 - 5x + 8x - 20$
$= 2x^2 + 3x - 20$

15. $8x^2 + 10x - 3$; grouping number: −24
$8x^2 + 10x - 3 = 8x^2 + 12x - 2x - 3$
$= 4x(2x + 3) - (2x + 3)$
$= (2x + 3)(4x - 1)$

Check: $(2x+3)(4x-1) = 8x^2 - 2x + 12x - 3$
$$= 8x^2 + 10x - 3$$

17. $6x^2 - 5x - 6$; grouping number: -36
$$6x^2 - 5x - 6 = 6x^2 - 9x + 4x - 6$$
$$= 3x(2x-3) + 2(2x-3)$$
$$= (2x-3)(3x+2)$$

19. $10x^2 + 3x - 1$; grouping number -10
$$10x^2 + 3x - 1 = 10x^2 + 5x - 2x - 1$$
$$= 5x(2x+1) - 1(2x+1)$$
$$= (5x-1)(2x+1)$$

21. $7x^2 - 5x - 18$; grouping number: -126
$$7x^2 - 5x - 18 = 7x^2 - 14x + 9x - 18$$
$$= 7x(x-2) + 9(x-2)$$
$$= (7x+9)(x-2)$$

23. $9y^2 - 13y + 4$; grouping number: 36
$$9y^2 - 13y + 4 = 9y^2 - 9y - 4y + 4$$
$$= 9y(y-1) - 4(y-1)$$
$$= (y-1)(9y-4)$$

25. $5a^2 - 13a - 6$; grouping number: -30
$$5a^2 - 13a - 6 = 5a^2 - 15a + 2a - 6$$
$$= 5a(a-3) + 2(a-3)$$
$$= (a-3)(5a+2)$$

27. $14x^2 + 17x - 6$; grouping number -84
$$14x^2 + 17x - 6 = 14x^2 + 21x - 4x - 6$$
$$= 7x(2x+3) - 2(2x+3)$$
$$= (7x-2)(2x+3)$$

29. $15x^2 + 4x - 4$; grouping number: -60
$$15x^2 + 4x - 4 = 15x^2 + 10x - 6x - 4$$
$$= 5x(3x+2) - 2(3x+2)$$
$$= (3x+2)(5x-2)$$

31. $12x^2 + 28x + 15$; grouping number: 180
$$12x^2 + 28x + 15 = 12x^2 + 18x + 10x + 15$$
$$= 6x(2x+3) + 5(2x+3)$$
$$= (6x+5)(2x+3)$$

33. $12x^2 - 16x - 3$; grouping number: -36
$$12x^2 - 16x - 3 = 12x^2 + 2x - 18x - 3$$
$$= 2x(6x+1) - 3(6x+1)$$
$$= (6x+1)(2x-3)$$

35. $3x^4 - 14x^2 - 5$; grouping number -15
$$3x^4 - 14x^2 - 5 = 3x^4 + x^2 - 15x^2 - 5$$
$$= x^2(3x^2+1) - 5(3x^2+1)$$
$$= (x^2-5)(3x^2+1)$$

37. $2x^2 + 11xy + 15y^2$; grouping number: 30
$$2x^2 + 11xy + 15y^2 = 2x^2 + 6xy + 5xy + 15y^2$$
$$= 2x(x+3y) + 5y(x+3y)$$
$$= (x+3y)(2x+5y)$$

39. $5x^2 + 16xy - 16y^2$; grouping number: -80
$$5x^2 + 16xy - 16y^2 = 5xy^2 + 20xy - 4xy - 16y^2$$
$$= 5x(x+4y) - 4y(x+4y)$$
$$= (x+4y)(5x-4y)$$

41. $10x^2 - 25x - 15 = 5(2x^2 - 5x - 3)$
$$= 5(2x^2 - 6x + x - 3)$$
$$= 5[2x(x-3) + 1(x-3)]$$
$$= 5(2x+1)(x-3)$$

43. $6x^3 + 9x^2 - 60x = 3x(2x^2 + 3x - 20)$
$$= 3x(2x^2 + 8x - 5x - 20)$$
$$= 3x[2x(x+4) - 5(x+4)]$$
$$= 3x(2x-5)(x+4)$$

45. $5x^2 + 3x - 2 = 5x^2 + 5x - 2x - 2$
$$= 5x(x+1) - 2(x+1)$$
$$= (x+1)(5x-2)$$

47. $12x^2 - 38x + 20 = 2(6x^2 - 19x + 10)$
$$= 2(6x^2 - 15x - 4x + 10)$$
$$= 2[3x(2x-5) - 2(2x-5)]$$
$$= 2(2x-5)(3x-2)$$

49. $12x^3 - 20x^2 + 3x = x(12x^2 - 20x + 3)$
$$= x(12x^2 - 18x - 2x + 3)$$
$$= x[6x(2x-3) - (2x-3)]$$
$$= x(2x-3)(6x-1)$$

51. $8x^2 + 24x - 14 = 2(4x^2 + 12x - 7)$
$$= 2(4x^2 - 2x + 14x - 7)$$
$$= 2[2x(2x - 1) + 7(2x - 1)]$$
$$= 2(2x - 1)(2x + 7)$$

Cumulative Review

53. $(x_1, y_1) = (-1, 6), \ (x_2, y_2) = (2, 4)$
$$m = \frac{y_2 - y_1}{x_2 - x_1} = \frac{4 - 6}{2 - (-1)} = \frac{-2}{3} = -\frac{2}{3}$$

54.
$$\frac{x}{3} - \frac{x}{5} = \frac{7}{15}$$
$$15\left(\frac{x}{3}\right) - 15\left(\frac{x}{5}\right) = 15\left(\frac{7}{15}\right)$$
$$5x - 3x = 7$$
$$2x = 7$$
$$x = \frac{7}{2}$$

55. a. Add the heights of the "Last year" bars.
$8 + 4.6 + 1.9 + 1.9 + 1.2 = 17.6$
17,600,000 travelers visited these states last year.

   b. 1,900,000 overseas travelers visited Hawaii last year.
$$\frac{1,900,000}{17,600,000} \approx 0.108$$
Approximately 10.8% visited Hawaii.

56. total = $5.9 + 6.4 + 6.0 + 2.7 + 2.4 = 23.4$
23,400,000 overseas travelers visited these states five years ago. 6,400,000 visited California.
$$\frac{6,400,000}{23,400,000} \approx 0.274$$
Approximately 27.4% visited California.

57. a. $6.4 - 4.6 = 1.8$
The difference is 1.8 million.

   b. 1.8 is what percent of 6.4?
$$\frac{1.8}{6.4} \approx 0.281 = 28.1\%$$
The decrease is approximately 28.1% of the number of visitors from five years ago.

58. a. $8 - 5.9 = 2.1$
The difference is 2.1 million.

   b. 2.1 is what percent of 5.9?
$$\frac{2.1}{5.9} \approx 0.356 = 35.6\%$$
The increase is approximately 35.6% of the number of visitors from five years ago.

Quick Quiz 6.4

1. $12x^2 + 16x - 3$; grouping number -36
$$12x^2 + 16x - 3 = 12x^2 + 18x - 2x - 3$$
$$= 6x(2x + 3) - (2x + 3)$$
$$= (6x - 1)(2x + 3)$$

2. $10x^2 - 21x + 9$; grouping number 90
$$10x^2 - 21x + 9 = 10x^2 - 15x - 6x + 9$$
$$= 5x(2x - 3) - 3(2x - 3)$$
$$= (5x - 3)(2x - 3)$$

3. $6x^3 - 3x^2 - 30x = 3x(2x^2 - x - 10)$
grouping number -20
$$6x^3 - 3x^2 - 30x = 3x[2x^2 + 4x - 5x - 10]$$
$$= 3x[2x(x + 2) - 5(x + 2)]$$
$$= 3x(2x - 5)(x + 2)$$

4. Answers may vary. Possible solution:
First step is to factor out the greatest common factor.
$$2x(5x^2 + 9xy - 2y^2)$$
Next step is to factor the inside expression using grouping.
Grouping number -10
$$2x[5x^2 + 10xy - xy - 2y^2]$$
$$= 2x[5x(x + 2y) - y(x + 2y)]$$
$$= 2x(5x - y)(x + 2y)$$
Lastly, check the solution by multiplying the factors.

Use Math to Save Money

1. $\$5100 + \$3800 + \$3200 = \$12,100$
Megan's total current balance is $12,100.

2. $\$5500 + \$4000 + \$3500 = \$13,000$
Megan's total credit limit is $13,000.

3. $\dfrac{\$12,100}{\$13,000} \approx 0.93 = 93\%$
Megan's debt-to-credit ratio is 93%.

4. $0.5 \times \$13,000 = \$6500$
50% of her total credit limit is $6500.

5. $12,100 - $6500 = $5600
She needs to pay $5600.

6. $0.33 \times $13,000 = $4290
$6500 - $4290 = $2210
She needs to pay $2210 to reach 33% in the third year.

How Am I Doing? Sections 6.1–6.4
(Available online through MyMathLab.)

1. $6xy - 15z + 21 = 3(2xy - 5z + 7)$

2. $30x^2 - 45xy - 10x = 5x(6x - 9y - 2)$

3. $7(4x - 5) - b(4x - 5) = (4x - 5)(7 - b)$

4. $2x(8y + 3z) - 5y(8y + 3z) = (8y + 3z)(2x - 5y)$

5. $18 + 3x - 6y - xy = 3(6 + x) - y(6 + x)$
$\qquad = (6 + x)(3 - y)$

6. $15x - 9xb + 20w - 12bw = 3x(5 - 3b) + 4w(5 - 3b)$
$\qquad\qquad = (5 - 3b)(3x + 4w)$

7. $x^3 - 5x^2 - 3x + 15 = x^2(x - 5) - 3(x - 5)$
$\qquad\qquad = (x - 5)(x^2 - 3)$

8. $7a + 21b + 2ab + 6b^2 = 7(a + 3b) + 2b(a + 3b)$
$\qquad\qquad = (a + 3b)(7 + 2b)$

9. $x^2 - 15x + 56 = x^2 - 7x - 8x + 56$
$\qquad\qquad = x(x - 7) - 8(x - 7)$
$\qquad\qquad = (x - 7)(x - 8)$

10. $x^2 + 12x - 64 = x^2 + 16x - 4x - 64$
$\qquad\qquad = x(x + 16) - 4(x + 16)$
$\qquad\qquad = (x + 16)(x - 4)$

11. $a^2 + 10ab + 21b^2$; product: 21, sum: 10, + signs
$a^2 + 10ab + 21b^2 = a^2 + 3ab + 7ab + 21b^2$
$\qquad\qquad = a(a + 3b) + 7b(a + 3b)$
$\qquad\qquad = (a + 7b)(a + 3b)$

12. $7x^2 - 14x - 245 = 7(x^2 - 2x - 35)$
$\qquad\qquad = 7(x^2 - 7x + 5x - 35)$
$\qquad\qquad = 7[x(x - 7) + 5(x - 7)]$
$\qquad\qquad = 7(x - 7)(x + 5)$

13. $12x^2 + 17x - 5$; product: -60, sum: 17
$12x^2 + 17x - 5 = 12x^2 + 20x - 3x - 5$
$\qquad\qquad = 4x(3x + 5) - (3x + 5)$
$\qquad\qquad = (4x - 1)(3x + 5)$

14. $3x^2 - 23x + 14 = 3x^2 - 21x - 2x + 14$
$\qquad\qquad = 3x(x - 7) - 2(x - 7)$
$\qquad\qquad = (x - 7)(3x - 2)$

15. $6x^2 + 17xy + 12y^2 = 6x^2 + 9xy + 8xy + 12y^2$
$\qquad\qquad = 3x(2x + 3y) + 4y(2x + 3y)$
$\qquad\qquad = (2x + 3y)(3x + 4y)$

16. $14x^3 - 20x^2 - 16x = 2x(7x^2 - 10x - 8)$
$\qquad\qquad = 2x(7x^2 - 14x + 4x - 8)$
$\qquad\qquad = 2x[7x(x - 2) + 4(x - 2)]$
$\qquad\qquad = 2x(x - 2)(7x + 4)$

6.5 Exercises

1. $100x^2 - 1 = (10x)^2 - (1)^2 = (10x + 1)(10x - 1)$

3. $81x^2 - 16 = (9x)^2 - (4)^2 = (9x + 4)(9x - 4)$

5. $x^2 - 49 = (x)^2 - (7)^2 = (x + 7)(x - 7)$

7. $25x^2 - 81 = (5x)^2 - (9)^2 = (5x + 9)(5x - 9)$

9. $x^2 - 25 = (x)^2 - (5)^2 = (x + 5)(x - 5)$

11. $1 - 16x^2 = (1)^2 - (4x)^2 = (1 + 4x)(1 - 4x)$

13. $16x^2 - 49y^2 = (4x)^2 - (7y)^2$
$\qquad\qquad = (4x + 7y)(4x - 7y)$

15. $36x^2 - 169y^2 = (6x)^2 - (13y)^2$
$\qquad\qquad = (6x + 13y)(6x - 13y)$

17. $100x^2 - 81 = (10x)^2 - (9)^2 = (10x + 9)(10x - 9)$

19. $25a^2 - 81b^2 = (5a)^2 - (9b)^2 = (5a + 9b)(5a - 9b)$

21. $9x^2 + 6x + 1 = (3x)^2 + 2(3x)(1) + (1)^2 = (3x + 1)^2$

23. $y^2 - 10y + 25 = y^2 - 2(y)(5) + (5)^2 = (y - 5)^2$

25. $36x^2 - 60x + 25 = (6x)^2 - 2(6x)(5) + (5)^2$
$$= (6x - 5)^2$$

27. $49x^2 + 28x + 4 = (7x)^2 + 2(7x)(2) + 2^2$
$$= (7x + 2)^2$$

29. $x^2 + 14x + 49 = (x)^2 + 2(x)(7) + (7)^2 = (x + 7)^2$

31. $25x^2 - 40x + 16 = (5x)^2 - 2(5x)(4) + 4^2$
$$= (5x - 4)^2$$

33. $81x^2 + 36xy + 4y^2 = (9x)^2 + 2(9x)(2y) + (2y)^2$
$$= (9x + 2y)^2$$

35. $9x^2 - 30xy + 25y^2 = (3x)^2 - 2(3x)(5y) + (5y)^2$
$$= (3x - 5y)^2$$

37. $16a^2 + 72ab + 81b^2 = (4a)^2 + 2(4a)(9b) + (9b)^2$
$$= (4a + 9b)^2$$

39. $49x^2 - 42xy + 9y^2 = (7x)^2 - 2(7x)(3y) + (3y)^2$
$$= (7x - 3y)^2$$

41. $64x^2 + 80x + 25 = (8x)^2 + 2(8x)(5) + (5)^2$
$$= (8x + 5)^2$$

43. $144x^2 - 1 = (12x)^2 - (1)^2 = (12x - 1)(12x + 1)$

45. $x^4 - 16 = (x^2)^2 - (4)^2$
$$= (x^2 + 4)(x^2 - 4)$$
$$= (x^2 + 4)(x^2 - 2^2)$$
$$= (x^2 + 4)(x + 2)(x - 2)$$

47. $9x^4 - 24x^2 + 16 = (3x^2)^2 - 2(3x^2)(4) + (4)^2$
$$= (3x^2 - 4)^2$$

49. You cannot factor $9x^2 + 1$ because there are no combinations of the product of two binomials that give $9x^2 + 1$.

51. $16 = 4^2$
$24 = (2)(4)(3)$
$c = 3^2$
$c = 9$
There is only one answer.

53. $16x^2 - 36 = 4(4x^2 - 9) = 4(2x + 3)(2x - 3)$

55. $147x^2 - 3y^2 = 3(49x^2 - y^2)$
$$= 3[(7x)^2 - y^2]$$
$$= 3(7x - y)(7x + y)$$

57. $16x^2 - 16x + 4 = 4(4x^2 - 4x + 1)$
$$= 4[(2x)^2 - 2(2x)(1) + (1)^2]$$
$$= 4(2x - 1)^2$$

59. $98x^2 + 84x + 18 = 2(49x^2 + 42x + 9) = 2(7x + 3)^2$

61. $x^2 + 16x + 63 = (x + 9)(x + 7)$

63. $2x^2 + 5x - 3 = (2x - 1)(x + 3)$

65. $12x^2 - 27 = 3(4x^2 - 9)$
$$= 3[(2x)^2 - (3)^2]$$
$$= 3(2x + 3)(2x - 3)$$

67. $9x^2 + 42x + 49 = (3x + 7)^2$

69. $36x^2 - 36x + 9 = 9(4x^2 - 4x + 1)$
$$= 9[(2x)^2 - 2(2x)(1) + (1)^2]$$
$$= 9(2x - 1)^2$$

71. $2x^2 - 32x + 126 = 2(x^2 - 16x + 63)$
$$= 2(x - 9)(x - 7)$$

Cumulative Review

73.
$$\begin{array}{r}
x^2 + 3x + 4 \\
x - 2 \overline{\smash{)}\ x^3 + \ x^2 - 2x - 11} \\
\underline{x^3 - 2x^2} \\
3x^2 - 2x \\
\underline{3x^2 - 6x} \\
4x - 11 \\
\underline{4x - 8} \\
-3
\end{array}$$

$(x^3 + x^2 - 2x - 11) \div (x - 2) = x^2 + 3x + 4 - \dfrac{3}{x - 2}$

74.

$$3x+4\overline{\smash{\big)}\begin{array}{l}2x^2+x-5\\6x^3+11x^2-11x-20\end{array}}$$

$$\underline{6x^3+\ 8x^2}$$

$$3x^2-11x$$

$$\underline{3x^2+\ 4x}$$

$$-15x-20$$

$$\underline{-15x-20}$$

$$(6x^3+11x^2-11x-20)\div(3x+4)=2x^2+x-5$$

75. Daily diet = $0.02(150)=3$ ounces
$0.4(3)=1.2$ ounces of greens
$0.35(3)=1.05$ ounces of bulk vegetables
$0.25(3)=0.75$ ounce of fruit

76. Daily diet = $0.03(120)=3.6$ ounces
$0.4(3.6)=1.44$ ounces of greens
$0.35(3.6)=1.26$ ounces of bulk vegetables
$0.25(3.6)=0.9$ ounce of fruit

Quick Quiz 6.5

1. $49x^2-81y^2=(7x)^2-(9y)^2$
$\qquad\qquad\quad=(7x-9y)(7x+9y)$

2. $9x^2-48x+64=(3x)^2-2(3x)(8)+(8)^2$
$\qquad\qquad\qquad\ =(3x-8)^2$

3. $162x^2-200=2(81x^2-100)$
$\qquad\qquad\quad=2[(9x)^2-(10)^2]$
$\qquad\qquad\quad=2(9x-10)(9x+10)$

4. Answers may vary. Possible solution:
First factor out the common factor of 2.

$2(12x^2+60x+75)$

Next use grouping to factor inside expression.
Grouping number 900

$2(12x^2+30x+30x+75)$
$=2[6x(2x+5)+15(2x+5)]$
$=2(6x+15)(2x+5)$

Lastly, check by multiplying.

6.6 Exercises

1. $3x^2-6xy+5x=x(3x-6y+5)$

Check: $x(3x-6y+5)=3x^2-6xy+5x$

3. $16x^2-25y^2=(4x)^2-(5y)^2$
$\qquad\qquad\quad=(4x+5y)(4x-5y)$

Check:

$(4x+5y)(4x-5y)=(4x)^2-(5y)^2=16x^2-25y^2$

5. x^2+64 cannot be factored. It is prime.

7. $x^2+8x+15=(x+5)(x+3)$

Check:

$(x+5)(x+3)=x^2+3x+5x+15=x^2+8x+15$

9. $15x^2+7x-2=(5x-1)(3x+2)$

Check: $(5x-1)(3x+2)=15x^2+10x-3x-2$
$\qquad\qquad\qquad\qquad\ =15x^2+7x-2$

11. $ax-3cx+3ay-9cy=x(a-3c)+3y(a-3c)$
$\qquad\qquad\qquad\qquad\ =(a-3c)(x+3y)$

Check: $(x+3y)(a-3c)=ax-3cx+3ay-9cy$

13. $y^2+14y+49=(y)^2+2(y)(7)+(7)^2=(y+7)^2$

15. $4x^2-12x+9=(2x)^2-2(2x)(3)+3^2=(2x-3)^2$

17. $2x^2-11x+12=(2x-3)(x-4)$

19. $x^2-3xy-70y^2=(x-10y)(x+7y)$

21. $ax-5a+3x-15=a(x-5)+3(x-5)$
$\qquad\qquad\qquad\quad=(a+3)(x-5)$

23. $16x-4x^3=4x(4-x^2)$
$\qquad\qquad\quad=4x[(2)^2-(x)^2]$
$\qquad\qquad\quad=4x(2-x)(2+x)$

25. $2x^2+3x-36$ cannot be factored. It is prime.

27. $3xyz^2-6xyz-9xy=3xy(z^2-2z-3)$
$\qquad\qquad\qquad\qquad\ =3xy(z-3)(z+1)$

29. $3x^2+6x-105=3(x^2+2x-35)$
$\qquad\qquad\qquad\ =3(x+7)(x-5)$

31. $5x^3y^3-10x^2y^3+5xy^3=5xy^3(x^2-2x+1)$
$\qquad\qquad\qquad\qquad\quad=5xy^3[x^2-2(x)(1)+1^2]$
$\qquad\qquad\qquad\qquad\quad=5xy^3(x-1)^2$

33. $7x^2 - 2x^4 + 4 = -1(2x^4 - 7x^2 - 4)$
$$= -1(2x^2 + 1)(x^2 - 4)$$
$$= -1(2x^2 + 1)(x + 2)(x - 2)$$

35. $6x^2 - 3x + 2$ is prime.

37. $5x^4 - 5x^2 + 10x^3 y - 10xy$
$$= 5x(x^3 - x + 2x^2 y - 2y)$$
$$= 5x[x(x^2 - 1) + 2y(x^2 - 1)]$$
$$= 5x(x + 2y)(x^2 - 1)$$
$$= 5x(x + 2y)(x + 1)(x - 1)$$

39. $5x^2 + 10xy - 30y = 5(x^2 + 2xy - 6y)$

41. $30x^3 + 3x^2 y - 6xy^2 = 3x(10x^2 + xy - 2y^2)$
$$= 3x(2x + y)(5x - 2y)$$

43. $30x^2 - 38x + 12 = 2(15x^2 - 19x + 6)$
$$= 2(15x^2 - 9x - 10x + 6)$$
$$= 2[3x(5x - 3) - 2(5x - 3)]$$
$$= 2(5x - 3)(3x - 2)$$

45. A polynomial that cannot be factored by the methods of this chapter is called <u>prime</u>.

Cumulative Review

47. Let x = his previous salary.
$$x - 0.14x = 24,080$$
$$0.86x = 24,080$$
$$\frac{0.86x}{0.86} = \frac{24,080}{0.86}$$
$$x = 28,000$$
His previous salary was $28,000.

48. $6 \times 13 = 78$
78 strains of virus have been killed.
$294 + 78 = 372$
There were 372 strains of virus 6 hours ago.

49. $\frac{1}{2}x - y = 7$ (1)
$-3x + 2y = -22$ (2)
Multiply (1) by 2 to eliminate y.
$$x - 2y = 14$$
$$\underline{-3x + 2y = -22}$$
$$-2x = -8$$
$$x = 4$$
Substitute 4 for x in (1).

$$\frac{1}{2}(4) - y = 7$$
$$2 - y = 7$$
$$-y = 5$$
$$y = -5$$
$(4, -5)$

50. $y = 2x + 3$

x	y
0	3
−1	1
−2	−1

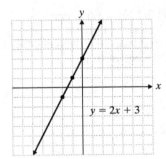

Quick Quiz 6.6

1. $6x^2 - 17x + 12$; product: 72, sum: −17, − signs
$$6x^2 - 17x + 12 = 6x^2 - 8x - 9x + 12$$
$$= 2x(3x - 4) - 3(3x - 4)$$
$$= (2x - 3)(3x - 4)$$

2. $60x^2 - 9x - 6 = 3(20x^2 - 3x - 2)$
grouping number: −40
$$= 3(20x^2 + 5x - 8x - 2)$$
$$= 3[5x(4x + 1) - 2(4x + 1)]$$
$$= 3(5x - 2)(4x + 1)$$

3. $25x^2 + 49$
prime

4. Answers may vary. Possible solution:
$2x^2 + 6xw - 5x - 15w$
The first step is to group terms and factor out common factors.
$2x(x + 3w) - 5(x + 3w)$
Next, factor out $(x + 3w)$.
$(2x - 5)(x + 3w)$
Finally, check by multiplying.

6.7 Exercises

1. $x^2 - 4x - 12 = 0$
$(x+2)(x-6) = 0$
$x+2 = 0 \qquad\qquad x-6 = 0$
$\quad x = -2 \qquad\qquad\qquad x = 6$
Check:
$(-2)^2 - 4(-2) - 12 \stackrel{?}{=} 0$
$\qquad\quad 4 + 8 - 12 \stackrel{?}{=} 0$
$\qquad\qquad\qquad\quad 0 = 0$

$6^2 - 4(6) - 12 \stackrel{?}{=} 0$
$36 - 24 - 12 \stackrel{?}{=} 0$
$\qquad\qquad\quad 0 = 0$

3. $x^2 + 14x + 24 = 0$
$(x+12)(x+2) = 0$
$x+12 = 0 \qquad\qquad x+2 = 0$
$\quad x = -12 \qquad\qquad\quad x = -2$
Check:
$(-12)^2 + 14(-12) + 24 \stackrel{?}{=} 0$
$\qquad 144 - 168 + 24 \stackrel{?}{=} 0$
$\qquad\qquad\qquad\qquad 0 = 0$

$(-2)^2 + 14(-2) + 24 \stackrel{?}{=} 0$
$\qquad\quad 4 - 28 + 24 \stackrel{?}{=} 0$
$\qquad\qquad\qquad\quad 0 = 0$

5. $2x^2 - 7x + 6 = 0$
$2x^2 - 3x - 4x + 6 = 0$
$x(2x-3) - 2(2x-3) = 0$
$\quad (2x-3)(x-2) = 0$
$2x-3 = 0 \qquad\qquad x-2 = 0$
$\quad x = \dfrac{3}{2} \qquad\qquad\qquad x = 2$
Check:
$2\left(\dfrac{3}{2}\right)^2 - 7\left(\dfrac{3}{2}\right) + 6 \stackrel{?}{=} 0$
$\qquad \dfrac{9}{2} - \dfrac{21}{2} + \dfrac{12}{2} \stackrel{?}{=} 0$
$\qquad\qquad\qquad\qquad 0 = 0$

$2(2)^2 - 7(2) + 6 \stackrel{?}{=} 0$
$\qquad 8 - 14 + 6 \stackrel{?}{=} 0$
$\qquad\qquad\qquad\quad 0 = 0$

7. $6x^2 - 13x = -6$
$6x^2 - 13x + 6 = 0$
$(3x-2)(2x-3) = 0$
$3x-2 = 0 \qquad\qquad 2x-3 = 0$
$\quad 3x = 2 \qquad\qquad\qquad 2x = 3$
$\quad x = \dfrac{2}{3} \qquad\qquad\qquad x = \dfrac{3}{2}$
Check:
$6\left(\dfrac{2}{3}\right)^2 - 13\left(\dfrac{2}{3}\right) \stackrel{?}{=} -6 \qquad 6\left(\dfrac{3}{2}\right)^2 - 13\left(\dfrac{3}{2}\right)^2 \stackrel{?}{=} -6$
$\qquad\quad \dfrac{8}{3} - \dfrac{26}{3} \stackrel{?}{=} -6 \qquad\qquad\qquad \dfrac{27}{2} - \dfrac{39}{2} \stackrel{?}{=} -6$
$\qquad\qquad\qquad -6 = -6 \qquad\qquad\qquad\qquad\qquad -6 = -6$

9. $x^2 + 13x = 0$
$x(x+13) = 0$
$x = 0 \qquad\qquad\qquad x+13 = 0$
$\qquad\qquad\qquad\qquad\quad x = -13$
Check:
$(-13)^2 + (13)(-13) \stackrel{?}{=} 0 \quad 0^2 + 13(0) \stackrel{?}{=} 0$
$\qquad\quad 169 - 169 \stackrel{?}{=} 0 \qquad\qquad\quad 0 \stackrel{?}{=} 0$
$\qquad\qquad\qquad\quad 0 = 0$

11. $8x^2 = 72$
$8x^2 - 72 = 0$
$8(x^2 - 9) = 0$
$8(x-3)(x+3) = 0$
$x-3 = 0 \qquad\qquad x+3 = 0$
$\quad x = 3 \qquad\qquad\qquad x = -3$
Check:
$8(-3)^2 \stackrel{?}{=} 72 \qquad\qquad 8(3)^2 \stackrel{?}{=} 72$
$\quad 72 = 72 \qquad\qquad\qquad 72 = 72$

13. $5x^2 + 3x = 8x$
$5x^2 - 5x = 0$
$5x(x-1) = 0$
$5x = 0 \qquad\qquad x-1 = 0$
$\quad x = 0 \qquad\qquad\quad x = 1$
Check:
$5(0)^2 + 3(0) \stackrel{?}{=} 8(0) \qquad 5(1)^2 + 3(1) \stackrel{?}{=} 8(1)$
$\qquad 0 + 0 \stackrel{?}{=} 0 \qquad\qquad\qquad 5 + 3 \stackrel{?}{=} 8$
$\qquad\qquad 0 = 0 \qquad\qquad\qquad\qquad\quad 8 = 8$

15.
$$6x^2 = 16x - 8$$
$$6x^2 - 16x + 8 = 0$$
$$2(3x^2 - 8x + 4) = 0$$
$$2(3x - 2)(x - 2) = 0$$

$3x - 2 = 0$ \qquad\qquad $x - 2 = 0$

$3x = 2$ \qquad\qquad\quad $x = 2$

$$x = \frac{2}{3}$$

Check:

$$6\left(\frac{2}{3}\right)^2 \overset{?}{=} 16\left(\frac{2}{3}\right) - 8 \qquad 6(2)^2 \overset{?}{=} 16(2) - 8$$
$$\qquad\qquad\qquad\qquad\qquad 24 \overset{?}{=} 32 - 8$$
$$\frac{24}{9} \overset{?}{=} \frac{32}{3} - \frac{24}{3} \qquad\qquad 24 = 24$$
$$\frac{8}{3} = \frac{8}{3}$$

17.
$$(x - 5)(x + 2) = -4(x + 1)$$
$$x^2 + 2x - 5x - 10 = -4x - 4$$
$$x^2 + x - 6 = 0$$
$$(x + 3)(x - 2) = 0$$

$x + 3 = 0$ \qquad\qquad $x - 2 = 0$

$x = -3$ \qquad\qquad\quad $x = 2$

Check:
$$(-3 - 5)(-3 + 2) \overset{?}{=} -4(-3 + 1)$$
$$(-8)(-1) \overset{?}{=} -4(-2)$$
$$8 = 8$$
$$(2 - 5)(2 + 2) \overset{?}{=} -4(2 + 1)$$
$$-3(4) \overset{?}{=} -4(3)$$
$$-12 = -12$$

19.
$$9x^2 + x + 1 = -5x$$
$$9x^2 + 6x + 1 = 0$$
$$(3x + 1)^2 = 0$$
$$3x + 1 = 0$$
$$3x = -1$$
$$x = -\frac{1}{3}$$

Check: $9\left(-\dfrac{1}{3}\right)^2 + \left(-\dfrac{1}{3}\right) + 1 \overset{?}{=} -5\left(-\dfrac{1}{3}\right)$

$$9\left(\frac{1}{9}\right) - \frac{1}{3} + 1 \overset{?}{=} \frac{5}{3}$$
$$1 - \frac{1}{3} + 1 \overset{?}{=} \frac{5}{3}$$
$$\frac{5}{3} = \frac{5}{3}$$

21.
$$\frac{x^2}{4} + \frac{5x}{4} + 2 = 2$$
$$x^2 + 5x + 8 = 8$$
$$x^2 + 5x = 0$$
$$x(x + 5) = 0$$

$x = 0$ \qquad or \qquad $x + 5 = 0$

$\qquad\qquad\qquad\qquad x = -5$

Check: $\dfrac{0^2}{4} + \dfrac{5(0)}{4} + 2 \overset{?}{=} 2$

$$2 = 2$$
$$\frac{(-5)^2}{4} + \frac{5(-5)}{4} + 2 \overset{?}{=} 2$$
$$\frac{25}{4} - \frac{25}{4} + 2 \overset{?}{=} 2$$
$$2 = 2$$

23.
$$\frac{x^2 + 10x}{8} = -2$$
$$x^2 + 10x = -16$$
$$x^2 + 10x + 16 = 0$$
$$(x + 8)(x + 2) = 0$$

$x + 8 = 0$ \qquad\qquad $x + 2 = 0$

$x = -8$ \qquad\qquad\quad $x = -2$

Check:
$$\frac{(-8)^2 + 10(-8)}{8} \overset{?}{=} -2 \qquad \frac{(-2)^2 + 10(-2)}{8} \overset{?}{=} -2$$
$$\frac{64 - 80}{8} \overset{?}{=} -2 \qquad\qquad \frac{4 - 20}{8} \overset{?}{=} -2$$
$$-2 = -2 \qquad\qquad\qquad\qquad -2 = -2$$

25.
$$\frac{10x^2 - 25x}{12} = 5$$
$$10x^2 - 25x = 60$$
$$10x^2 - 25x - 60 = 0$$
$$5(2x^2 - 5x - 12) = 0$$
$$5(2x + 3)(x - 4) = 0$$

$2x + 3 = 0$ \qquad\qquad $x - 4 = 0$

$x = -\dfrac{3}{2}$ \qquad\qquad $x = 4$

Check:

$$\frac{10\left(-\frac{3}{2}\right)-25\left(-\frac{3}{2}\right)}{12} \stackrel{?}{=} 5$$

$$\frac{\frac{90}{4}+\frac{150}{4}}{12} \stackrel{?}{=} 5$$

$$\frac{60}{4} \stackrel{?}{=} 5$$

$$5 = 5$$

$$\frac{10(4)^2-25(4)}{12} \stackrel{?}{=} 5$$

$$\frac{160-100}{12} \stackrel{?}{=} 5$$

$$\frac{60}{4} \stackrel{?}{=} 5$$

$$5 = 5$$

27. An equation in the form $ax^2 + bx = 0$ can always be solved by factoring out x.

29. $x = $ length

$$\frac{1}{2}x + 3 = \text{width}$$

$$x\left(\frac{1}{2}x+3\right) = 140$$

$$\frac{x^2}{2} + 3x = 140$$

$$x^2 + 6x = 280$$

$$x^2 + 6x - 280 = 0$$

$$(x+20)(x-14) = 0$$

$$x + 20 = 0 \qquad\qquad x - 14 = 0$$

$$x = -20 \qquad\qquad x = 14$$

Width not negative $\quad \frac{1}{2}x + 3 = 10$

The length is 14 m and the width is 10 m.

31. $G = \dfrac{x^2 - 3x + 2}{2}, \quad x = 13$

$$G = \frac{13^2 - 3(13) + 2}{2}$$

$$G = \frac{169 - 39 + 2}{2}$$

$$G = 66$$

There are 66 possible groups.

33. $G = \dfrac{x^2 - 3x + 2}{2}$

$$72 = x^2 - 3x + 2$$

$$0 = x^2 - 3x - 70$$

$$0 = (x - 10)(x + 7)$$

$$x - 10 = 0 \qquad\qquad x + 7 = 0$$

$$x = 10 \qquad\qquad x = -7$$

$$\text{Not possible}$$

There are 10 students.

35. $S = -5t^2 + vt + h$

$$v = 13 \qquad\qquad\qquad h = 6$$

$$S = -5t^2 + 13t + 6$$

When the ball hits the ground $S = 0$.

$$0 = -5t^2 + 13t + 6$$

$$0 = 5t^2 - 13t - 6$$

$$0 = (5t + 2)(t - 3)$$

$$5t + 2 = 0 \qquad\qquad t - 3 = 0$$

$$t = -\frac{2}{5} \qquad\qquad t = 3$$

Time can't be negative. The ball will hit the ground at $t = 3$ sec.

$$S = -5(2)^2 + 13(2) + 6 = -20 + 26 + 6 = 12$$

After 2 seconds the ball is 12 m from the ground.

37. $x = 70$

$$T = 0.5[(70)^2 - 70]$$

$$= 0.5(4900 - 70)$$

$$= 2415 \text{ telephone calls}$$

These 70 people could make 2415 telephone calls between them.

39. $H = $ number of handshakes

$$H = 0.5(17^2 - 17)$$

$$H = 0.5(289 - 17)$$

$$H = 136$$

136 handshakes will take place.

Cumulative Review

41. $(2x^2y^3)(-5x^3y) = (2)(-5)x^{2+3}y^{3+1} = -10x^5y^4$

42. $(3a^4b^5)(4a^6b^8) = (3)(4)a^{4+6}b^{5+8} = 12a^{10}b^{13}$

43. $\dfrac{21a^5b^{10}}{-14ab^{12}} = \dfrac{3a^{5-4}}{-2b^{12-10}} = -\dfrac{3a^4}{2b^2}$

44. $\dfrac{18x^3y^6}{54x^8y^{10}} = \dfrac{1}{3x^{8-3}y^{10-6}} = \dfrac{1}{3x^5y^4}$

Quick Quiz 6.7

1. $15x^2 - 8x + 1 = 0$
   grouping number 15
   $15x^2 - 8x + 1 = 15x^2 - 5x - 3x + 1$
   $\qquad\qquad\qquad = 5x(3x-1) - (3x-1)$
   $\qquad\qquad\qquad = (5x-1)(3x-1)$

   $\begin{array}{ll} 5x - 1 = 0 & \quad 3x - 1 = 0 \\ 5x = 1 & \quad 3x = 1 \\ x = \dfrac{1}{5} & \quad x = \dfrac{1}{3} \end{array}$

2. $4 + x(x-2) = 7$
   $4 + x^2 - 2x = 7$
   $x^2 - 2x - 3 = 0$
   $(x-3)(x+1) = 0$

   $\begin{array}{ll} x - 3 = 0 & \quad x + 1 = 0 \\ x = 3 & \quad x = -1 \end{array}$

3. $\qquad\qquad 4x^2 = 9x + 9$
   $4x^2 - 9x - 9 = 0$
   grouping number -36
   $4x^2 - 9x - 9 = 4x^2 - 12x + 3x - 9$
   $\qquad\qquad\qquad = 4x(x-3) + 3(x-3)$
   $\qquad\qquad\qquad = (4x+3)(x-3)$

   $\begin{array}{ll} 4x + 3 = 0 & \quad x - 3 = 0 \\ 4x = -3 & \quad x = 3 \\ x = -\dfrac{3}{4} & \end{array}$

4. Answers may vary. Possible solution:
   Let x = width of rectangle, then
   length = $(2x + 3)$.
   $A = $ (width)(length)
   $65 = x(2x+3)$
   $65 = 2x^2 + 3x$
   $0 = 2x^2 + 3x - 65$
   $0 = (2x+13)(x-5)$

   $\begin{array}{ll} 2x + 13 = 0 & \quad x - 5 = 0 \\ 2x = -13 & \quad x = 5 \\ x = -\dfrac{13}{2} & \end{array}$

   x cannot be negative in this case, because it
   describes a length. So, $x = 5$.
   length = $2(5) + 3 = 13$ feet
   width = 5 feet

Career Exploration Problems

1. Rectangular piece:
   width = w
   length = $3w - 3$
   Area = length $\cdot$ width
   $126 = (3w-3)w$
   $126 = 3w^2 - 3w$
   $0 = 3w^2 - 3w - 126$
   $0 = 3(w^2 - w - 42)$
   $0 = 3(w-7)(w+6)$

   $\begin{array}{ll} w - 7 = 0 & \quad w + 6 = 0 \\ w = 7 & \quad w = -6 \end{array}$

   Eliminate negative dimensions.
   The width is 7 feet and the length is
   $3(7) - 3 = 18$ ft.
   Triangular piece:
   base = b
   height = $b - 1$
   Area = $\dfrac{1}{2} \cdot$ base $\cdot$ height
   $10 = \dfrac{1}{2}b(b-1)$
   $20 = b(b-1)$
   $0 = b^2 - b - 20$
   $0 = (b-5)(b+4)$

   $\begin{array}{ll} b - 5 = 0 & \quad b + 4 = 0 \\ b = 5 & \quad b = -4 \end{array}$

   Eliminate negative dimensions.
   The base is 5 feet and the height is $5 - 1 = 4$ feet.

2. base = b
   height = $2b - 2$
   Area = $\dfrac{1}{2} \cdot$ base $\cdot$ height
   $132 = \dfrac{1}{2}b(2b-2)$
   $132 = b^2 - b$
   $0 = b^2 - b - 132$
   $0 = (b-12)(b+11)$

   $\begin{array}{ll} b - 12 = 0 & \quad b + 11 = 0 \\ b = 12 & \quad b = -11 \end{array}$

   The base is 12 feet and the height is
   $2(12) - 2 = 22$ feet.

3. $132 \text{ ft}^2 \times \dfrac{1 \text{ yd}^2}{9 \text{ ft}^2} = \dfrac{132}{9} \text{ yd}^2 = \dfrac{44}{3} \text{ yd}^2$

   The sail's area is $\dfrac{44}{3}$ square yards.

　　　　　　Copyright © 2017 Pearson Education, Inc.

You Try It

1. a. $5a^2 - 15a = 5a(a) - 5a(3) = 5a(a-3)$

   b. $4x^2 - 8xy + 4x = 4x(x) - 4x(2y) + 4x(1)$
   $$= 4x(x - 2y + 1)$$

   c. $6x^4 - 18x^2 = 6x^2(x^2) - 6x^2(3)$
   $$= 6x^2(x^2 - 3)$$

2. $3ax^2 - 12x^2 - 8 + 2a = 3ax^2 - 12x^2 + 2a - 8$
   $$= 3x^2(a-4) + 2(a-4)$$
   $$= (3x^2 + 2)(a-4)$$

3. a. $x^2 + 9x + 18$
   Product: 18; sum: 9; 6 and 3
   $x^2 + 9x + 18 = (x+6)(x+3)$

   b. $x^2 + 2x - 35$
   Product: −35; sum: 2; −5 and 7
   $x^2 + 2x - 35 = (x-5)(x+7)$

   c. $3x^2 - 9x - 12 = 3(x^2 - 3x - 4)$
   $$= 3(x-4)(x+1)$$

4. $8x^2 + 6x - 9$
   Product: −72; sum: 6; −6 and 12
   $8x^2 + 6x - 9 = 8x^2 - 6x + 12x - 9$
   $$= 2x(4x-3) + 3(4x-3)$$
   $$= (4x-3)(2x+3)$$

5. a. $9x^2 - 16y^2 = (3x)^2 - (4y)^2$
   $$= (3x+4y)(3x-4y)$$

   b. $81x^4 - 1 = (9x^2)^2 - (1)^2$
   $$= (9x^2 + 1)(9x^2 - 1)$$
   $$= (9x^2 + 1)[(3x)^2 - (1)^2]$$
   $$= (9x^2 + 1)(3x+1)(3x-1)$$

   c. $16a^2 + 24a + 9 = (4a)^2 + 2(4a)(3) + (3)^2$
   $$= (4a+3)^2$$

   d. $4x^2 - 20xy + 25y^2$
   $$= (2x)^2 - 2(2x)(5y) + (5y)^2$$
   $$= (2x-5y)^2$$

6. a. $4x^2 + 4x - 24 = 4(x^2 + x - 6)$
   $$= 4(x+3)(x-2)$$

   b. $3x^3 + 7x^2 + 2x = x(3x^2 + 7x + 2)$
   $$= x(3x+1)(x+2)$$

   c. $9x^3 - 64x = x(9x^2 - 64)$
   $$= x[(3x)^2 - (8)^2]$$
   $$= x(3x+8)(3x-8)$$

   d. $48x^2 - 24x + 3 = 3(16x^2 - 8x + 1)$
   $$= 3[(4x)^2 - 2(4x)(1) + (1)^2]$$
   $$= 3(4x-1)^2$$

7. a. $x^2 + 4$ is prime because it is the sum of two squares.

   b. $x^2 + x + 2$ is prime because there are no two factors of 2 that add to 1.

8. $$2x^2 - x = 3$$
   $$2x^2 - x - 3 = 0$$
   $$(2x-3)(x+1) = 0$$

$2x - 3 = 0$	$x + 1 = 0$
$2x = 3$	$x = -1$
$x = \dfrac{3}{2}$	

9. $w = $ width
   $2w + 3 = $ length
   $$w(2w+3) = 90$$
   $$2w^2 + 3w - 90 = 0$$
   $$(2w+15)(w-6) = 0$$

$2w + 15 = 0$	$w - 6 = 0$
$2w = -15$	$w = 6$
$w = -\dfrac{15}{2}$	

   Discard the negative solution.
   $2w + 3 = 2(6) + 3 = 15$
   The length is 15 feet, and the width is 6 feet.

Chapter 6 Review Problems

1. $12x^3 - 20x^2 y = 4x^2(3x - 5y)$

2. $10x^3 - 35x^3 y = 5x^3(2 - 7y)$

3. $24x^3 y - 8x^2 y^2 - 16x^3 y^3 = 8x^2 y(3x - y - 2xy^2)$

4. $3a^3 + 6a^2 - 9ab + 12a = 3a(a^2 + 2a - 3b + 4)$

5. $2a(a+3b) - 5(a+3b) = (a+3b)(2a-5)$

6. $15x^3y + 6xy^2 + 3xy = 3xy(5x^2 + 2y + 1)$

7. $2ax + 5a - 8x - 20 = 2ax - 8x + 5a - 20$
$$= 2x(a-4) + 5(a-4)$$
$$= (2x+5)(a-4)$$

8. $a^2 - 4ab + 7a - 28b = a(a-4b) + 7(a-4b)$
$$= (a+7)(a-4b)$$

9. $x^2y + 3y - 2x^2 - 6 = y(x^2+3) - 2(x^2+3)$
$$= (x^2+3)(y-2)$$

10. $30ax - 15ay + 42x - 21y$
$$= 3(10ax - 5ay + 14x - 7y)$$
$$= 3[5a(2x-y) + 7(2x-y)]$$
$$= 3(2x-y)(5a+7)$$

11. $15x^2 - 3x + 10x - 2 = 3x(5x-1) + 2(5x-1)$
$$= (5x-1)(3x+2)$$

12. $30w^2 - 18w + 5wz - 3z = 6w(5w-3) + z(5w-3)$
$$= (5w-3)(6w+z)$$

13. $x^2 + 6x - 27 = (x+9)(x-3)$

14. $x^2 + 9x - 10 = (x+10)(x-1)$

15. $x^2 + 14x + 48 = (x+6)(x+8)$

16. $x^2 + 8xy + 15y^2 = (x+3y)(x+5y)$

17. $x^4 + 13x^2 + 42 = (x^2+6)(x^2+7)$

18. $x^4 - 2x^2 - 35 = (x^2-7)(x^2+5)$

19. $6x^2 + 30x + 36 = 6(x^2+5x+6) = 6(x+2)(x+3)$

20. $2x^2 - 28x + 96 = 2(x^2-14x+48)$
$$= 2(x-8)(x-6)$$

21. $4x^2 + 7x - 15 = (4x-5)(x+3)$

22. $12x^2 + 11x - 5 = 12x^2 - 4x + 15x - 5$
$$= 4x(3x-1) + 5(3x-1)$$
$$= (3x-1)(4x+5)$$

23. $2x^2 - x - 3 = (2x-3)(x+1)$

24. $3x^2 + 2x - 8 = (3x-4)(x+2)$

25. $20x^2 + 48x - 5 = (10x-1)(2x+5)$

26. $20x^2 + 21x - 5 = 20x^2 - 4x + 25x - 5$
$$= 4x(5x-1) + 5(5x-1)$$
$$= (5x-1)(4x+5)$$

27. $6x^2 + 4x - 10 = 2(3x^2 + 2x - 5)$
$$= 2(3x+5)(x-1)$$

28. $6x^2 - 4x - 10 = 2(3x^2 - 2x - 5)$
$$= 2(3x^2 - 5x + 3x - 5)$$
$$= 2[x(3x-5) + 1(3x-5)]$$
$$= 2(3x-5)(x+1)$$

29. $4x^2 - 26x + 30 = 2(2x^2 - 13x + 15)$
$$= 2(2x-3)(x-5)$$

30. $4x^2 - 20x - 144 = 4(x^2 - 5x - 36)$
$$= 4(x-9)(x+4)$$

31. $12x^2 + xy - 6y^2 = (4x+3y)(3x-2y)$

32. $6x^2 + 5xy - 25y^2 = (3x-5y)(2x+5y)$

33. $49x^2 - y^2 = (7x)^2 - (y)^2 = (7x+y)(7x-y)$

34. $16x^2 - 36y^2 = 4(4x^2 - 9y^2)$
$$= 4[(2x)^2 - (3y)^2]$$
$$= 4(2x-3y)(2x+3y)$$

35. $y^2 - 36x^2 = y^2 - (6x)^2 = (y+6x)(y-6x)$

36. $9y^2 - 25x^2 = (3y)^2 - (5x)^2 = (3y+5x)(3y-5x)$

37. $36x^2 + 12x + 1 = (6x)^2 + 2(6x)(1) + 1^2 = (6x+1)^2$

38. $25x^2 - 20x + 4 = (5x)^2 - 2(5x)(2) + 2^2$
$$= (5x-2)^2$$

39. $16x^2 - 24xy + 9y^2 = (4x)^2 - 2(4x)(3y) + (3y)^2$
$= (4x - 3y)^2$

40. $49x^2 - 28xy + 4y^2 = (7x)^2 - 2(7x)(2y) + (2y)^2$
$= (7x - 2y)^2$

41. $2x^2 - 32 = 2(x^2 - 16)$
$= 2[(x)^2 - (4)^2]$
$= 2(x - 4)(x + 4)$

42. $3x^2 - 27 = 3(x^2 - 9)$
$= 3[(x)^2 - (3)^2]$
$= 3(x - 3)(x + 3)$

43. $28x^2 + 140x + 175 = 7(4x^2 + 20x + 25)$
$= 7[(2x)^2 + 2(2x)(5) + 5^2]$
$= 7(2x + 5)^2$

44. $72x^2 - 192x + 128 = 8(9x^2 - 24x + 16)$
$= 8(3x - 4)^2$

45. $4x^2 - 9y^2 = (2x)^2 - (3y)^2 = (2x + 3y)(2x - 3y)$

46. $x^2 + 13x - 30 = (x + 15)(x - 2)$

47. $9x^2 - 9x - 4 = 9x^2 + 3x - 12x - 4$
$= 3x(3x + 1) - 4(3x + 1)$
$= (3x - 4)(3x + 1)$

48. $50x^3y^2 + 30x^2y^2 - 10x^2y^2 = 50x^3y^2 + 20x^2y^2$
$= 10x^2y^2(5x + 2)$

49. $3x^2 - 18x + 27 = 3(x^2 - 6x + 9) = 3(x - 3)^2$

50. $25x^3 - 60x^2 + 36x = x(25x^2 - 60x + 36)$
$= x[(5x)^2 - 2(5x)(6) + 6^2]$
$= x(5x - 6)^2$

51. $4x^2 - 13x - 12 = (4x + 3)(x - 4)$

52. $3x^3a^3 - 11x^4a^2 - 20x^5a$
$= x^3a(3a^2 - 11xa - 20x^2)$
$= x^3a(3a^2 + 4xa - 15xa - 20x^2)$
$= x^3a[a(3a + 4x) - 5x(3a + 4x)]$
$= x^3a(3a + 4x)(a - 5x)$

53. $12a^2 + 14ab - 10b^2 = 2(6a^2 + 7ab - 5b^2)$
$= 2(3a + 5b)(2a - b)$

54. $121a^2 + 66ab + 9b^2 = (11a)^2 + 2(11a)(3b) + (3b)^2$
$= (11a + 3b)^2$

55. $7a - 7 - ab + b = 7(a - 1) - b(a - 1)$
$= (a - 1)(7 - b)$

56. $3x^3 - 3x + 5yx^2 - 5y = 3x(x^2 - 1) + 5y(x^2 - 1)$
$= (3x + 5y)(x^2 - 1)$
$= (3x + 5y)(x + 1)(x - 1)$

57. $18b - 42 + 3bc - 7c = 18b + 3bc - 42 - 7c$
$= 3b(6 + c) - 7(6 + c)$
$= (3b - 7)(6 + c)$

58. $10b + 16 - 24x - 15bx = 10b - 15bx + 16 - 24x$
$= 5b(2 - 3x) + 8(2 - 3x)$
$= (5b + 8)(2 - 3x)$

59. $5xb - 35x + 4by - 28y = 5x(b - 7) + 4y(b - 7)$
$= (b - 7)(5x + 4y)$

60. $x^4 - 81y^{12} = (x^2)^2 - (9y^6)^2$
$= (x^2 + 9y^6)(x^2 - 9y^6)$
$= (x^2 + 9y^6)(x + 3y^3)(x - 3y^3)$

61. $6x^4 - x^2 - 15 = 6x^4 - 10x^2 + 9x^2 - 15$
$= 2x^2(3x^2 - 5) + 3(3x^2 - 5)$
$= (3x^2 - 5)(2x^2 + 3)$

62. $28yz - 16xyz + x^2yz = yz(28 - 16x + x^2)$
$= yz(14 - x)(2 - x)$

63. $12x^3 + 17x^2 + 6x = x(12x^2 + 17x + 6)$
$= x(12x^2 + 8x + 9x + 6)$
$= x[4x(3x + 2) + 3(3x + 2)]$
$= x(3x + 2)(4x + 3)$

64. $12w^2 - 12w + 3 = 3(4w^2 - 4w + 1)$
$$= 3[(2w)^2 - 2(2w)(1) + 1^2]$$
$$= 3(2w - 1)^2$$

65. $4y^3 + 10y^2 - 6y = 2y(2y^2 + 5y - 3)$
$$= 2y(2y - 1)(y + 3)$$

66. $9x^4 - 144 = 9(x^4 - 16)$
$$= 9[(x^2)^2 - (4)^2]$$
$$= 9(x^2 + 4)(x^2 - 4)$$
$$= 9(x^2 + 4)(x + 2)(x - 2)$$

67. $x^2 - 6x + 12$ is prime.

68. $8x^2 - 19x - 6$ is prime.

69. $8y^5 + 4y^3 - 60y = 4y(2y^4 + y^2 - 15)$
$$= 4y(2y^2 - 5)(y^2 + 3)$$

70. $16x^4y^2 - 56x^2y + 49$
$$= (4x^2y)^2 - 2(4x^2y)(7) + (7)^2$$
$$= (4x^2y - 7)^2$$

71. $2ax + 5a - 10b - 4bx = 2ax + 5a - 4bx - 10b$
$$= a(2x + 5) - 2b(2x + 5)$$
$$= (2x + 5)(a - 2b)$$

72. $2x^3 - 9 + x^2 - 18x = 2x^3 + x^2 - 18x - 9$
$$= x^2(2x + 1) - 9(2x + 1)$$
$$= (2x + 1)(x^2 - 9)$$
$$= (2x + 1)(x^2 - 3^2)$$
$$= (2x + 1)(x - 3)(x + 3)$$

73. $x^2 + x - 20 = 0$
$(x + 5)(x - 4) = 0$
$x + 5 = 0 \qquad\qquad x - 4 = 0$
$\quad x = -5 \qquad\qquad\quad x = 4$

74. $2x^2 + 11x - 6 = 0$
$2x^2 + 12x - x - 6 = 0$
$2x(x + 6) - 1(x + 6) = 0$
$(2x - 1)(x + 6) = 0$
$2x - 1 = 0 \qquad\qquad x + 6 = 0$
$\quad 2x = 1 \qquad\qquad\quad x = -6$
$\quad\quad x = \dfrac{1}{2}$

75. $7x^2 = 15x + x^2$
$7x^2 - x^2 - 15x = 0$
$6x^2 - 15x = 0$
$3x(2x - 5) = 0$
$3x = 0 \qquad\qquad 2x - 5 = 0$
$\ x = 0 \qquad\qquad\quad 2x = 5$
$\qquad\qquad\qquad\qquad x = \dfrac{5}{2}$

76. $5x^2 - x = 4x^2 + 12$
$x^2 - x = 12$
$x^2 - x - 12 = 0$
$(x - 4)(x + 3) = 0$
$x - 4 = 0 \qquad\qquad x + 3 = 0$
$\quad x = 4 \qquad\qquad\quad x = -3$

77. $2x^2 + 9x - 5 = 0$
$(2x - 1)(x + 5) = 0$
$2x - 1 = 0 \qquad\qquad x + 5 = 0$
$\qquad x = \dfrac{1}{2} \qquad\qquad\ x = -5$

78. $x^2 + 11x + 24 = 0$
$(x + 8)(x + 3) = 0$
$x + 8 = 0 \qquad\qquad x + 3 = 0$
$\quad x = -8 \qquad\qquad\quad x = -3$

79. $x^2 + 14x + 45 = 0$
$(x + 9)(x + 5) = 0$
$x + 9 = 0 \qquad\qquad x + 5 = 0$
$\quad x = -9 \qquad\qquad\quad x = -5$

80.
$$5x^2 = 7x + 6$$
$$5x^2 - 7x - 6 = 0$$
$$(5x + 3)(x - 2) = 0$$

$5x + 3 = 0$ 　　　　 $x - 2 = 0$
$5x = -3$ 　　　　　 $x = 2$
$$x = -\frac{3}{5}$$

81.
$$3x^2 + 6x = 2x^2 - 9$$
$$x^2 + 6x + 9 = 0$$
$$(x + 3)^2 = 0$$
$$x + 3 = 0$$
$$x = -3$$

82.
$$4x^2 + 9x - 9 = 0$$
$$(4x - 3)(x + 3) = 0$$

$4x - 3 = 0$ 　　　　 $x + 3 = 0$
$4x = 3$ 　　　　　　 $x = -3$
$$x = \frac{3}{4}$$

83.
$$5x^2 - 11x + 2 = 0$$
$$(5x - 1)(x - 2) = 0$$

$5x - 1 = 0$ 　　　　 $x - 2 = 0$
$$x = \frac{1}{5}$$ 　　　　　 $x = 2$

84. b = base of triangle
a = altitude = $b + 5$

$$\text{area} = \frac{1}{2}ba$$
$$25 = \frac{1}{2}b(b + 5)$$
$$25 = \frac{1}{2}(b^2 + 5b)$$
$$50 = b^2 + 5b$$
$$0 = b^2 + 5b - 50$$
$$0 = (b + 10)(b - 5)$$

$b + 10 = 0$ 　　　　　 $b - 5 = 0$
$b = -10$ not possible 　 $b = 5$

The base = 5 inches and the
altitude = 5 + 5 = 10 inches.

85. w = width of rectangle
l = length = $2w - 4$
Area = $wl = w(2w - 4)$
$$30 = w(2w - 4)$$
$$30 = 2w^2 - 4w$$
$$0 = 2w^2 - 4w - 30$$
$$0 = 2(w^2 - 2w - 15)$$
$$0 = 2(w - 5)(w + 3)$$

$w - 5 = 0$ 　　　　　 $w + 3 = 0$
$w = 5$ feet 　　　　 $w = -3$ not possible
width = 5
length = 2(5) − 4 = 6
The width is 5 feet and the length is 6 feet.

86. $h = -16t^2 + 80t + 96$
$$0 = -16t^2 + 80t + 96$$
$$0 = -16(t^2 - 5t - 6)$$
$$0 = -16(t - 6)(t + 1)$$

$t - 6 = 0$ 　　　　　 $t + 1 = 0$
$t = 6$ 　　　　　　 $t = -1$
Since the time must be positive, $t = 6$ seconds.

87.
$$480 = -5x^2 + 100x$$
$$5x^2 - 100x + 480 = 0$$
$$5(x^2 - 20x + 96) = 0$$
$$5(x - 12)(x - 8) = 0$$

$x - 12 = 0$ 　or　 $x - 8 = 0$
$x = 12$ 　　　　　 $x = 8$
The current is 12 amperes or 8 amperes.

How Am I Doing? Chapter 6 Test

1. $x^2 + 12x - 28 = (x + 14)(x - 2)$

2. $16x^2 - 81 = (4x)^2 - 9^2 = (4x + 9)(4x - 9)$

3. $10x^2 + 27x + 5 = 10x^2 + 2x + 25x + 5$
$$= 2x(5x + 1) + 5(5x + 1)$$
$$= (5x + 1)(2x + 5)$$

4. $9a^2 - 30a + 25 = (3a)^2 - 2(3a)(5) + (5)^2$
$$= (3a - 5)^2$$

5. $7x - 9x^2 + 14xy = x(7 - 9x + 14y)$

6. $10xy + 15by - 8x - 12b = 5y(2x + 3b) - 4(2x + 3b)$
$$= (2x + 3b)(5y - 4)$$

7. $6x^3 - 20x^2 + 16x = 2x(3x^2 - 10x + 8)$
$$= 2x(3x^2 - 6x - 4x + 8)$$
$$= 2x[3x(x-2) - 4(x-2)]$$
$$= 2x(x-2)(3x-4)$$

8. $5a^2c - 11ac + 2c = c(5a^2 - 11a + 2)$
$$= c(5a - 1)(a - 2)$$

9. $81x^2 - 100 = (9x)^2 - (10)^2 = (9x + 10)(9x - 10)$

10. $9x^2 - 15x + 4 = (3x - 1)(3x - 4)$

11. $20x^2 - 45 = 5(4x^2 - 9)$
$$= 5[(2x)^2 - 3^2]$$
$$= 5(2x + 3)(2x - 3)$$

12. $36x^2 + 1 = (6x)^2 + (1)^2$
It is prime.

13. $3x^3 + 11x^2 + 10x = x(3x^2 + 11x + 10)$
$$= x(3x + 5)(x + 2)$$

14. $60xy^2 - 20x^2y - 45y^3$
$$= -5y(-12xy + 4x^2 + 9y^2)$$
$$= -5y(4x^2 - 12xy + 9y^2)$$
$$= -5y[(2x)^2 - 2(2x)(3y) + (3y)^2]$$
$$= -5y(2x - 3y)^2$$

15. $81x^2 - 1 = (9x)^2 - (1)^2 = (9x + 1)(9x - 1)$

16. $81y^4 - 1 = (9y^2 + 1)(9y^2 - 1)$
$$= (9y^2 + 1)(3y + 1)(3y - 1)$$

17. $2ax + 6a - 5x - 15 = 2a(x + 3) - 5(x + 3)$
$$= (x + 3)(2a - 5)$$

18. $aw^2 - 8b + 2bw^2 - 4a = aw^2 - 4a + 2bw^2 - 8b$
$$= a(w^2 - 4) + 2b(w^2 - 4)$$
$$= (w^2 - 4)(a + 2b)$$
$$= (w - 2)(w + 2)(a + 2b)$$

19. $3x^2 - 3x - 90 = 3(x^2 - x - 30) = 3(x - 6)(x + 5)$

20. $2x^3 - x^2 - 15x = x(2x^2 - x - 15)$
$$= x(2x^2 - 6x + 5x - 15)$$
$$= x[2x(x - 3) + 5(x - 3)]$$
$$= x(x - 3)(2x + 5)$$

21. $x^2 + 14x + 45 = 0$
$(x + 9)(x + 5) = 0$
$x + 9 = 0 \qquad\qquad x + 5 = 0$
$\quad x = -9 \qquad\qquad\quad x = -5$

22. $14 + 3x(x + 2) = -7x$
$14 + 3x^2 + 6x = -7x$
$3x^2 + 13x + 14 = 0$
$(3x + 7)(x + 2) = 0$
$3x + 7 = 0 \qquad\qquad x + 2 = 0$
$\quad 3x = -7 \qquad\qquad\quad x = -2$
$\quad x = -\dfrac{7}{3}$

23. $\quad 2x^2 + x - 10 = 0$
$(2x + 5)(x - 2) = 0$
$2x + 5 = 0 \qquad\qquad x - 2 = 0$
$\quad x = -\dfrac{5}{2} \qquad\qquad\quad x = 2$

24. $x^2 - 3x - 28 = 0$
$(x - 7)(x + 4) = 0$
$x - 7 = 0 \qquad\qquad x + 4 = 0$
$\quad x = 7 \qquad\qquad\quad x = -4$

25. $x = $ width
$2x - 1 = $ length
$$x(2x - 1) = 91$$
$$2x^2 - x = 91$$
$$2x^2 - x - 91 = 0$$
$$(2x + 13)(x - 7) = 0$$
$2x + 13 = 0 \qquad\qquad x - 7 = 0$
$\quad 2x = -13 \qquad\qquad\quad x = 7$
$\quad x = -\dfrac{13}{2}$
Since the width cannot be negative, $x = 7$ and
$2x - 1 = 2(7) - 1 = 13$.
The width = 7 miles, and the length = 13 miles.

Cumulative Test for Chapters 0–6

1. $\dfrac{11}{15} - \dfrac{7}{10} = \dfrac{22}{30} - \dfrac{21}{30} = \dfrac{1}{30}$

2. $-\dfrac{5}{3} + \dfrac{1}{2} + \dfrac{5}{6} = -\dfrac{5 \cdot 2}{3 \cdot 2} + \dfrac{1 \cdot 3}{2 \cdot 3} + \dfrac{5}{6}$

$= -\dfrac{10}{6} + \dfrac{3}{6} + \dfrac{5}{6}$

$= \dfrac{-2}{6}$

$= -\dfrac{1}{3}$

3. $\left(-4\dfrac{1}{2}\right) \div \left(5\dfrac{1}{4}\right) = \left(-\dfrac{9}{2}\right) \div \left(\dfrac{21}{4}\right)$

$= -\dfrac{9}{2} \cdot \dfrac{4}{21}$

$= -\dfrac{\cancel{3} \cdot 3 \cdot \cancel{2} \cdot 2}{\cancel{2} \cdot \cancel{3} \cdot 7}$

$= -\dfrac{6}{7}$

4. 6% of $1842.5 = 0.06(1842.5) = 110.55$

5. $\dfrac{11,904}{x} = 0.96$

$11,904 = 0.96x$

$12,400 = x$

VBM has 12,400 employees.

6. $3a^2 + ab - 4b^2 = 3(5)^2 + (5)(-1) - 4(-1)^2$

$= 3(25) + (5)(-1) - 4(1)$

$= 75 - 5 - 4$

$= 66$

7. $7x(3x-4) - 5x(2x-3) - (3x)^2$

$= 21x^2 - 28x - 10x^2 + 15x - 9x^2$

$= 2x^2 - 13x$

8. $\dfrac{2}{3}x + 6 = 4(x-11)$

$3\left(\dfrac{2}{3}x\right) + 3(6) = 3(4)(x-11)$

$2x + 18 = 12x - 132$

$18 = 10x - 132$

$150 = 10x$

$15 = x$

9. $7x - 3(4-2x) = 14x - (3-x)$

$7x - 12 + 6x = 14x - 3 + x$

$13x - 12 = 15x - 3$

$-9 = 2x$

$-\dfrac{9}{2} = x$

10. $2 - 5x > 17$

$2 - 2 - 5x > 17 - 2$

$-5x > 15$

$\dfrac{-5x}{-5} < \dfrac{15}{-5}$

$x < -3$

11. $\dfrac{1}{3}x - \dfrac{1}{2}y = 4$

$\dfrac{3}{8}x - \dfrac{1}{4}y = 7$

Clear the fractions.

$2x - 3y = 24 \quad (1)$

$3x - 2y = 56 \quad (2)$

Multiply (1) by -3 and (2) by 2.

$-6x + 9y = -72$

$\underline{6x - 4y = 112}$

$5y = 40$

$y = 8$

Substitute 8 for y in (1).

$2x - 3(8) = 24$

$2x = 48$

$x = 24$

$(24, 8)$

12. $10x - 5y = 45 \quad (1)$

$3x - 8y = 7 \quad (2)$

Multiply (1) by -8 and (2) by 5.

$-80x + 40y = -360$

$\underline{15x - 40y = 35}$

$-65x = -325$

$x = 5$

Substitute 5 for x in (2).

$3(5) - 8y = 7$

$-8y = -8$

$y = 1$

$(5, 1)$

13. r = speed of boat in still water
s = speed of the stream

$$r - s = \frac{6}{3} \quad (1)$$

$$r + s = \frac{6}{1.5} \quad (2)$$

Multiply (1) by -1.

$$-r + s = -2$$
$$\underline{r + s = 4}$$
$$2s = 2$$
$$s = 1$$

Speed of the stream is 1 mph.

14. $(6, -1), (-4, -2)$

$$m = \frac{y_2 - y_1}{x_2 - x_1} = \frac{-2 - (-1)}{-4 - 6} = \frac{-2 + 1}{-10} = \frac{-1}{-10} = \frac{1}{10}$$

15. $m = -\dfrac{3}{4}, (x_1, y_1) = (2, 5)$

$$y - y_1 = m(x - x_1)$$
$$y - 5 = -\frac{3}{4}(x - 2)$$
$$y = -\frac{3}{4}x + \frac{6}{4} + 5$$
$$y = -\frac{3}{4}x + \frac{13}{2}$$

16. $4x - 8y = 10$
$$-8y = -4x + 10$$
$$y = \frac{1}{2}x - \frac{5}{4}$$

x	y	(x, y)
0	$\frac{1}{2}(0) - \frac{5}{4} = -\frac{5}{4}$	$\left(0, -\frac{5}{4}\right)$
$\frac{5}{2}$	$\frac{1}{2}\left(\frac{5}{2}\right) - \frac{5}{4} = \frac{5}{4} - \frac{5}{4} = 0$	$\left(\frac{5}{2}, 0\right)$
4	$\frac{1}{2}(4) - \frac{5}{4} = 2 - \frac{5}{4} = \frac{8}{4} - \frac{5}{4} = \frac{3}{4}$	$\left(4, \frac{3}{4}\right)$

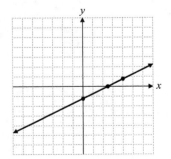

17. $(4x - 5)(5x + 1) = 20x^2 + 4x - 25x - 5$
$$= 20x^2 - 21x - 5$$

18. $(3x - 5)^2 = (3x)^2 - 2(3x)(5) + 5^2$
$$= 9x^2 - 30x + 25$$

19. $(-4x^4 y^5)(5xy^3) = (-4)(5)x^{4+1}y^{5+3} = -20x^5 y^8$

20. $(-2x^3 y^2 z^0)^4 = (-2x^3 y^2)^4$
$$= (-2)^4 (x^3)^4 (y^2)^4$$
$$= 16x^{12} y^8$$

21. $\dfrac{9x^{-3} y^{-4}}{w^2 z^{-8}} = \dfrac{9z^8}{w^2 x^3 y^4}$

22. $0.00056 = 5.6 \times 10^{-4}$

23. $121x^2 - 64y^2 = (11x)^2 - (8y)^2$
$$= (11x + 8y)(11x - 8y)$$

24. $4x + 120 - 80x^2 = -80x^2 + 4x + 120$
$$= -4(20x^2 - x - 30)$$
$$= -4(20x^2 + 24x - 25x - 30)$$
$$= -4[4x(5x + 6) - 5(5x + 6)]$$
$$= -4(5x + 6)(4x - 5)$$

25. $16x^3 + 40x^2 + 25x = x(16x^2 + 40x + 25)$
$$= x[(4x)^2 + 2(4x)(5) + 5^2]$$
$$= x(4x + 5)^2$$

26. $16x^4 - b^4 = (4x^2)^2 - (b^2)^2$
$$= (4x^2 - b^2)(4x^2 + b^2)$$
$$= [(2x)^2 - (b)^2](4x^2 + b^2)$$
$$= (2x - b)(2x + b)(4x^2 + b^2)$$

27. $2ax - 4bx + 3a - 6b = 2x(a - 2b) + 3(a - 2b)$
$$= (a - 2b)(2x + 3)$$

28. $x^2 + 5x - 36 = 0$
$$(x - 4)(x + 9) = 0$$
$$x - 4 = 0 \qquad\qquad x + 9 = 0$$
$$x = 4 \qquad\qquad\quad x = -9$$

Chapter 7

7.1 Exercises

1. $\dfrac{4x-24y}{x-6y} = \dfrac{4(x-6y)}{x-6y} = 4$

3. $\dfrac{6x+18}{x^2+3x} = \dfrac{6(x+3)}{x(x+3)} = \dfrac{6}{x}$

5. $\dfrac{9x^2+6x+1}{1-9x^2} = \dfrac{(3x+1)(3x+1)}{(1+3x)(1-3x)} = \dfrac{3x+1}{1-3x}$

7. $\dfrac{3a^2b(a-2b)}{6ab^2} = \dfrac{a(a-2b)}{2b}$

9. $\dfrac{x^2+x-2}{x^2-x} = \dfrac{(x+2)(x-1)}{x(x-1)} = \dfrac{x+2}{x}$

11. $\dfrac{x^2-3x-10}{3x^2+5x-2} = \dfrac{(x-5)(x+2)}{(3x-1)(x+2)} = \dfrac{x-5}{3x-1}$

13. $\dfrac{x^2+4x-21}{x^3-49x} = \dfrac{(x+7)(x-3)}{x(x^2-49)}$

$\qquad = \dfrac{(x+7)(x-3)}{x(x+7)(x-7)}$

$\qquad = \dfrac{x-3}{x(x-7)}$

15. $\dfrac{3x^2-11x-4}{x^2+x-20} = \dfrac{(3x+1)(x-4)}{(x+5)(x-4)} = \dfrac{3x+1}{x+5}$

17. $\dfrac{3x^2-8x+5}{4x^2-5x+1} = \dfrac{(3x-5)(x-1)}{(4x-1)(x-1)} = \dfrac{3x-5}{4x-1}$

19. $\dfrac{5x^2-27x+10}{5x^2+3x-2} = \dfrac{(5x-2)(x-5)}{(5x-2)(x+1)} = \dfrac{x-5}{x+1}$

21. $\dfrac{12-3x}{5x^2-20x} = \dfrac{-3(-4+x)}{5x(x-4)} = \dfrac{-3}{5x} = -\dfrac{3}{5x}$

23. $\dfrac{2x^2-7x-15}{25-x^2} = \dfrac{(2x+3)(x-5)}{(5-x)(5+x)}$

$\qquad = \dfrac{-2x-3}{x+5}$

$\qquad = \dfrac{-2x-3}{5+x}$ or $-\dfrac{2x+3}{5+x}$

25. $\dfrac{(3x+4)^2}{9x^2+9x-4} = \dfrac{(3x+4)(3x+4)}{(3x+4)(3x-1)} = \dfrac{3x+4}{3x-1}$

27. $\dfrac{3x^2+13x-10}{20-x-x^2} = \dfrac{(3x-2)(x+5)}{(5+x)(4-x)} = \dfrac{3x-2}{4-x}$

29. $\dfrac{a^2+ab-6b^2}{3a^2+8ab-3b^2} = \dfrac{(a+3b)(a-2b)}{(3a-b)(a+3b)} = \dfrac{a-2b}{3a-b}$

Cumulative Review

31. $(3x-7)^2 = (3x)^2 - 2(3x)(7) + (7)^2$

$\qquad = 9x^2 - 42x + 49$

32. $(7x+6y)(7x-6y) = (7x)^2 - (6y)^2$

$\qquad\qquad = 49x^2 - 36y^2$

33. $(2x+3)(x-4) = 2x^2 - 8x + 3x - 12$

$\qquad\qquad = 2x^2 - 5x - 12$

34. $(2x+3)(x-4)(x-2)$

$\quad = (2x+3)(x^2-6x+8)$

$\quad = 2x^3 - 12x^2 + 16x + 3x^2 - 18x + 24$

$\quad = 2x^3 - 9x^2 - 2x + 24$

35. $\dfrac{2a^2}{7} + \dfrac{3b}{2} + 3a^2 - \dfrac{3b}{4} = \dfrac{2}{7}a^2 + \dfrac{3}{2}b + 3a^2 - \dfrac{3}{4}b$

$\qquad = \left(\dfrac{2}{7}+3\right)a^2 + \left(\dfrac{3}{2}-\dfrac{3}{4}\right)b$

$\qquad = \left(\dfrac{2}{7}+\dfrac{21}{7}\right)a^2 + \left(\dfrac{6}{4}-\dfrac{3}{4}\right)b$

$\qquad = \dfrac{23}{7}a^2 + \dfrac{3}{4}b$

36. $\dfrac{-35}{12} \div \dfrac{5}{14} = -\dfrac{35}{12} \cdot \dfrac{14}{5}$

$\qquad\qquad = -\dfrac{5 \cdot 7 \cdot 2 \cdot 7}{2 \cdot 6 \cdot 5}$

$\qquad\qquad = -\dfrac{7 \cdot 7}{6}$

$\qquad\qquad = -\dfrac{49}{6}$ or $-8\dfrac{1}{6}$

37. $\dfrac{4\frac{7}{8}}{3} = \dfrac{\frac{39}{8}}{\frac{3}{1}} = \dfrac{39}{8} \div \dfrac{3}{1} = \dfrac{39}{8} \cdot \dfrac{1}{3} = \dfrac{13}{8} = 1\dfrac{5}{8}$

Each lot will be $1\dfrac{5}{8}$ acres.

38. $17.5 \text{ lb} \cdot \dfrac{22 \text{ min}}{\text{lb}} = 385$

The cooking time is 385 minutes, or 6 hours, 25 minutes.

Quick Quiz 7.1

1. $\dfrac{x^3 + 3x^2}{x^3 - 2x^2 - 15x} = \dfrac{x^2(x+3)}{x(x^2 - 2x - 15)}$

$\qquad\qquad = \dfrac{x^2(x+3)}{x(x+3)(x-5)}$

$\qquad\qquad = \dfrac{x}{x-5}$

2. $\dfrac{6 - 2ab}{ab^2 - 3b} = \dfrac{2(3 - ab)}{b(ab - 3)} = \dfrac{2(3-ab)}{-b(3-ab)} = -\dfrac{2}{b}$

3. $\dfrac{8x^2 + 6x - 5}{16x^2 + 40x + 25} = \dfrac{8x^2 - 4x + 10x - 5}{16x^2 + 20x + 20x + 25}$

$\qquad\qquad = \dfrac{4x(2x-1) + 5(2x-1)}{4x(4x+5) + 5(4x+5)}$

$\qquad\qquad = \dfrac{(4x+5)(2x-1)}{(4x+5)(4x+5)}$

$\qquad\qquad = \dfrac{2x-1}{4x+5}$

4. Answers may vary. Possible solution: Completely factoring both numerator and denominator is the only way to see what factors are shared, and may consequently be eliminated. In this case, it can be seen that y and $(x - y)$ are common factors.

7.2 Exercises

1. Before multiplying rational expressions, we should always first try to <u>factor the numerators and denominators completely and divide out any common factors</u>.

3. $\dfrac{4x+12}{x-4} \cdot \dfrac{x^2 + x - 20}{x^2 + 6x + 9} = \dfrac{4(x+3)}{x-4} \cdot \dfrac{(x+5)(x-4)}{(x+3)(x+3)}$

$\qquad\qquad = \dfrac{4(x+5)}{x+3}$

5. $\dfrac{24x^3}{4x^2 - 36} \cdot \dfrac{2x^2 + 6x}{16x^2} = \dfrac{3(8x^3)}{4(x^2 - 9)} \cdot \dfrac{2x(x+3)}{2 \cdot 8x^2}$

$\qquad\qquad = \dfrac{3 \cdot 8x^3}{4(x+3)(x-3)} \cdot \dfrac{2x(x+3)}{2 \cdot 8x^2}$

$\qquad\qquad = \dfrac{3x^2}{4(x-3)}$

7. $\dfrac{x^2 + 3x - 10}{x^2 + x - 20} \cdot \dfrac{x^2 - 3x - 4}{x^2 + 4x + 3}$

$\qquad = \dfrac{(x-2)(x+5)}{(x-4)(x+5)} \cdot \dfrac{(x-4)(x+1)}{(x+3)(x+1)}$

$\qquad = \dfrac{x-2}{x+3}$

9. $\dfrac{x+6}{x-8} \div \dfrac{x+5}{x^2 - 6x - 16} = \dfrac{x+6}{x-8} \cdot \dfrac{x^2 - 6x - 16}{x+5}$

$\qquad\qquad = \dfrac{x+6}{x-8} \cdot \dfrac{(x-8)(x+2)}{x+5}$

$\qquad\qquad = \dfrac{(x+6)(x+2)}{x+5}$

11. $(5x+4) \div \dfrac{25x^2 - 16}{5x^2 + 11x - 12}$

$\qquad = \dfrac{5x+4}{1} \cdot \dfrac{5x^2 + 11x - 12}{25x^2 - 16}$

$\qquad = \dfrac{5x+4}{1} \cdot \dfrac{(x+3)(5x-4)}{(5x+4)(5x-4)}$

$\qquad = x+3$

13. $\dfrac{3x^2+12xy+12y^2}{x^2+4xy+3y^2} \div \dfrac{4x+8y}{x+y}$

$= \dfrac{3x^2+12xy+12y^2}{x^2+4xy+3y^2} \cdot \dfrac{x+y}{4x+8y}$

$= \dfrac{3(x+2y)(x+2y)}{(x+3y)(x+y)} \cdot \dfrac{x+y}{4(x+2y)}$

$= \dfrac{3(x+2y)}{4(x+3y)}$

15. $\dfrac{(x+5)^2}{3x^2-7x+2} \cdot \dfrac{x^2-4x+4}{x+5}$

$= \dfrac{(x+5)(x+5)}{(3x-1)(x-2)} \cdot \dfrac{(x-2)(x-2)}{x+5}$

$= \dfrac{(x+5)(x-2)}{3x-1}$

17. $\dfrac{x^2+x-30}{10-2x} \div \dfrac{x^2+4x-12}{5x+15}$

$= \dfrac{x^2+x-30}{10-2x} \cdot \dfrac{5x+15}{x^2+4x-12}$

$= \dfrac{(x+6)(x-5)}{-2(x-5)} \cdot \dfrac{5(x+3)}{(x+6)(x-2)}$

$= -\dfrac{5(x+3)}{2(x-2)}$

19. $\dfrac{y^2+4y-12}{y^2+2y-24} \cdot \dfrac{y^2-16}{y^2+2y-8}$

$= \dfrac{(y+6)(y-2)}{(y+6)(y-4)} \cdot \dfrac{(y+4)(y-4)}{(y+4)(y-2)}$

$= 1$

21. $\dfrac{x^2+10x+24}{2x^2+13x+6} \cdot \dfrac{2x^2-5x-3}{x^2+5x-24}$

$= \dfrac{(x+6)(x+4)}{(2x+1)(x+6)} \cdot \dfrac{(2x+1)(x-3)}{(x+8)(x-3)}$

$= \dfrac{x+4}{x+8}$

Cumulative Review

23. $6x^2+3x-18 = 5x-2+6x^2$

$3x-18 = 5x-2$

$-18 = 2x-2$

$-16 = 2x$

$-8 = x$

24. $\dfrac{3}{4}+\dfrac{1}{2}-\dfrac{4}{7} = \dfrac{3\cdot7}{4\cdot7}+\dfrac{1\cdot14}{2\cdot14}-\dfrac{4\cdot4}{7\cdot4}$

$= \dfrac{21+14-16}{28}$

$= \dfrac{19}{28}$

25. Area of road $= 8981(90)$
Area of sidewalk $= 8981(10.5)$
Difference in area $= 8981(90)-8981(10.5)$
$\qquad\qquad\qquad = 8981(79.5)$
Difference in cost $= 8981(79.5)x = \$713{,}989.5x$

26. Harold: length $= x$, width $= x$
George: length $= x+3$, width $= x-2$
$A = LW$
$36 = (x+3)(x-2)$
$36 = x^2+x-6$
$0 = x^2+x-42$
$0 = (x+7)(x-6)$
$\begin{array}{ll} x+7=0 & x-6=0 \\ \quad x=-7 & \quad x=6 \end{array}$
Not possible
The dimensions of their gardens are as follows:
Harold: 6 ft by 6 ft
George: 9 ft by 4 ft

Quick Quiz 7.2

1. $\dfrac{2x-10}{x-4} \cdot \dfrac{x^2+5x+4}{x^2-4x-5} = \dfrac{2(x-5)}{x-4} \cdot \dfrac{(x+4)(x+1)}{(x-5)(x+1)}$

$= \dfrac{2(x+4)}{x-4}$

2. $\dfrac{3x^2-13x-10}{3x^2+2x} \cdot \dfrac{x^2-25x}{x^2-25}$

$= \dfrac{(3x+2)(x-5)}{x(3x+2)} \cdot \dfrac{x(x-25)}{(x-5)(x+5)}$

$= \dfrac{x-25}{x+5}$

3. $\dfrac{2x^2-18}{3x^2+3x} \div \dfrac{x^2+6x+9}{x^2+4x+3}$

$= \dfrac{2x^2-18}{3x^2+3x} \cdot \dfrac{x^2+4x+3}{x^2+6x+9}$

$= \dfrac{2(x-3)(x+3)}{3x(x+1)} \cdot \dfrac{(x+3)(x+1)}{(x+3)(x+3)}$

$= \dfrac{2(x-3)}{3x}$

4. Answers may vary. Possible solution:
The first step is to change the operation from division to multiplication, by changing the operator and inverting the second fraction. Secondly, all terms must be factored completely. Next, common factors in the numerators and denominators may be canceled. Lastly the multiplication operation is performed.

7.3 Exercises

1. The LCD would be a product that contains each factor. However, any repeated factor in any one denominator must be repeated the greatest number of times it occurs in any one denominator. So the LCD would be

$(x+5)(x+3)^2$.

3. $\dfrac{3x+2}{5+2x}+\dfrac{x}{2x+5}=\dfrac{3x+2+x}{2x+5}=\dfrac{4x+2}{2x+5}=\dfrac{2(2x+1)}{2x+5}$

5. $\dfrac{3x}{x+3}-\dfrac{x+5}{x+3}=\dfrac{3x-(x+5)}{x+3}$

$=\dfrac{3x-x-5}{x+3}$

$=\dfrac{2x-5}{x+3}$

7. $\dfrac{8x+3}{5x+7}-\dfrac{6x+10}{5x+7}=\dfrac{8x+3-(6x+10)}{5x+7}$

$=\dfrac{8x+3-6x-10}{5x+7}$

$=\dfrac{2x-7}{5x+7}$

9. $3a^2b^3$, ab^2

$\text{LCD}=3a^2b^3$

11. $18x^2y^5=2\cdot3^2x^2y^5$

$30x^3y^3=2\cdot3\cdot5x^3y^3$

$\text{LCD}=2\cdot3^2\cdot5x^3y^5=90x^3y^5$

13. $2x-6=2(x-3)$

$9x-27=9(x-3)$

$\text{LCD}=2\cdot9(x-3)=18(x-3)$

15. $x+3=x+3$

$x^2-9=(x+3)(x-3)$

$\text{LCD}=(x+3)(x-3)=x^2-9$

17. $3x^2+14x-5=(3x-1)(x+5)$

$9x^2-6x+1=(3x-1)(3x-1)$

$\text{LCD}=(x+5)(3x-1)^2$

19. $\dfrac{7}{ab}+\dfrac{3}{b}=\dfrac{7}{ab}+\dfrac{3}{b}\cdot\dfrac{a}{a}=\dfrac{7}{ab}+\dfrac{3a}{ab}=\dfrac{7+3a}{ab}$

21. $\dfrac{3}{x+7}+\dfrac{8}{x^2-49}=\dfrac{3}{x+7}\cdot\dfrac{x-7}{x-7}+\dfrac{8}{(x+7)(x-7)}$

$=\dfrac{3x-21+8}{(x+7)(x-7)}$

$=\dfrac{3x-13}{(x+7)(x-7)}$

23. $\dfrac{4y}{y+1}+\dfrac{y}{y-1}=\dfrac{4y}{y+1}\cdot\dfrac{y-1}{y-1}+\dfrac{y}{y-1}\cdot\dfrac{y+1}{y+1}$

$=\dfrac{4y^2-4y}{(y+1)(y-1)}+\dfrac{y^2+y}{(y+1)(y-1)}$

$=\dfrac{4y^2-4y+y^2+y}{(y+1)(y-1)}$

$=\dfrac{5y^2-3y}{(y+1)(y-1)}$

$=\dfrac{y(5y-3)}{(y+1)(y-1)}$

25. $\dfrac{6}{5a}+\dfrac{5}{3a+2}=\dfrac{6}{5a}\cdot\dfrac{3a+2}{3a+2}+\dfrac{5}{3a+2}\cdot\dfrac{5a}{5a}$

$=\dfrac{6(3a+2)+5\cdot5a}{5a(3a+2)}$

$=\dfrac{18a+12+25a}{5a(3a+2)}$

$=\dfrac{43a+12}{5a(3a+2)}$

27. $\dfrac{2}{3xy}+\dfrac{1}{6yz}=\dfrac{2}{3xy}\cdot\dfrac{2z}{2z}+\dfrac{1}{6yz}\cdot\dfrac{x}{x}$

$=\dfrac{4z}{6xyz}+\dfrac{x}{6xyz}$

$=\dfrac{4z+x}{6xyz}$

29. $\dfrac{5x+6}{x-3} - \dfrac{x-2}{2x-6} = \dfrac{5x+6}{x-3} \cdot \dfrac{2}{2} - \dfrac{x-2}{2(x-3)}$

$\qquad = \dfrac{2(5x+6) - (x-2)}{2(x-3)}$

$\qquad = \dfrac{10x+12 - x+2}{2(x-3)}$

$\qquad = \dfrac{9x+14}{2(x-3)}$

31. $\dfrac{3x}{x^2-25} - \dfrac{2}{x+5} = \dfrac{3x}{(x+5)(x-5)} - \dfrac{2}{x+5} \cdot \dfrac{x-5}{x-5}$

$\qquad = \dfrac{3x - 2(x-5)}{(x+5)(x-5)}$

$\qquad = \dfrac{3x - 2x + 10}{(x+5)(x-5)}$

$\qquad = \dfrac{x+10}{(x+5)(x-5)}$

33. $\dfrac{a+3b}{2} - \dfrac{a-b}{5} = \dfrac{a+3b}{2} \cdot \dfrac{5}{5} - \dfrac{a-b}{5} \cdot \dfrac{2}{2}$

$\qquad = \dfrac{5(a+3b) - 2(a-b)}{2 \cdot 5}$

$\qquad = \dfrac{5a + 15b - 2a + 2b}{10}$

$\qquad = \dfrac{3a + 17b}{10}$

35. $\dfrac{8}{2x-3} - \dfrac{6}{x+2} = \dfrac{8}{2x-3} \cdot \dfrac{x+2}{x+2} - \dfrac{6}{x+2} \cdot \dfrac{2x-3}{2x-3}$

$\qquad = \dfrac{8x+16}{(2x-3)(x+2)} - \dfrac{12x-18}{(2x-3)(x+2)}$

$\qquad = \dfrac{8x+16 - (12x-18)}{(2x-3)(x+2)}$

$\qquad = \dfrac{8x+16 - 12x + 18}{(2x-3)(x+2)}$

$\qquad = \dfrac{-4x+34}{(2x-3)(x+2)}$

$\qquad = \dfrac{-2(2x-17)}{(2x-3)(x+2)}$

37. $\dfrac{x}{x^2+2x-3} - \dfrac{x}{x^2-5x+4}$

$\qquad = \dfrac{x}{(x+3)(x-1)} - \dfrac{x}{(x-4)(x-1)}$

$\qquad = \dfrac{x}{(x+3)(x-1)} \cdot \dfrac{x-4}{x-4} - \dfrac{x}{(x-4)(x-1)} \cdot \dfrac{x+3}{x+3}$

$\qquad = \dfrac{x^2-4x}{(x+3)(x-1)(x-4)} - \dfrac{x^2+3x}{(x+3)(x-1)(x-4)}$

$\qquad = \dfrac{x^2-4x - (x^2+3x)}{(x+3)(x-1)(x-4)}$

$\qquad = \dfrac{x^2-4x - x^2 - 3x}{(x+3)(x-1)(x-4)}$

$\qquad = \dfrac{-7x}{(x+3)(x-1)(x-4)}$

39. $\dfrac{3}{x^2+9x+20} + \dfrac{1}{x^2+10x+24}$

$\qquad = \dfrac{3}{(x+4)(x+5)} \cdot \dfrac{x+6}{x+6} + \dfrac{1}{(x+4)(x+6)} \cdot \dfrac{x+5}{x+5}$

$\qquad = \dfrac{3x+18 + x+5}{(x+4)(x+5)(x+6)}$

$\qquad = \dfrac{4x+23}{(x+4)(x+5)(x+6)}$

41. $\dfrac{3x-8}{x^2-5x+6} + \dfrac{x+2}{x^2-6x+8}$

$\qquad = \dfrac{3x-8}{(x-2)(x-3)} + \dfrac{x+2}{(x-4)(x-2)}$

$\qquad = \dfrac{3x-8}{(x-2)(x-3)} \cdot \dfrac{x-4}{x-4} + \dfrac{x+2}{(x-4)(x-2)} \cdot \dfrac{x-3}{x-3}$

$\qquad = \dfrac{3x^2-20x+32}{(x-2)(x-3)(x-4)} + \dfrac{x^2-x-6}{(x-2)(x-3)(x-4)}$

$\qquad = \dfrac{3x^2-20x+32 + x^2 - x - 6}{(x-2)(x-3)(x-4)}$

$\qquad = \dfrac{4x^2-21x+26}{(x-2)(x-3)(x-4)}$

$\qquad = \dfrac{(x-2)(4x-13)}{(x-2)(x-3)(x-4)}$

$\qquad = \dfrac{4x-13}{(x-3)(x-4)}$

43. $\dfrac{6x}{y-2x} - \dfrac{5x}{2x-y} = \dfrac{6x}{-(2x-y)} - \dfrac{5x}{2x-y}$

$\qquad\qquad = \dfrac{-6x-5x}{2x-y}$

$\qquad\qquad = \dfrac{-11x}{2x-y}$

$\qquad\qquad = \dfrac{11x}{y-2x}$

45. $\dfrac{3y}{8y^2+2y-1} - \dfrac{5y}{2y^2-9y-5}$

$= \dfrac{3y}{(4y-1)(2y+1)} - \dfrac{5y}{(2y+1)(y-5)}$

$= \dfrac{3y}{(4y-1)(2y+1)} \cdot \dfrac{y-5}{y-5} - \dfrac{5y}{(2y+1)(y-5)} \cdot \dfrac{4y-1}{4y-1}$

$= \dfrac{3y^2-15y}{(4y-1)(2y+1)(y-5)} - \dfrac{20y^2-5y}{(4y-1)(2y+1)(y-5)}$

$= \dfrac{3y^2-15y-(20y^2-5y)}{(4y-1)(2y+1)(y-5)}$

$= \dfrac{3y^2-15y-20y^2+5y}{(4y-1)(2y+1)(y-5)}$

$= \dfrac{-17y^2-10y}{(4y-1)(2y+1)(y-5)}$

47. $\dfrac{2x}{2x^2-5x-3} + \dfrac{3}{x-3}$

$= \dfrac{2x}{(2x+1)(x-3)} + \dfrac{3}{x-3}$

$= \dfrac{2x}{(2x+1)(x-3)} + \dfrac{3}{x-3} \cdot \dfrac{2x+1}{2x+1}$

$= \dfrac{2x}{(2x+1)(x-3)} + \dfrac{6x+3}{(2x+1)(x-3)}$

$= \dfrac{2x+6x+3}{(2x+1)(x-3)}$

$= \dfrac{8x+3}{(2x+1)(x-3)}$

Cumulative Review

49. $\dfrac{1}{3}(x-2) + \dfrac{1}{2}(x+3) = \dfrac{1}{4}(3x+1)$

$12\left[\dfrac{1}{3}(x-2) + \dfrac{1}{2}(x+3)\right] = 12\left(\dfrac{1}{4}\right)(3x+1)$

$\qquad 4(x-2)+6(x+3) = 3(3x+1)$

$\qquad 4x-8+6x+18 = 9x+3$

$\qquad\qquad 10x+10 = 9x+3$

$\qquad\qquad\quad x+10 = 3$

$\qquad\qquad\qquad\quad x = -7$

50. $\qquad 4.8-0.6x = 0.8(x-1)$

$\quad 10(4.8-0.6x) = 10[0.8(x-1)]$

$\qquad 48-6x = 8(x-1)$

$\qquad 48-6x = 8x-8$

$\qquad 48-14x = -8$

$\qquad\quad -14x = -56$

$\qquad\qquad\quad x = 4$

51. $\qquad x-\dfrac{1}{5}x > \dfrac{1}{2}+\dfrac{1}{10}x$

$10\left(x-\dfrac{1}{5}x\right) > 10\left(\dfrac{1}{2}+\dfrac{1}{10}x\right)$

$\qquad 10x-2x > 5+x$

$\qquad\quad 8x > 5+x$

$\qquad\quad 7x > 5$

$\qquad\qquad x > \dfrac{5}{7}$

52. $(3x^3y^4)^4 = 3^4 x^{3\cdot4} y^{4\cdot4} = 81x^{12}y^{16}$

53. Let x = number of days.

$\qquad 2.75(2x) > 90$

$\qquad\quad 5.5x > 90$

$\qquad\qquad x > \dfrac{90}{5.5}$

$\qquad\qquad x > 16\dfrac{4}{11}$

It is cheaper if you use the subway at least 17 days per month.

54. $5.3\% - 0.04\% = 5.26\%$

$0.0526(5,475,000) = 287,985$

287,985 more people spoke Swedish than Sámi.

Quick Quiz 7.3

1. $\dfrac{3}{x^2-2x-8}+\dfrac{2}{x-4}$

    $=\dfrac{3}{(x-4)(x+2)}+\dfrac{2}{x-4}\cdot\dfrac{x+2}{x+2}$

    $=\dfrac{3+2x+4}{(x-4)(x+2)}$

    $=\dfrac{7+2x}{(x-4)(x+2)}$

2. $\dfrac{2x+y}{xy}-\dfrac{b-y}{by}=\dfrac{2x+y}{xy}\cdot\dfrac{b}{b}-\dfrac{b-y}{by}\cdot\dfrac{x}{x}$

    $=\dfrac{2bx+by}{bxy}-\dfrac{bx-xy}{bxy}$

    $=\dfrac{2bx+by-bx+xy}{bxy}$

    $=\dfrac{bx+by+xy}{bxy}$

3. $\dfrac{2}{x^2-9}+\dfrac{3}{x^2+7x+12}$

    $=\dfrac{2}{(x-3)(x+3)}\cdot\dfrac{x+4}{x+4}+\dfrac{3}{(x+4)(x+3)}\cdot\dfrac{x-3}{x-3}$

    $=\dfrac{2x+8+3x-9}{(x+4)(x+3)(x-3)}$

    $=\dfrac{5x-1}{(x+4)(x+3)(x-3)}$

4. Answers may vary. Possible solution: First factor each denominator completely. The LCD will be the product containing each different factor. If a factor occurs more than once in any one denominator, the LCD will contain that factor repeated the greatest number of times that it occurs in any one denominator.

Use Math to Save Money

1. 15% of $500 = 0.15 × $500 = $75
    Adam will save $75 today.

2. 25% of $500 = 0.25 × $500 = $125
    Using simple interest, Adam would pay $125 in interest if he carries the balance for a year.

3. $\dfrac{\$500}{\$5000}=0.10=10\%$

    $500 is 10% of the available credit on his existing card.

4. 48 × $9 = $432
    Adam would pay $432 more over the life of the loan.

How Am I Doing? Sections 7.1–7.3
(Available online through MyMathLab.)

1. $\dfrac{8x-48}{x^2-6x}=\dfrac{8(x-6)}{x(x-6)}=\dfrac{8}{x}$

2. $\dfrac{2x^2-7x-15}{x^2-12x+35}=\dfrac{(2x+3)(x-5)}{(x-5)(x-7)}=\dfrac{2x+3}{x-7}$

3. $\dfrac{y^2+2y+1}{2x^2-2x^2y^2}=\dfrac{(y+1)(y+1)}{-2x^2(y^2-1)}$

    $=\dfrac{(y+1)(y+1)}{-2x^2(y-1)(y+1)}$

    $=\dfrac{y+1}{2x^2(1-y)}$

4. $\dfrac{5x^2-23x+12}{5x^2+7x-6}=\dfrac{(5x-3)(x-4)}{(5x-3)(x+2)}=\dfrac{x-4}{x+2}$

5. $\dfrac{12a^2}{2x+10}\cdot\dfrac{8x+40}{16a^3}=\dfrac{12a^2}{2(x+5)}\cdot\dfrac{8(x+5)}{16a^3}$

    $=\dfrac{6a^2}{x+5}\cdot\dfrac{x+5}{2a^3}$

    $=\dfrac{6a^2}{2a^3}$

    $=\dfrac{3}{a}$

6. $\dfrac{x-5}{x^2+5x-14}\cdot\dfrac{x^2+12x+35}{15-3x}$

    $=\dfrac{x-5}{(x+7)(x-2)}\cdot\dfrac{(x+7)(x+5)}{3(5-x)}$

    $=-\dfrac{x+5}{3(x-2)}$

7.
$$\frac{x^2-9}{2x+6} \div \frac{2x^2-5x-3}{4x^2-1}$$
$$=\frac{x^2-9}{2x+6} \cdot \frac{4x^2-1}{2x^2-5x-3}$$
$$=\frac{(x+3)(x-3)}{2(x+3)} \cdot \frac{(2x+1)(2x-1)}{(2x+1)(x-3)}$$
$$=\frac{2x-1}{2}$$

8.
$$\frac{3a^2+7a+2}{4a^2+11a+6} \div \frac{6a^2-13a-5}{16a^2-9}$$
$$=\frac{3a^2+7a+2}{4a^2+11a+6} \cdot \frac{16a^2-9}{6a^2-13a-5}$$
$$=\frac{(3a+1)(a+2)}{(4a+3)(a+2)} \cdot \frac{(4a+3)(4a-3)}{(2a-5)(3a+1)}$$
$$=\frac{4a-3}{2a-5}$$

9.
$$\frac{x-3y}{xy}-\frac{4a-y}{ay}=\frac{x-3y}{xy} \cdot \frac{a}{a}-\frac{4a-y}{ay} \cdot \frac{x}{x}$$
$$=\frac{a(x-3y)-x(4a-y)}{axy}$$
$$=\frac{ax-3ay-4ax+xy}{axy}$$
$$=\frac{xy-3ax-3ay}{axy}$$

10.
$$\frac{7}{2x-4}+\frac{-14}{x^2-4}$$
$$=\frac{7}{2(x-2)} \cdot \frac{x+2}{x+2}+\frac{-14}{(x+2)(x-2)} \cdot \frac{2}{2}$$
$$=\frac{7(x+2)-14 \cdot 2}{2(x+2)(x-2)}$$
$$=\frac{7x+14-28}{2(x+2)(x-2)}$$
$$=\frac{7x-14}{2(x+2)(x-2)}$$
$$=\frac{7(x-2)}{2(x+2)(x-2)}$$
$$=\frac{7}{2(x+2)}$$

11.
$$\frac{2x}{x^2+10x+21}+\frac{x-3}{x+7}$$
$$=\frac{2x}{(x+3)(x+7)}+\frac{x-3}{x+7} \cdot \frac{x+3}{x+3}$$
$$=\frac{2x+(x-3)(x+3)}{(x+3)(x+7)}$$
$$=\frac{2x+x^2-9}{(x+3)(x+7)}$$
$$=\frac{x^2+2x-9}{(x+3)(x+7)}$$

12.
$$\frac{4}{x^2-2x-3}-\frac{5x}{x^2+5x+4}$$
$$=\frac{4}{(x-3)(x+1)} \cdot \frac{x+4}{x+4}-\frac{5x}{(x+4)(x+1)} \cdot \frac{x-3}{x-3}$$
$$=\frac{4x+16-5x^2+15x}{(x-3)(x+1)(x+4)}$$
$$=\frac{-5x^2+19x+16}{(x-3)(x+1)(x+4)}$$

7.4 Exercises

1.
$$\frac{\frac{5}{x}}{\frac{4}{x}+\frac{3}{x^2}}=\frac{\frac{5}{x}}{\frac{4}{x}+\frac{3}{x^2}} \cdot \frac{x^2}{x^2}=\frac{\frac{5}{x} \cdot x^2}{\frac{4}{x} \cdot x^2+\frac{3}{x^2} \cdot x^2}=\frac{5x}{4x+3}$$

3.
$$\frac{\frac{4}{a}+\frac{1}{b}}{\frac{5}{ab}}=\frac{\frac{4}{a}+\frac{1}{b}}{\frac{5}{ab}} \cdot \frac{ab}{ab}=\frac{\frac{4}{a} \cdot ab+\frac{1}{b} \cdot ab}{\frac{5}{ab} \cdot ab}=\frac{4b+a}{5}$$

5.
$$\frac{\frac{x}{6}-\frac{1}{3}}{\frac{2}{3x}+\frac{5}{6}}=\frac{\frac{x}{6}(6x)-\frac{1}{3}(6x)}{\frac{2}{3x}(6x)+\frac{5}{6}(6x)}=\frac{x^2-2x}{4+5x}=\frac{x(x-2)}{4+5x}$$

7.
$$\frac{\frac{7}{5x}-\frac{1}{x}}{\frac{3}{5}+\frac{2}{x}}=\frac{\frac{7}{5x}(5x)-\frac{1}{x}(5x)}{\frac{3}{5}(5x)+\frac{2}{x}(5x)}=\frac{7-5}{3x+10}=\frac{2}{3x+10}$$

9.
$$\frac{\frac{5}{x}+\frac{3}{y}}{3x+5y}=\frac{\frac{5}{x}+\frac{3}{y}}{3x+5y} \cdot \frac{xy}{xy}$$
$$=\frac{\frac{5}{x} \cdot xy+\frac{3}{y} \cdot xy}{(3x+5y)xy}$$
$$=\frac{5y+3x}{(3x+5y)xy}$$
$$=\frac{1}{xy}$$

11. $\dfrac{4-\frac{1}{x^2}}{2+\frac{1}{x}} = \dfrac{4-\frac{1}{x^2}}{2+\frac{1}{x}} \cdot \dfrac{x^2}{x^2}$

$= \dfrac{4 \cdot x^2 - \frac{1}{x^2} \cdot x^2}{2x^2 + \frac{1}{x} \cdot x^2}$

$= \dfrac{4x^2 - 1}{2x^2 + x}$

$= \dfrac{(2x+1)(2x-1)}{x(2x+1)}$

$= \dfrac{2x-1}{x}$

13. $\dfrac{\frac{2}{x+6}}{\frac{2}{x-6}-\frac{2}{x^2-36}} = \dfrac{\frac{2}{x+6}}{\frac{2}{x-6}-\frac{2}{(x-6)(x+6)}} \cdot \dfrac{(x-6)(x+6)}{(x-6)(x+6)}$

$= \dfrac{\frac{2(x-6)(x+6)}{(x+6)}}{\frac{2(x-6)(x+6)}{(x-6)} - \frac{2(x-6)(x+6)}{(x-6)(x+6)}}$

$= \dfrac{2(x-6)}{2(x+6)-2}$

$= \dfrac{2(x-6)}{2[(x+6)-1]}$

$= \dfrac{x-6}{x+6-1}$

$= \dfrac{x-6}{x+5}$

15. $\dfrac{a+\frac{3}{a}}{\frac{a^2+2}{3a}} = \dfrac{a+\frac{3}{a}}{\frac{a^2+2}{3a}} \cdot \dfrac{3a}{3a}$

$= \dfrac{a(3a) + \left(\frac{3}{a}\right)(3a)}{\left(\frac{a^2+2}{3a}\right)(3a)}$

$= \dfrac{3a^2 + 9}{a^2 + 2}$

$= \dfrac{3(a^2+3)}{a^2 + 2}$

17. $\dfrac{\frac{3}{x-3}}{\frac{1}{x^2-9}+\frac{2}{x+3}}$

$= \dfrac{\frac{3}{x-3}}{\frac{1}{x^2-9}+\frac{2}{x+3}} \cdot \dfrac{(x+3)(x-3)}{(x+3)(x-3)}$

$= \dfrac{\left(\frac{3}{x-3}\right)\frac{(x+3)(x-3)}{1}}{\frac{1}{(x+3)(x-3)}\frac{(x+3)(x-3)}{1} + \left(\frac{2}{x+3}\right)\frac{(x+3)(x-3)}{1}}$

$= \dfrac{3(x+3)}{1+2(x-3)}$

$= \dfrac{3(x+3)}{1+2x-6}$

$= \dfrac{3x+9}{2x-5}$

$= \dfrac{3(x+3)}{2x-5}$

19. $\dfrac{\frac{3}{x-1}+4}{\frac{3}{x-1}-4} = \dfrac{\frac{3}{x-1}+4}{\frac{3}{x-1}-4} \cdot \dfrac{x-1}{x-1}$

$= \dfrac{\frac{3}{x-1} \cdot (x-1) + 4(x-1)}{\frac{3}{x-1} \cdot (x-1) - 4(x-1)}$

$= \dfrac{3+4(x-1)}{3-4(x-1)}$

$= \dfrac{3+4x-4}{3-4x+4}$

$= \dfrac{4x-1}{7-4x}$

21. No expression in any denominator can be allowed to be zero, since division by zero is undefined. So −3, 5, and 0 are not allowable replacements for the variable x.

23. $\dfrac{x+5y}{x-6y} \div \left(\dfrac{1}{5y} - \dfrac{1}{x+5y} \right)$

$= \dfrac{x+5y}{x-6y} \div \left(\dfrac{1}{5y} \cdot \dfrac{x+5y}{x+5y} - \dfrac{1}{x+5y} \cdot \dfrac{5y}{5y} \right)$

$= \dfrac{x+5y}{x-6y} \div \left[\dfrac{x+5y}{5y(x+5y)} - \dfrac{5y}{5y(x+5y)} \right]$

$= \dfrac{x+5y}{x-6y} \div \dfrac{x+5y-5y}{5y(x+5y)}$

$= \dfrac{x+5y}{x-6y} \div \dfrac{x}{5y(x+5y)}$

$= \dfrac{x+5y}{x-6y} \cdot \dfrac{5y(x+5y)}{x}$

$= \dfrac{5y(x+5y)^2}{x(x-6y)}$

Cumulative Review

25. $5x+6y=8$

$6y=-5x+8$

$y=\dfrac{-5x+8}{6}=-\dfrac{5}{6}x+\dfrac{4}{3}$

$m=-\dfrac{5}{6}$; y-intercept $\left(0, \dfrac{4}{3} \right)$

26. $7+x<11+5x$

$7<11+4x$

$-4<4x$

$\dfrac{-4}{4}<\dfrac{4x}{4}$

$-1<x$ or $x>-1$

27. $2x-9=\dfrac{x}{2}$

$2(2x-9)=2\left(\dfrac{x}{2} \right)$

$4x-18=x$

$3x=18$

$x=6$

The number is 6.

28. $x=$ salary last year

$0.05x=$ raise

$x+0.05x=25,200$

$1.05x=25,200$

$x=24,000$

Isabella's salary was $24,000 last year.

Quick Quiz 7.4

1. $\dfrac{\frac{a}{4b}-\frac{1}{3}}{\frac{5}{4b}-\frac{4}{a}}=\dfrac{\frac{a}{4b}-\frac{1}{3}}{\frac{5}{4b}-\frac{4}{a}} \cdot \dfrac{12ab}{12ab}$

$= \dfrac{\frac{a}{4b} \cdot 12ab - \frac{1}{3} \cdot 12ab}{\frac{5}{46} \cdot 12ab - \frac{4}{a} \cdot 12ab}$

$= \dfrac{3a^2-4ab}{15a-48b}$

$= \dfrac{a(3a-4b)}{3(5a-16b)}$

2. $\dfrac{a+b}{\frac{1}{a}+\frac{1}{b}}=\dfrac{a \cdot ab + b \cdot ab}{\frac{1}{a} \cdot ab + \frac{1}{b} \cdot ab}$

$= \dfrac{a^2b+ab^2}{b+a}$

$= \dfrac{ab(a+b)}{b+a}$

$= ab$

3. $\dfrac{\frac{10}{x^2-25}}{\frac{3}{x+5}+\frac{2}{x-5}}=\dfrac{\frac{10}{(x+5)(x-5)}}{\frac{3}{x+5}+\frac{2}{x-5}} \cdot \dfrac{(x+5)(x-5)}{(x+5)(x-5)}$

$= \dfrac{\frac{10(x+5)(x-5)}{(x+5)(x-5)}}{\frac{3(x+5)(x-5)}{x+5}+\frac{2(x+5)(x-5)}{x-5}}$

$= \dfrac{10}{3(x-5)+2(x+5)}$

$= \dfrac{10}{3x-15+2x+10}$

$= \dfrac{10}{5x-5}$

$= \dfrac{10}{5(x-1)}$

$= \dfrac{2}{x-1}$

4. Answers may vary. Possible solution:
The first step is to find the LCD.

$x-3=\qquad (x-3)$

$2x-6=2 \cdot (x-3)$

$x+5=\qquad\qquad (x+5)$

LCD $= 2 \quad (x-3)(x+5)$

Next, multiply the numerator and denominator of the complex fraction by the LCD. Lastly, cancel common factors to eliminate the denominators of each individual fraction.

Copyright © 2017 Pearson Education, Inc.

7.5 Exercises

1.
$$\frac{7}{x} + \frac{3}{4} = \frac{-2}{x}$$
$$4x\left(\frac{7}{x}\right) + 4x\left(\frac{3}{4}\right) = 4x\left(-\frac{2}{x}\right)$$
$$28 + 3x = -8$$
$$3x = -36$$
$$x = -12$$

Check: $\quad \dfrac{7}{-12} + \dfrac{3}{4} \stackrel{?}{=} \dfrac{-2}{-12}$

$$\frac{7}{-12} + \frac{9}{12} \stackrel{?}{=} \frac{1}{6}$$
$$\frac{2}{12} \stackrel{?}{=} \frac{1}{6}$$
$$\frac{1}{6} = \frac{1}{6}$$

3.
$$\frac{3}{x} - \frac{5}{4} = \frac{1}{2x}$$
$$4x\left(\frac{3}{x}\right) - 4x\left(\frac{5}{4}\right) = 4x\left(\frac{1}{2x}\right)$$
$$12 - 5x = 2$$
$$-5x = -10$$
$$x = 2$$

Check: $\dfrac{3}{2} - \dfrac{5}{4} \stackrel{?}{=} \dfrac{1}{2(2)}$

$$\frac{6}{4} - \frac{5}{4} \stackrel{?}{=} \frac{1}{4}$$
$$\frac{1}{4} = \frac{1}{4}$$

5.
$$\frac{5x+3}{3x} = \frac{7}{3} - \frac{9}{x}$$
$$3x\left(\frac{5x+3}{3x}\right) = 3x\left(\frac{7}{3}\right) - 3x\left(\frac{9}{x}\right)$$
$$5x + 3 = 7x - 27$$
$$3 = 2x - 27$$
$$30 = 2x$$
$$15 = x$$

Check: $\dfrac{5(15)+3}{3(15)} \stackrel{?}{=} \dfrac{7}{3} - \dfrac{9}{15}$

$$\frac{78}{45} \stackrel{?}{=} \frac{35}{15} - \frac{9}{15}$$
$$\frac{26}{15} = \frac{26}{15}$$

7.
$$\frac{x+5}{3x} = \frac{1}{2}$$
$$6x \cdot \frac{x+5}{3x} = 6x \cdot \frac{1}{2}$$
$$2(x+5) = 3x$$
$$2x + 10 = 3x$$
$$-x = -10$$
$$x = 10$$

Check: $\dfrac{10+5}{3(10)} \stackrel{?}{=} \dfrac{1}{2}$

$$\frac{15}{30} \stackrel{?}{=} \frac{1}{2}$$
$$\frac{1}{2} = \frac{1}{2}$$

9.
$$\frac{6}{3x-5} = \frac{3}{2x}$$
$$2x(3x-5)\left(\frac{6}{3x-5}\right) = 2x(3x-5)\left(\frac{3}{2x}\right)$$
$$12x = 9x - 15$$
$$3x = -15$$
$$x = -5$$

Check: $\dfrac{6}{3(-5)-5} \stackrel{?}{=} \dfrac{3}{2(-5)}$

$$-\frac{6}{20} \stackrel{?}{=} -\frac{3}{10}$$
$$-\frac{3}{10} = -\frac{3}{10}$$

11.
$$\frac{2}{2x+5} = \frac{4}{x-4}$$
$$(2x+5)(x-4)\left(\frac{2}{2x+5}\right) = (2x+5)(x-4)\left(\frac{4}{x-4}\right)$$
$$2(x-4) = 4(2x+5)$$
$$2x - 8 = 8x + 20$$
$$-8 = 6x + 20$$
$$-28 = 6x$$
$$-\frac{28}{6} = x$$
$$-\frac{14}{3} = x \text{ or } x = -4\frac{2}{3}$$

Check: $\dfrac{2}{2\left(-\frac{14}{3}\right)+5} \overset{?}{=} \dfrac{4}{-\frac{14}{3}-4}$

$\dfrac{2}{-\frac{28}{3}+\frac{15}{3}} \overset{?}{=} \dfrac{4}{-\frac{14}{3}-\frac{12}{3}}$

$\dfrac{2}{-\frac{13}{3}} \overset{?}{=} \dfrac{4}{-\frac{26}{3}}$

$-\dfrac{6}{13} = -\dfrac{6}{13}$

13.

$$\dfrac{2}{x}+\dfrac{x}{x+1}=1$$

$$x(x+1)\left(\dfrac{2}{x}\right)+x(x+1)\left(\dfrac{x}{x+1}\right)=x(x+1)(1)$$

$$(x+1)(2)+x^2 = x^2+x$$

$$2x+2+x^2 = x^2+x$$

$$2x+2 = x$$

$$2 = -x$$

$$-2 = x$$

Check: $\dfrac{2}{-2}+\dfrac{-2}{-2+1} \overset{?}{=} 1$

$-1+2 \overset{?}{=} 1$

$1 = 1$

15.

$$\dfrac{85-4x}{x}=7-\dfrac{3}{x}$$

$$x\left(\dfrac{85-4x}{x}\right)=7x-x\left(\dfrac{3}{x}\right)$$

$$85-4x = 7x-3$$

$$85 = 11x-3$$

$$88 = 11x$$

$$8 = x$$

Check: $\dfrac{85-4(8)}{8} \overset{?}{=} 7-\dfrac{3}{8}$

$\dfrac{53}{8} \overset{?}{=} \dfrac{56}{8}-\dfrac{3}{8}$

$\dfrac{53}{8} = \dfrac{53}{8}$

17.

$$\dfrac{1}{x+4}-2=\dfrac{3x-2}{x+4}$$

$$(x+4)\left(\dfrac{1}{x+4}\right)-(x+4)\cdot 2 = (x+4)\cdot\dfrac{3x-2}{x+4}$$

$$1-(2x+8) = 3x-2$$

$$1-2x-8 = 3x-2$$

$$-5x = 5$$

$$x = -1$$

Check: $\dfrac{1}{-1+4} - 2 \overset{?}{=} \dfrac{3(-1)-2}{-1+4}$

$$\dfrac{1}{3} - \dfrac{6}{3} \overset{?}{=} -\dfrac{5}{3}$$

$$-\dfrac{5}{3} = -\dfrac{5}{3}$$

19.
$$\dfrac{2}{x-6} - 5 = \dfrac{2(x-5)}{x-6}$$

$$(x-6)\left(\dfrac{2}{x-6}\right) - 5(x-6) = (x-6)\left[\dfrac{2(x-5)}{x-6}\right]$$

$$2 - 5x + 30 = 2x - 10$$
$$-5x + 32 = 2x - 10$$
$$42 = 7x$$
$$6 = x$$

$x = 6$ makes the denominators zero so it is an extraneous solution. There is no solution.

21.
$$\dfrac{2}{x+1} - \dfrac{1}{x-1} = \dfrac{2x}{x^2-1}$$

$$\dfrac{2}{x+1} - \dfrac{1}{x-1} = \dfrac{2x}{(x+1)(x-1)}$$

$$(x+1)(x-1)\left(\dfrac{2}{x+1}\right) - (x+1)(x-1)\left(\dfrac{1}{x-1}\right) = (x+1)(x-1)\left(\dfrac{2x}{(x+1)(x-1)}\right)$$

$$(x-1)(2) - (x+1) = 2x$$
$$2x - 2 - x - 1 = 2x$$
$$x - 3 = 2x$$
$$-3 = x$$

Check: $\dfrac{2}{-3+1} - \dfrac{1}{-3-1} \overset{?}{=} \dfrac{2(-3)}{(-3)^2-1}$

$$-1 + \dfrac{1}{4} \overset{?}{=} -\dfrac{6}{8}$$

$$-\dfrac{3}{4} = -\dfrac{3}{4}$$

23.
$$\dfrac{x+2}{x^2-x-12} = \dfrac{1}{x+3} - \dfrac{1}{x-4}$$

$$\dfrac{x+2}{(x+3)(x-4)} = \dfrac{1}{x+3} - \dfrac{1}{x-4}$$

$$(x+3)(x-4)\left(\dfrac{x+2}{(x+3)(x-4)}\right) = (x+3)(x-4)\left(\dfrac{1}{x+3}\right) - (x+3)(x-4)\left(\dfrac{1}{x-4}\right)$$

$$x + 2 = x - 4 - x - 3$$
$$x + 2 = -7$$
$$x = -9$$

Check: $\dfrac{-9+2}{(-9)^2-(-9)-12} \overset{?}{=} \dfrac{1}{-9+3}-\dfrac{1}{-9-4}$

$$\dfrac{-7}{78} \overset{?}{=} \dfrac{1}{-6}-\dfrac{1}{-13}$$

$$\dfrac{-7}{78} \overset{?}{=} \dfrac{-13}{78}+\dfrac{6}{78}$$

$$\dfrac{-7}{78}=\dfrac{-7}{78}$$

25.

$$\dfrac{2x}{x+4}-\dfrac{8}{x-4}=\dfrac{2x^2+32}{x^2-16}$$

$$\dfrac{2x}{x+4}-\dfrac{8}{x-4}=\dfrac{2x^2+32}{(x+4)(x-4)}$$

$$(x+4)(x-4)\left(\dfrac{2x}{x+4}\right)-(x+4)(x-4)\left(\dfrac{8}{x-4}\right)=(x+4)(x-4)\left(\dfrac{2x^2+32}{(x+4)(x-4)}\right)$$

$$(x-4)(2x)-(x+4)(8)=2x^2+32$$

$$2x^2-8x-8x-32=2x^2+32$$

$$2x^2-16x-32=2x^2+32$$

$$-16x-32=32$$

$$-16x=64$$

$$x=-4$$

Since $x=-4$ makes the first denominator 0, $x=-4$ is extraneous and there is no solution.

27.

$$\dfrac{4}{x^2-1}+\dfrac{7}{x+1}=\dfrac{5}{x-1}$$

$$\dfrac{4}{(x+1)(x-1)}+\dfrac{7}{x+1}=\dfrac{5}{x-1}$$

$$(x+1)(x-1)\dfrac{4}{(x+1)(x-1)}+(x+1)(x-1)\dfrac{7}{x+1}=(x+1)(x-1)\dfrac{5}{x-1}$$

$$4+(x-1)(7)=(x+1)(5)$$

$$4+7x-7=5x+5$$

$$2x-3=5$$

$$2x=8$$

$$x=4$$

Check: $\dfrac{4}{(4)^2-1}+\dfrac{7}{4+1} \overset{?}{=} \dfrac{5}{4-1}$

$$\dfrac{4}{15}+\dfrac{7}{5} \overset{?}{=} \dfrac{5}{3}$$

$$\dfrac{25}{15} \overset{?}{=} \dfrac{5}{3}$$

$$\dfrac{5}{3}=\dfrac{5}{3}$$

29.

$$\frac{x+11}{x^2-5x+4}+\frac{3}{x-1}=\frac{5}{x-4}$$

$$\frac{x+11}{(x-4)(x-1)}+\frac{3}{x-1}=\frac{5}{x-4}$$

$$(x-4)(x-1)\left(\frac{x+11}{(x-4)(x-1)}\right)+(x-4)(x-1)\left(\frac{3}{x-1}\right)=(x-4)(x-1)\left(\frac{5}{x-4}\right)$$

$$x+11+(x-4)(3)=(x-1)(5)$$
$$x+11+3x-12=5x-5$$
$$4x-1=5x-5$$
$$-1=x-5$$
$$4=x$$

Since $x=4$ makes a denominator zero, $x=4$ is extraneous and there is no solution.

31. $x+6\neq0$, so $x\neq-6$

$3x+1\neq0$, so $x\neq-\dfrac{1}{3}$

$3x^2+19x+6\neq0$
$(x+6)(3x+1)\neq0$

$$x+6\neq0 \qquad\qquad 3x+1\neq0$$
$$x\neq-6 \qquad\qquad\quad x\neq-\frac{1}{3}$$

$x=-6$ and $x=-\dfrac{1}{3}$ are not allowable replacements for x.

Cumulative Review

33. $8x^2-2x-1=(4x+1)(2x-1)$

34. $5(x-2)=8-(3+x)$
$$5x-10=8-3-x$$
$$6x=15$$
$$x=\frac{5}{2}\text{ or }2\frac{1}{2}\text{ or }2.5$$

35. Let $w=$ width.
length $=2w-8$
Perimeter $=2(\text{length})+2(\text{width})$
$$44=2(2w-8)+2w$$
$$44=4w-16+2w$$
$$44=6w-16$$
$$-6w=-60$$
$$w=10$$
$2w-8=2(10)-8=20-8=12$
The dimensions are as follows:
width $=10$ in.
length $=12$ in.

36. $\{(7, 3), (2, 2), (-2, 0), (2, -2), (7, -3)\}$
Domain = $\{7, 2, -2\}$
Range = $\{3, 2, 0, -2, -3\}$
Not a function because two different ordered pairs have the same first element.

Quick Quiz 7.5

1.
$$\frac{3}{4x} - \frac{5}{6x} = 2 - \frac{1}{2x}$$
$$12x \cdot \frac{3}{4x} - 12x \cdot \frac{5}{6x} = 12x \cdot 2 - 12x \cdot \frac{1}{2x}$$
$$9 - 10 = 24x - 6$$
$$6 - 1 = 24x$$
$$x = \frac{5}{24}$$

2.
$$\frac{x}{x-1} - \frac{2}{x} = \frac{1}{x-1}$$
$$x(x-1) \cdot \frac{x}{x-1} - x(x-1) \cdot \frac{2}{x} = x(x-1) \cdot \frac{1}{x-1}$$
$$x^2 - 2(x-1) = x$$
$$x^2 - 2x + 2 = x$$
$$x^2 - 3x + 2 = 0$$
$$(x-2)(x-1) = 0$$
$$x - 2 = 0 \qquad\qquad x - 1 = 0$$
$$x = 2 \qquad\qquad\quad x = 1$$
$x = 1$ is extraneous because it makes denominator 0. $x = 2$ is the solution.

3.
$$\frac{6}{x^2 - 2x - 8} + \frac{5}{x+2} = \frac{1}{x-4}$$
$$(x-4)(x+2) \cdot \frac{6}{(x-4)(x+2)} + (x-4)(x+2) \cdot \frac{5}{x+2} = (x-4)(x+2) \cdot \frac{1}{x-4}$$
$$6 + 5(x-4) = x + 2$$
$$6 + 5x - 20 = x + 2$$
$$5x - 14 = x + 2$$
$$4x = 16$$
$$x = 4$$
$x = 4$ is extraneous because it makes denominator 0, so there is no solution.

4. Answers may vary. Possible solution:
To find the LCD first factor each denominator, then multiply one instance of each factor.
$$x^2 - 9 = \qquad (x-3)(x+3)$$
$$3x - 9 = \qquad 3\,(x-3)$$
$$2x + 6 = 2 \qquad\quad (x+3)$$
$$2x^2 - 18 = 2 \quad\ (x-3)(x+3)$$
$$\text{LCD} = 2 \cdot 3 \cdot (x-3)(x+3)$$

7.6 Exercises

1. $\dfrac{5}{11} = \dfrac{8}{x}$

   $5 \cdot x = 11 \cdot 8$

   $5x = 88$

   $x = \dfrac{88}{5}$ or $17\dfrac{3}{5}$ or 17.6

3. $\dfrac{x}{17} = \dfrac{12}{5}$

   $x \cdot 5 = 17 \cdot 12$

   $5x = 204$

   $x = \dfrac{204}{5}$ or $40\dfrac{4}{5}$ or 40.8

5. $\dfrac{9.1}{8.4} = \dfrac{x}{6}$

   $9.1 \cdot 6 = 8.4 \cdot x$

   $54.6 = 8.4x$

   $x = 6.5$

7. $\dfrac{7}{x} = \dfrac{40}{130}$

   $x \cdot 40 = 7 \cdot 130$

   $40x = 910$

   $x = \dfrac{910}{40} = \dfrac{91}{4}$ or 22.75 or $22\dfrac{3}{4}$

9. a. x = New Zealand dollars received

   $\dfrac{1.3 \text{ N.Z.}}{1.0 \text{ U.S.}} = \dfrac{x}{500 \text{ U.S.}}$

   $1 \cdot x = 1.3 \cdot 500$

   $x = 650$

   Robyn received 650 New Zealand dollars.

b. x = New Zealand dollars received

   $\dfrac{1.15 \text{ N.Z.}}{1.00 \text{ U.S.}} = \dfrac{x}{500 \text{ U.S.}}$

   $1 \cdot x = 1.15 \cdot 500$

   $x = 575$

   $650 - 575 = 75$

   She would have received 75 New Zealand dollars less.

11. x = speed limit in miles per hour

   $\dfrac{x}{90} = \dfrac{62}{100}$

   $100x = 5580$

   $x = 55.8 \approx 56$

   The limit is 56 miles per hour.

13. x = miles from base of mountain

   $\dfrac{x}{\frac{3}{4}} = \dfrac{136}{3\frac{1}{2}}$

   $3\dfrac{1}{2}x = \dfrac{3}{4}(136)$

   $\dfrac{7}{2}x = 102$

   $x \approx 29.1$

   He is 29 miles away.

15. $x = 20$, $y = 29$, $m = 13$

   $\dfrac{n}{m} = \dfrac{y}{x}$

   $\dfrac{n}{13} = \dfrac{29}{20}$

   $20n = 377$

   $n = \dfrac{377}{20}$ or $18\dfrac{17}{20}$ inches

17. $x = 175$, $n = 40$, $m = 35$

   $\dfrac{y}{n} = \dfrac{x}{m}$

   $\dfrac{y}{40} = \dfrac{175}{35}$

   $35y = 7000$

   $y = 200$ meters

19. $a = 5$, $d = 8$, $g = 7$

   $\dfrac{a}{g} = \dfrac{d}{k}$

   $\dfrac{5}{7} = \dfrac{8}{k}$

   $5k = 56$

   $k = \dfrac{56}{5}$ or $11\dfrac{1}{5}$ ft

21. $b = 20$, $h = 24$, $d = 32$

   $\dfrac{k}{d} = \dfrac{h}{b}$

   $\dfrac{k}{32} = \dfrac{24}{20}$

   $20k = 32 \cdot 24$

   $k = \dfrac{768}{20} = \dfrac{192}{5}$ or $38\dfrac{2}{5}$ m

23. $x = $ length

$$\frac{30}{x} = \frac{5}{8}$$

$$5x = 8(30)$$

$$x = \frac{240}{5}$$

$$x = 48 \text{ inches}$$

25. $x = $ flower height

$$\frac{x}{13} = \frac{5}{3}$$

$$3x = 65$$

$$x = \frac{65}{3} \approx 22 \text{ inches}$$

Total height = 13 + 22 = 35 inches

27. $x = $ amount of acceleration in 11 seconds

$$\frac{x}{11} = \frac{3}{2}$$

$$2x = 33$$

$$x = 16.5$$

$$45 + 16.5 = 61.5$$

He will be traveling at 61.5 miles per hour.

29.

	D	R	$T = \frac{D}{R}$
Helicopter	1050	s	$\frac{1050}{s}$
Airliner	1250	$s + 40$	$\frac{1250}{s+40}$

$$\frac{1250}{s+40} = \frac{1050}{s}$$

$$1050s = 1050s + 42{,}000$$

$$200s = 42{,}000$$

$$s = 210$$

$$s + 40 = 250$$

The speeds are as follows:
commuter airliner, 250 kilometers/hr;
helicopter, 210 kilometers/hr

31. a. $x = $ cost of 1 ounce in glass jar

$$\frac{x}{1} = \frac{1.29}{7\frac{1}{2}}$$

$$7\frac{1}{2}x = 1.29 \cdot 1$$

$$x = \frac{1.29}{7.5} = 0.172$$

It costs about $0.17 per ounce.

b. $x = $ cost of 1 ounce in plastic tub

$$\frac{x}{1} = \frac{2.19}{16}$$

$$16x = 2.19 \cdot 1$$

$$x = \frac{2.19}{16} \approx 0.137$$

It costs about $0.14 per ounce.

c. $x = $ cost of 40-oz bucket

$$\frac{x}{40} = \frac{2.19}{16}$$

$$16x = 2.19 \cdot 40$$

$$16x = 87.6$$

$$x = \frac{87.6}{16} = 5.475$$

The price would be $5.48.

33. $x = $ time to rake together

$$\frac{1}{6} + \frac{1}{8} = \frac{1}{x}$$

$$24x\left(\frac{1}{6}\right) + 24x\left(\frac{1}{8}\right) = 24x\left(\frac{1}{x}\right)$$

$$4x + 3x = 24$$

$$7x = 24$$

$$x = \frac{24}{7} = 3\frac{3}{7}$$

It will take $3\frac{3}{7}$ hours or 3 hours 26 minutes.

Cumulative Review

35. $0.000892465 = 8.92465 \times 10^{-4}$

36. $6.83 \times 10^9 = 6{,}830{,}000{,}000$

37. $\dfrac{x^{-3}y^{-2}}{z^4 w^{-8}} = \dfrac{w^8}{x^3 y^2 z^4}$

38. $\left(\dfrac{2}{3}\right)^{-3} = \left(\dfrac{3}{2}\right)^3 = \dfrac{3^3}{2^3} = \dfrac{27}{8}$ or $3\dfrac{3}{8}$

Quick Quiz 7.6

1. $\dfrac{16.5}{2.1} = \dfrac{x}{7}$

$$2.1 \cdot x = 16.5 \cdot 7$$

$$2.1x = 115.5$$

$$x = 55$$

2. x = height of tree

$$\frac{x}{34} = \frac{6}{8.5}$$

$$8.5x = 34 \cdot 6$$

$$8.5x = 204$$

$$x = 24$$

The tree is 24 ft tall.

3. x = expected on-time departures

$$\frac{164}{205} = \frac{x}{215}$$

$$205x = 164 \cdot 215$$

$$205x = 35,260$$

$$x = 172$$

He would expect 172 on-time departures.

4. Answers may vary. Possible solution:
Since 18 gallons corresponds to 396 miles, the

known ratio of miles to gallons is $\dfrac{18}{396}$. He

should use the equation $\dfrac{18}{396} = \dfrac{x}{450}$.

Career Exploration Problems

1. n = the number of treadmills produced daily
$\$22,000$ = the fixed daily costs.
$\$350n$ = the cost of producing n number of treadmills.
The cost for producing n treadmills is
$C = 350n + 22,000$.
Therefore, the average treadmill cost model is:

$$C_{\text{ave}} = \frac{350n + 22,000}{n}$$

2. h = number of hours to produce one shipment
Line A produces one shipment in 18 hours, so it

produces $\dfrac{1}{18}$ of the treadmills in 1 hour.

Line B produces one shipment in 14 hours, so it

produces $\dfrac{1}{14}$ of the treadmills in 1 hour.

$$\frac{1}{18} + \frac{1}{14} = \frac{1}{h}$$

$$252h\left(\frac{1}{18}\right) + 252h\left(\frac{1}{14}\right) = 252h\left(\frac{1}{h}\right)$$

$$14h + 18h = 252$$

$$32h = 252$$

$$h \approx 7.8$$

It will take about 8 hours for the two lines to produce one shipment of treadmills.

3. $C_{\text{ave}} = \dfrac{350n + 22,000}{n}$

$$\text{average cost per hour} = \frac{\frac{350n + 22,000}{n}}{8}$$

$$= \frac{350n + 22,000}{n} \div 8$$

$$= \frac{350n + 22,000}{n} \cdot \frac{1}{8}$$

$$= \frac{350n + 22,000}{8n}$$

For 50 treadmills:

$$\frac{\$350(50) + \$22,000}{8(50)} = \$98.75 \text{ per hour}$$

For 100 treadmills:

$$\frac{\$350(100) + \$22,000}{8(100)} = \$71.25 \text{ per hour}$$

You Try It

1.
$$\frac{6x^2 - 12x - 90}{3x^2 - 27} = \frac{6(x^2 - 2x - 15)}{3(x^2 - 9)}$$
$$= \frac{6(x-5)(x+3)}{3(x+3)(x-3)}$$
$$= \frac{2(x-5)}{x-3}$$

2.
$$\frac{x^2 - 4xy - 5y^2}{2x^2 - 9xy - 5y^2} \cdot \frac{4x^2 - y^2}{4x^2 - 4xy + y^2}$$
$$= \frac{(x-5y)(x+y)}{(2x+y)(x-5y)} \cdot \frac{(2x+y)(2x-y)}{(2x-y)(2x-y)}$$
$$= \frac{x+y}{2x-y}$$

3.
$$\frac{2x^2 + 3x - 20}{8x + 8} \div \frac{x^2 - 16}{4x^2 - 12x - 16}$$
$$= \frac{2x^2 + 3x - 20}{8x + 8} \cdot \frac{4x^2 - 12x - 16}{x^2 - 16}$$
$$= \frac{(2x-5)(x+4)}{8(x+1)} \cdot \frac{4(x-4)(x+1)}{(x+4)(x-4)}$$
$$= \frac{2x-5}{2}$$

4. $\dfrac{x+2}{2x+6} + \dfrac{x}{x^2-9}$

   $= \dfrac{x+2}{2(x+3)} \cdot \dfrac{x-3}{x-3} + \dfrac{x}{(x+3)(x-3)} \cdot \dfrac{2}{2}$

   $= \dfrac{x^2-x-6+2x}{2(x+3)(x-3)}$

   $= \dfrac{x^2+x-6}{2(x+3)(x-3)}$

   $= \dfrac{(x+3)(x-2)}{2(x+3)(x-3)}$

   $= \dfrac{x-2}{2(x-3)}$

5. $\dfrac{9x}{x+3} - \dfrac{3x-18}{x+3} = \dfrac{9x-(3x-18)}{x+3}$

   $= \dfrac{6x+18}{x+3}$

   $= \dfrac{6(x+3)}{x+3}$

   $= 6$

6. $\dfrac{\dfrac{x}{x-3}+\dfrac{2}{x+3}}{\dfrac{1}{x-3}+\dfrac{3}{x^2-9}} = \dfrac{\dfrac{x(x+3)}{(x-3)(x+3)}+\dfrac{2(x-3)}{(x+3)(x-3)}}{\dfrac{1(x+3)}{(x+3)(x-3)}+\dfrac{3}{(x+3)(x-3)}}$

   $= \dfrac{\dfrac{x^2+3x+2x-6}{(x+3)(x-3)}}{\dfrac{x+3+3}{(x+3)(x-3)}}$

   $= \dfrac{x^2+5x-6}{(x+3)(x-3)} \cdot \dfrac{(x+3)(x-3)}{x+6}$

   $= \dfrac{(x+6)(x-1)}{(x+3)(x-3)} \cdot \dfrac{(x+3)(x-3)}{x+6}$

   $= x-1$

7. $\dfrac{5x}{x^2-16} = \dfrac{5}{x+4}$

   $(x+4)(x-4)\left(\dfrac{5x}{(x+4)(x-4)}\right) = (x+4)(x-4)\left(\dfrac{5}{x+4}\right)$

   $5x = 5(x-4)$

   $5x = 5x-20$

   $0 = -20$

   There is no solution.

8. x = length of room

$$\frac{2}{10.5} = \frac{3}{x}$$

$$2 \cdot x = 10.5 \cdot 3$$

$$2x = 31.5$$

$$x = 15.75$$

The room will be 15.75 ft long.

Chapter 7 Review Problems

1. $\dfrac{bx}{bx - by} = \dfrac{bx}{b(x-y)} = \dfrac{x}{x-y}$

2. $\dfrac{4x - 4y}{5y - 5x} = \dfrac{4(x-y)}{-5(x-y)} = -\dfrac{4}{5}$

3. $\dfrac{x^3 - 4x^2}{x^3 - x^2 - 12x} = \dfrac{x^2(x-4)}{x(x-4)(x+3)} = \dfrac{x}{x+3}$

4. $\dfrac{2x^2 + 7x - 15}{25 - x^2} = \dfrac{2x^2 + 10x - 3x - 15}{25 - x^2}$

$$= \dfrac{2x(x+5) - 3(x+5)}{(5-x)(5+x)}$$

$$= \dfrac{(2x-3)(x+5)}{(5-x)(5+x)}$$

$$= \dfrac{2x-3}{5-x}$$

5. $\dfrac{2x^2 - 2xy - 24y^2}{2x^2 + 5xy - 3y^2} = \dfrac{2(x^2 - xy - 12y^2)}{2x^2 + 5xy - 3y^2}$

$$= \dfrac{2(x-4y)(x+3y)}{(2x-y)(x+3y)}$$

$$= \dfrac{2(x-4y)}{2x-y}$$

6. $\dfrac{4 - y^2}{3y^2 + 5y - 2} = \dfrac{(2+y)(2-y)}{(3y-1)(y+2)} = \dfrac{2-y}{3y-1}$

7. $\dfrac{5x^3 - 10x^2}{25x^4 + 5x^3 - 30x^2} = \dfrac{5x^2(x-2)}{5x^2(5x+6)(x-1)}$

$$= \dfrac{x-2}{(5x+6)(x-1)}$$

8. $\dfrac{16x^2 - 4y^2}{4x - 2y} = \dfrac{4(2x+y)(2x-y)}{2(2x-y)}$

$$= 2(2x+y)$$

$$= 4x + 2y$$

9. $\dfrac{2x^2 + 6x}{3x^2 - 27} \cdot \dfrac{x^2 + 3x - 18}{4x^2 - 4x}$

$$= \dfrac{2x(x+3)}{3(x+3)(x-3)} \cdot \dfrac{(x+6)(x-3)}{4x(x-1)}$$

$$= \dfrac{x+6}{6(x-1)}$$

10. $\dfrac{y^2 + 8y + 16}{5y^2 + 20y} \div \dfrac{y^2 + 7y + 12}{2y^2 + 5y - 3}$

$$= \dfrac{y^2 + 8y + 16}{5y^2 + 20y} \cdot \dfrac{2y^2 + 5y - 3}{y^2 + 7y + 12}$$

$$= \dfrac{(y+4)(y+4)}{5y(y+4)} \cdot \dfrac{(2y-1)(y+3)}{(y+4)(y+3)}$$

$$= \dfrac{2y-1}{5y}$$

11. $\dfrac{6y^2 + 13y - 5}{9y^2 + 3y} \div \dfrac{4y^2 + 20y + 25}{12y^2}$

$$= \dfrac{6y^2 + 13y - 5}{9y^2 + 3y} \cdot \dfrac{12y^2}{4y^2 + 20y + 25}$$

$$= \dfrac{(3y-1)(2y+5)}{3y(3y+1)} \cdot \dfrac{12y^2}{(2y+5)(2y+5)}$$

$$= \dfrac{4y(3y-1)}{(3y+1)(2y+5)}$$

12. $\dfrac{3xy^2 + 12y^2}{2x^2 - 11x + 5} \div \dfrac{2xy + 8y}{8x^2 + 2x - 3}$

$$= \dfrac{3xy^2 + 12y^2}{2x^2 - 11x + 5} \cdot \dfrac{8x^2 + 2x - 3}{2xy + 8y}$$

$$= \dfrac{3y^2(x+4)}{(2x-1)(x-5)} \cdot \dfrac{(2x-1)(4x+3)}{2y(x+4)}$$

$$= \dfrac{3y(4x+3)}{2(x-5)}$$

13. $\dfrac{x^2 - 5xy - 24y^2}{2x^2 - 2xy - 24y^2} \cdot \dfrac{4x^2 + 4xy - 24y^2}{x^2 - 10xy + 16y^2}$

$$= \dfrac{(x-8y)(x+3y)}{2(x-4y)(x+3y)} \cdot \dfrac{4(x+3y)(x-2y)}{(x-8y)(x-2y)}$$

$$= \dfrac{2(x+3y)}{x-4y}$$

14. $\dfrac{2x^2+10x+2}{8x-8}\cdot\dfrac{3x-3}{4x^2+20x+4}$

$=\dfrac{2(x^2+5x+1)}{8(x-1)}\cdot\dfrac{3(x-1)}{4(x^2+5x+1)}$

$=\dfrac{3}{16}$

15. $\dfrac{6}{y+2}+\dfrac{2}{3y}=\dfrac{6}{y+2}\cdot\dfrac{3y}{3y}+\dfrac{2}{3y}\cdot\dfrac{y+2}{y+2}$

$=\dfrac{18y+2(y+2)}{3y(y+2)}$

$=\dfrac{18y+2y+4}{3y(y+2)}$

$=\dfrac{20y+4}{3y(y+2)}$

$=\dfrac{4(5y+1)}{3y(y+2)}$

16. $3+\dfrac{2}{x+1}+\dfrac{1}{x}=\dfrac{3}{1}\cdot\dfrac{x(x+1)}{x(x+1)}+\dfrac{2}{x+1}\cdot\dfrac{x}{x}+\dfrac{1}{x}\cdot\dfrac{x+1}{x+1}$

$=\dfrac{3x(x+1)+2x+(x+1)}{x(x+1)}$

$=\dfrac{3x^2+3x+2x+x+1}{x(x+1)}$

$=\dfrac{3x^2+6x+1}{x(x+1)}$

17. $\dfrac{7}{x+2}+\dfrac{3}{x-4}=\dfrac{7}{x+2}\cdot\dfrac{x-4}{x-4}+\dfrac{3}{x-4}\cdot\dfrac{x+2}{x+2}$

$=\dfrac{7x-28}{(x+2)(x-4)}+\dfrac{3x+6}{(x+2)(x-4)}$

$=\dfrac{7x-28+3x+6}{(x+2)(x-4)}$

$=\dfrac{10x-22}{(x+2)(x-4)}$

$=\dfrac{2(5x-11)}{(x+2)(x-4)}$

18. $\dfrac{2}{x^2-9}+\dfrac{x}{x+3}=\dfrac{2}{(x+3)(x-3)}+\dfrac{x}{x+3}\cdot\dfrac{x-3}{x-3}$

$=\dfrac{2+x^2-3x}{(x+3)(x-3)}$

$=\dfrac{(x-2)(x-1)}{(x+3)(x-3)}$

19. $\dfrac{x}{y}+\dfrac{3}{2y}+\dfrac{1}{y+2}$

$=\dfrac{x}{y}\cdot\dfrac{2(y+2)}{2(y+2)}+\dfrac{3}{2y}\cdot\dfrac{y+2}{y+2}+\dfrac{1}{y+2}\cdot\dfrac{2y}{2y}$

$=\dfrac{2x(y+2)+3y+6+2y}{2y(y+2)}$

$=\dfrac{2xy+4x+5y+6}{2y(y+2)}$

20. $\dfrac{4}{a}+\dfrac{2}{b}+\dfrac{3}{a+b}$

$=\dfrac{4}{a}\cdot\dfrac{b(a+b)}{b(a+b)}+\dfrac{2}{b}\cdot\dfrac{a(a+b)}{a(a+b)}+\dfrac{3}{a+b}\cdot\dfrac{ab}{ab}$

$=\dfrac{4ab+4b^2}{ab(a+b)}+\dfrac{2a^2+2ab}{ab(a+b)}+\dfrac{3ab}{ab(a+b)}$

$=\dfrac{4ab+4b^2+2a^2+2ab+3ab}{ab(a+b)}$

$=\dfrac{2a^2+9ab+4b^2}{ab(a+b)}$

$=\dfrac{(2a+b)(a+4b)}{ab(a+b)}$

21. $\dfrac{3x+1}{3x}-\dfrac{1}{x}=\dfrac{3x+1}{3x}-\dfrac{1}{x}\cdot\dfrac{3}{3}$

$=\dfrac{3x+1-3}{3x}$

$=\dfrac{3x-2}{3x}$

22. $\dfrac{x+4}{x+2}-\dfrac{1}{2x}=\dfrac{x+4}{x+2}\cdot\dfrac{2x}{2x}-\dfrac{1}{2x}\cdot\dfrac{x+2}{x+2}$

$=\dfrac{2x^2+8x}{2x(x+2)}-\dfrac{x+2}{2x(x+2)}$

$=\dfrac{2x^2+8x-(x+2)}{2x(x+2)}$

$=\dfrac{2x^2+8x-x-2}{2x(x+2)}$

$=\dfrac{2x^2+7x-2}{2x(x+2)}$

23. $\dfrac{27}{x^2-81}+\dfrac{3}{2(x+9)}$

$=\dfrac{27}{(x+9)(x-9)}\cdot\dfrac{2}{2}+\dfrac{3}{2(x+9)}\cdot\dfrac{x-9}{x-9}$

$=\dfrac{54+3x-27}{2(x+9)(x-9)}$

$=\dfrac{3x+27}{2(x+9)(x-9)}$

$=\dfrac{3(x+9)}{2(x+9)(x-9)}$

$=\dfrac{3}{2(x-9)}$

24. $\dfrac{1}{x^2+7x+10}-\dfrac{x}{x+5}$

$=\dfrac{1}{(x+2)(x+5)}-\dfrac{x}{x+5}\cdot\dfrac{x+2}{x+2}$

$=\dfrac{1-x^2-2x}{(x+2)(x+5)}$

25. $\dfrac{\frac{4}{3y}-\frac{2}{y}}{\frac{1}{2y}+\frac{1}{y}}=\dfrac{\frac{4}{3y}-\frac{2}{y}}{\frac{1}{2y}+\frac{1}{y}}\cdot\dfrac{6y}{6y}$

$=\dfrac{\frac{4}{3y}\cdot 6y-\frac{2}{y}\cdot 6y}{\frac{1}{2y}\cdot 6y+\frac{1}{y}\cdot 6y}$

$=\dfrac{8-12}{3+6}$

$=\dfrac{-4}{9}$

$=-\dfrac{4}{9}$

26. $\dfrac{\frac{5}{x}+\frac{1}{2x}}{\frac{x}{4}+x}=\dfrac{\frac{5}{x}+\frac{1}{2x}}{\frac{x}{4}+\frac{x}{1}}\cdot\dfrac{4x}{4x}$

$=\dfrac{\frac{5}{x}\cdot 4x+\frac{1}{2x}\cdot 4x}{\frac{x}{4}\cdot 4x+\frac{x}{1}\cdot 4x}$

$=\dfrac{20+2}{x^2+4x^2}$

$=\dfrac{22}{5x^2}$

27. $\dfrac{w-\frac{4}{w}}{1+\frac{2}{w}}=\dfrac{w-\frac{4}{w}}{1+\frac{2}{w}}\cdot\dfrac{w}{w}$

$=\dfrac{w^2-4}{w+2}$

$=\dfrac{(w+2)(w-2)}{w+2}$

$=w-2$

28. $\dfrac{1-\frac{w}{w-1}}{1+\frac{w}{1-w}}=\dfrac{1-\frac{w}{w-1}}{1+\frac{w}{-(w-1)}}=\dfrac{1-\frac{w}{w-1}}{1-\frac{w}{w-1}}=1$

29. $\dfrac{1+\frac{1}{y^2-1}}{\frac{1}{y+1}-\frac{1}{y-1}}=\dfrac{1+\frac{1}{(y+1)(y-1)}}{\frac{1}{y+1}-\frac{1}{y-1}}\cdot\dfrac{(y+1)(y-1)}{(y+1)(y-1)}$

$=\dfrac{(y+1)(y-1)+1}{y-1-(y+1)}$

$=\dfrac{y^2-1+1}{y-1-y-1}$

$=\dfrac{y^2}{-2}$

$=-\dfrac{y^2}{2}$

30. $\dfrac{\frac{1}{y}+\frac{1}{x+y}}{1+\frac{2}{x+y}}=\dfrac{\frac{1}{y}+\frac{1}{x+y}}{1+\frac{2}{x+y}}\cdot\dfrac{y(x+y)}{y(x+y)}$

$=\dfrac{x+y+y}{y(x+y)+2y}$

$=\dfrac{x+2y}{xy+y^2+2y}$

$=\dfrac{x+2y}{y(x+y+2)}$

31. $\dfrac{\frac{1}{a+b}-\frac{1}{a}}{b}=\dfrac{\frac{a-(a+b)}{a(a+b)}}{b}$

$=\dfrac{a-a-b}{a(a+b)}\cdot\dfrac{1}{b}$

$=\dfrac{-b}{a(a+b)}\cdot\dfrac{1}{b}$

$=-\dfrac{1}{a(a+b)}$

32. $\dfrac{\frac{2}{a+b} - \frac{3}{b}}{\frac{1}{a+b}} = \dfrac{\frac{2}{a+b} - \frac{3}{b}}{\frac{1}{a+b}} \cdot \dfrac{b(a+b)}{b(a+b)}$

$\qquad = \dfrac{2b - 3(a+b)}{b}$

$\qquad = \dfrac{2b - 3a - 3b}{b}$

$\qquad = \dfrac{-3a - b}{b}$ or $-\dfrac{3a+b}{b}$

33.

$$\frac{8a-1}{6a+8} = \frac{3}{4}$$

$$\frac{8a-1}{2(3a+4)} = \frac{3}{4}$$

$$4(3a+4)\left[\frac{8a-1}{2(3a+4)}\right] = 4(3a+4)\left(\frac{3}{4}\right)$$

$$2(8a-1) = 3(3a+4)$$

$$16a - 2 = 9a + 12$$

$$7a - 2 = 12$$

$$7a = 14$$

$$a = 2$$

34.

$$\frac{8}{a-3} = \frac{12}{a+3}$$

$$(a-3)(a+3)\frac{8}{a-3} = (a-3)(a+3)\frac{12}{a+3}$$

$$8(a+3) = 12(a-3)$$

$$8a + 24 = 12a - 36$$

$$60 = 4a$$

$$15 = a$$

35.

$$\frac{2x-1}{x} - \frac{1}{2} = -2$$

$$2x\left(\frac{2x-1}{x}\right) - 2x\left(\frac{1}{2}\right) = -2(2x)$$

$$4x - 2 - x = -4x$$

$$7x = 2$$

$$x = \frac{2}{7}$$

36.

$$\frac{5}{4} - \frac{1}{2x} = \frac{1}{x} + 2$$

$$4x\left(\frac{5}{4}\right) - 4x\left(\frac{1}{2x}\right) = 4x\left(\frac{1}{x}\right) + 4x(2)$$

$$5x - 2 = 4 + 8x$$

$$-2 = 4 + 3x$$

$$-6 = 3x$$

$$-2 = x$$

37.
$$\frac{7}{8x} - \frac{3}{4} = \frac{1}{4x} + \frac{1}{2}$$
$$8x\left(\frac{7}{8x}\right) - 8x\left(\frac{3}{4}\right) = 8x\left(\frac{1}{4x}\right) + 8x\left(\frac{1}{2}\right)$$
$$7 - 6x = 2 + 4x$$
$$7 = 2 + 10x$$
$$5 = 10x$$
$$\frac{1}{2} = x$$

38.
$$\frac{3}{y-3} = \frac{3}{2} + \frac{y}{y-3}$$
$$2(y-3)\left(\frac{3}{y-3}\right) = 2(y-3)\left(\frac{3}{2}\right) + 2(y-3)\left(\frac{y}{y-3}\right)$$
$$6 = 3(y-3) + 2y$$
$$6 = 3y - 9 + 2y$$
$$6 = 5y - 9$$
$$15 = 5y$$
$$3 = y$$
Since $y = 3$ causes a denominator in the original equation to equal 0, there is no solution.

39.
$$\frac{3x}{x^2-4} - \frac{2}{x+2} = -\frac{4}{x-2}$$
$$(x+2)(x-2) \cdot \frac{3x}{(x+2)(x-2)} - (x+2)(x-2) \cdot \frac{2}{x+2} = (x+2)(x-2) \cdot \frac{-4}{x-2}$$
$$3x - 2(x-2) = -4(x+2)$$
$$3x - 2x + 4 = -4x - 8$$
$$5x = -12$$
$$x = -\frac{12}{5} \text{ or } -2\frac{2}{5} \text{ or } -2.4$$

40.
$$\frac{3y-1}{3y} - \frac{6}{5y} = \frac{1}{y} - \frac{4}{15}$$
$$15y\left(\frac{3y-1}{3y}\right) - 15y\left(\frac{6}{5y}\right) = 15y\left(\frac{1}{y}\right) - 15y\left(\frac{4}{15}\right)$$
$$5(3y-1) - 3(6) = 15 - 4y$$
$$15y - 5 - 18 = 15 - 4y$$
$$19y = 38$$
$$y = 2$$

41.
$$\frac{y+18}{y^2-16} = \frac{y}{y+4} - \frac{y}{y-4}$$

$$\frac{y+18}{(y+4)(y-4)} = \frac{y}{y+4} - \frac{y}{y-4}$$

$$(y+4)(y-4) \cdot \frac{y+18}{(y+4)(y-4)} = (y+4)(y-4) \cdot \frac{y}{y+4} - (y+4)(y-4) \cdot \frac{y}{y-4}$$

$$y+18 = (y-4)(y) - (y+4)(y)$$

$$y+18 = y^2 - 4y - y^2 - 4y$$

$$y+18 = -8y$$

$$18 = -9y$$

$$-2 = y$$

42.
$$\frac{4}{x^2-1} = \frac{2}{x-1} + \frac{2}{x+1}$$

$$(x+1)(x-1) \cdot \frac{4}{(x+1)(x-1)} = (x+1)(x-1) \cdot \frac{2}{x-1} + (x+1)(x-1) \cdot \frac{2}{x+1}$$

$$4 = (x+1)(2) + (x-1)(2)$$

$$4 = 2x + 2 + 2x - 2$$

$$4 = 4x$$

$$1 = x$$

Since $x = 1$ causes a denominator in the original equation to equal 0, there is no solution.

43.
$$\frac{3y+1}{y^2-y} - \frac{3}{y-1} = \frac{4}{y}$$

$$y(y-1) \cdot \frac{3y+1}{y(y-1)} - y(y-1) \cdot \frac{3}{y-1} = y(y-1) \cdot \frac{4}{y}$$

$$\frac{3y+1}{y(y-1)} - \frac{3}{y-1} \cdot \frac{y}{y} = \frac{4}{y} \cdot \frac{y-1}{y-1}$$

$$3y+1-3y = 4(y-1)$$

$$3y+1-3y = 4y-4$$

$$-4y = -5$$

$$y = \frac{5}{4} \text{ or } 1\frac{1}{4} \text{ or } 1.25$$

44.
$$\frac{3}{y-2} + \frac{4}{3y+2} = \frac{1}{2-y}$$

$$(y-2)(3y+2) \cdot \frac{3}{y-2} + (y-2)(3y+2) \cdot \frac{4}{3y+2} = (y-2)(3y+2) \cdot \frac{1}{2-y}$$

$$9y+6+4y-8 = -3y-2$$

$$10y = 0$$

$$y = 0$$

45.
$$\frac{x}{4} = \frac{7}{10}$$
$$x \cdot 10 = 4 \cdot 7$$
$$10x = 28$$
$$x = \frac{28}{10}$$
$$x = \frac{14}{5} = 2\frac{4}{5} = 2.8$$

46.
$$\frac{8}{5} = \frac{2}{x}$$
$$8 \cdot x = 5 \cdot 2$$
$$8x = 10$$
$$x = \frac{5}{4} = 1.25 = 1\frac{1}{4}$$

47.
$$\frac{33}{10} = \frac{x}{8}$$
$$33 \cdot 8 = 10 \cdot x$$
$$264 = 10x$$
$$26.4 = x \text{ or } x = \frac{132}{5} \text{ or } 26\frac{2}{5}$$

48.
$$\frac{16}{x} = \frac{24}{9}$$
$$16 \cdot 9 = x \cdot 24$$
$$144 = 24x$$
$$6 = x$$

49.
$$\frac{13.5}{0.6} = \frac{360}{x}$$
$$13.5x = 0.6(360)$$
$$13.5x = 216$$
$$x = 16$$

50.
$$\frac{2\frac{1}{2}}{3\frac{1}{4}} = \frac{7}{x}$$
$$\frac{\frac{5}{2}}{\frac{13}{4}} = \frac{7}{x}$$
$$\frac{5}{2}x = \frac{91}{4}$$
$$x = \frac{91}{10} \text{ or } 9\frac{1}{10} \text{ or } 9.1$$

51. x = gallons to cover 400 square feet
$$\frac{x}{400} = \frac{5}{240}$$
$$240x = 2000$$
$$x = \frac{2000}{240} = 8\frac{1}{3}$$
8.3 gallons of paint are needed.

52.
$$\frac{3}{100} = \frac{5}{x}$$
$$3x = 5(100)$$
$$3x = 500$$
$$\frac{3x}{3} = \frac{500}{3}$$
$$x = \frac{500}{3} = 166\frac{2}{3}$$
She can make 167 cookies.

53. x = distance from Houston to Dallas
$$\frac{4}{640} = \frac{1.5}{x}$$
$$4x = 640 \cdot 1.5$$
$$4x = 960$$
$$x = 240$$
Distance from Houston to Dallas is 240 miles.

54.

	D	R	$T = \frac{D}{R}$
Train	180	$s + 20$	$\frac{180}{s+20}$
Car	120	s	$\frac{120}{s}$

$$\frac{180}{s+20} = \frac{120}{s}$$
$$180s = 120s + 2400$$
$$60s = 2400$$
$$s = 40$$
$$s + 20 = 60$$
Car's speed is 40 mph. Train's speed is 60 mph.

55.
$$\frac{5.75}{3} = \frac{x}{95}$$
$$(5.75)(95) = 3x$$
$$546.25 = 3x$$
$$\frac{546.25}{3} = \frac{3x}{3}$$
$$182.1 = x$$
The peak of the canyon is 182 feet tall.

56. x = height of building

$$\frac{x}{450} = \frac{8}{3}$$
$$3x = 3600$$
$$x = 1200$$

The office building is 1200 feet tall.

57. x = time in hours for both people to wash windows when working together

$$\frac{1}{4} + \frac{1}{6} = \frac{1}{x}$$
$$12x \cdot \frac{1}{4} + 12x \cdot \frac{1}{6} = 12x \cdot \frac{1}{x}$$
$$3x + 2x = 12$$
$$5x = 12$$
$$x = 2\frac{2}{5}$$

Working together it will take them $2\frac{2}{5}$ hours or

2 hours, 24 minutes.

58. x = time to plow together

$$\frac{1}{20} + \frac{1}{30} = \frac{1}{x}$$
$$60x\left(\frac{1}{20}\right) + 60x\left(\frac{1}{30}\right) = 60x\left(\frac{1}{x}\right)$$
$$3x + 2x = 60$$
$$5x = 60$$
$$x = 12$$

Working together it will take them 12 hours.

59. $\dfrac{a^2 + 2a - 8}{6a^2 - 3a^3} = \dfrac{(a+4)(a-2)}{-3a^2(a-2)} = -\dfrac{a+4}{3a^2}$

60. $\dfrac{4a^3 + 20a^2}{2a^2 + 13a + 15} = \dfrac{4a^2(a+5)}{(2a+3)(a+5)} = \dfrac{4a^2}{2a+3}$

61. $\dfrac{x^2 - y^2}{x^2 + 4xy + 3y^2} \cdot \dfrac{x^2 + xy - 6y^2}{x^2 + xy - 2y^2}$

$= \dfrac{(x+y)(x-y)}{(x+y)(x+3y)} \cdot \dfrac{(x+3y)(x-2y)}{(x+2y)(x-y)}$

$= \dfrac{x - 2y}{x + 2y}$

62. $\dfrac{x}{x+3} + \dfrac{9x+18}{x^2+3x} = \dfrac{x}{x+3} \cdot \dfrac{x}{x} + \dfrac{9x+18}{x(x+3)}$

$= \dfrac{x^2 + 9x + 18}{x(x+3)}$

$= \dfrac{(x+6)(x+3)}{x(x+3)}$

$= \dfrac{x+6}{x}$

63. $\dfrac{x-30}{x^2-5x} + \dfrac{x}{x-5} = \dfrac{x-30}{x(x-5)} + \dfrac{x}{x-5} \cdot \dfrac{x}{x}$

$= \dfrac{x - 30 + x^2}{x(x-5)}$

$= \dfrac{x^2 + x - 30}{x(x-5)}$

$= \dfrac{(x+6)(x-5)}{x(x-5)}$

$= \dfrac{x+6}{x}$

64. $\dfrac{a+b}{ax+ay} - \dfrac{a+b}{bx+by} = \dfrac{a+b}{a(x+y)} \cdot \dfrac{b}{b} - \dfrac{a+b}{b(x+y)} \cdot \dfrac{a}{a}$

$= \dfrac{b(a+b) - a(a+b)}{ab(x+y)}$

$= \dfrac{ab + b^2 - a^2 - ab}{ab(x+y)}$

$= \dfrac{b^2 - a^2}{ab(x+y)}$

$= \dfrac{(b-a)(b+a)}{ab(x+y)}$

65. $\dfrac{\frac{5}{3x} + \frac{2}{9x}}{\frac{3}{x} + \frac{8}{3x}} = \dfrac{\frac{5}{3x} + \frac{2}{9x}}{\frac{3}{x} + \frac{8}{3x}} \cdot \dfrac{9x}{9x}$

$= \dfrac{\frac{5(9x)}{3x} + \frac{2(9x)}{9x}}{\frac{3(9x)}{x} + \frac{8(9x)}{3x}}$

$= \dfrac{15 + 2}{27 + 24}$

$= \dfrac{17}{51}$

$= \dfrac{1}{3}$

66. $\dfrac{\frac{4}{5y} - \frac{8}{y}}{y + \frac{y}{5}} = \dfrac{\frac{4}{5y} - \frac{8}{y}}{y + \frac{y}{5}} \cdot \dfrac{5y}{5y}$

$= \dfrac{\frac{4(5y)}{5y} - \frac{8(5y)}{y}}{y(5y) + \frac{y(5y)}{5}}$

$= \dfrac{4 - 40}{5y^2 + y^2}$

$= \dfrac{-36}{6y^2}$

$= -\dfrac{6}{y^2}$

67. $\dfrac{x - 3y}{x + 2y} \div \left(\dfrac{2}{y} - \dfrac{12}{x + 3y} \right)$

$= \dfrac{\frac{x-3y}{x+2y}}{\frac{2}{y} - \frac{12}{x+3y}}$

$= \dfrac{\frac{x-3y}{x+2y}}{\frac{2}{y} - \frac{12}{x+3y}} \cdot \dfrac{y(x+2y)(x+3y)}{y(x+2y)(x+3y)}$

$= \dfrac{y(x-3y)(x+3y)}{2(x+2y)(x+3y) - 12y(x+2y)}$

$= \dfrac{y(x-3y)(x+3y)}{2(x+2y)[x+3y-6y]}$

$= \dfrac{y(x-3y)(x+3y)}{2(x+2y)(x-3y)}$

$= \dfrac{y(x+3y)}{2(x+2y)}$

68. $\dfrac{7}{x+2} = \dfrac{4}{x-4}$

$7(x-4) = 4(x+2)$

$7x - 28 = 4x + 8$

$3x = 36$

$x = 12$

69. $\dfrac{2x-1}{3x-8} = \dfrac{5}{8}$

$8(2x-1) = 5(3x-8)$

$16x - 8 = 15x - 40$

$x = -32$

70. $2 + \dfrac{4}{b-1} = \dfrac{4}{b^2 - b}$

$2b(b-1) + b(b-1)\left(\dfrac{4}{b-1} \right) = b(b-1)\left(\dfrac{4}{b(b-1)} \right)$

$2b^2 - 2b + 4b = 4$

$2b^2 + 2b = 4$

$2b^2 + 2b - 4 = 0$

$2(b^2 + b - 2) = 0$

$2(b+2)(b-1) = 0$

$b + 2 = 0 \qquad\qquad b - 1 = 0$

$b = -2 \qquad\qquad\quad b = 1$

Check: $2 + \dfrac{4}{1-1} \overset{?}{=} \dfrac{4}{1^2 - 1}$

$2 + \dfrac{4}{0} \overset{?}{=} \dfrac{4}{0}$

$b = 1$ does not check.

Solution: $b = -2$

How Am I Doing? Chapter 7 Test

1. $\dfrac{2ac + 2ad}{3a^2 c + 3a^2 d} = \dfrac{2a(c+d)}{3a^2(c+d)} = \dfrac{2}{3a}$

2. $\dfrac{8x^2 - 2x^2 y^2}{y^2 + 4y + 4} = \dfrac{2x^2(2-y)(2+y)}{(y+2)^2} = \dfrac{2x^2(2-y)}{y+2}$

3. $\dfrac{x^2 + 2x}{2x - 1} \cdot \dfrac{10x^2 - 5x}{12x^3 + 24x^2} = \dfrac{x(x+2)}{2x-1} \cdot \dfrac{5x(2x-1)}{12x^2(x+2)}$

$= \dfrac{5}{12}$

4. $\dfrac{x + 2y}{12y^2} \cdot \dfrac{4y}{x^2 + xy - 2y^2} = \dfrac{x+2y}{12y^2} \cdot \dfrac{4y}{(x+2y)(x-y)}$

$= \dfrac{1}{3y(x-y)}$

5. $\dfrac{2a^2 - 3a - 2}{a^2 + 5a + 6} \div \dfrac{a^2 - 5a + 6}{a^2 - 9}$

$= \dfrac{2a^2 - 3a - 2}{a^2 + 5a + 6} \cdot \dfrac{a^2 - 9}{a^2 - 5a + 6}$

$= \dfrac{(2a+1)(a-2)}{(a+2)(a+3)} \cdot \dfrac{(a+3)(a-3)}{(a-2)(a-3)}$

$= \dfrac{2a+1}{a+2}$

6. $\dfrac{1}{a^2-a-2}+\dfrac{3}{a-2}=\dfrac{1}{(a-2)(a+1)}+\dfrac{3}{a-2}\cdot\dfrac{a+1}{a+1}$

$\qquad\qquad\qquad=\dfrac{1+3(a+1)}{(a-2)(a+1)}$

$\qquad\qquad\qquad=\dfrac{1+3a+3}{(a-2)(a+1)}$

$\qquad\qquad\qquad=\dfrac{3a+4}{(a-2)(a+1)}$

7. $\dfrac{x-y}{xy}-\dfrac{a-y}{ay}=\dfrac{x-y}{xy}\cdot\dfrac{a}{a}-\dfrac{a-y}{ay}\cdot\dfrac{x}{x}$

$\qquad\qquad\qquad=\dfrac{ax-ay-ax+xy}{axy}$

$\qquad\qquad\qquad=\dfrac{xy-ay}{axy}$

$\qquad\qquad\qquad=\dfrac{y(x-a)}{axy}$

$\qquad\qquad\qquad=\dfrac{x-a}{ax}$

8. $\dfrac{3x}{x^2-3x-18}-\dfrac{x-4}{x-6}$

$\quad=\dfrac{3x}{(x-6)(x+3)}-\dfrac{x-4}{x-6}$

$\quad=\dfrac{3x}{(x-6)(x+3)}-\dfrac{x-4}{x-6}\cdot\dfrac{x+3}{x+3}$

$\quad=\dfrac{3x}{(x-6)(x+3)}-\dfrac{x^2-x-12}{(x-6)(x+3)}$

$\quad=\dfrac{3x-(x^2-x-12)}{(x-6)(x+3)}$

$\quad=\dfrac{3x-x^2+x+12}{(x-6)(x+3)}$

$\quad=\dfrac{-x^2+4x+12}{(x-6)(x+3)}$

$\quad=\dfrac{-(x-6)(x+2)}{(x-6)(x+3)}$

$\quad=-\dfrac{x+2}{x+3}$

9. $\dfrac{\frac{x}{3y}-\frac{1}{2}}{\frac{4}{3y}-\frac{2}{x}}=\dfrac{\frac{x}{3y}-\frac{1}{2}}{\frac{4}{3y}-\frac{2}{x}}\cdot\dfrac{6xy}{6xy}$

$\qquad\quad=\dfrac{\frac{x}{3y}\cdot 6xy-\frac{1}{2}\cdot 6xy}{\frac{4}{3y}\cdot 6xy-\frac{2}{x}\cdot 6xy}$

$\qquad\quad=\dfrac{2x^2-3xy}{8x-12y}$

$\qquad\quad=\dfrac{x(2x-3y)}{4(2x-3y)}$

$\qquad\quad=\dfrac{x}{4}$

10. $\dfrac{\frac{6}{b}-4}{\frac{5}{bx}-\frac{10}{3x}}=\dfrac{\frac{6}{b}-4}{\frac{5}{bx}-\frac{10}{3x}}\cdot\dfrac{3bx}{3bx}$

$\qquad\quad=\dfrac{\frac{6(3bx)}{b}-4(3bx)}{\frac{5(3bx)}{bx}-\frac{10(3bx)}{3x}}$

$\qquad\quad=\dfrac{18x-12bx}{15-10b}$

$\qquad\quad=\dfrac{6x(3-2b)}{5(3-2b)}$

$\qquad\quad=\dfrac{6x}{5}$

11. $\dfrac{2x^2+3xy-9y^2}{4x^2+13xy+3y^2}=\dfrac{(2x-3y)(x+3y)}{(4x+y)(x+3y)}=\dfrac{2x-3y}{4x+y}$

12. $\dfrac{1}{x+4}-\dfrac{2}{x^2+6x+8}$

$\quad=\dfrac{1}{x+4}-\dfrac{2}{(x+4)(x+2)}$

$\quad=\dfrac{1}{x+4}\cdot\dfrac{x+2}{x+2}-\dfrac{2}{(x+4)(x+2)}$

$\quad=\dfrac{x+2}{(x+4)(x+2)}-\dfrac{2}{(x+4)(x+2)}$

$\quad=\dfrac{x+2-2}{(x+4)(x+2)}$

$\quad=\dfrac{x}{(x+4)(x+2)}$

13.
$$\frac{4}{3x} - \frac{5}{2x} = 5 - \frac{1}{6x}$$

$$6x\left(\frac{4}{3x}\right) - 6x\left(\frac{5}{2x}\right) = 6x(5) - 6x\left(\frac{1}{6x}\right)$$

$$8 - 15 = 30x - 1$$

$$-7 = 30x - 1$$

$$-6 = 30x$$

$$\frac{-6}{30} = x$$

$$-\frac{1}{5} = x$$

Check: $\dfrac{4}{3\left(-\frac{1}{5}\right)} - \dfrac{5}{2\left(-\frac{1}{5}\right)} \overset{?}{=} 5 - \dfrac{1}{6\left(-\frac{1}{5}\right)}$

$$-\frac{20}{3} + \frac{25}{2} \overset{?}{=} 5 + \frac{5}{6}$$

$$-\frac{40}{6} + \frac{75}{6} \overset{?}{=} \frac{30}{6} + \frac{5}{6}$$

$$\frac{35}{6} = \frac{35}{6}$$

14.
$$\frac{x-3}{x-2} = \frac{2x^2 - 15}{x^2 + x - 6} - \frac{x+1}{x+3}$$

$$\frac{x-3}{x-2} = \frac{2x^2 - 15}{(x+3)(x-2)} - \frac{x+1}{x+3}$$

$$(x-2)(x+3)\left(\frac{x-3}{x-2}\right) = (x-2)(x+3)\left[\frac{2x^2 - 15}{(x+3)(x-2)}\right] - (x-2)(x+3)\left(\frac{x+1}{x+3}\right)$$

$$(x+3)(x-3) = 2x^2 - 15 - (x-2)(x+1)$$

$$x^2 - 9 = 2x^2 - 15 - (x^2 - x - 2)$$

$$x^2 - 9 = 2x^2 - 15 - x^2 + x + 2$$

$$x^2 - 9 = x^2 + x - 13$$

$$-9 = x - 13$$

$$4 = x$$

Check: $\dfrac{4-3}{4-2} \overset{?}{=} \dfrac{2(4)^2 - 15}{4^2 + 4 - 6} - \dfrac{4+1}{4+3}$

$$\frac{1}{2} \overset{?}{=} \frac{17}{14} - \frac{5}{7}$$

$$\frac{7}{14} \overset{?}{=} \frac{17}{14} - \frac{10}{14}$$

$$\frac{7}{14} = \frac{7}{14}$$

15.

$$3 - \frac{7}{x+3} = \frac{x-4}{x+3}$$

$$(x+3)(3) - (x+3)\left(\frac{7}{x+3}\right) = (x+3)\left(\frac{x-4}{x+3}\right)$$

$$3x + 9 - 7 = x - 4$$

$$2x = -6$$

$$x = -3$$

Since $x = -3$ causes a denominator in the original equation to equal 0, there is no solution.

16.

$$\frac{3}{3x-5} = \frac{7}{5x+4}$$

$$3(5x+4) = 7(3x-5)$$

$$15x + 12 = 21x - 35$$

$$12 = 6x - 35$$

$$47 = 6x$$

$$\frac{47}{6} = x$$

Check: $\dfrac{3}{3\left(\frac{47}{6}\right)-5} \stackrel{?}{=} \dfrac{7}{5\left(\frac{47}{6}\right)+4}$

$$\frac{3}{\frac{111}{6}} \stackrel{?}{=} \frac{7}{\frac{259}{6}}$$

$$\frac{18}{111} \stackrel{?}{=} \frac{42}{259}$$

$$\frac{6}{37} = \frac{6}{37}$$

17.

$$\frac{9}{x} = \frac{13}{5}$$

$$45 = 13x$$

$$\frac{45}{13} = x$$

18.

$$\frac{9.3}{2.5} = \frac{x}{10}$$

$$2.5x = 9.3 \cdot 10$$

$$2.5x = 93$$

$$x = 37.2$$

19. x = on-time flights

$$\frac{x}{200} = \frac{113}{150}$$

$$150x = 113(200)$$

$$150x = 22,600$$

$$x = 151$$

151 flights can be expected to be on time.

20. x = cost of wood for 92 days

$$\frac{x}{92} = \frac{100}{25}$$

$$25x = 9200$$

$$x = 368$$

It will cost \$368.

21. $\dfrac{\text{height}}{\text{shadow}}$:

$$\frac{6}{7} = \frac{87}{x}$$

$$6x = 609$$

$$x = 101.5$$

The rope bridge should be 102 feet long.

Chapter 8

1. $\left(\dfrac{3xy^{-1}}{z^2}\right)^4 = \dfrac{3^4 x^4 y^{-4}}{z^8} = \dfrac{81x^4}{y^4 z^8}$

3. $\left(\dfrac{2a^2 b}{3b^{-1}}\right)^3 = \dfrac{2^3 a^{2\cdot3} b^3}{3^3 b^{-1\cdot3}}$

$= \dfrac{8a^6 b^3}{27b^{-3}}$

$= \dfrac{8a^6 b^{3+3}}{27}$

$= \dfrac{8a^6 b^6}{27}$

5. $\left(\dfrac{2x^2}{y}\right)^{-3} = \dfrac{2^{-3} x^{-6}}{y^{-3}} = \dfrac{y^3}{8x^6}$

7. $\left(\dfrac{3xy^{-2}}{y^3}\right)^{-2} = \dfrac{3^{-2} x^{-2} y^4}{y^{-6}} = \dfrac{y^{4+6}}{3^2 x^2} = \dfrac{y^{10}}{9x^2}$

9. $(x^{3/4})^2 = x^{6/4} = x^{3/2}$

11. $(y^{12})^{2/3} = y^{12\cdot\frac{2}{3}} = y^8$

13. $\dfrac{x^{3/5}}{x^{1/5}} = x^{\frac{3}{5}-\frac{1}{5}} = x^{2/5}$

15. $\dfrac{x^{8/9}}{x^{2/9}} = x^{\frac{8}{9}-\frac{2}{9}} = x^{6/9} = x^{2/3}$

17. $\dfrac{a^2}{a^{1/4}} = a^{\frac{8}{4}-\frac{1}{4}} = a^{7/4}$

19. $x^{1/7} \cdot x^{3/7} = x^{\frac{1}{7}+\frac{3}{7}} = x^{4/7}$

21. $a^{3/8} \cdot a^{1/2} = a^{\frac{3}{8}+\frac{4}{8}} = a^{7/8}$

23. $y^{3/5} \cdot y^{-1/10} = y^{\frac{6}{10}-\frac{1}{10}} = y^{5/10} = y^{1/2}$

25. $x^{-3/4} = \dfrac{1}{x^{3/4}}$

27. $a^{-5/6} b^{1/3} = \dfrac{b^{1/3}}{a^{5/6}}$

29. $6^{-1/2} = \dfrac{1}{6^{1/2}}$

31. $3a^{-1/3} = \dfrac{3}{a^{1/3}}$

33. $(27)^{5/3} = (3^3)^{5/3} = 3^{3\cdot\frac{5}{3}} = 3^5 = 243$

35. $(4)^{3/2} = (2^2)^{3/2} = 2^{2\cdot3/2} = 2^3 = 8$

37. $(-8)^{5/3} = ((-2)^3)^{5/3} = (-2)^{3\cdot5/3} = (-2)^5 = -32$

39. $(-27)^{2/3} = ((-3)^3)^{2/3} = (-3)^{3\cdot2/3} = (-3)^2 = 9$

41. $(x^{1/4} y^{-1/3})(x^{3/4} y^{1/2}) = x^{\frac{1}{4}+\frac{3}{4}} y^{-\frac{2}{6}+\frac{3}{6}} = xy^{1/6}$

43. $(7x^{1/3} y^{1/4})(-2x^{1/4} y^{-1/6}) = 7(-2)x^{\frac{1}{3}+\frac{1}{4}} y^{\frac{1}{4}-\frac{1}{6}}$

$= -14x^{\frac{3}{12}+\frac{4}{12}} y^{\frac{3}{12}-\frac{2}{12}}$

$= -14x^{7/12} y^{1/12}$

45. $6^2 \cdot 6^{-2/3} = 6^{\frac{6}{3}-\frac{2}{3}} = 6^{4/3}$

47. $\dfrac{2x^{1/5}}{x^{-1/2}} = 2x^{\frac{2}{10}+\frac{5}{10}} = 2x^{7/10}$

49. $\dfrac{-20x^2 y^{-1/5}}{5x^{-1/2} y} = -\dfrac{4x^{2+\frac{1}{2}}}{y^{1+\frac{1}{5}}} = -\dfrac{4x^{5/2}}{y^{6/5}}$

51. $\left(\dfrac{8a^2 b^6}{a^{-1} b^3}\right)^{1/3} = (8a^3 b^3)^{1/3} = 8^{1/3} a^{3\cdot\frac{1}{3}} b^{3\cdot\frac{1}{3}} = 2ab$

53. $(-4x^{1/4} y^{5/2} z^{1/2})^2 = 16x^{2/4} y^{10/2} z^{2/2}$

$= 16x^{1/2} y^5 z$

55. $x^{2/3}(x^{4/3} - x^{1/5}) = x^{\frac{2}{3}+\frac{4}{3}} - x^{\frac{2}{3}+\frac{1}{5}}$

$= x^{6/3} - x^{\frac{10}{15}+\frac{3}{15}}$

$= x^2 - x^{13/15}$

57. $m^{7/8}(m^{-1/2} + 2m) = m^{\frac{7}{8}-\frac{4}{8}} + 2m^{\frac{7}{8}+\frac{8}{8}}$

$= m^{3/8} + 2m^{15/8}$

59. $(8)^{-1/3} = (2^3)^{-1/3} = 2^{3(-1/3)} = 2^{-1} = \dfrac{1}{2}$

61. $(64)^{-2/3} = (4^3)^{-2/3} = 4^{3(-2/3)} = 4^{-2} = \dfrac{1}{4^2} = \dfrac{1}{16}$

63. $(81)^{3/4} + (25)^{1/2} = (3^4)^{3/4} + (5^2)^{1/2}$

$= 3^{4 \cdot 3/4} + 5^{2 \cdot 1/2}$

$= 3^3 + 5$

$= 27 + 5$

$= 32$

65. $3y^{1/2} + y^{-1/2} = 3y^{1/2} \cdot \dfrac{y^{1/2}}{y^{1/2}} + \dfrac{1}{y^{1/2}}$

$= \dfrac{3y}{y^{1/2}} + \dfrac{1}{y^{1/2}}$

$= \dfrac{3y+1}{y^{1/2}}$

67. $x^{-1/3} + 6^{4/3} = \dfrac{1}{x^{1/3}} + 6^{4/3}$

$= \dfrac{1}{x^{1/3}} + \dfrac{6^{4/3} x^{1/3}}{x^{1/3}}$

$= \dfrac{1 + 6^{4/3} x^{1/3}}{x^{1/3}}$

69. $10a^{5/4} - 4a^{8/5} = 2a^{4/4} \cdot 5a^{1/4} - 2a^{5/5} \cdot 2a^{3/5}$

$= 2a(5a^{1/4} - 2^{3/5})$

71. $12x^{4/3} - 3x^{5/2} = 3x^{3/3} \cdot 4x^{1/3} - 3x^{2/2} \cdot x^{3/2}$

$= 3x(4x^{1/3} - x^{3/2})$

73. $x^a \cdot x^{1/4} = x^{-1/8} \Rightarrow x^{a+\frac{1}{4}} = x^{-1/8}$

$a + \dfrac{1}{4} = -\dfrac{1}{8}$

$a = -\dfrac{3}{8}$

75. $r = 0.62(V)^{1/3}$

$r = 0.62(27)^{1/3} = 0.62(3) = 1.86$

The radius is 1.86 meters.

77. $r = \left(\dfrac{3V}{\pi h}\right)^{1/2}$

$r = \left(\dfrac{3(314)}{3.14(12)}\right)^{1/2} = (25)^{1/2} = 5$

The radius is 5 feet.

Cumulative Review

79. $-4(x+1) = \dfrac{1}{3}(3 - 2x)$

$-12(x+1) = 3 - 2x$

$-12x - 12 = 3 - 2x$

$-10x = 15$

$x = -\dfrac{3}{2}$

80. $\dfrac{2}{3}x + 4 = -\dfrac{1}{2}(x-1)$

$6\left(\dfrac{2}{3}x + 4\right) = 6\left[-\dfrac{1}{2}(x-1)\right]$

$4x + 24 = -3(x-1)$

$4x + 24 = -3x + 3$

$7x = -21$

$x = -3$

Quick Quiz 8.1

1. $(-4x^{2/3}y^{1/4})(3x^{1/6}y^{1/2}) = (-4)(3)x^{\frac{2}{3}+\frac{1}{6}}y^{\frac{1}{4}+\frac{1}{2}}$

$= -12x^{\frac{4}{6}+\frac{1}{6}}y^{\frac{1}{4}+\frac{2}{4}}$

$= -12x^{5/6}y^{3/4}$

2. $\dfrac{16x^4}{8x^{2/3}} = 2x^{4-\frac{2}{3}} = 2x^{\frac{12}{3}-\frac{2}{3}} = 2x^{10/3}$

3. $(25x^{1/4})^{3/2} = (5^2 x^{2/8})^{3/2}$

$= (5x^{1/8})^{2(3/2)}$

$= (5x^{1/8})^3$

$= 5^3 x^{3 \cdot \frac{1}{8}}$

$= 125x^{3/8}$

4. Answers may vary. Possible answer: Change the exponents to have equal denominators, then add the numerators over the common denominator. This is the combined, simplified exponent for x.

8.2 Exercises

1. A square root of a number is a value that when multiplied by itself is equal to the original number.

3. One answer is $\sqrt[3]{-8} = -2$ because $(-2)(-2)(-2) = -8$.

5. $\sqrt{100} = 10$ because $10^2 = 100$.

7. $\sqrt{16} + \sqrt{81} = 4 + 9 = 13$

9. $-\sqrt{\dfrac{1}{9}} = -\dfrac{1}{3}$

11. $\sqrt{-36}$ is not a real number.

13. $\sqrt{0.25} = 0.5$ because $0.5^2 = 0.25$.

15. $f(x) = \sqrt{4x + 12}$
$f(-3) = \sqrt{4(-3) + 12} = \sqrt{0} = 0$
$f(-1) = \sqrt{4(-1) + 12} = \sqrt{8} \approx 2.8$
$f(1) = \sqrt{4(1) + 12} = \sqrt{16} = 4$
$f(3) = \sqrt{4(3) + 12} = \sqrt{24} \approx 4.9$
The domain is $4x + 12 \geq 0$
$\qquad\qquad\qquad 4x \geq -12$
$\qquad\qquad\qquad x \geq -3$
The domain is all real numbers x where $x \geq -3$.

17. $f(x) = \sqrt{0.5x - 5}$
$f(10) = \sqrt{0.5(10) - 5} = \sqrt{0} = 0$
$f(12) = \sqrt{0.5(12) - 5} = \sqrt{1} = 1$
$f(18) = \sqrt{0.5(18) - 5} = \sqrt{4} = 2$
$f(20) = \sqrt{0.5(20) - 5} = \sqrt{5} \approx 2.2$
The domain is $0.5x - 5 \geq 0$
$\qquad\qquad\qquad 0.5x \geq 5$
$\qquad\qquad\qquad x \geq 10$
The domain is all real numbers x where $x \geq 10$.

19. $f(x) = \sqrt{x - 1}$
$f(1) = \sqrt{1 - 1} = \sqrt{0} = 0$
$f(2) = \sqrt{2 - 1} = \sqrt{1} = 1$
$f(5) = \sqrt{5 - 1} = \sqrt{4} = 2$
$f(10) = \sqrt{10 - 1} = \sqrt{9} = 3$

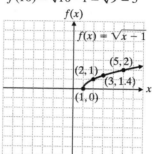

21. $f(x) = \sqrt{3x + 9}$
$f(-3) = \sqrt{3(-3) + 9} = \sqrt{0} = 0$
$f\left(-\dfrac{8}{3}\right) = \sqrt{3\left(-\dfrac{8}{3}\right) + 9} = \sqrt{1} = 1$
$f\left(-\dfrac{5}{3}\right) = \sqrt{3\left(-\dfrac{5}{3}\right) + 9} = \sqrt{4} = 2$
$f(0) = \sqrt{3(0) + 9} = \sqrt{9} = 3$

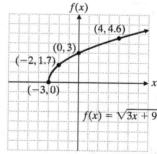

23. $\sqrt[3]{64} = 4$ because $4^3 = 64$.

25. $\sqrt[3]{-1000} = -10$ because $(-10)^3 = -1000$.

27. $\sqrt[4]{16} = \sqrt[4]{2^4} = 2$

29. $\sqrt[4]{81} = \sqrt[4]{3^4} = 3$

31. $\sqrt[5]{(8)^5} = 8$

33. $\sqrt[8]{(5)^8} = 5$

35. $\sqrt[3]{-\dfrac{1}{8}} = \sqrt[3]{\left(-\dfrac{1}{2}\right)^3} = -\dfrac{1}{2}$

37. $\sqrt[3]{y} = y^{1/3}$

39. $\sqrt[5]{m^3} = m^{3/5}$

41. $\sqrt[4]{2a} = (2a)^{1/4}$

43. $\sqrt[7]{(a+b)^3} = (a+b)^{3/7}$

45. $\sqrt{\sqrt[3]{x}} = (x^{1/3})^{1/2} = x^{1/6}$

47. $\left(\sqrt[6]{3x}\right)^5 = ((3x)^{1/6})^5 = (3x)^{5/6}$

49. $\sqrt[6]{(12)^6} = 12$

51. $\sqrt[3]{x^{12}y^3} = \sqrt[3]{(x^4 y)^3} = x^4 y$

53. $\sqrt{36x^8 y^4} = \sqrt{(6x^4 y^2)^2} = 6x^4 y^2$

55. $\sqrt[4]{16a^8 b^4} = \sqrt[4]{(2a^2 b)^4} = 2a^2 b$

57. $y^{4/7} = \sqrt[7]{y^4}$ or $\sqrt[7]{y}^4$

59. $7^{-2/3} = \dfrac{1}{7^{2/3}} = \dfrac{1}{\sqrt[3]{7^2}} = \dfrac{1}{\sqrt[3]{49}}$

61. $(a+5b)^{3/4} = \sqrt[4]{(a+5b)^3}$ or $\sqrt[4]{a+5b}^3$

63. $(-x)^{3/5} = \sqrt[5]{(-x)^3}$ or $\sqrt[5]{-x}^3$

65. $(2xy)^{3/5} = \sqrt[5]{(2xy)^3} = \sqrt[5]{8x^3 y^3}$

67. $9^{3/2} = (3)^{2(3/2)} = 3^3 = 27$

69. $\left(\dfrac{4}{25}\right)^{1/2} = \sqrt{\dfrac{4}{25}} = \dfrac{2}{5}$

71. $\left(\dfrac{1}{8}\right)^{-1/3} = (8^{-1})^{-1/3} = 8^{-1(-1/3)} = 8^{1/3} = \sqrt[3]{8} = 2$

73. $(64x^4)^{-1/2} = (8x^2)^{2(-1/2)} = (8x^2)^{-1} = \dfrac{1}{8x^2}$

75. $\sqrt{121x^4} = \sqrt{(11x^2)^2} = 11x^2$

77. $\sqrt{144a^6 b^{24}} = \sqrt{(12a^3 b^{12})^2} = 12a^3 b^{12}$

79. $\sqrt{25x^2} = \sqrt{(5x)^2} = |5x| = 5|x|$

81. $\sqrt[3]{-8x^6} = \sqrt[3]{(-2x^2)^3} = -2x^2$

83. $\sqrt[4]{x^4 y^{24}} = \sqrt[4]{(xy^6)^4} = |xy^6| = |x|y^6$

85. $\sqrt[4]{a^{12} b^4} = \sqrt[4]{(a^3 b)^4} = |a^3 b|$

87. $\sqrt{25x^{12} y^4} = \sqrt{(5x^6 y^2)^2} = |5x^6 y^2| = 5x^6 y^2$

Cumulative Review

89. $-5x + 2y = 6$
$5x - 2y = -6$
$5x = 2y - 6$
$x = \dfrac{2y-6}{5}$

90. $x = \dfrac{2}{3}y + 4$
$3(x) = 3\left(\dfrac{2}{3}y + 4\right)$
$3x = 2y + 12$
$3x - 12 = 2y$
$\dfrac{3x-12}{2} = y$ or $y = \dfrac{3x}{2} - 6$

Quick Quiz 8.2

1. $\left(\dfrac{4}{25}\right)^{3/2} = \left(\dfrac{2^2}{5^2}\right)^{3/2} = \left(\dfrac{2}{5}\right)^{2(3/2)} = \left(\dfrac{2}{5}\right)^3 = \dfrac{8}{125}$

2. $\sqrt[3]{-64} = -4$ because $(-4)^3 = -64$.

3. $\sqrt{121x^{10}y^{12}} = \sqrt{(11x^5 y^6)^2} = 11x^5 y^6$

4. Answers may vary. Possible solution:
Factor the coefficient completely to identify the fourth root. Remove the fourth root from under the radical. Divide the exponents of the variables by 4 to remove the variables from under the radical.

8.3 Exercises

1. $\sqrt{8} = \sqrt{4 \cdot 2} = \sqrt{4}\sqrt{2} = 2\sqrt{2}$

3. $\sqrt{18} = \sqrt{9 \cdot 2} = \sqrt{9}\sqrt{2} = 3\sqrt{2}$

5. $\sqrt{28} = \sqrt{4 \cdot 7} = \sqrt{4}\sqrt{7} = 2\sqrt{7}$

7. $\sqrt{50} = \sqrt{25}\sqrt{2} = 5\sqrt{2}$

9. $\sqrt{9x^2} = \sqrt{9}\sqrt{x^2} = 3x$

11. $\sqrt{40a^6 b^7} = \sqrt{4 \cdot 10 a^6 b^6 b}$
$= \sqrt{4a^6 b^6}\sqrt{10b}$
$= 2a^3 b^3 \sqrt{10b}$

13. $\sqrt{90x^3 yz^4} = \sqrt{9 \cdot 10 x^2 xyz^4}$
$= \sqrt{9x^2 z^4}\sqrt{10xy}$
$= 3xz^2 \sqrt{10xy}$

15. $\sqrt[3]{8} = \sqrt[3]{2^3} = 2$

17. $\sqrt[3]{40} = \sqrt[3]{8 \cdot 5} = \sqrt[3]{8}\sqrt[3]{5} = 2\sqrt[3]{5}$

19. $\sqrt[3]{54a^2} = \sqrt[3]{27 \cdot 2a^2} = \sqrt[3]{27}\sqrt[3]{2a^2} = 3\sqrt[3]{2a^2}$

21. $\sqrt[3]{27a^5 b^9} = \sqrt[3]{27a^3 b^9 \cdot a^2}$
$= \sqrt[3]{27a^3 b^9}\sqrt[3]{a^2}$
$= 3ab^3 \sqrt[3]{a^2}$

23. $\sqrt[3]{40x^{12} y^{13}} = \sqrt[3]{8 \cdot 5 x^{12} y^{12} y}$
$= \sqrt[3]{8x^{12} y^{12}}\sqrt[3]{5y}$
$= 2x^4 y^4 \sqrt[3]{5y}$

25. $\sqrt[4]{81kp^{23}} = \sqrt[4]{3^4 kp^{20} p^3}$
$= \sqrt[4]{3^4 p^{20}}\sqrt[4]{kp^3}$
$= 3p^5 \sqrt[4]{kp^3}$

27. $\sqrt[5]{-32x^5 y^6} = \sqrt[5]{(-2)^5 x^5 y^5 y}$
$= \sqrt[5]{(-2)^5 x^5 y^5}\sqrt[5]{y}$
$= -2xy\sqrt[5]{y}$

29. $\sqrt[4]{1792} = a\sqrt[4]{7}$
$\sqrt[4]{256 \cdot 7} = a\sqrt[4]{7}$
$\sqrt[4]{4^4 \cdot 7} = a\sqrt[4]{7}$
$4\sqrt[4]{7} = a\sqrt[4]{7}$
$a = 4$

31. $4\sqrt{5} + 8\sqrt{5} = 12\sqrt{5}$

33. $4\sqrt{3} + \sqrt{7} - 5\sqrt{7} = 4\sqrt{3} - 4\sqrt{7}$

35. $3\sqrt{32} - \sqrt{2} = 3\sqrt{16 \cdot 2} - \sqrt{2} = 12\sqrt{2} - \sqrt{2} = 11\sqrt{2}$

37. $4\sqrt{12} + \sqrt{27} = 4\sqrt{4 \cdot 3} + \sqrt{9 \cdot 3}$
$= 8\sqrt{3} + 3\sqrt{3}$
$= 11\sqrt{3}$

39. $\sqrt{8} + \sqrt{50} - 2\sqrt{72} = \sqrt{4 \cdot 2} + \sqrt{25 \cdot 2} - 2\sqrt{36 \cdot 2}$
$= 2\sqrt{2} + 5\sqrt{2} - 12\sqrt{2}$
$= -5\sqrt{2}$

41. $\sqrt{48} - 2\sqrt{27} + \sqrt{12} = \sqrt{16 \cdot 3} - 2\sqrt{9 \cdot 3} + \sqrt{4 \cdot 3}$
$= 4\sqrt{3} - 6\sqrt{3} + 2\sqrt{3}$
$= 0$

43. $-2\sqrt{24} + 3\sqrt{6} + \sqrt{54} = -2\sqrt{4 \cdot 6} + 3\sqrt{6} + \sqrt{9 \cdot 6}$
$= -4\sqrt{6} + 3\sqrt{6} + 3\sqrt{6}$
$= 2\sqrt{6}$

45. $3\sqrt{48x} - 2\sqrt{12x} = 3\sqrt{16 \cdot 3x} - 2\sqrt{4 \cdot 3x}$
$= 12\sqrt{3x} - 4\sqrt{3x}$
$= 8\sqrt{3x}$

47. $5\sqrt{2x} + 2\sqrt{18x} + 2\sqrt{32x}$
$= 5\sqrt{2x} + 2\sqrt{9 \cdot 2x} + 2\sqrt{16 \cdot 2x}$
$= 5\sqrt{2x} + 6\sqrt{2x} + 8\sqrt{2x}$
$= 19\sqrt{2x}$

49. $\sqrt{44} - 3\sqrt{63x} + 4\sqrt{28x}$
$= \sqrt{4 \cdot 11} - 3\sqrt{9 \cdot 7x} + 4\sqrt{4 \cdot 7x}$
$= 2\sqrt{11} - 9\sqrt{7x} + 8\sqrt{7x}$
$= 2\sqrt{11} - \sqrt{7x}$

51. $\sqrt{200x^3} - x\sqrt{32x} = \sqrt{100x^2 \cdot 2x} - x\sqrt{16 \cdot 2x}$
$$= 10x\sqrt{2x} - 4x\sqrt{2x}$$
$$= 6x\sqrt{2x}$$

53. $\sqrt[3]{16} + 3\sqrt[3]{54} = \sqrt[3]{2^3 \cdot 2} + 3\sqrt[3]{3^3 \cdot 2}$
$$= 2\sqrt[3]{2} + 9\sqrt[3]{2}$$
$$= 11\sqrt[3]{2}$$

55. $4\sqrt[3]{x^4 y^3} - 3\sqrt[3]{xy^5} = 4\sqrt[3]{x^3 y^3 \cdot x} - 3\sqrt[3]{y^3 \cdot xy^2}$
$$= 4xy\sqrt[3]{x} - 3y\sqrt[3]{xy^2}$$

57. $\sqrt{48} + \sqrt{27} + \sqrt{75}$
$$= 6.92820323 + 5.196152423 + 8.660254038$$
$$\approx 20.7846097$$
$12\sqrt{3} \approx 20.7846097$ which shows
$\sqrt{48} + \sqrt{27} + \sqrt{75} = 12\sqrt{3}$.

59. $I = \sqrt{\dfrac{P}{R}} = \sqrt{\dfrac{825}{15}} = \sqrt{55} \approx 7.416$
The current is approximately 7.416 amps.

61. $T = 2\pi\sqrt{\dfrac{L}{32}} \approx 2(3.14)\sqrt{\dfrac{8}{32}} \approx 3.14$
The period of the pendulum is approximately 3.14 seconds.

Cumulative Review

63. $16x^3 - 56x^2 y + 49xy^2$
$$= x(16x^2 - 56xy + 49y^2)$$
$$= x[(4x)^2 - 2(4x)(7y) + (7y)^2]$$
$$= x(4x - 7y)^2$$

64. $81x^2 y - 25y = y(81x^2 - 25)$
$$= y[(9x)^2 - 5^2]$$
$$= y(9x + 5)(9x - 5)$$

65. $(2x+3)(2x-3) = (2x)^2 - 3^2 = 4x^2 - 9$

66. $(2x+3)^2 = (2x)^2 + 2(2x)(3) + 3^2$
$$= 4x^2 + 12x + 9$$

Quick Quiz 8.3

1. $\sqrt{120x^7 y^8} = \sqrt{4 \cdot 30xx^6 y^8}$
$$= \sqrt{4x^6 y^8}\sqrt{30x}$$
$$= 2x^3 y^4 \sqrt{30x}$$

2. $\sqrt[3]{16x^{15} y^{10}} = \sqrt[3]{2y \cdot 2^3 x^{15} y^9}$
$$= \sqrt[3]{2^3 x^{15} y^9}\sqrt[3]{2y}$$
$$= 2x^5 y^3 \sqrt[3]{2y}$$

3. $2\sqrt{75} + 3\sqrt{48} - 4\sqrt{27}$
$$= 2\sqrt{25 \cdot 3} + 3\sqrt{16 \cdot 3} - 4\sqrt{9 \cdot 3}$$
$$= 10\sqrt{3} + 12\sqrt{3} - 12\sqrt{3}$$
$$= 10\sqrt{3}$$

4. Answers may vary. Possible solution: Completely factor the coefficient of the radicand, and express as its primes to their respective powers. Divide exponents of the radicand coefficient primes and variables by the index. The results are moved outside the radical. The remainders remain inside.

8.4 Exercises

1. $\sqrt{5}\sqrt{7} = \sqrt{5 \cdot 7} = \sqrt{35}$

3. $(4\sqrt{7})(-2\sqrt{3}) = -8\sqrt{7 \cdot 3} = -8\sqrt{21}$

5. $(3\sqrt{10})(-4\sqrt{2}) = -12\sqrt{20} = -12\sqrt{4 \cdot 5} = -24\sqrt{5}$

7. $(-3\sqrt{y})(\sqrt{5x}) = -3\sqrt{5xy}$

9. $(3x\sqrt{2x})(-2\sqrt{10xy}) = -6x\sqrt{20x^2 y}$
$$= -6x\sqrt{4x^2 \cdot 5y}$$
$$= -12x^2 \sqrt{5y}$$

11. $5\sqrt{a}(3\sqrt{b} - 5) = 15\sqrt{ab} - 25\sqrt{a}$

13. $-3\sqrt{a}(\sqrt{2b} + 2\sqrt{5}) = -3\sqrt{2ab} - 6\sqrt{5a}$

15. $-\sqrt{a}(\sqrt{a} - 2\sqrt{b}) = -a + 2\sqrt{ab}$

17. $7\sqrt{x}(2\sqrt{3} - 5\sqrt{x}) = 14\sqrt{3x} - 35x$

19. $(3 - \sqrt{2})(8 + \sqrt{2}) = 24 + 3\sqrt{2} - 8\sqrt{2} - 2$
$$= 24 - 5\sqrt{2} - 2$$
$$= 22 - 5\sqrt{2}$$

21. $\left(2\sqrt{3}+\sqrt{2}\right)\left(2\sqrt{3}-4\sqrt{2}\right)$
$= 4\cdot 3 - 8\sqrt{6} + 2\sqrt{6} - 4\cdot 2$
$= 12 - 6\sqrt{6} - 8$
$= 4 - 6\sqrt{6}$

23. $\left(\sqrt{7}+4\sqrt{5x}\right)\left(2\sqrt{7}+3\sqrt{5x}\right)$
$= 2\cdot 7 + 3\sqrt{35x} + 8\sqrt{35x} + 12(5x)$
$= 14 + 11\sqrt{35x} + 60x$

25. $\left(\sqrt{3}+2\sqrt{2}\right)\left(\sqrt{5}+\sqrt{3}\right) = \sqrt{15} + 3 + 2\sqrt{10} + 2\sqrt{6}$

27. $\left(\sqrt{5}-2\sqrt{6}\right)^2 = \left(\sqrt{5}\right)^2 - 2\cdot 2\sqrt{5}\sqrt{6} + \left(2\sqrt{6}\right)^2$
$= 5 - 4\sqrt{30} + 24$
$= 29 - 4\sqrt{30}$

29. $\left(9-2\sqrt{b}\right)^2 = 9^2 - 2(9)2\sqrt{b} + 2^2\cdot b$
$= 81 - 36\sqrt{b} + 4b$

31. $\left(\sqrt{3x+4}+3\right)^2 = \left(\sqrt{3x+4}\right)^2 + 2\cdot 3\sqrt{3x+4} + 3^2$
$= 3x + 4 + 6\sqrt{3x+4} + 9$
$= 3x + 13 + 6\sqrt{3x+4}$

33. $\left(\sqrt[3]{x^2}\right)\left(3\sqrt[3]{4x} - 4\sqrt[3]{x^5}\right) = 3\sqrt[3]{4x^3} - 4\sqrt[3]{x^7}$
$= 3x\sqrt[3]{4} - 4\sqrt[3]{x^6\cdot x}$
$= 3x\sqrt[3]{4} - 4x^2\sqrt[3]{x}$

35. $\left(\sqrt[3]{3}+\sqrt[3]{2}\right)\left(\sqrt[3]{9}-\sqrt[3]{4}\right) = \sqrt[3]{27} - \sqrt[3]{12} + \sqrt[3]{18} - \sqrt[3]{8}$
$= 3 - \sqrt[3]{12} + \sqrt[3]{18} - 2$
$= 1 + \sqrt[3]{18} - \sqrt[3]{12}$

37. $\sqrt{\dfrac{49}{25}} = \dfrac{\sqrt{49}}{\sqrt{25}} = \dfrac{7}{5}$

39. $\sqrt{\dfrac{12x}{49y^6}} = \dfrac{\sqrt{4\cdot 3x}}{\sqrt{49y^6}} = \dfrac{2\sqrt{3x}}{7y^3}$

41. $\sqrt[3]{\dfrac{8x^5y^6}{27}} = \dfrac{\sqrt[3]{8x^3y^6\cdot x^2}}{\sqrt[3]{27}} = \dfrac{2xy^2\sqrt[3]{x^2}}{3}$

43. $\dfrac{\sqrt[3]{5y^8}}{\sqrt[3]{27x^3}} = \dfrac{\sqrt[3]{5y^6y^2}}{3x} = \dfrac{y^2\sqrt[3]{5y^2}}{3x}$

45. $\dfrac{3}{\sqrt{2}} = \dfrac{3}{\sqrt{2}}\cdot\dfrac{\sqrt{2}}{\sqrt{2}} = \dfrac{3\sqrt{2}}{2}$

47. $\sqrt{\dfrac{4}{3}} = \dfrac{\sqrt{4}}{\sqrt{3}} = \dfrac{\sqrt{4}}{\sqrt{3}}\cdot\dfrac{\sqrt{3}}{\sqrt{3}} = \dfrac{2\sqrt{3}}{3}$

49. $\dfrac{1}{\sqrt{5y}} = \dfrac{1}{\sqrt{5y}}\cdot\dfrac{\sqrt{5y}}{\sqrt{5y}} = \dfrac{\sqrt{5y}}{5y}$

51. $\dfrac{\sqrt{14a}}{\sqrt{2y}} = \dfrac{\sqrt{14a}}{\sqrt{2y}}\cdot\dfrac{\sqrt{2y}}{\sqrt{2y}}$
$= \dfrac{\sqrt{28ay}}{2y}$
$= \dfrac{\sqrt{4\cdot 7ay}}{2y}$
$= \dfrac{2\sqrt{7ay}}{2y}$
$= \dfrac{\sqrt{7ay}}{y}$

53. $\dfrac{\sqrt{2}}{\sqrt{6x}} = \dfrac{\sqrt{2}}{\sqrt{6x}}\cdot\dfrac{\sqrt{6x}}{\sqrt{6x}}$
$= \dfrac{\sqrt{12x}}{6x}$
$= \dfrac{\sqrt{4\cdot 3x}}{6x}$
$= \dfrac{2\sqrt{3x}}{6x}$
$= \dfrac{\sqrt{3x}}{3x}$

55. $\dfrac{x}{\sqrt{5}-\sqrt{2}} = \dfrac{x}{\sqrt{5}-\sqrt{2}}\cdot\dfrac{\sqrt{5}+\sqrt{2}}{\sqrt{5}+\sqrt{2}}$
$= \dfrac{x\left(\sqrt{5}+\sqrt{2}\right)}{5-2}$
$= \dfrac{x\left(\sqrt{5}+\sqrt{2}\right)}{3}$

57. $\dfrac{2y}{\sqrt{6}+\sqrt{5}} = \dfrac{2y}{\sqrt{6}+\sqrt{5}}\cdot\dfrac{\sqrt{6}-\sqrt{5}}{\sqrt{6}-\sqrt{5}}$
$= \dfrac{2y\left(\sqrt{6}-\sqrt{5}\right)}{6-5}$
$= 2y\left(\sqrt{6}-\sqrt{5}\right)$

59. $\dfrac{\sqrt{y}}{\sqrt{6}+\sqrt{2y}} = \dfrac{\sqrt{y}}{\sqrt{6}+\sqrt{2y}} \cdot \dfrac{\sqrt{6}-\sqrt{2y}}{\sqrt{6}-\sqrt{2y}}$

$= \dfrac{\sqrt{6y}-\sqrt{2y^2}}{6-2y}$

$= \dfrac{\sqrt{6y}-y\sqrt{2}}{6-2y}$

61. $\dfrac{\sqrt{5}+\sqrt{3}}{\sqrt{5}-\sqrt{3}} = \dfrac{\sqrt{5}+\sqrt{3}}{\sqrt{5}-\sqrt{3}} \cdot \dfrac{\sqrt{5}+\sqrt{3}}{\sqrt{5}+\sqrt{3}}$

$= \dfrac{5+2\sqrt{15}+3}{5-3}$

$= \dfrac{8+2\sqrt{15}}{2}$

$= \dfrac{2\left(4+\sqrt{15}\right)}{2}$

$= 4+\sqrt{15}$

63. $\dfrac{\sqrt{3x}-2\sqrt{y}}{\sqrt{3x}+\sqrt{y}} = \dfrac{\sqrt{3x}-2\sqrt{y}}{\sqrt{3x}+\sqrt{y}} \cdot \dfrac{\sqrt{3x}-\sqrt{y}}{\sqrt{3x}-\sqrt{y}}$

$= \dfrac{3x-3\sqrt{3xy}+2y}{3x-y}$

65. $2\sqrt{32}-\sqrt{72}+3\sqrt{18}$

$= 2\sqrt{16\cdot 2}-\sqrt{36\cdot 2}+3\sqrt{9\cdot 2}$

$= 8\sqrt{2}-6\sqrt{2}+9\sqrt{2}$

$= 11\sqrt{2}$

67. $\left(3\sqrt{2}-5\sqrt{3}\right)\left(\sqrt{2}+2\sqrt{3}\right)$

$= 3\cdot 2+6\sqrt{6}-5\sqrt{6}-10\cdot 3$

$= 6+\sqrt{6}-30$

$= -24+\sqrt{6}$

69. $\dfrac{9}{\sqrt{8x}} = \dfrac{9}{\sqrt{8x}} \cdot \dfrac{\sqrt{8x}}{\sqrt{8x}} = \dfrac{9\sqrt{4\cdot 2x}}{8x} = \dfrac{18\sqrt{2x}}{8x} = \dfrac{9\sqrt{2x}}{4x}$

71. $\dfrac{\sqrt{5}+1}{\sqrt{5}+2} = \dfrac{\sqrt{5}+1}{\sqrt{5}+2} \cdot \dfrac{\sqrt{5}-2}{\sqrt{5}-2}$

$= \dfrac{5-2\sqrt{5}+\sqrt{5}-2}{5-4}$

$= 5-2\sqrt{5}+\sqrt{5}-2$

$= 3-\sqrt{5}$

73. $\dfrac{\sqrt{6}}{2\sqrt{3}-\sqrt{2}} = 1.194938299...$

$\dfrac{\sqrt{3}+3\sqrt{2}}{5} = 1.194938299...$

The decimal approximations are the same. The student worked correctly.

75. $\dfrac{\sqrt{2}+3\sqrt{5}}{4} = \dfrac{\sqrt{2}+3\sqrt{5}}{4} \cdot \dfrac{\sqrt{2}-3\sqrt{5}}{\sqrt{2}-3\sqrt{5}}$

$= \dfrac{2-9(5)}{4\left(\sqrt{2}-3\sqrt{5}\right)}$

$= \dfrac{-43}{4\left(\sqrt{2}-3\sqrt{5}\right)}$

77. $C = 1.25\dfrac{\left(12+\sqrt{5}\right)\left(\sqrt{80}\right)}{2} \approx 79.58$

The cost to fertilize the lawn is $79.58.

79. $A = LW = \left(\sqrt{x}+5\right)\left(\sqrt{x}+3\right)$

$A = x+8\sqrt{x}+15$

The area is $\left(x+8\sqrt{x}+15\right)$ mm^2.

Cumulative Review

81. $-2x+3y = 21$ (1)

$3x+2y = 1$ (2)

Multiply (1) by 3, (2) by 2 to eliminate x.

$-6x+9y = 63$

$\underline{6x+4y = 2}$

$13y = 65$

$y = 5$

Substitute 5 for y in (2).

$3x+2(5) = 1$

$3x = -9$

$x = -3$

$(-3, 5)$

82. $2x+3y-z = 8$ (1)

$-x+2y+3z = -14$ (2)

$3x-y-z = 10$ (3)

Solve (3) for z.

$z = 3x-y-10$ (4)

Substitute $3x-y-10$ for z in (1), (2).

$2x+3y-(3x-y-10) = 8$

$-x+2y+3(3x-y-10) = -14$

Simplify above equations.

$-x + 4y = -2$ (5)
$8x - y = 16$ (6)
Multiply top equation by 8 to eliminate x.
$-8x + 32y = -16$
$\underline{8x - y = 16}$
$31y = 0$
$y = 0$
Substitute 0 for y in (5).
$-x + 4(0) = -2$
$x = 2$
Substitute 2 for x, 0 for y in (4).
$z = (3)(2) - 0 - 10 = -4$
$(2, 0, -4)$

83. $24\% + 24\% = 48\%$
48% of U.S. households earned $49,999 or less.

84. 10.1% of $123,000,000 = 0.101 \cdot 123,000,000$
$= 12,423,000$
12,423,000 households earned $150,000 or more.

85. 48% of $123,000,000 = 0.48 \cdot 123,000,000$
$= 59,040,000$
59,040,000 households earned $49,999 or less.

Quick Quiz 8.4

1. $\left(2\sqrt{3} - \sqrt{5}\right)\left(3\sqrt{3} + 2\sqrt{5}\right)$
$= 6(3) + 4\sqrt{15} - 3\sqrt{15} - 2(5)$
$= 18 + \sqrt{15} - 10$
$= 8 + \sqrt{15}$

2. $\dfrac{9}{\sqrt{3x}} = \dfrac{9}{\sqrt{3x}} \cdot \dfrac{\sqrt{3x}}{\sqrt{3x}} = \dfrac{9\sqrt{3x}}{3x} = \dfrac{3\sqrt{3x}}{x}$

3. $\dfrac{1 + 2\sqrt{5}}{4 - \sqrt{5}} = \dfrac{1 + 2\sqrt{5}}{4 - \sqrt{5}} \cdot \dfrac{4 + \sqrt{5}}{4 + \sqrt{5}}$
$= \dfrac{4 + \sqrt{5} + 8\sqrt{5} + 2(5)}{16 - 5}$
$= \dfrac{14 + 9\sqrt{5}}{11}$

4. Answers may vary. Possible solution:
To rationalize the denominator, the numerator and the denominator must be multiplied by the conjugate of the denominator. In this case, the conjugate of the denominator is $3\sqrt{2} + 2\sqrt{3}$. The product will not contain a radical in the denominator.

Use Math to Save Money

1. Total deposits:
$200 + 150.50 + 120.25 + 50 + 25 = 545.75$
The total of his deposits was $545.75.

2. Total checks:
$238.50 + 75 + 200 + 28.56 + 36 = 578.06$
The total of his checks was $578.06.

3. $578.06 \geq 545.75$
He spent more than he deposited, but the $300.50 would help to cover his expenses.

4. $\$300.50 + \$545.75 - \$578.06 = \$268.19$
His balance would be $268.19.

5. Eventually he will be in debt.

6. Answers will vary.

7. Answers will vary.

How Am I Doing? Sections 8.1–8.4
(Available online through MyMathLab.)

1. $(-3x^{1/4}y^{1/2})(-2x^{-1/2}y^{1/3}) = 6x^{\frac{1}{4}-\frac{1}{2}}y^{\frac{1}{2}+\frac{1}{3}}$
$= 6x^{-1/4}y^{5/6}$
$= \dfrac{6y^{5/6}}{x^{1/4}}$

2. $(-a^{-2/3}b^{1/2})^6 = (-1)^6(a^{-2/3})^6(b^{1/2})^6$
$= 1 \cdot a^{-4}b^3$
$= \dfrac{b^3}{a^4}$

3. $\dfrac{-24x^{-1}y}{3x^{-4}y^{2/3}} = -8x^{-1+4}y^{1-2/3} = -8x^3y^{1/3}$

4. $\left(\dfrac{27x^2y^{-5}}{x^{-4}y^4}\right)^{2/3} = \left(\dfrac{3^3x^6}{y^9}\right)^{2/3}$

$= \left[\left(\dfrac{3^3x^6}{y^9}\right)^{1/3}\right]^2$

$= \left(\dfrac{3x^2}{y^3}\right)^2$

$= \dfrac{3^2x^4}{y^6}$

$= \dfrac{9x^4}{y^6}$

5. $27^{-4/3} = \dfrac{1}{(3^3)^{4/3}} = \dfrac{1}{3^4} = \dfrac{1}{81}$

6. $\sqrt[5]{-243} = \sqrt[5]{(-3)^5} = -3$

7. $\sqrt{169} + \sqrt[3]{-64} = 13 - 4 = 9$

8. $\sqrt{64a^8y^{16}} = \sqrt{(8a^4y^8)^2} = 8a^4y^8$

9. $\sqrt[3]{27a^{12}b^6c^{15}} = \sqrt[3]{27}\sqrt[3]{a^{12}}\sqrt[3]{b^6}\sqrt[3]{c^{15}} = 3a^4b^2c^5$

10. $\left(\sqrt[6]{4x}\right)^5 = [(4x)^{1/6}]^5 = (4x)^{5/6}$

11. $\sqrt[4]{16x^{20}y^{28}} = \sqrt[4]{(2x^5y^7)^4} = 2x^5y^7$

12. $\sqrt[3]{32x^8y^{15}} = \sqrt[3]{8 \cdot 4x^2x^6y^{15}}$

$= \sqrt[3]{8}\sqrt[3]{x^6}\sqrt[3]{y^{15}}\sqrt[3]{4x^2}$

$= 2x^2y^5\sqrt[3]{4x^2}$

13. $\sqrt{44} - 2\sqrt{99} + 7\sqrt{11} = \sqrt{4 \cdot 11} - 2\sqrt{9 \cdot 11} + 7\sqrt{11}$

$= 2\sqrt{11} - 2 \cdot 3\sqrt{11} + 7\sqrt{11}$

$= 2\sqrt{11} - 6\sqrt{11} + 7\sqrt{11}$

$= 3\sqrt{11}$

14. $3\sqrt{48y^3} - 2\sqrt[3]{16} + 3\sqrt[3]{54} - 5y\sqrt{12y}$

$= 3\sqrt{16y^2 \cdot 3y} - 2\sqrt[3]{8 \cdot 2} + 3\sqrt[3]{27 \cdot 2} - 5y\sqrt{4 \cdot 3y}$

$= 12y\sqrt{3y} - 4\sqrt[3]{2} + 9\sqrt[3]{2} - 10y\sqrt{3y}$

$= 2y\sqrt{3y} + 5\sqrt[3]{2}$

15. $\left(5\sqrt{2} - 3\sqrt{5}\right)\left(\sqrt{8} + 3\sqrt{5}\right)$

$= 5\sqrt{16} + 15\sqrt{10} - 3\sqrt{40} - 9(5)$

$= 20 + 15\sqrt{10} - 6\sqrt{10} - 45$

$= -25 + 9\sqrt{10}$

16. $\dfrac{5}{\sqrt{18x}} = \dfrac{5}{\sqrt{18x}} \cdot \dfrac{\sqrt{18x}}{\sqrt{18x}}$

$= \dfrac{5\sqrt{18x}}{18x}$

$= \dfrac{5\sqrt{9 \cdot 2x}}{18x}$

$= \dfrac{15\sqrt{2x}}{18x}$

$= \dfrac{5\sqrt{2x}}{6x}$

17. $\dfrac{\sqrt{2} + \sqrt{3}}{\sqrt{2} - \sqrt{3}} = \dfrac{\sqrt{2} + \sqrt{3}}{\sqrt{2} - \sqrt{3}} \cdot \dfrac{\sqrt{2} + \sqrt{3}}{\sqrt{2} + \sqrt{3}}$

$= \dfrac{2 + 2\sqrt{6} + 3}{2 - 3}$

$= \dfrac{5 + 2\sqrt{6}}{-1}$

$= -5 - 2\sqrt{6}$

8.5 Exercises

1. Isolate one of the radicals on one side of the equation.

3. $\sqrt{8x+1} = 5$

$\left(\sqrt{8x+1}\right)^2 = 5^2$

$8x + 1 = 25$

$8x = 24$

$x = 3$

Check: $\sqrt{8(3)+1} \stackrel{?}{=} 5$, $5 = 5$

$x = 3$ is the solution.

5. $\sqrt{4x-3} - 3 = 0$

$\sqrt{4x-3} = 3$

$\left(\sqrt{4x-3}\right)^2 = 3^2$

$4x - 3 = 9$

$4x = 12$

$x = 3$

Check: $\sqrt{4(3)-3} - 3 \stackrel{?}{=} 0$, $\sqrt{9} - 3 \stackrel{?}{=} 0$, $0 = 0$

7.
$$y+1 = \sqrt{5y-1}$$
$$(y+1)^2 = \left(\sqrt{5y-1}\right)^2$$
$$y^2 + 2y + 1 = 5y - 1$$
$$y^2 - 3y + 2 = 0$$
$$(y-2)(y-1) = 0$$
$$y = 2 \qquad\qquad y = 1$$
Check: $2+1 \overset{?}{=} \sqrt{5(2)-1}$, $3 = 3$
$$1+1 \overset{?}{=} \sqrt{5(1)-1}, \ 2 = 2$$
$y = 2$, $y = 1$ is the solution.

9.
$$2x = \sqrt{11x+3}$$
$$(2x)^2 = \left(\sqrt{11x+3}\right)^2$$
$$4x^2 = 11x + 3$$
$$4x^2 - 11x - 3 = 0$$
$$(x-3)(4x+1) = 0$$
$$x - 3 = 0 \qquad\qquad 4x + 1 = 0$$
$$x = 3 \qquad\qquad 4x = -1$$
$$\qquad\qquad\qquad x = -\frac{1}{4}$$

Check:
$$2(3) \overset{?}{=} \sqrt{11 \cdot 3 + 3}$$
$$6 = 6$$
$$2\left(-\frac{1}{4}\right) \overset{?}{=} \sqrt{11 \cdot \left(-\frac{1}{4}\right) + 3}$$
$$-\frac{1}{2} \neq \frac{1}{2}$$
$x = 3$ only

11. $2 = 5 + \sqrt{2x+1} \Rightarrow \sqrt{2x+1} = -3$
No solution since $\sqrt{2x+1} \geq 0$.

13.
$$y - \sqrt{y-3} = 5$$
$$y - 5 = \sqrt{y-3}$$
$$(y-5)^2 = \left(\sqrt{y-3}\right)^2$$
$$y^2 - 10y + 25 = y - 3$$
$$y^2 - 11y + 28 = 0$$
$$(y-7)(y-4) = 0$$
$$y = 7, \ y = -4$$
Check: $4 - \sqrt{4-3} \overset{?}{=} 5$, $3 \neq 5$
$$7 - \sqrt{7-3} \overset{?}{=} 5, \ 5 = 5$$
$y = 7$ is the only solution.

15.
$$y = \sqrt{y+3} - 3$$
$$y + 3 = \sqrt{y+3}$$
$$(y+3)^2 = \left(\sqrt{y+3}\right)^2$$
$$y^2 + 6y + 9 = y + 3$$
$$y^2 + 5y + 6 = 0$$
$$(y+2)(y+3) = 0$$
$$y = -2, \ y = -3$$
Check: $-2 \overset{?}{=} \sqrt{-2+3} - 3$, $-2 = -2$
$$-3 \overset{?}{=} \sqrt{-3+3} - 3, \ -3 = -3$$
$y = -2$, $y = -3$ is the solution.

17. $x - 2\sqrt{x-3} = 3$
$$x - 3 = 2\sqrt{x-3}$$
$$(x-3)^2 = \left(2\sqrt{x-3}\right)^2$$
$$x^2 - 6x + 9 = 4x - 12$$
$$x^2 - 10x + 21 = 0$$
$$(x-7)(x-3) = 0$$
$$x = 7, \ x = 3$$
Check: $7 - 2\sqrt{7-3} \overset{?}{=} 3$, $3 = 3$
$$3 - 2\sqrt{3-3} \overset{?}{=} 3, \ 3 = 3$$
$x = 7$, $x = 3$ is the solution.

19. $\sqrt{3x^2 - x} = x$
$$\left(\sqrt{3x^2 - x}\right)^2 = x^2$$
$$3x^2 - x = x^2$$
$$2x^2 - x = 0$$
$$x(2x-1) = 0$$
$$x = 0, \ x = \frac{1}{2}$$
Check: $\sqrt{3(0)^2 - 0} \overset{?}{=} 0$, $0 = 0$
$$\sqrt{3\left(\frac{1}{2}\right)^2 - \frac{1}{2}} \overset{?}{=} \frac{1}{2}, \ \frac{1}{2} = \frac{1}{2}$$
$x = 0$, $x = \frac{1}{2}$ is the solution.

21. $\sqrt[3]{2x+3} = 2$
$$\left(\sqrt[3]{2x+3}\right)^3 = 2^3$$
$$2x + 3 = 8$$
$$2x = 5$$
$$x = \frac{5}{2}$$

Check: $\sqrt[3]{2\left(\dfrac{5}{2}\right)+3} \overset{?}{=} 2$

$$2 = 2$$

$x = \dfrac{5}{2}$ is the solution.

23. $\sqrt[3]{4x-1} = 3$

$\left(\sqrt[3]{4x-1}\right)^3 = 3^3$

$4x-1 = 27$

$4x = 28$

$x = 7$

Check: $\sqrt[3]{4(7)-1} \overset{?}{=} 3$, $3 = 3$

$x = 7$ is the solution.

25. $\sqrt{x+4} = 1+\sqrt{x-3}$

$\left(\sqrt{x+4}\right)^2 = \left(1+\sqrt{x-3}\right)^2$

$x+4 = 1+2\sqrt{x-3}+x-3$

$6 = 2\sqrt{x-3}$

$3 = \sqrt{x-3}$

$3^2 = \left(\sqrt{x-3}\right)^2$

$9 = x-3$

$x = 12$

Check: $\sqrt{12+4} \overset{?}{=} 1+\sqrt{12-3}$, $4 = 4$

$x = 12$ is the solution.

27. $\sqrt{9x+1} = 1+\sqrt{7x}$

$\left(\sqrt{9x+1}\right)^2 = \left(1+\sqrt{7x}\right)^2$

$9x+1 = 1+2\sqrt{7x}+7x$

$2x = 2\sqrt{7x}$

$x = \sqrt{7x}$

$x^2 = \left(\sqrt{7x}\right)^2$

$x^2 = 7x$

$x^2 - 7x = 0$

$x(x-7) = 0$

$x = 0 \ \ \text{or} \ \ x - 7 = 0$

$x = 7$

Check:

$\sqrt{9(0)+1} \overset{?}{=} 1+\sqrt{7(0)}$, $\sqrt{1} \overset{?}{=} 1$, $1 = 1$

$\sqrt{9(7)+1} \overset{?}{=} 1+\sqrt{7(7)}$; $\sqrt{64} \overset{?}{=} 1+\sqrt{49}$, $8 = 8$

$x = 0$, $x = 7$ is the solution.

29. $\sqrt{x+6} = 1+\sqrt{x+2}$

$\left(\sqrt{x+6}\right)^2 = \left(1+\sqrt{x+2}\right)^2$

$x+6 = 1+2\sqrt{x+2}+x+2$

$3 = 2\sqrt{x+2}$

$3^2 = \left(2\sqrt{x+2}\right)^2$

$9 = 4(x+2)$

$4x+8 = 9$

$4x = 1$

$x = \dfrac{1}{4}$

Check: $\sqrt{\dfrac{1}{4}+6} \overset{?}{=} 1+\sqrt{\dfrac{1}{4}+2}$, $\dfrac{5}{2} = \dfrac{5}{2}$

$x = \dfrac{1}{4}$ is the solution.

31. $\sqrt{3x+13} = 1+\sqrt{x+4}$

$\left(\sqrt{3x+13}\right)^2 = \left(1+\sqrt{x+4}\right)^2$

$3x+13 = 1+2\sqrt{x+4}+x+4$

$2x+8 = 2\sqrt{x+4}$

$x+4 = \sqrt{x+4}$

$(x+4)^2 = \left(\sqrt{x+4}\right)^2$

$x^2+8x+16 = x+4$

$x^2+7x+12 = 0$

$(x+3)(x+4) = 0$

$x = -3, \ x = -4$

Check: $\sqrt{3(-3)+13} \overset{?}{=} 1+\sqrt{-3+4}$, $2 = 2$

$\sqrt{3(-4)+13} \overset{?}{=} 1+\sqrt{-4+4}$, $1 = 1$

$x = -3$, $x = -4$ is the solution.

33. $\sqrt{2x+9}-\sqrt{x+1} = 2$

$\sqrt{2x+9} = 2+\sqrt{x+1}$

$\left(\sqrt{2x+9}\right)^2 = \left(2+\sqrt{x+1}\right)^2$

$2x+9 = 4+4\sqrt{x+1}+x+1$

$x+4 = 4\sqrt{x+1}$

$(x+4)^2 = \left(4\sqrt{x+1}\right)^2$

$x^2+8x+16 = 16x+16$

$x^2-8x = 0$

$x(x-8) = 0$

$x = 0, \ x = 8$

Check: $\sqrt{2(0)+9} - \sqrt{0+1} \stackrel{?}{=} 2, \, 2 = 2$

$\qquad\quad \sqrt{2(8)+9} - \sqrt{8+1} \stackrel{?}{=} 2, \, 2 = 2$

$x = 0, \, x = 8$ is the solution.

35. $\sqrt{4x+6} = \sqrt{x+1} - \sqrt{x+5}$

$\left(\sqrt{4x+6}\right)^2 = \left(\sqrt{x+1} - \sqrt{x+5}\right)^2$

$\qquad 4x+6 = x+1 - 2\sqrt{x+1}\sqrt{x+5} + x+5$

$\qquad\quad 2x = 2\sqrt{x^2+6x+5}$

$\qquad\quad\; x = \sqrt{x^2+6x+5}$

$\qquad\quad x^2 = \left(\sqrt{x^2+6x+5}\right)^2$

$\qquad\quad x^2 = x^2 + 6x + 5$

$\qquad\quad\; 0 = 6x + 5$

$\qquad\quad\; x = -\dfrac{5}{6}$

Check: $\sqrt{4\left(-\dfrac{5}{6}\right)+6} \stackrel{?}{=} \sqrt{-\dfrac{5}{6}+1} - \sqrt{-\dfrac{5}{6}+5}$

$\qquad\qquad\quad \dfrac{4\sqrt{6}}{6} \neq \dfrac{-4\sqrt{6}}{6}$

No solution

37. $2\sqrt{x} - \sqrt{x-5} = \sqrt{2x-2}$

$\left(2\sqrt{x} - \sqrt{x-5}\right)^2 = \left(\sqrt{2x-2}\right)^2$

$4x - 4\sqrt{x}\sqrt{x-5} + x - 5 = 2x - 2$

$\qquad\quad 4\sqrt{x}\sqrt{x-5} = 3x - 3$

$\qquad \left(4\sqrt{x}\sqrt{x-5}\right)^2 = (3x-3)^2$

$\qquad\qquad 16x^2 - 80x = 9x^2 - 18x + 9$

$\qquad\quad 7x^2 - 62x - 9 = 0$

$\qquad\quad (7x+1)(x-9) = 0$

$x = -\dfrac{1}{7}, \, x = 9$

$x = -\dfrac{1}{7}$ does not check since it gives the square root of a negative.

Check: $2\sqrt{9} - \sqrt{9-5} \stackrel{?}{=} \sqrt{2(9)-2}, \, 4 = 4$

$x = 9$ is the solution.

39. $\qquad\qquad x = \sqrt{4.28x - 3.15}$

$\qquad\qquad x^2 = \left(\sqrt{4.28x - 3.15}\right)^2$

$\qquad\qquad x^2 = 4.28x - 3.15$

$x^2 - 4.28x + 3.15 = 0$

$x \approx 3.3357, \, x \approx 0.9443$

Check: $3.3357 \overset{?}{=} \sqrt{4.28(3.3357) - 3.15}$
$\qquad\qquad 3.3357 \approx 3.3357$
$\qquad\qquad 0.9443 \overset{?}{=} \sqrt{4.28(0.9443) - 3.15}$
$\qquad\qquad 0.9443 \approx 0.9442$
$x \approx 3.3357$, $x \approx 0.9443$ is the solution.

41. a. $V = 2\sqrt{3S} \Rightarrow V^2 = 4(3S) \Rightarrow S = \dfrac{V^2}{12}$

b. $S = \dfrac{V^2}{12} = \dfrac{(30)^2}{12} = 75$
The skid mark is 75 feet.

43. $\qquad\qquad 0.11y + 1.25 = \sqrt{3.7625 + 0.22x}$
$\qquad\qquad (0.11y + 1.25)^2 = \left(\sqrt{3.7625 + 0.22x}\right)^2$
$0.0121y^2 + 0.275y + 1.5625 = 3.7625 + 0.22x$
$\qquad\qquad 0.22x = 0.0121y^2 + 0.275y - 2.2$
$\qquad\qquad x = 0.055y^2 + 1.25y - 10$

45. $\sqrt{x^2 - 3x + c} = x - 2$
$x = 3$: $\sqrt{3^2 - 3(3) + c} = 3 - 2$
$\qquad\qquad \sqrt{c} = 1$
$\qquad\qquad c = 1$

Cumulative Review

47. $(4^3 x^6)^{2/3} = 4^{3 \cdot 2/3} x^{6 \cdot 2/3} = 4^2 x^4 = 16x^4$

48. $(2^{-3} x^{-6})^{1/3} = 2^{-3 \cdot \frac{1}{3}} x^{-6 \cdot \frac{1}{3}} = 2^{-1} x^{-2} = \dfrac{1}{2x^2}$

49. $\sqrt[3]{-216x^6 y^9} = \sqrt[3]{(-6x^2 y^3)^3} = -6x^2 y^3$

50. $\sqrt[5]{-32x^{15} y^5} = \sqrt[5]{(-2)^5 (x^3)^5 y^5} = -2x^3 y$

51. w = speed of current
$(12 + w) \cdot 4 = (12 - w) \cdot 5$
$\qquad 48 + 4w = 60 - 5w$
$\qquad\qquad 9w = 12$
$\qquad\qquad w = \dfrac{12}{9} \approx 1.33$
The speed of the current is approximately 1.33 mph.

52. $d = \text{dogs}$
$c = \text{cats}$
$$c + d = 28 \quad (1)$$
$$55c + 68d = 1748 \quad (2)$$
Solve (1) for c.
$$c = 28 - d \quad (3)$$
Substitute $28 - d$ for c in (2).
$$55(28 - d) + 68d = 1748$$
$$13d = 208$$
$$d = 16$$
Substitute 16 for d in (3).
$$c = 28 - 16 = 12$$
There were 16 dogs and 12 cats examined.

Quick Quiz 8.5

1. $\sqrt{5x - 4} = x$
$$\left(\sqrt{5x - 4}\right)^2 = x^2$$
$$5x - 4 = x^2$$
$$x^2 - 5x + 4 = 0$$
$$(x - 4)(x - 1) = 0$$

$x - 4 = 0 \qquad\qquad x - 1 = 0$
$\quad x = 4 \qquad\qquad\qquad x = 1$

Check:
$$\sqrt{5(4) - 4} \stackrel{?}{=} 4$$
$$4 = 4$$
$$\sqrt{5(1) - 4} \stackrel{?}{=} 1$$
$$1 = 1$$
$x = 4$, $x = 1$ is the solution.

2. $x = 3 - \sqrt{2x - 3}$
$$x - 3 = -\sqrt{2x - 3}$$
$$3 - x = \sqrt{2x - 3}$$
$$(3 - x)^2 = \left(\sqrt{2x - 3}\right)^2$$
$$9 - 6x + x^2 = 2x - 3$$
$$x^2 - 8x + 12 = 0$$
$$(x - 2)(x - 6) = 0$$

$x - 2 = 0 \qquad\qquad x - 6 = 0$
$\quad x = 2 \qquad\qquad\qquad x = 6$

Check:
$$2 \stackrel{?}{=} 3 - \sqrt{2(2) - 3}$$
$$2 \stackrel{?}{=} 3 - 1$$
$$2 = 2$$
$$6 \stackrel{?}{=} 3 - \sqrt{2(6) - 3}$$
$$6 \stackrel{?}{=} 3 - 3$$
$$6 \neq 0$$
$x = 2$ is the only solution.

3. $4 - \sqrt{x - 4} = \sqrt{2x - 1}$
$$\left(4 - \sqrt{x - 4}\right)^2 = \left(\sqrt{2x - 1}\right)^2$$
$$16 - 8\sqrt{x - 4} + x - 4 = 2x - 1$$
$$8\sqrt{x - 4} = 13 - x$$
$$\left(8\sqrt{x - 4}\right)^2 = (13 - x)^2$$
$$64(x - 4) = 169 - 26x + x^2$$
$$64x - 256 = 169 - 26x + x^2$$
$$x^2 - 90x + 425 = 0$$
$$(x - 85)(x - 5) = 0$$

$x - 85 = 0 \qquad\qquad x - 5 = 0$
$\quad x = 85 \qquad\qquad\qquad x = 5$

Check:
$$4 - \sqrt{85 - 4} \stackrel{?}{=} \sqrt{2(85) - 1}$$
$$4 - 9 \stackrel{?}{=} \sqrt{169}$$
$$-5 \neq 13$$
$$4 - \sqrt{5 - 4} \stackrel{?}{=} \sqrt{2(5) - 1}$$
$$4 - 1 \stackrel{?}{=} \sqrt{9}$$
$$3 = 3$$
$x = 5$ is the only solution.

4. Answers may vary. Possible solution: Substitute the found values for x back into the original equation to test for validity.

8.6 Exercises

1. No; there is no real number that, when squared, will equal -9.

3. No; to be equal, the real parts must be equal, and the imaginary parts must be equal. $2 \neq 3$ and $3i \neq 2i$.

5. $\sqrt{-25} = \sqrt{25}\sqrt{-1} = 5i$

7. $\sqrt{-28} = \sqrt{4 \cdot 7} \cdot \sqrt{-1} = 2i\sqrt{7}$

9. $\sqrt{-\dfrac{25}{4}} = \sqrt{\dfrac{25}{4}}\sqrt{-1} = \dfrac{5}{2}i$

11. $-\sqrt{-81} = -\sqrt{81}\sqrt{-1} = -9i$

13. $2 + \sqrt{-3} = 2 + \sqrt{3}\sqrt{-1} = 2 + i\sqrt{3}$

15. $-2.8 + \sqrt{-16} = -2.8 + \sqrt{16}\sqrt{-1} = -2.8 + 4i$

17. $-3 + \sqrt{-24} = -3 + \sqrt{4 \cdot 6}\sqrt{-1} = -3 + 2i\sqrt{6}$

19. $\left(\sqrt{-5}\right)\left(\sqrt{-2}\right) = \sqrt{5}\sqrt{-1}\sqrt{2}\sqrt{-1}$
$$= i\sqrt{5} \cdot i\sqrt{2}$$
$$= i^2\sqrt{10}$$
$$= -\sqrt{10}$$

21. $\left(\sqrt{-36}\right)\left(\sqrt{-4}\right) = \left(\sqrt{36}\sqrt{-1}\right)\left(\sqrt{4}\sqrt{-1}\right)$
$$= (6i)(2i)$$
$$= 12i^2$$
$$= -12$$

23. $x - 3i = 5 + yi$
$x = 5,\ y = -3$

25. $1.3 - 2.5yi = x - 5i,\ x = 1.3$
$$-2.5y = -5$$
$$y = 2$$

27. $23 + yi = 17 - x + 3i$
$$23 = 17 - x$$
$$x = -6$$
$$y = 3$$

29. $(1 + 8i) + (-6 + 3i) = 1 - 6 + (8 + 3)i = -5 + 11i$

31. $\left(-\dfrac{3}{2} + \dfrac{1}{2}i\right) + \left(\dfrac{5}{2} - \dfrac{3}{2}i\right) = -\dfrac{3}{2} + \dfrac{5}{2} + \left(\dfrac{1}{2} - \dfrac{3}{2}\right)i$
$$= 1 - i$$

33. $(2.8 - 0.7i) - (1.6 - 2.8i)$
$$= 2.8 - 1.6 + (-0.7 + 2.8)i$$
$$= 1.2 + 2.1i$$

35. $(2i)(7i) = 14i^2 = -14$

37. $(-7i)(6i) = -42i^2 = (-42)(-1) = 42$

39. $(2 + 3i)(2 - i) = 4 - 2i + 6i - 3i^2$
$$= 4 + 4i - 3(-1)$$
$$= 4 + 4i + 3$$
$$= 7 + 4i$$

41. $9i - 3(-2 + i) = 9i + 6 - 3i = 6 + 6i$

43. $2i(5i - 6) = 10i^2 - 12i = -10 - 12i$

45. $\left(\dfrac{1}{2} + i\right)^2 = \dfrac{1}{4} + \dfrac{1}{2}i + \dfrac{1}{2}i + i^2 = \dfrac{1}{4} + i - 1 = -\dfrac{3}{4} + i$

47. $\left(i\sqrt{3}\right)\left(i\sqrt{7}\right) = i^2\sqrt{21} = -\sqrt{21}$

49. $\left(3 + \sqrt{-2}\right)\left(4 + \sqrt{-5}\right) = \left(3 + i\sqrt{2}\right)\left(4 + i\sqrt{5}\right)$
$$= 12 + 3i\sqrt{5} + 4i\sqrt{2} + i^2\sqrt{10}$$
$$= 12 + 3i\sqrt{5} + 4i\sqrt{2} - \sqrt{10}$$
$$= 12 - \sqrt{10} + \left(3\sqrt{5} + 4\sqrt{2}\right)i$$

51. $i^{17} = (i^4)^4 \cdot i = 1^4 \cdot i = i$

53. $i^{24} = (i^4)^6 = 1^6 = 1$

55. $i^{46} = (i^4)^{11} \cdot i^2 = 1^{11}(-1) = -1$

57. $i^{47} = i^{44}i^3 = (i^4)^{11}i^3 = 1^{11} \cdot (-i) = -i$

59. $i^{30} + i^{28} = (i^4)^7 \cdot i^2 + (i^4)^7$
$$= 1^7(-1) + 1^7$$
$$= -1 + 1$$
$$= 0$$

61. $i^{100} - i^7 = (i^4)^{25} - i^4 i^3 = 1^{25} - 1 \cdot (-i) = 1 + i$

63. $\dfrac{2+i}{3-i} = \dfrac{2+i}{3-i} \cdot \dfrac{3+i}{3+i}$
$$= \dfrac{6 + 5i + i^2}{9 + 1}$$
$$= \dfrac{6 + 5i - 1}{10}$$
$$= \dfrac{5(1+i)}{10}$$
$$= \dfrac{1+i}{2} \text{ or } \dfrac{1}{2} + \dfrac{i}{2}$$

65. $\dfrac{i}{1+4i} = \dfrac{i}{1+4i} \cdot \dfrac{1-4i}{1-4i}$
$$= \dfrac{i - 4i^2}{1 - 16i^2}$$
$$= \dfrac{i + 4}{1 + 16}$$
$$= \dfrac{4+i}{17} \text{ or } \dfrac{4}{17} + \dfrac{i}{17}$$

67. $\dfrac{5-2i}{6i} = \dfrac{5-2i}{6i} \cdot \dfrac{-6i}{-6i}$

$= \dfrac{-30i+12i^2}{-36i^2}$

$= \dfrac{-12-30i}{36}$

$= \dfrac{-2-5i}{6}$ or $-\dfrac{1}{3} - \dfrac{5i}{6}$

69. $\dfrac{2}{i} = \dfrac{2}{i} \cdot \dfrac{i}{i} = \dfrac{2i}{i^2} = \dfrac{2i}{-1} = -2i$

71. $\dfrac{7}{5-6i} = \dfrac{7}{5-6i} \cdot \dfrac{5+6i}{5+6i}$

$= \dfrac{35+42i}{25+36}$

$= \dfrac{35+42i}{61}$ or $\dfrac{35}{61} + \dfrac{42i}{61}$

73. $\dfrac{5-2i}{3+2i} = \dfrac{5-2i}{3+2i} \cdot \dfrac{3-2i}{3-2i}$

$= \dfrac{15-10i-6i+4i^2}{9+4}$

$= \dfrac{15-16i-4}{13}$

$= \dfrac{11-16i}{13}$ or $\dfrac{11}{13} - \dfrac{16i}{13}$

75. $\sqrt{-98} = \sqrt{98}\sqrt{-1} = \sqrt{49}\sqrt{2}\sqrt{-1} = 7i\sqrt{2}$

77. $(8-5i)-(-1+3i) = 8-5i+1-3i = 9-8i$

79. $(3i-1)(5i-3) = 15i^2 - 9i - 5i + 3$

$= 15i^2 - 14i + 3$

$= -15 - 14i + 3$

$= -12 - 14i$

81. $\dfrac{2-3i}{2+i} = \dfrac{2-3i}{2+i} \cdot \dfrac{2-i}{2-i}$

$= \dfrac{4-2i-6i+3i^2}{4+1}$

$= \dfrac{4-8i-3}{5}$

$= \dfrac{1-8i}{5}$ or $\dfrac{1}{5} - \dfrac{8i}{5}$

83. $Z = \dfrac{V}{I}$

$= \dfrac{3+2i}{3i}$

$= \dfrac{3+2i}{3i} \cdot \dfrac{-3i}{-3i}$

$= \dfrac{-9i-6i^2}{9}$

$= \dfrac{6-9i}{9}$

$= \dfrac{2-3i}{3}$ or $\dfrac{2}{3} - i$

Cumulative Review

85. $x+3+2x-5+4x+2 = 105$

$7x = 105$

$x = 15$

$x + 3 = 18$

$2x - 5 = 25$

$4x + 2 = 62$

18 hours producing juice in glass bottles, 25 hours producing juice in cans, and 62 hours producing juice in plastic bottles.

86. $C = 60[1850(0.93) - 120] = 96{,}030$
The net cost to the bank is \$96,030.

Quick Quiz 8.6

1. $(6-7i)(3+2i) = 18 + 12i - 21i - 14i^2$

$= 18 - 9i + 14$

$= 32 - 9i$

2. $\dfrac{4+3i}{1-2i} = \dfrac{4+3i}{1-2i} \cdot \dfrac{1+2i}{1+2i}$

$= \dfrac{4+8i+3i+6i^2}{1-4i^2}$

$= \dfrac{4+11i-6}{1+4}$

$= \dfrac{-2+11i}{5}$ or $-\dfrac{2}{5} + \dfrac{11i}{5}$

3. $i^{33} = i \cdot i^{32} = i \cdot (i^4)^8 = i(1)^8 = i \cdot (1) = i$

4. Answers may vary. Possible solution:

Multiply by FOIL method. Replace i^2 with -1. Combine like terms.

8.7 Exercises

1. Answers may vary. A person's weekly paycheck varies as the number of hours worked, $y = kx$ where y is the weekly salary, k is the hourly salary, and x is the number of hours worked.

3. If y varies inversely with x, we write the equation $y = \dfrac{k}{x}$.

5. $y = kx$, $20 = k \cdot 25$, $k = \dfrac{4}{5}$

   $y = \dfrac{4}{5}x$, $y = \dfrac{4}{5}(65) = 52$

7. $p = kd$, $21 = k \cdot 50$, $k = \dfrac{21}{50}$

   $p = \dfrac{21}{50}d$, $p = \dfrac{21}{50} \cdot 170 = 71.4$

   The pressure would be 71.4 pounds per square inch.

9. $d = ks^2$, $40 = k \cdot 30^2$, $k = \dfrac{2}{45}$

   $d = \dfrac{2}{45}s^2$, $d = \dfrac{2}{45} \cdot 60^2 = 160$

   It will take 160 feet to stop.

11. $y = \dfrac{k}{x^2}$, $5 = \dfrac{k}{3^2}$, $k = 45$

    $y = \dfrac{45}{x^2}$, $y = \dfrac{45}{0.3^2} = \dfrac{45}{0.09} = 500$

13. g = gallons sold
    p = price in dollars

    $g = \dfrac{k}{p}$

    $k = gp = 2800(3.10) = 8680$

    $g = \dfrac{8680}{p}$

    When $p = \$2.90$, $g = \dfrac{8680}{2.90} \approx 2993$.

    He could expect to sell approximately 2943 gallons.

15. v = number of volunteers
    t = time in hours

    $t = \dfrac{k}{v}$

    When $v = 39$, $t = 6$.
    $k = tv = 39(6) = 234$

    $t = \dfrac{234}{v}$, $t = \dfrac{234}{60} = 3.9$

    Cleanup will take 3.9 hours.

17. $w = \dfrac{k}{l}$, $900 = \dfrac{k}{8}$, $k = 7200$

    $w = \dfrac{7200}{l}$, $w = \dfrac{7200}{18} = 400$

    The beam can safely support 400 pounds.

19. t = time in minutes
    r = radius of pipe in inches

    $t = \dfrac{k}{r^2}$

    When $r = 2.5$, $t = 6$.
    $k = tr^2 = 6(2.5)^2 = 37.5$

    $t = \dfrac{37.5}{r^2}$, $t = \dfrac{37.5}{(3.5)^2} \approx 3.1$

    The tub would fill in approximately 3.1 minutes.

Cumulative Review

21. $3x^2 - 8x + 4 = 0$
    $(3x - 2)(x - 2) = 0$
    $3x - 2 = 0$ or $x - 2 = 0$
           $x = \dfrac{2}{3}$ $x = 2$

22. $4x^2 = -28x + 32$
    $4x^2 + 28x - 32 = 0$
         $x^2 + 7x - 8 = 0$
         $(x + 8)(x - 1) = 0$
    $x + 8 = 0$ or $x - 1 = 0$
       $x = -8$ $x = 1$

23. $503.47 = 1.0945p$
       $460 = p$
    The original price was \$460.

24. $\dfrac{7.5}{3} = \dfrac{x}{22}$

$3x = 165$

$x = 55$

It will take 55 gallons of paint.

Quick Quiz 8.7

1. $y = \dfrac{k}{x}$

When $y = 9$, $x = 3$.

$k = yx = 9(3) = 27$

$y = \dfrac{27}{x}$, $y = \dfrac{27}{6} = 4.5$

2. $y = \dfrac{kx}{z^2}$

When $x = 3$, $z = 5$, $y = 6$.

$k = \dfrac{yz^2}{x} = \dfrac{6(5)^2}{3} = 50$

$y = \dfrac{50x^2}{z^2}$, $y = \dfrac{50(6)}{(10)^2} = 3$

3. d = distance to stop in feet

v = speed traveling in miles per hour

$d = kv^2$

When $v = 50$, $d = 80$.

$k = \dfrac{d}{v^2} = \dfrac{80}{(50)^2} = 0.032$

$d = 0.032v^2$, $d = 0.032(65)^2 = 135.2$

Distance to stop will be 135.2 feet.

4. Answers may vary. Possible solution:
Writing the equation is always the first step. In this case:

$y = k\sqrt{x}$

Next use the known values of x and y to find the numerical value of k.

Career Exploration Problems

1. Use the Pythagorean theorem, $a^2 + b^2 = c^2$.
For the first section, $a = 170$ and $b = 38$.

$170^2 + 38^2 = c^2$

$28,900 + 1444 = c^2$

$30,344 = c^2$

$174.2 \approx c$

For the second section, $c = 84$ and $b = 27$.

$a^2 + 27^2 = 84^2$

$a^2 + 729 = 7056$

$a^2 = 6327$

$a \approx 79.5$

$174.2 + 79.5 = 253.7$

The length of the shoreline is approximately 254 feet.

2. $V = \dfrac{1}{2}\left(20\sqrt{3}\right)\left(6\sqrt{50} + 7\sqrt{8}\right)$

$= 10\sqrt{3}\left(6\sqrt{25 \cdot 2} + 7\sqrt{4 \cdot 2}\right)$

$= 10\sqrt{3}\left(30\sqrt{2} + 14\sqrt{2}\right)$

$= 10\sqrt{3}\left(44\sqrt{2}\right)$

$= 440\sqrt{6}$

≈ 1077.78

Her computation of the volume is approximately 1078 cubic units.

3. $L = \dfrac{k}{d^2}$

When $d = 25$, $L = 3900$.

$k = Ld^2 = 3900(25)^2 = 2,437,500$

$L = \dfrac{2,437,500}{d^2}$

20-foot pole:

$L = \dfrac{2,437,500}{20^2} = \dfrac{2,437,500}{400} = 6093.75$

30-foot pole:

$L = \dfrac{2,437,500}{30^2} = \dfrac{2,437,500}{900} \approx 2708.3$

The light would be 6093.75 lumens if the light source were mounted on a 20-foot pole, and 2708.3 lumens on a 30-foot pole.

4. Use the Pythagorean theorem with $a = 13$ and $c = 40.5$.

$13^2 + b^2 = 40.5^2$

$169 + b^2 = 1640.25$

$b^2 = 1471.25$

$b \approx 38.4$

The horizontal distance of the roadway is approximately 38.4 feet.

You Try It

1. **a.** $(x^{-2/5})^{-1/2} = x^{-2/5(-1/2)} = x^{1/5}$

    b. $(2a^{-3}b^{-1/4})^{1/3} = 2^{1/3}(a^{-3})^{1/3}(b^{-1/4})^{1/3}$
    $$= 2^{1/3}a^{-1}b^{-1/12}$$

    c. $\left(\dfrac{5x^{-3}}{4^{-2}y^2}\right)^{1/4} = \dfrac{5^{1/4}(x^{-3})^{1/4}}{(4^{-2})^{1/4}(y^2)^{1/4}} = \dfrac{5^{1/4}\,x^{-3/4}}{4^{-1/2}\,y^{1/2}}$

2. $(-a^{1/2})(4a^{1/3}) = -1 \cdot 4a^{1/2+1/3}$
    $$= -4a^{3/6+2/6}$$
    $$= -4a^{5/6}$$

3. $\dfrac{12x^{7/12}}{-2x^{1/3}} = -6x^{7/12-1/3}$
    $$= -6x^{7/12-4/12}$$
    $$= -6x^{3/12}$$
    $$= -6x^{1/4}$$

4. **a.** $5a^{-3} = 5 \cdot \dfrac{1}{a^3} = \dfrac{5}{a^3}$

    b. $\dfrac{3a^{-2}}{6a^{-6}} = \dfrac{3}{6}a^{-2-(-6)} = \dfrac{1}{2}a^{-2+6} = \dfrac{a^4}{2}$

    c. $2^{-5} = \dfrac{1}{2^5} = \dfrac{1}{32}$

5. $x^{1/2}(x^{2/3} - x^{1/2}) = x^{1/2+1/3} - x^{1/2+1/2}$
    $$= x^{3/6+4/6} - x^1$$
    $$= x^{7/6} - x$$

6. $(-5x^{2/3})^0 = 1$

7. **a.** $\sqrt[4]{16} = 2$ because $2^4 = 16$.

    b. $\sqrt[5]{-1} = -1$ because $(-1)^5 = -1$.

    c. $\sqrt[6]{-64}$ is not a real number.

8. **a.** $x^{4/5} = \sqrt[5]{x^4}$ or $\sqrt[5]{x}^{\,4}$

    b. $\sqrt[4]{v^9} = v^{9/4}$

    c. $27^{4/3} = \left(\sqrt[3]{27}\right)^4 = 3^4 = 81$

9. **a.** $\sqrt[4]{x^4} = |x|$

    b. $\sqrt[7]{y^7} = y$

10. $\sqrt[3]{-64m^{18}} = \sqrt[3]{(-4m^6)^3} = -4m^6$

11. **a.** $\sqrt{24x^5} = \sqrt{4x^4 \cdot 6x} = \sqrt{4x^4}\,\sqrt{6x} = 2x^2\sqrt{6x}$

    b. $\sqrt[3]{54r^4s^9} = \sqrt[3]{27r^3s^9 \cdot 2r}$
    $$= \sqrt[3]{27r^3s^9}\,\sqrt[3]{2r}$$
    $$= 3rs^3\sqrt[3]{2r}$$

12. $3\sqrt{72} + 4\sqrt{18} = 3\sqrt{36}\sqrt{2} + 4\sqrt{9}\sqrt{2}$
    $$= 3 \cdot 6\sqrt{2} + 4 \cdot 3\sqrt{2}$$
    $$= 18\sqrt{2} + 12\sqrt{2}$$
    $$= 30\sqrt{2}$$

13. **a.** $\left(\sqrt{6}\right)\left(3\sqrt{5}\right) = 3\sqrt{6 \cdot 5} = 3\sqrt{30}$

    b. $3\sqrt{3}\left(2\sqrt{6} - \sqrt{15}\right) = 6\sqrt{18} - 3\sqrt{45}$
    $$= 6\sqrt{9}\sqrt{2} - 3\sqrt{9}\sqrt{5}$$
    $$= 18\sqrt{2} - 9\sqrt{5}$$

    c. $\left(\sqrt{3} - \sqrt{5}\right)\left(2\sqrt{3} + \sqrt{5}\right)$
    $$= 2\sqrt{9} + \sqrt{15} - 2\sqrt{15} - \sqrt{25}$$
    $$= 6 - \sqrt{15} - 5$$
    $$= 1 - \sqrt{15}$$

14. $\sqrt[4]{\dfrac{3}{16}} = \dfrac{\sqrt[4]{3}}{\sqrt[4]{16}} = \dfrac{\sqrt[4]{3}}{2}$

15. **a.** $\dfrac{3}{\sqrt{6}} = \dfrac{3}{\sqrt{6}} \cdot \dfrac{\sqrt{6}}{\sqrt{6}} = \dfrac{3\sqrt{6}}{6} = \dfrac{\sqrt{6}}{2}$

b. $\dfrac{4}{\sqrt{2}-\sqrt{3}} = \dfrac{4}{\sqrt{2}-\sqrt{3}} \cdot \dfrac{\sqrt{2}+\sqrt{3}}{\sqrt{2}+\sqrt{3}}$

$\qquad = \dfrac{4\sqrt{2}+4\sqrt{3}}{\left(\sqrt{2}\right)^2 - \left(\sqrt{3}\right)^2}$

$\qquad = \dfrac{4\sqrt{2}+4\sqrt{3}}{-1}$

$\qquad = -4\sqrt{2}-4\sqrt{3}$

16. $5+\sqrt{3x-11} = x$

$\qquad \sqrt{3x-11} = x-5$

$\qquad \left(\sqrt{3x-11}\right)^2 = (x-5)^2$

$\qquad\qquad 3x-11 = x^2 -10x+25$

$\qquad\qquad\quad 0 = x^2 -13x+36$

$\qquad\qquad\quad 0 = (x-4)(x-9)$

$x = 4, x = 9$

Check: $5+\sqrt{3(4)-11} \overset{?}{=} 4,\ 6 \ne 4$

$\qquad\quad 5+\sqrt{3(9)-11} \overset{?}{=} 9,\ 9 = 9$

$x = 9$ is the only solution.

17. a. $\sqrt{-100} = \sqrt{-1}\sqrt{100} = 10i$

    b. $\sqrt{-24} = \sqrt{-1}\sqrt{24} = i\sqrt{4}\sqrt{6} = 2i\sqrt{6}$

18. a. $(8+i)+(3-7i) = 8+3+(1-7)i = 11-6i$

    b. $(4+3i)-(5+i) = 4-5+(3-1)i = -1+2i$

19. $(1-3i)(3+2i) = 3+2i-9i-6i^2$

$\qquad\qquad\qquad\quad = 3-7i-6(-1)$

$\qquad\qquad\qquad\quad = 3-7i+6$

$\qquad\qquad\qquad\quad = 9-7i$

20. $i^{36} = (i^4)^9 = 1^9 = 1$

21. $\dfrac{4-5i}{2+i} = \dfrac{4-5i}{2+i} \cdot \dfrac{2-i}{2-i}$

$\qquad = \dfrac{8-4i-10i+5i^2}{2^2 -i^2}$

$\qquad = \dfrac{8-14i+5(-1)}{4-(-1)}$

$\qquad = \dfrac{8-14i-5}{4+1}$

$\qquad = \dfrac{3-14i}{5}$ or $\dfrac{3}{5}-\dfrac{14}{5}i$

22. $y = kx^2$

$\qquad 16 = k(2)^2$

$\qquad 16 = k \cdot 4$

$\qquad\ \ 4 = k$

$\qquad\ \ y = 4x^2$

$\qquad\ \ y = 4(-3)^2 = 4 \cdot 9 = 36$

23. $w = \dfrac{k}{z}$

$\qquad 5 = \dfrac{k}{0.5}$

$\qquad 2.5 = k$

$\qquad w = \dfrac{2.5}{z}$

$\qquad w = \dfrac{2.5}{5} = 0.5$

Chapter 8 Review Problems

1. $(3xy^{1/2})(5x^2 y^{-3}) = (3 \cdot 5)x^{1+3}y^{\frac{1}{2}-3}$

$\qquad\qquad\qquad\qquad = 15x^3 y^{-5/2}$

$\qquad\qquad\qquad\qquad = \dfrac{15x^3}{y^{5/2}}$

2. $(16a^6 b^5)^{1/2} = (4a^3 b^{5/2})^{2(1/2)} = 4a^3 b^{5/2}$

3. $3^{1/2} \cdot 3^{1/6} = 3^{\frac{3}{6}+\frac{1}{6}} = 3^{4/6} = 3^{2/3}$

4. $\dfrac{6x^{2/3}y^{1/10}}{12x^{1/6}y^{-1/5}} = \dfrac{x^{\frac{2}{3}-\frac{1}{6}}y^{\frac{1}{10}+\frac{1}{5}}}{2} = \dfrac{x^{1/2}y^{3/10}}{2}$

5. $(2x^{-1/5}y^{1/10}z^{4/5})^{-5} = 2^{-5}x^{-\frac{1}{5}(-5)}y^{\frac{1}{10}(-5)}z^{\frac{4}{5}(-5)}$

$\qquad\qquad\qquad\qquad\quad = \dfrac{xy^{-1/2}z^{-4}}{2^5}$

$\qquad\qquad\qquad\qquad\quad = \dfrac{x}{32y^{1/2}z^4}$

6. $\left(\dfrac{49a^3 b^6}{a^{-7}b^4}\right)^{1/2} = (49a^{10}b^2)^{1/2}$

$\qquad\qquad\qquad\qquad = 49^{1/2}a^{10(1/2)}b^{2(1/2)}$

$\qquad\qquad\qquad\qquad = 7a^5 b$

7. $\left(\dfrac{8a^4}{a^{-2}}\right)^{1/3} = (8a^{(4+2)})^{1/3}$

   $= (8a^6)^{1/3}$

   $= 8^{1/3} a^{6 \cdot \frac{1}{3}}$

   $= 2a^2$

8. $(4^{5/3})^{6/5} = 4^{\frac{5 \cdot 6}{3 \cdot 5}} = 4^{30/15} = 4^2 = 16$

9. $2x^{1/3} + x^{-2/3} = 2x^{1/3} + \dfrac{1}{x^{2/3}}$

   $= \dfrac{2x^{1/3}}{1} \cdot \dfrac{x^{2/3}}{x^{2/3}} + \dfrac{1}{x^{2/3}}$

   $= \dfrac{2x^{\frac{1}{3}+\frac{2}{3}} + 1}{x^{2/3}}$

   $= \dfrac{2x + 1}{x^{2/3}}$

10. $6x^{3/2} - 9x^{1/2} = 3x \cdot 2x^{1/2} - 3x \cdot 3x^{-1/2}$

    $= 3x(2x^{1/2} - 3x^{-1/2})$

11. $-\sqrt{16} = -4$

12. $\sqrt[5]{-32} = \sqrt[5]{(-2)^5} = -2$

13. $-\sqrt{\dfrac{1}{25}} = -\dfrac{1}{5}$

14. $\sqrt{0.04} = 0.2$ because $0.2^2 = 0.04$.

15. $\sqrt[4]{-256}$ is not a real number.

16. $\sqrt[3]{-\dfrac{1}{8}} = -\dfrac{1}{2}$ because $\left(-\dfrac{1}{2}\right)^3 = -\dfrac{1}{8}$.

17. $64^{2/3} = (4^3)^{2/3} = 4^{3 \cdot \frac{2}{3}} = 4^2 = 16$

18. $125^{4/3} = (5^3)^{4/3} = 5^{3 \cdot \frac{4}{3}} = 5^4 = 625$

19. $\sqrt{49x^4 y^{10} z^2} = \sqrt{7^2 (x^2)^2 (y^5)^2 (z)^2} = 7x^2 y^5 z$

20. $\sqrt[3]{64a^{12}b^{30}} = \sqrt[3]{4^3 (a^4)^3 (b^{10})^3} = 4a^4 b^{10}$

21. $\sqrt[3]{-8a^{12}b^{15}c^{21}} = \sqrt[3]{(-2)^3 (a^4)^3 (b^5)^3 (c^7)^3}$
    $= -2a^4 b^5 c^7$

22. $\sqrt{49x^{22}y^2} = \sqrt{7^2 (x^{11})^2 y^2} = 7x^{11} y$

23. $\sqrt[5]{a^2} = a^{2/5}$

24. $\sqrt{2b} = (2b)^{1/2}$

25. $\sqrt[3]{5a} = (5a)^{1/3}$

26. $\left(\sqrt[5]{xy}\right)^7 = ((xy)^{1/5})^7 = (xy)^{7/5}$

27. $m^{1/2} = \sqrt[2]{m^1} = \sqrt{m}$

28. $y^{3/5} = \sqrt[5]{y^3}$ or $\sqrt[5]{y}^3$

29. $(3z)^{2/3} = \sqrt[3]{(3z)^2} = \sqrt[3]{9z^2}$

30. $(2x)^{3/7} = \sqrt[7]{(2x)^3} = \sqrt[7]{8x^3}$

31. $16^{3/4} = (2^4)^{3/4} = 2^{4 \cdot \frac{3}{4}} = 2^3 = 8$

32. $(-27)^{2/3} = ((-3)^3)^{2/3} = (-3)^{3 \cdot \frac{2}{3}} = (-3)^2 = 9$

33. $\left(\dfrac{1}{9}\right)^{1/2} = (3^{-2})^{1/2} = 3^{-2 \cdot \frac{1}{2}} = 3^{-1} = \dfrac{1}{3^1} = \dfrac{1}{3}$

34. $(0.49)^{1/2} = ((0.7)^2)^{1/2} = 0.7^{2 \cdot \frac{1}{2}} = 0.7^1 = 0.7$

35. $\left(\dfrac{1}{36}\right)^{-1/2} = (6^{-2})^{-1/2} = 6^{-2 \cdot \frac{-1}{2}} = 6^1 = 6$

36. $(25a^2 b^4)^{3/2} = (5^2)^{3/2} (a^2)^{3/2} (b^4)^{3/2}$
    $= 5^{2 \cdot \frac{3}{2}} a^{2 \cdot \frac{3}{2}} b^{4 \cdot \frac{3}{2}}$
    $= 5^3 a^3 b^6$
    $= 125a^3 b^6$

37. $\sqrt{50} + 2\sqrt{32} - \sqrt{8} = \sqrt{25 \cdot 2} + 2\sqrt{16 \cdot 2} - \sqrt{4 \cdot 2}$
    $= 5\sqrt{2} + 8\sqrt{2} - 2\sqrt{2}$
    $= 11\sqrt{2}$

38. $\sqrt{28}-4\sqrt{7}+5\sqrt{63}=\sqrt{4\cdot7}-4\sqrt{7}+5\sqrt{9\cdot7}$
$\qquad\qquad\qquad\qquad = 2\sqrt{7}-4\sqrt{7}+15\sqrt{7}$
$\qquad\qquad\qquad\qquad = 13\sqrt{7}$

39. $2\sqrt{12}-\sqrt{48}+5\sqrt{75}=2\sqrt{4\cdot3}-\sqrt{16\cdot3}+5\sqrt{25\cdot3}$
$\qquad\qquad\qquad\qquad = 2\cdot2\sqrt{3}-4\sqrt{3}+5\cdot5\sqrt{3}$
$\qquad\qquad\qquad\qquad = 4\sqrt{3}-4\sqrt{3}+25\sqrt{3}$
$\qquad\qquad\qquad\qquad = 25\sqrt{3}$

40. $\sqrt{125x^3}+x\sqrt{45x}=\sqrt{25x^2\cdot5x}+x\sqrt{9\cdot5x}$
$\qquad\qquad\qquad\qquad = 5x\sqrt{5x}+3x\sqrt{5x}$
$\qquad\qquad\qquad\qquad = 8x\sqrt{5x}$

41. $2\sqrt{32x}-5x\sqrt{2}+\sqrt{18x}$
$\qquad = 2\sqrt{16\cdot2x}-5x\sqrt{2}+\sqrt{9\cdot2x}$
$\qquad = 8\sqrt{2x}-5x\sqrt{2}+3\sqrt{2x}$
$\qquad = 11\sqrt{2x}-5x\sqrt{2}$

42. $3\sqrt[3]{16}-4\sqrt[3]{54}=3\sqrt[3]{2^3\cdot2}-4\sqrt[3]{3^3\cdot2}$
$\qquad\qquad\qquad\quad = 3\cdot2\sqrt[3]{2}-4\cdot3\sqrt[3]{2}$
$\qquad\qquad\qquad\quad = 6\sqrt[3]{2}-12\sqrt[3]{2}$
$\qquad\qquad\qquad\quad = -6\sqrt[3]{2}$

43. $\left(5\sqrt{12}\right)\left(3\sqrt{6}\right)=15\sqrt{72}$
$\qquad\qquad\qquad\qquad = 15\sqrt{36\cdot2}$
$\qquad\qquad\qquad\qquad = 15\cdot6\sqrt{2}$
$\qquad\qquad\qquad\qquad = 90\sqrt{2}$

44. $\left(-2\sqrt{15}\right)\left(4x\sqrt{3}\right)=-8x\sqrt{45}$
$\qquad\qquad\qquad\qquad = -8x\sqrt{9\cdot5}$
$\qquad\qquad\qquad\qquad = -24x\sqrt{5}$

45. $3\sqrt{x}\left(2\sqrt{8x}-3\sqrt{48}\right)=3\sqrt{x}\left(4\sqrt{2x}-12\sqrt{3}\right)$
$\qquad\qquad\qquad\qquad\qquad = 12x\sqrt{2}-36\sqrt{3x}$

46. $\sqrt{3a}\left(4-\sqrt{21a}\right)=4\sqrt{3a}-\sqrt{3\cdot21a^2}$
$\qquad\qquad\qquad\qquad = 4\sqrt{3a}-\sqrt{3^2a^2\cdot7}$
$\qquad\qquad\qquad\qquad = 4\sqrt{3a}-3a\sqrt{7}$

47. $2\sqrt{7b}\left(\sqrt{ab}-b\sqrt{3bc}\right)=2\sqrt{7ab^2}-2b\sqrt{21b^2c}$
$\qquad\qquad\qquad\qquad\qquad\quad = 2b\sqrt{7a}-2b^2\sqrt{21c}$

48. $\left(5\sqrt{2}+\sqrt{3}\right)\left(\sqrt{2}-2\sqrt{3}\right)$
$\qquad = 5\cdot2-10\sqrt{6}+\sqrt{6}-2\cdot3$
$\qquad = 10-9\sqrt{6}-6$
$\qquad = 4-9\sqrt{6}$

49. $\left(2\sqrt{5}-3\sqrt{6}\right)^2=20-12\sqrt{30}+54=74-12\sqrt{30}$

50. $\left(\sqrt[3]{2x}+\sqrt[3]{6}\right)\left(\sqrt[3]{4x^2}-\sqrt[3]{y}\right)$
$\qquad = \sqrt[3]{8x^3}-\sqrt[3]{2xy}+\sqrt[3]{24x^2}-\sqrt[3]{6y}$
$\qquad = 2x-\sqrt[3]{2xy}+2\sqrt[3]{3x^2}-\sqrt[3]{6y}$

51. $f(x)=\sqrt{4x+16}$

    a. $f(12)=\sqrt{4(12)+16}=\sqrt{64}=8$

    b. Domain is all real numbers x where $4x+16\ge0$ or $x\ge-4$.

52. $f(x)=\sqrt{36-3x}$

    a. $f(9)=\sqrt{36-3(9)}=\sqrt{9}=3$

    b. Domain is all real numbers x where $36-3x\ge0$ or $x\le12$.

53. $\sqrt{\dfrac{6y^2}{x}}=\dfrac{\sqrt{6y^2}}{\sqrt{x}}\cdot\dfrac{\sqrt{x}}{\sqrt{x}}=\dfrac{\sqrt{6xy^2}}{\sqrt{x^2}}=\dfrac{y\sqrt{6x}}{x}$

54. $\dfrac{3}{\sqrt{5y}}=\dfrac{3}{\sqrt{5y}}\cdot\dfrac{\sqrt{5y}}{\sqrt{5y}}=\dfrac{3\sqrt{5y}}{5y}$

55. $\dfrac{3\sqrt{7x}}{\sqrt{21x}}=\dfrac{3\sqrt{7x}}{\sqrt{3}\cdot\sqrt{7x}}=\dfrac{3}{\sqrt{3}}\cdot\dfrac{\sqrt{3}}{\sqrt{3}}=\dfrac{3\sqrt{3}}{3}=\sqrt{3}$

56. $\dfrac{2}{\sqrt{6}-\sqrt{5}}=\dfrac{2}{\sqrt{6}-\sqrt{5}}\cdot\dfrac{\sqrt{6}+\sqrt{5}}{\sqrt{6}+\sqrt{5}}$
$\qquad = \dfrac{2\sqrt{6}+2\sqrt{5}}{6-5}$
$\qquad = 2\sqrt{6}+2\sqrt{5}$

57. $\dfrac{\sqrt{x}}{3\sqrt{x}+\sqrt{y}}=\dfrac{\sqrt{x}}{3\sqrt{x}+\sqrt{y}}\cdot\dfrac{3\sqrt{x}-\sqrt{y}}{3\sqrt{x}-\sqrt{y}}=\dfrac{3x-\sqrt{xy}}{9x-y}$

58.
$$\frac{\sqrt{5}}{\sqrt{7}-3} = \frac{\sqrt{5}}{\sqrt{7}-3} \cdot \frac{\sqrt{7}+3}{\sqrt{7}+3}$$
$$= \frac{\sqrt{35}+3\sqrt{5}}{7-9}$$
$$= -\frac{\sqrt{35}+3\sqrt{5}}{2}$$

59.
$$\frac{2\sqrt{3}+\sqrt{6}}{\sqrt{3}+2\sqrt{6}} = \frac{2\sqrt{3}+\sqrt{6}}{\sqrt{3}+2\sqrt{6}} \cdot \frac{\sqrt{3}-2\sqrt{6}}{\sqrt{3}-2\sqrt{6}}$$
$$= \frac{6-4\sqrt{18}+\sqrt{18}-12}{3-24}$$
$$= \frac{-6-3\sqrt{9\cdot 2}}{-21}$$
$$= \frac{-6-9\sqrt{2}}{-21}$$
$$= \frac{2+3\sqrt{2}}{7}$$

60.
$$\frac{2xy}{\sqrt[3]{16xy^5}} = \frac{2xy}{\sqrt[3]{16xy^5}} \cdot \frac{\sqrt[3]{4x^2y}}{\sqrt[3]{4x^2y}}$$
$$= \frac{2xy\sqrt[3]{4x^2y}}{\sqrt[3]{4^3 x^3 y^6}}$$
$$= \frac{2xy\sqrt[3]{4x^2y}}{4xy^2}$$
$$= \frac{\sqrt[3]{4x^2y}}{2y}$$

61.
$$\sqrt{-16}+\sqrt{-45} = \sqrt{16}\sqrt{-1}+\sqrt{45}\sqrt{-1}$$
$$= 4i+i\sqrt{9\cdot 5}$$
$$= 4i+3i\sqrt{5}$$

62.
$$2x-3i+5 = yi-2+\sqrt{6}$$
$$2x+5 = -2+\sqrt{6}$$
$$2x = -7+\sqrt{6}$$
$$x = \frac{-7+\sqrt{6}}{2}$$
$$y = -3$$

63.
$$(-12-6i)+(3-5i) = -12-6i+3-5i$$
$$= -12+3+(-6-5)i$$
$$= -9-11i$$

64. $(2-i)-(12-3i) = 2-i-12+3i = -10+2i$

65.
$$(5-2i)(3+3i) = 15+15i-6i-6i^2$$
$$= 15+15i-6i+6$$
$$= 21+9i$$

66.
$$(6-2i)^2 = (6-2i)(6-2i)$$
$$= 36-12i-12i+4i^2$$
$$= 36-24i-4$$
$$= 32-24i$$

67. $2i(3+4i) = 6i+8i^2 = -8+6i$

68. $3-4(2+i) = 3-8-4i = -5-4i$

69. $i^{34} = (i^4)^8 \cdot i^2 = 1^8(-1) = -1$

70. $i^{65} = (i^4)^{16} \cdot i = 1^{16} \cdot i = i$

71.
$$\frac{7-2i}{3+4i} = \frac{7-2i}{3+4i} \cdot \frac{3-4i}{3-4i}$$
$$= \frac{21-6i-28i+8i^2}{9+16}$$
$$= \frac{21-34i-8}{25}$$
$$= \frac{13-34i}{25} \text{ or } \frac{13}{25}-\frac{34i}{25}$$

72.
$$\frac{5-2i}{1-3i} = \frac{5-2i}{1-3i} \cdot \frac{1+3i}{1+3i}$$
$$= \frac{5-2i+15i-6i^2}{1+9}$$
$$= \frac{5+13i+6}{10}$$
$$= \frac{11+13i}{10} \text{ or } \frac{11}{10}+\frac{13i}{10}$$

73.
$$\frac{4-3i}{5i} = \frac{4-3i}{5i} \cdot \frac{-5i}{-5i}$$
$$= \frac{-20i+15i^2}{-25i^2}$$
$$= \frac{-20i-15}{-25(-1)}$$
$$= \frac{-15-20i}{25}$$
$$= \frac{5(-3-4i)}{5\cdot 5}$$
$$= \frac{-3-4i}{5} \text{ or } -\frac{3}{5}-\frac{4i}{5}$$

74. $\dfrac{12}{3-5i} = \dfrac{12}{3-5i} \cdot \dfrac{3+5i}{3+5i}$

$\qquad = \dfrac{36+60i}{9+25}$

$\qquad = \dfrac{36+60i}{34}$

$\qquad = \dfrac{18+30i}{17}$ or $\dfrac{18}{17}+\dfrac{30i}{17}$

75. $\sqrt{3x-2} = 5$

$\qquad \left(\sqrt{3x-2}\right)^2 = 5^2$

$\qquad\qquad 3x-2 = 25$

$\qquad\qquad\quad 3x = 27$

$\qquad\qquad\quad\ x = 9$

Check: $\sqrt{3(9)-2} \stackrel{?}{=} 5,\ 5 = 5$

$x = 9$ is the solution.

76. $\sqrt[3]{3x-1} = 2$

$\qquad \left(\sqrt[3]{3x-1}\right)^3 = 2^3$

$\qquad\qquad 3x-1 = 8$

$\qquad\qquad\quad 3x = 9$

$\qquad\qquad\quad\ x = 3$

Check: $\sqrt[3]{3(3)-1} \stackrel{?}{=} 2,\ 2 = 2$

$x = 3$ is the solution.

77. $\qquad\sqrt{2x+1} = 2x-5$

$\qquad \left(\sqrt{2x+1}\right)^2 = (2x-5)^2$

$\qquad\qquad 2x+1 = 4x^2 -20x+25$

$4x^2 -22x+24 = 0$

$2x-11x+12 = 0$

$(x-4)(2x-3) = 0$

$x = 4,\ \ x = \dfrac{3}{2}$

Check: $\sqrt{2(4)+1} \stackrel{?}{=} 2(4)-5,\ 3 = 3$

$\qquad\sqrt{2\left(\dfrac{3}{2}\right)+1} \stackrel{?}{=} 2\left(\dfrac{3}{2}\right)-5,\ \sqrt{4} \neq -2$

$x = 4$ is the solution.

78. $1+\sqrt{3x+1} = x$

$\qquad \sqrt{3x+1} = x-1$

$\qquad \left(\sqrt{3x+1}\right)^2 = (x-1)^2$

$\qquad\qquad 3x+1 = x^2 -2x+1$

$\qquad\qquad x^2 -5x = 0$

$\qquad\qquad x(x-5) = 0$

$\qquad x = 0,\ x = 5$

Check: $1+\sqrt{3(0)+1} \stackrel{?}{=} 0,\ 2 \neq 0$

$\qquad\qquad 1+\sqrt{3(5)+1} \stackrel{?}{=} 5,\ 5 = 5$

$x = 5$ is the solution.

79. $\sqrt{3x+1} -\sqrt{2x-1} = 1$

$\qquad \sqrt{3x+1} = \sqrt{2x-1}+1$

$\qquad \left(\sqrt{3x+1}\right)^2 = \left(\sqrt{2x-1}+1\right)^2$

$\qquad\qquad 3x+1 = 2x-1+2\sqrt{2x-1}+1$

$\qquad\qquad x+1 = 2\sqrt{2x-1}$

$\qquad\qquad (x+1)^2 = \left(2\sqrt{2x-1}\right)^2$

$\qquad\qquad x^2 +2x+1 = 8x-4$

$\qquad\qquad x^2 -6x+5 = 0$

$\qquad\qquad (x-5)(x-1) = 0$

$x = 5,\ x = 1$

Check: $\sqrt{3(5)+1} -\sqrt{2(5)-1} \stackrel{?}{=} 1,\ 1 = 1$

$\qquad\ \sqrt{3(1)+1} -\sqrt{2(1)-1} \stackrel{?}{=} 1,\ 1 = 1$

$x = 5,\ x = 1$ is the solution.

80. $\qquad\sqrt{7x+2} = \sqrt{x+3}+\sqrt{2x-1}$

$\qquad \left(\sqrt{7x+2}\right)^2 = \left[\sqrt{x+3}+\sqrt{2x-1}\right]^2$

$\qquad\qquad 7x+2 = x+3+2\sqrt{x+3}\sqrt{2x-1}+2x-1$

$\qquad\qquad 4x = 2\sqrt{x+3}\sqrt{2x-1}$

$\qquad\qquad 2x = \sqrt{x+3}\sqrt{2x-1}$

$\qquad\qquad (2x)^2 = \left(\sqrt{x+3}\sqrt{2x-1}\right)^2$

$\qquad\qquad 4x^2 = 2x^2 +5x-3$

$\qquad 2x^2 -5x+3 = 0$

$\qquad (x-1)(2x-3) = 0$

$\qquad x = 1,\ x = \dfrac{3}{2}$

Check:

$\sqrt{7(1)+2} \overset{?}{=} \sqrt{1+3} + \sqrt{2(1)-1}, \ 3 = 3$

$\sqrt{7\left(\dfrac{3}{2}\right)+2} \overset{?}{=} \sqrt{\left(\dfrac{3}{2}\right)+3} + \sqrt{2\left(\dfrac{3}{2}\right)-1}$

$\dfrac{5}{\sqrt{2}} = \dfrac{5}{\sqrt{2}}$

$x = 1, \ x = \dfrac{3}{2}$ is the solution.

81. $y = kx$

When $y = 11, \ x = 4, \ k = \dfrac{y}{x} = \dfrac{11}{4}.$

$y = \dfrac{11}{4}x$

When $x = 6, \ y = \dfrac{11}{4}(6) = 16.5.$

82. c = calories consumed
g = fat grams
$g = kc$

When $c = 2000, \ g = 18, \ k = \dfrac{g}{c} = \dfrac{18}{2000} = 0.009.$

$g = 0.009c$
When $c = 2500$: $g = 0.009(2500) = 22.5.$
She should consume a maximum of 22.5 grams of fat.

83. $t = k\sqrt{d}, \ 2 = k\sqrt{64}, \ k = \dfrac{1}{4}$

$t = \dfrac{1}{4}\sqrt{d}, \ t = \dfrac{1}{4}\sqrt{196} = \dfrac{14}{4} = 3.5$

It will take 3.5 seconds.

84. $y = \dfrac{k}{x}, \ 8 = \dfrac{k}{3}, \ k = 24$

$y = \dfrac{24}{x}, \ y = \dfrac{24}{48} = 0.5$

85. $y = \dfrac{kx}{z^2}$

When $x = 10$ and $z = 5, \ y = 20$:

$k = \dfrac{y(z)^2}{x} = \dfrac{20(5)^2}{10} = 50$

$y = \dfrac{50x}{z^2}$

When $x = 8$ and $z = 2, \ y = \dfrac{50(8)}{(2)^2} = 100.$

86. $V = khr^2, \ 135 = k(5)(3)^2, \ k = 3$

$V = 3r^2h, \ V = 3(4)^2(9) = 432$

The capacity is $432 \ \text{cm}^3.$

How Am I Doing? Chapter 8 Test

1. $(2x^{1/2}y^{1/3})(-3x^{1/3}y^{1/6}) = -6x^{\frac{1}{2}+\frac{1}{3}}y^{\frac{1}{3}+\frac{1}{6}}$
$= -6x^{5/6}y^{1/2}$

2. $\dfrac{7x^3}{4x^{3/4}} = \dfrac{7x^{3-\frac{3}{4}}}{4} = \dfrac{7x^{9/4}}{4}$

3. $(8x^{1/3})^{3/2} = 8^{3/2}x^{\frac{1}{3}\cdot\frac{3}{2}} = 8^{3/2}x^{1/2}$
or $(8x^{1/3})^{3/2} = (2^3x^{1/3})^{3/2}$
$= (2^3)^{3/2}(x^{1/3})^{3/2}$
$= 2^{9/2}x^{1/2}$
$= 2^4 2^{1/2}x^{1/2}$
$= 16(2x)^{1/2}$

4. $\left(\dfrac{4}{9}\right)^{3/2} = \left(\left(\dfrac{2}{3}\right)^2\right)^{3/2} = \left(\dfrac{2}{3}\right)^{2\cdot\frac{3}{2}} = \left(\dfrac{2}{3}\right)^3 = \dfrac{8}{27}$

5. $\sqrt[5]{-32} = \sqrt[5]{(-2)^5} = -2$

6. $8^{-2/3} = \dfrac{1}{(8^{1/3})^2} = \dfrac{1}{2^2} = \dfrac{1}{4}$

7. $16^{5/4} = (16^{1/4})^5 = 2^5 = 32$

8. $\sqrt{75a^4b^9} = \sqrt{25a^4b^8 \cdot 3b} = 5a^2b^4\sqrt{3b}$

9. $\sqrt{49a^4b^{10}} = \sqrt{7^2(a^2)^2(b^5)^2} = 7a^2b^5$

10. $\sqrt[3]{54m^3n^5} = \sqrt[3]{3^3 m^3 n^3 \cdot 2n^2} = 3mn\sqrt[3]{2n^2}$

11. $3\sqrt{24} - \sqrt{18} + \sqrt{50} = 3\sqrt{4\cdot6} - \sqrt{9\cdot2} + \sqrt{25\cdot2}$
$= 6\sqrt{6} - 3\sqrt{2} + 5\sqrt{2}$
$= 6\sqrt{6} + 2\sqrt{2}$

12. $\sqrt{40x} - \sqrt{27x} + 2\sqrt{12x}$
$= \sqrt{4 \cdot 10x} - \sqrt{9 \cdot 3x} + 2\sqrt{4 \cdot 3x}$
$= 2\sqrt{10x} - 3\sqrt{3x} + 4\sqrt{3x}$
$= 2\sqrt{10x} + \sqrt{3x}$

13. $\left(-3\sqrt{2y}\right)\left(5\sqrt{10xy}\right) = -15\sqrt{20xy^2}$
$= -15\sqrt{4y^2 \cdot 5x}$
$= -30y\sqrt{5x}$

14. $2\sqrt{3}\left(3\sqrt{6} - 5\sqrt{2}\right) = 6\sqrt{18} - 10\sqrt{6}$
$= 6\sqrt{9 \cdot 2} - 10\sqrt{6}$
$= 18\sqrt{2} - 10\sqrt{6}$

15. $\left(5\sqrt{3} - \sqrt{6}\right)\left(2\sqrt{3} + 3\sqrt{6}\right)$
$= 5 \cdot 2 \cdot 3 + 5 \cdot 3\sqrt{3 \cdot 6} - 2\sqrt{3 \cdot 6} - 3 \cdot 6$
$= 30 + 15\sqrt{18} - 2\sqrt{18} - 18$
$= 12 + 13\sqrt{9 \cdot 2}$
$= 12 + 39\sqrt{2}$

16. $\dfrac{30}{\sqrt{5x}} = \dfrac{30}{\sqrt{5x}} \cdot \dfrac{\sqrt{5x}}{\sqrt{5x}} = \dfrac{30\sqrt{5x}}{5x} = \dfrac{6\sqrt{5x}}{x}$

17. $\sqrt{\dfrac{xy}{3}} = \dfrac{\sqrt{xy}}{\sqrt{3}} \cdot \dfrac{\sqrt{3}}{\sqrt{3}} = \dfrac{\sqrt{3xy}}{3}$

18. $\dfrac{1 + 2\sqrt{3}}{3 - \sqrt{3}} = \dfrac{1 + 2\sqrt{3}}{3 - \sqrt{3}} \cdot \dfrac{3 + \sqrt{3}}{3 + \sqrt{3}}$
$= \dfrac{3 + \sqrt{3} + 6\sqrt{3} + 2(3)}{9 - 3}$
$= \dfrac{9 + 7\sqrt{3}}{6}$

19. $\sqrt{3x - 2} = x$
$\left(\sqrt{3x - 2}\right)^2 = x^2$
$3x - 2 = x^2$
$x^2 - 3x + 2 = 0$
$(x - 2)(x - 1) = 0$
$x = 2, x = 1$
Check: $\sqrt{3(2) - 2} \stackrel{?}{=} 2$, $2 = 2$
$\quad\quad\quad \sqrt{3(1) - 2} \stackrel{?}{=} 1$, $1 = 1$
$x = 2, x = 1$ is the solution.

20. $5 + \sqrt{x + 15} = x$
$\sqrt{x + 15} = x - 5$
$\left(\sqrt{x + 15}\right)^2 = (x - 5)^2$
$x + 15 = x^2 - 10x + 25$
$x^2 - 11x + 10 = 0$
$(x - 10)(x - 11) = 0$
$x = 10, x = 11$
Check: $5 + \sqrt{10 + 15} \stackrel{?}{=} 10$, $10 = 10$
$\quad\quad\quad 5 + \sqrt{11 + 15} \stackrel{?}{=} 11$, $5 + \sqrt{26} \neq 11$
$x = 10$ is the solution.

21. $5 - \sqrt{x - 2} = \sqrt{x + 3}$
$\left(5 - \sqrt{x - 2}\right)^2 = \left(\sqrt{x + 3}\right)^2$
$25 - 10\sqrt{x - 2} + x - 2 = x + 3$
$20 = 10\sqrt{x - 2}$
$2 = \sqrt{x - 2}$
$2^2 = \left(\sqrt{x - 2}\right)^2$
$4 = x - 2$
$x = 6$
Check: $5 - \sqrt{6 - 2} \stackrel{?}{=} \sqrt{6 + 3}$, $3 = 3$
$x = 6$ is the solution.

22. $(8 + 2i) - 3(2 - 4i) = 8 + 2i - 6 + 12i = 2 + 14i$

23. $i^{18} + \sqrt{-16} = (i^4)^4 \cdot i^2 + \sqrt{16}\sqrt{-1}$
$= 1^4(-1) + 4i$
$= -1 + 4i$

24. $(3 - 2i)(4 + 3i) = 12 + i - 6i^2 = 12 + i + 6 = 18 + i$

25. $\dfrac{2 + 5i}{1 - 3i} = \dfrac{2 + 5i}{1 - 3i} \cdot \dfrac{1 + 3i}{1 + 3i}$
$= \dfrac{2 + 11i + 15i^2}{1 + 9}$
$= \dfrac{2 + 11i - 15}{10}$
$= \dfrac{-13 + 11i}{10}$ or $-\dfrac{13}{10} + \dfrac{11i}{10}$

26. $(6 + 3i)^2 = 36 + 36i + 9i^2 = 36 + 36i - 9 = 27 + 36i$

27. $i^{43} = (i^4)^{10} \cdot i^3 = 1^{10} \cdot (-i) = -i$

28. $y = \dfrac{k}{x}$, $9 = \dfrac{k}{2}$, $k = 18$

    $y = \dfrac{18}{x}$, $y = \dfrac{18}{6} = 3$

29. $y = k \cdot \dfrac{x}{z^2}$, $3 = k \cdot \dfrac{8}{4^2}$, $k = 6$

    $y = 6 \cdot \dfrac{x}{z^2}$, $y = 6 \cdot \dfrac{5}{6^2} = \dfrac{5}{6}$

30. $d = kv^2$, $30 = k \cdot 30^2$, $k = \dfrac{1}{30}$

    $d = \dfrac{1}{30} \cdot v^2$, $d = \dfrac{1}{30} \cdot 50^2 \approx 83.3$

The car's stopping distance is about 83.3 feet.

Chapter 9

1. $x^2 = 100$

$x = \pm\sqrt{100}$

$x = \pm 10$

3. $4x^2 - 12 = 0$

$4x^2 = 12$

$x^2 = 3$

$x = \pm\sqrt{3}$

5. $2x^2 - 80 = 0$

$2x^2 = 80$

$x^2 = 40$

$x = \pm\sqrt{40} = \pm\sqrt{4 \cdot 10}$

$x = \pm 2\sqrt{10}$

7. $x^2 = -81$

$x = \pm\sqrt{-81}$

$x = \pm 9i$

9. $x^2 + 16 = 0$

$x^2 = -16$

$x = \pm\sqrt{-16}$

$x = \pm 4i$

11. $(x-3)^2 = 12$

$x - 3 = \pm\sqrt{12} = \pm\sqrt{4 \cdot 3}$

$x - 3 = \pm 2\sqrt{3}$

$x = 3 \pm 2\sqrt{3}$

13. $(x+9)^2 = 21$

$x + 9 = \pm\sqrt{21}$

$x = -9 \pm\sqrt{21}$

15. $(2x+1)^2 = 7$

$2x + 1 = \pm\sqrt{7}$

$2x = -1 \pm\sqrt{7}$

$x = \dfrac{-1 \pm\sqrt{7}}{2}$

17. $(4x-3)^2 = 36$

$4x - 3 = \pm 6$

$4x = 3 \pm 6$

$x = \dfrac{3 \pm 6}{4}$

$x = \dfrac{9}{4}, \; x = -\dfrac{3}{4}$

19. $(2x+5)^2 = 49$

$2x + 5 = \pm 7$

$2x = -5 \pm 7$

$x = \dfrac{-5 \pm 7}{2}$

$x = 1, \; x = -6$

21. $2x^2 - 9 = 0$

$2x^2 = 9$

$x^2 = \dfrac{9}{2}$

$x = \pm\sqrt{\dfrac{9}{2}}$

$x = \pm\dfrac{3}{\sqrt{2}}$

$x = \pm\dfrac{3\sqrt{2}}{2}$

23. $x^2 + 10x + 5 = 0$

$x^2 + 10x + 25 = -5 + 25$

$(x+5)^2 = 20$

$x + 5 = \pm\sqrt{20} = \pm 2\sqrt{5}$

$x = -5 \pm 2\sqrt{5}$

25. $x^2 - 8x = 17$

$x^2 - 8x + 16 = 17 + 16$

$(x-4)^2 = 33$

$x - 4 = \pm\sqrt{33}$

$x = 4 \pm\sqrt{33}$

27. $x - 14x = -48$

$x^2 - 14x + 49 = -48 + 49$

$(x-7)^2 = 1$

$x - 7 = \pm 1$

$x = 7 \pm 1$

$x = 6, \; x = 8$

29. $\dfrac{x^2}{2}+\dfrac{3}{2}x=4$

$x^2+3x=8$

$x^2+3x+\dfrac{9}{4}=8+\left(\dfrac{3}{2}\right)^2$

$\left(x+\dfrac{3}{2}\right)^2=\dfrac{41}{4}$

$x+\dfrac{3}{2}=\pm\dfrac{\sqrt{41}}{2}$

$x=-\dfrac{3}{2}\pm\dfrac{\sqrt{41}}{2}$

$x=\dfrac{-3\pm\sqrt{41}}{2}$

31. $2y^2+10y=-11$

$y^2+5y=-\dfrac{11}{2}$

$y^2+5y+\dfrac{25}{4}=-\dfrac{11}{2}+\dfrac{25}{4}$

$\left(y+\dfrac{5}{2}\right)^2=\dfrac{3}{4}$

$y+\dfrac{5}{2}=\pm\dfrac{\sqrt{3}}{2}$

$y=-\dfrac{5}{2}\pm\dfrac{\sqrt{3}}{2}$

$y=\dfrac{-5\pm\sqrt{3}}{2}$

33. $3x^2+10x-2=0$

$3x^2+10x=2$

$x^2+\dfrac{10}{3}x=\dfrac{2}{3}$

$x^2+\dfrac{10}{3}x+\dfrac{25}{9}=\dfrac{2}{3}+\dfrac{25}{9}$

$\left(x+\dfrac{5}{3}\right)^2=\dfrac{31}{9}$

$x+\dfrac{5}{3}=\pm\dfrac{\sqrt{31}}{3}$

$x=-\dfrac{5}{3}\pm\dfrac{\sqrt{31}}{3}$

$x=\dfrac{-5\pm\sqrt{31}}{3}$

35. $x^2+4x-6=0$

$x^2+4x=6$

$x^2+4x+4=6+4$

$(x+2)^2=10$

$x+2=\pm\sqrt{10}$

$x=-2\pm\sqrt{10}$

37. $\dfrac{x^2}{2}-x=4$

$x^2-2x=8$

$x^2-2x+1=8+1$

$(x-1)^2=9$

$x-1=\pm3$

$x=1\pm3$

$x=4,\ x=-2$

39. $3x^2+1=x$

$3x^2-x=-1$

$x^2-\dfrac{1}{3}x=-\dfrac{1}{3}$

$x^2-\dfrac{1}{3}x+\dfrac{1}{36}=-\dfrac{1}{3}+\dfrac{1}{36}$

$\left(x-\dfrac{1}{6}\right)^2=-\dfrac{11}{36}$

$x-\dfrac{1}{6}=\pm\dfrac{i\sqrt{11}}{6}$

$x=\dfrac{1}{6}\pm\dfrac{i\sqrt{11}}{6}$

$x=\dfrac{1\pm i\sqrt{11}}{6}$

41. $x^2+2=x$

$x^2-x=-2$

$x^2-x+\dfrac{1}{4}=-2+\dfrac{1}{4}$

$\left(x-\dfrac{1}{2}\right)^2=-\dfrac{7}{4}$

$x-\dfrac{1}{2}=\pm\dfrac{i\sqrt{7}}{2}$

$x=\dfrac{1}{2}\pm\dfrac{i\sqrt{7}}{2}$

$x=\dfrac{1\pm i\sqrt{7}}{2}$

43.
$$2x^2 + 2 = 3x$$
$$2x^2 - 3x = -2$$
$$x^2 - \frac{3}{2}x = -1$$
$$x^2 - \frac{3}{2}x + \frac{9}{16} = -1 + \frac{9}{16}$$
$$\left(x - \frac{3}{4}\right)^2 = -\frac{7}{16}$$
$$x - \frac{3}{4} = \pm\frac{i\sqrt{7}}{4}$$
$$x = \frac{3}{4} \pm \frac{i\sqrt{7}}{4}$$
$$x = \frac{3 \pm i\sqrt{7}}{4}$$

45.
$$x^2 + 2x - 5 = 0$$
$$\left(-1 + \sqrt{6}\right)^2 + 2\left(-1 + \sqrt{6}\right) - 5 \stackrel{?}{=} 0$$
$$1 - 2\sqrt{6} + 6 - 2 + 2\sqrt{6} - 5 \stackrel{?}{=} 0$$
$$0 = 0 \checkmark$$

47. $(x-7)^2(8) = 200$
$$(x-7)^2 = 25$$
$$x - 7 = \pm 5$$
$$x = 7 \pm 5$$
Since the value of x must be greater than 7, $x = 12$.

49. $4t^2 = L$
$$4t^2 = 5$$
$$t^2 = 1.25$$
$$t = \pm\sqrt{1.25}$$
Since the time must be positive, $t = \sqrt{1.25} \approx 1.12$. The hang time is approximately 1.12 seconds.

Cumulative Review

51. $\sqrt{b^2 - 4ac} = \sqrt{4^2 - 4(3)(-4)}$
$$= \sqrt{16 + 48}$$
$$= \sqrt{64}$$
$$= 8$$

52. $\sqrt{b^2 - 4ac} = \sqrt{(-5)^2 - 4(2)(-3)}$
$$= \sqrt{25 + 24}$$
$$= \sqrt{49}$$
$$= 7$$

53. $5x^2 - 6x + 8 = 5(-2)^2 - 6(-2) + 8$
$$= 5(4) - 6(-2) + 8$$
$$= 20 + 12 + 8$$
$$= 40$$

54. $2x^2 + 3x - 5 = 2(-3)^2 + 3(-3) - 5$
$$= 2(9) + 3(-3) - 5$$
$$= 18 - 9 - 5$$
$$= 4$$

Quick Quiz 9.1

1. $(4x-3)^2 = 12$
$$4x - 3 = \pm\sqrt{12} = \pm 2\sqrt{3}$$
$$4x = 3 \pm 2\sqrt{3}$$
$$x = \frac{3 \pm 2\sqrt{3}}{4}$$

2. $x^2 - 8x = 28$
$$x^2 - 8x + 16 = 28 + 16$$
$$(x-4)^2 = 44$$
$$x - 4 = \pm\sqrt{44} = \pm 2\sqrt{11}$$
$$x = 4 \pm 2\sqrt{11}$$

3. $2x^2 + 10x = -11$
$$x^2 + 5x = -\frac{11}{2}$$
$$x^2 + 5x + \frac{25}{4} = -\frac{11}{2} + \frac{25}{4}$$
$$\left(x + \frac{5}{2}\right)^2 = \frac{3}{4}$$
$$x + \frac{5}{2} = \pm\frac{\sqrt{3}}{2}$$
$$x = -\frac{5}{2} \pm \frac{\sqrt{3}}{2}$$
$$x = \frac{-5 \pm \sqrt{3}}{2}$$

4. Answers may vary. Possible solution:

Divide the coefficient of x (1) by 2 to get $\frac{1}{2}$.

Since $\left(\frac{1}{2}\right)^2 = \frac{1}{4}$, add $\frac{1}{4}$ to both sides of the equation.

9.2 Exercises

1. Place the quadratic in standard form. Find a, b, and c. Substitute these values into the quadratic formula.

3. If the discriminant in the quadratic formula is zero, then the quadratic equation will have <u>one real</u> solution.

5. $x^2 + x - 5 = 0$
$a = 1$, $b = 1$, $c = -5$
$$x = \frac{-b \pm \sqrt{b^2 - 4ac}}{2a}$$
$$x = \frac{-1 \pm \sqrt{(1)^2 - 4(1)(-5)}}{2(1)}$$
$$x = \frac{-1 \pm \sqrt{21}}{2}$$

7. $2x^2 + 3x - 3 = 0$
$a = 2$, $b = 3$, $c = -3$
$$x = \frac{-b \pm \sqrt{b^2 - 4ac}}{2a}$$
$$x = \frac{-3 \pm \sqrt{3^2 - 4(2)(-3)}}{2(2)}$$
$$x = \frac{-3 \pm \sqrt{33}}{4}$$

9. $$x^2 = \frac{2}{3}x$$
$$x^2 - \frac{2}{3}x = 0$$
$$3x^2 - 2x = 0$$
$a = 3$, $b = -2$, $c = 0$
$$x = \frac{-b \pm \sqrt{b^2 - 4ac}}{2a}$$
$$x = \frac{-(-2) \pm \sqrt{(-2)^2 - 4(3)(0)}}{2(3)}$$
$$x = \frac{2 \pm \sqrt{4}}{6} = \frac{2 \pm 2}{6}$$
$$x = 0, \ x = \frac{2}{3}$$

11. $3x^2 - x - 2 = 0$
$a = 3$, $b = -1$, $c = -2$
$$x = \frac{-b \pm \sqrt{b^2 - 4ac}}{2a}$$
$$x = \frac{-(-1) \pm \sqrt{(-1)^2 - 4(3)(-2)}}{2(3)}$$
$$x = \frac{1 \pm \sqrt{25}}{6} = \frac{1 \pm 5}{6}$$
$$x = 1, \ x = -\frac{2}{3}$$

13. $4x^2 + 3x - 2 = 0$
$a = 4$, $b = 3$, $c = -2$
$$x = \frac{-b \pm \sqrt{b^2 - 4ac}}{2a}$$
$$x = \frac{-3 \pm \sqrt{(3)^2 - 4(4)(-2)}}{2(4)}$$
$$x = \frac{-3 \pm \sqrt{41}}{8}$$

15. $4x^2 + 1 = 7$
$4x^2 - 6 = 0$
$a = 4$, $b = 0$, $c = -6$
$$x = \frac{-b \pm \sqrt{b^2 - 4ac}}{2a}$$
$$x = \frac{-0 \pm \sqrt{(0)^2 - 4(4)(-6)}}{2(4)}$$
$$x = \pm\frac{\sqrt{96}}{8} = \pm\frac{4\sqrt{6}}{8}$$
$$x = \pm\frac{\sqrt{6}}{2}$$

17. $2x(x + 3) - 3 = 4x - 2$
$2x^2 + 6x - 4x - 1 = 0$
$2x^2 + 2x - 1 = 0$
$a = 2$, $b = 2$, $c = -1$
$$x = \frac{-b \pm \sqrt{b^2 - 4ac}}{2a}$$
$$x = \frac{-2 \pm \sqrt{(2)^2 - 4(2)(-1)}}{2(2)}$$
$$x = \frac{-2 \pm \sqrt{12}}{4} = \frac{-2 \pm 2\sqrt{3}}{4}$$
$$x = \frac{2\left(-1 \pm \sqrt{3}\right)}{4} = \frac{-1 \pm \sqrt{3}}{2}$$

19. $x(x+3)-2=3x+7$

$x^2+3x-2=3x+7$

$x^2-9=0$

$a=1,\ b=0,\ c=-9$

$x=\dfrac{-b\pm\sqrt{b^2-4ac}}{2a}$

$x=\dfrac{-0\pm\sqrt{(0)^2-4(1)(-9)}}{2(1)}$

$x=\pm\dfrac{\sqrt{36}}{2}=\pm\dfrac{6}{2}=\pm 3$

21. $(x-2)(x+1)=\dfrac{2x+3}{2}$

$x^2-x-2=\dfrac{2x+3}{2}$

$2x^2-2x-4=2x+3$

$2x^2-4x-7=0$

$a=2,\ b=-4,\ c=-7$

$x=\dfrac{-b\pm\sqrt{b^2-4ac}}{2a}$

$x=\dfrac{-(-4)\pm\sqrt{(-4)^2-4(2)(-7)}}{2(2)}$

$x=\dfrac{4\pm\sqrt{72}}{4}=\dfrac{4\pm 6\sqrt{2}}{4}$

$x=\dfrac{2\left(2\pm 3\sqrt{2}\right)}{4}=\dfrac{2\pm 3\sqrt{2}}{2}$

23. The LCD is $3x(x+2)$.

$$\dfrac{1}{x+2}+\dfrac{1}{x}=\dfrac{1}{3}$$

$$\dfrac{1}{x+2}(3x)(x+2)+\dfrac{1}{x}(3x)(x+2)=\dfrac{1}{3}(3x)(x+2)$$

$$3x+3(x+2)=x(x+2)$$

$$3x+3x+6=x^2+2x$$

$$x^2-4x-6=0$$

$a=1,\ b=-4,\ c=-6$

$x=\dfrac{-b\pm\sqrt{b^2-4ac}}{2a}$

$x=\dfrac{-(-4)\pm\sqrt{(-4)^2-4(1)(-6)}}{2(1)}$

$x=\dfrac{4\pm\sqrt{40}}{2}=\dfrac{4\pm 2\sqrt{10}}{2}$

$x=\dfrac{2\left(2\pm\sqrt{10}\right)}{2}=2\pm\sqrt{10}$

25. The LCD is $12y(y + 2)$.

$$\frac{1}{12} + \frac{1}{y} = \frac{2}{y+2}$$

$$\frac{1}{12}(12y)(y+2) + \frac{1}{y}(12y)(y+2) = \frac{2}{y+2}(12y)(y+2)$$

$$y(y+2) + 12(y+2) = 2(12y)$$

$$y^2 + 2y + 12y + 24 = 24y$$

$$y^2 - 10y + 24 = 0$$

$a = 1, b = -10, c = 24$

$$y = \frac{-b \pm \sqrt{b^2 - 4ac}}{2a}$$

$$y = \frac{-(-10) \pm \sqrt{(-10)^2 - 4(1)(24)}}{2(1)}$$

$$y = \frac{10 \pm \sqrt{4}}{2}$$

$$y = \frac{10 \pm 2}{2}$$

$$y = 6, \ y = 4$$

27.

$$x(x+4) = -12$$

$$x^2 + 4x + 12 = 0$$

$a = 1, b = 4, c = 12$

$$x = \frac{-b \pm \sqrt{b^2 - 4ac}}{2a}$$

$$x = \frac{-4 \pm \sqrt{(4)^2 - 4(1)(12)}}{2(1)}$$

$$x = \frac{-4 \pm \sqrt{-32}}{2} = \frac{-4 \pm 4i\sqrt{2}}{2}$$

$$x = \frac{2\left(-2 \pm 2i\sqrt{2}\right)}{2} = -2 \pm 2i\sqrt{2}$$

29.

$$2x^2 + 11 = 0$$

$a = 2, b = 0, c = 11$

$$x = \frac{-b \pm \sqrt{b^2 - 4ac}}{2a}$$

$$x = \frac{-0 \pm \sqrt{(0)^2 - 4(2)(11)}}{2(2)}$$

$$x = \pm \frac{\sqrt{-88}}{4}$$

$$x = \pm \frac{2i\sqrt{22}}{4}$$

$$x = \pm \frac{i\sqrt{22}}{2}$$

31. $3x^2 - 8x + 7 = 0$

$a = 3, b = -8, c = 7$

$$x = \frac{-b \pm \sqrt{b^2 - 4ac}}{2a}$$

$$x = \frac{-(-8) \pm \sqrt{(-8)^2 - 4(3)(7)}}{2(3)}$$

$$x = \frac{8 \pm \sqrt{-20}}{6} = \frac{8 \pm 2i\sqrt{5}}{6}$$

$$x = \frac{2\left(4 \pm i\sqrt{5}\right)}{6} = \frac{4 \pm i\sqrt{5}}{3}$$

33. $\qquad 3x^2 + 4x = 2$

$3x^2 + 4x - 2 = 0$

$a = 3, b = 4, c = -2$

$b^2 - 4ac = 4^2 - 4(3)(-2) = 40$

2 irrational roots

35. $2x^2 + 10x + 8 = 0$

$a = 2, b = 10, c = 8$

$b^2 - 4ac = 10^2 - 4(2)(8) = 36 = 6^2$

2 rational roots

37. $\qquad 9x^2 + 4 = 12x$

$9x^2 - 12x + 4 = 0$

$a = 9, b = -12, c = 4$

$b^2 - 4ac = (-12)^2 - 4(9)(4) = 0$

1 rational root

39. $7x(x-1) + 15 = 10$

$\qquad 7x^2 - 7x + 5 = 0$

$a = 7, b = -7, c = 5$

$b^2 - 4ac = (-7)^2 - 4(7)(5) = -91$

2 nonreal complex roots

41. 13, 2

$\qquad x = 13 \qquad\qquad\qquad x = 2$

$x - 13 = 0 \qquad\qquad\quad x - 2 = 0$

$(x - 13)(x - 2) = 0$

$x^2 - 15x + 26 = 0$

43. −7, −6

$\qquad x = -7 \qquad\qquad\qquad x = -6$

$x + 7 = 0 \qquad\qquad\quad x + 6 = 0$

$(x + 7)(x + 6) = 0$

$x^2 + 13x + 42 = 0$

45. $4i, -4i$

$\qquad x = 4i \qquad\qquad\qquad x = -4i$

$x - 4i = 0 \qquad\qquad\quad x + 4i = 0$

$(x - 4i)(x + 4i) = 0$

$\qquad\quad x^2 + 16 = 0$

47. $3, -\dfrac{5}{2}$

$\qquad x = 3 \qquad\qquad\qquad x = -\dfrac{5}{2}$

$x - 3 = 0 \qquad\qquad\quad 2x + 5 = 0$

$(x - 3)(2x + 5) = 0$

$\qquad\quad 2x^2 - x - 15 = 0$

49. $0.162x^2 + 0.094x - 0.485 = 0$

$a = 0.162, b = 0.094, c = -0.485$

$$x = \frac{-0.094 \pm \sqrt{0.094^2 - 4(0.162)(-0.485)}}{2(0.162)}$$

$$x = \frac{-0.094 \pm \sqrt{0.323116}}{0.324}$$

$x \approx 1.4643, \; x \approx -2.0445$

51. $p = -100x^2 + 4800x - 54{,}351 = 0$

$a = -100, b = 4800, c = -54{,}351$

$$x = \frac{-4800 \pm \sqrt{4800^2 - 4(-100)(-54{,}351)}}{2(-100)}$$

$$x = \frac{-4800 \pm \sqrt{1{,}299{,}600}}{-200}$$

$x = 18.3, \; x = 29.7$

Eighteen or thirty bikes per day will produce a zero profit.

Cumulative Review

53. $9x^2 - 6x + 3 - 4x - 12x^2 + 8 = -3x^2 - 10x + 11$

54. $3y(2 - y) + \dfrac{1}{5}(10y^2 - 15y) = 6y - 3y^2 + 2y^2 - 3y$

$\qquad\qquad\qquad\qquad\qquad\qquad = -y^2 + 3y$

Quick Quiz 9.2

1. $11x^2 - 9x - 1 = 0$
   $a = 11, b = -9, c = -1$
   $$x = \frac{-(-9) \pm \sqrt{(-9)^2 - 4(11)(-1)}}{2(11)}$$
   $$x = \frac{9 \pm \sqrt{125}}{22}$$
   $$x = \frac{9 \pm 5\sqrt{5}}{22}$$

2. The LCD is $4x^2$.
   $$\frac{3}{4} + \frac{5}{4x} = \frac{2}{x^2}$$
   $$\frac{3}{4}(4x^2) + \frac{5}{4x}(4x^2) = \frac{2}{x^2}(4x^2)$$
   $$3x^2 + 5x = 8$$
   $$3x^2 + 5x - 8 = 0$$
   $a = 3, b = 5, c = -8$
   $$x = \frac{-5 \pm \sqrt{(5)^2 - 4(3)(-8)}}{2(3)}$$
   $$x = \frac{-5 \pm \sqrt{121}}{6}$$
   $$x = \frac{-5 \pm 11}{6}$$
   $$x = 1, \ x = -\frac{8}{3}$$

3. $(x+2)(x+1) + (x-4)^2 = 9$
   $$x^2 + 3x + 2 + x^2 - 8x + 16 = 9$$
   $$2x^2 - 5x + 18 = 9$$
   $$2x^2 - 5x + 9 = 0$$
   $a = 2, b = -5, c = 9$
   $$x = \frac{-(-5) \pm \sqrt{(-5)^2 - 4(2)(9)}}{2(2)}$$
   $$x = \frac{5 \pm \sqrt{-47}}{4}$$
   $$x = \frac{5 \pm i\sqrt{47}}{4}$$

4. Answers may vary. Possible solution:
   Subtract 3 from both sides to put the equation in standard form. Identify the values of a, b, and c, and find the value of the discriminant, $b^2 - 4ac$. If the discriminant is a perfect square, there are two rational solutions; if it is a positive number that is not a perfect square, there are two irrational solutions; if it is zero, there is one rational solution; if it is negative, there are two nonreal complex solutions.

9.3 Exercises

1. $x^4 - 9x^2 + 20 = 0$
   $y = x^2$
   $$y^2 - 9y + 20 = 0$$
   $$(y-5)(y-4) = 0$$
   $y = 5, y = 4$
   $x^2 = 5 \qquad\qquad x^2 = 4$
   $x = \pm\sqrt{5} \qquad x = \pm 2$

3. $x^4 + x^2 - 12 = 0$
   $y = x^2$
   $$y^2 + y - 12 = 0$$
   $$(y-3)(y+4) = 0$$
   $y = 3, y = -4$
   $x^2 = 3 \qquad\qquad x^2 = -4$
   $x = \pm\sqrt{3} \qquad x = \pm 2i$

5. $4x^4 - x^2 - 3 = 0$
   $y = x^2$
   $$4y^2 - y - 3 = 0$$
   $$(y-1)(4y+3) = 0$$
   $$y = 1, -\frac{3}{4}$$
   $x^2 = 1 \qquad\qquad x^2 = -\frac{3}{4}$
   $x = \pm 1 \qquad\qquad x = \pm\sqrt{-\frac{3}{4}} = \pm i\sqrt{\frac{3}{4}}$
   $\qquad\qquad\qquad x = \pm\frac{i\sqrt{3}}{2}$

7. $x^6 - 7x^3 - 8 = 0$
   $y = x^3$
   $$y^2 - 7y - 8 = 0$$
   $$(y-8)(y+1) = 0$$
   $y = 8, y = -1$
   $x^3 = 8 \qquad\qquad x^3 = -1$
   $x = 2 \qquad\qquad x = -1$

9. $x^6 - 3x^3 = 0$

$y = x^3$

$y^2 - 3y = 0$

$y(y - 3) = 0$

$y = 0, \; y = 3$

$x^3 = 0 \qquad\qquad x^3 = 3$

$x = 0 \qquad\qquad\quad x = \sqrt[3]{3}$

11. $\qquad\qquad x^8 = 17x^4 - 16$

$x^8 - 17x^4 + 16 = 0$

$y = x^4$

$y^2 - 17y + 16 = 0$

$(y - 16)(y - 1) = 0$

$y = 16, \; y = 1$

$x^4 = 16 \qquad\qquad x^4 = 1$

$x = \pm 2 \qquad\qquad x = \pm 1$

13. $\qquad 3x^8 + 13x^4 = 10$

$3x^8 + 13x^4 - 10 = 0$

$y = x^4$

$3y^2 + 13y - 10 = 0$

$(3y - 2)(y + 5) = 0$

$y = \dfrac{2}{3}, \; y = -5$

$x^4 = \dfrac{2}{3} \qquad\qquad x^4 = -5$

$\qquad\qquad\qquad\qquad$ no real roots

$x = \pm\sqrt[4]{\dfrac{2}{3}} = \pm\dfrac{\sqrt[4]{54}}{3}$

15. $x^{2/3} + x^{1/3} - 12 = 0$

$y = x^{1/3}$

$y^2 + y - 12 = 0$

$(y + 4)(y - 3) = 0$

$y = -4, \; y = 3$

$x^{1/3} = -4 \qquad\qquad x^{1/3} = 3$

$x = (-4)^3 \qquad\qquad x = 3^3$

$x = -64 \qquad\qquad\; x = 27$

17. $12x^{2/3} + 5x^{1/3} - 2 = 0$

$y = x^{1/3}$

$12y^2 + 5y - 2 = 0$

$(4y - 1)(3y + 2) = 0$

$y = \dfrac{1}{4}, \; y = -\dfrac{2}{3}$

$x^{1/3} = \dfrac{1}{4} \qquad\qquad x^{1/3} = -\dfrac{2}{3}$

$x = \left(\dfrac{1}{4}\right)^3 \qquad\qquad x = \left(-\dfrac{2}{3}\right)^3$

$x = \dfrac{1}{64} \qquad\qquad\quad x = -\dfrac{8}{27}$

19. $2x^{1/2} - 5x^{1/4} - 3 = 0$

$y = x^{1/4}$

$2y^2 - 5y - 3 = 0$

$(2y + 1)(y - 3) = 0$

$y = -\dfrac{1}{2}, \; y = 3$

$x^{1/4} = -\dfrac{1}{2} \qquad\qquad\qquad x^{1/4} = 3$

$\qquad\qquad\qquad\qquad\qquad\qquad\quad x = 3^4 = 81$

$x = \left(-\dfrac{1}{2}\right)^4 = \dfrac{1}{16}$ (extraneous)

21. $4x^{1/2} + 11x^{1/4} - 3 = 0$

$y = x^{1/4}$

$4y^2 + 11y - 3 = 0$

$(4y - 1)(y + 3) = 0$

$y = \dfrac{1}{4}, \; y = -3$

$x^{1/4} = \dfrac{1}{4}$

$x = \left(\dfrac{1}{4}\right)^4 = \dfrac{1}{256}$

$x^{1/4} = -3$

$x = (-3)^4 = 81$ (extraneous)

23. $x^{2/5} - x^{1/5} - 2 = 0$

$y = x^{1/5}$

$y^2 - y - 2 = 0$

$(y - 2)(y + 1) = 0$

$y = 2, \; y = -1$

$x^{1/5} = 2 \qquad\qquad x^{1/5} = -1$

$x = 2^5 \qquad\qquad\; x = (-1)^5$

$x = 32 \qquad\qquad\; x = -1$

25.
$$x^6 - 5x^3 = 14$$
$$x^6 - 5x^3 - 14 = 0$$
$$y = x^3$$
$$y^2 - 5y - 14 = 0$$
$$(y - 7)(y + 2) = 0$$
$$y = 7, \; y = -2$$

$$x^3 = 7 \qquad\qquad x^3 = -2$$
$$x = \sqrt[3]{7} \qquad\qquad x = \sqrt[3]{-2}$$

27. $(x^2 + 3x)^2 - 2(x^2 + 3x) - 8 = 0$
$$y = x^2 + 3x$$
$$y^2 - 2y - 8 = 0$$
$$(y + 2)(y - 4) = 0$$
$$y = -2, \; y = 4$$
$$\qquad\quad y = -2$$
$$\qquad x^2 + 3x = -2$$
$$x^2 + 3x + 2 = 0$$
$$(x + 1)(x + 2) = 0$$
$$x + 1 = 0 \qquad\qquad x + 2 = 0$$
$$\quad x = -1 \qquad\qquad\quad x = -2$$
$$\qquad\quad y = 4$$
$$\qquad x^2 + 3x = 4$$
$$x^2 + 3x - 4 = 0$$
$$(x + 4)(x - 1) = 0$$
$$x + 4 = 0 \qquad\qquad x - 1 = 0$$
$$\quad x = -4 \qquad\qquad\quad x = 1$$

29. $x - 5x^{1/2} + 6 = 0$
$$y = x^{1/2}$$
$$y^2 - 5y + 6 = 0$$
$$(y - 3)(y - 2) = 0$$
$$y = 3, \; y = 2$$
$$x^{1/2} = 3 \qquad\qquad x^{1/2} = 2$$
$$x = 3^2 = 9 \qquad\qquad x = 2^2 = 4$$

31. $x^{-2} + 3x^{-1} = 0$
$$y = x^{-1}$$
$$y^2 + 3y = 0$$
$$y(y + 3) = 0$$
$$y = 0, \; y = -3$$
$$x^{-1} = -3 \qquad\qquad x^{-1} = 0, \text{ extraneous}$$
$$x = -\frac{1}{3}$$

33.
$$x^{-2} = 3x^{-1} + 10$$
$$x^{-2} - 3x^{-1} - 10 = 0$$
$$y = x^{-1}$$
$$y^2 - 3y - 10 = 0$$
$$(y - 5)(y + 2) = 0$$
$$y = 5, \; y = -2$$
$$x^{-1} = 5, \; x^{-1} = -2$$
$$x = \frac{1}{5}, \; x = -\frac{1}{2}$$

35.
$$15 - \frac{2x}{x - 1} = \frac{x^2}{x^2 - 2x + 1}$$
$$15 - \frac{2x}{x - 1} = \frac{x^2}{(x - 1)(x - 1)}$$
$$15(x - 1)^2 - 2x(x - 1) = x^2$$
$$15x^2 - 30x + 15 - 2x^2 + 2x = x^2$$
$$12x^2 - 28x + 15 = 0$$
$$(6x - 5)(2x - 3) = 0$$
$$x = \frac{5}{6}, \; x = \frac{3}{2}$$

Cumulative Review

37. $2x + 3y = 5$ **(1)**
    $-5x - 2y = 4$ **(2)**
Multiply equation **(1)** by 2 and equation **(2)** by 3.
$$4x + 6y = 10$$
$$\underline{-15x - 6y = 12}$$
$$-11x \qquad = 22$$
$$\qquad\quad x = -2$$

Replace x with -2 in equation **(1)**.
$$2(-2) + 3y = 5$$
$$-4 + 3y = 5$$
$$3y = 9$$
$$y = 3$$
The solution is $(-2, 3)$.

38. $\dfrac{5 + \frac{2}{x}}{\frac{7}{3x} - 1} = \dfrac{5 + \frac{2}{x}}{\frac{7}{3x} - 1} \cdot \dfrac{3x}{3x} = \dfrac{15x + 6}{7 - x}$

39. $3\sqrt{2}\left(\sqrt{5} - 2\sqrt{6}\right) = 3\sqrt{10} - 6\sqrt{12}$
$$= 3\sqrt{10} - 6\sqrt{4 \cdot 3}$$
$$= 3\sqrt{10} - 12\sqrt{3}$$

40. $\left(\sqrt{2}+\sqrt{6}\right)\left(3\sqrt{2}-2\sqrt{5}\right)$

$= 6-2\sqrt{10}+3\sqrt{12}-2\sqrt{30}$

$= 6-2\sqrt{10}+3\sqrt{4\cdot3}-2\sqrt{30}$

$= 6-2\sqrt{10}+6\sqrt{3}-2\sqrt{30}$

41. $\dfrac{508-386}{386}=\dfrac{122}{386}\approx0.316$

Less than H.S. diploma: 31.6%

$\dfrac{735-559}{559}=\dfrac{176}{559}\approx0.315$

H.S. graduate: 31.5%

$\dfrac{857-660}{660}=\dfrac{197}{660}\approx0.298$

Some college or associate's: 29.8%

$\dfrac{1371-1001}{1001}=\dfrac{370}{1001}\approx0.370$

Bachelor's or higher: 37.0%

42. Male earnings for 52 weeks = 52(857) = 44,564
Female earnings for 52 weeks = 52(660)
$\qquad\qquad\qquad\qquad\qquad = 34,320$

$44,564 - 34,320 = 10,244$

$\dfrac{10,244}{660}\approx15.5$

A female with some college or an associate's degree would have to work an extra 15.5 weeks to earn the same annual earnings as a male counterpart.

Quick Quiz 9.3

1. $x^4-18x^2+32=0$

$y=x^2$

$y^2-18y+32=0$

$(y-16)(y-2)=0$

$y=16, \ y=2$

$\begin{array}{ll} x^2=16 & x^2=2 \\ x=\pm4 & x=\pm\sqrt{2} \end{array}$

2. $2x^{-2}-3x^{-1}-20=0$

$y=x^{-1}$

$2y^2-3y-20=0$

$(2y+5)(y-4)=0$

$y=-\dfrac{5}{2}, \ y=4$

$\begin{array}{ll} x^{-1}=-\dfrac{5}{2} & x^{-1}=4 \\[2mm] x=-\dfrac{2}{5} & x=\dfrac{1}{4} \end{array}$

3. $x^{2/3}-4x^{1/3}-5=0$

$y=x^{1/3}$

$y^2-4y-5=0$

$(y-5)(y+1)=0$

$y=5, \ y=-1$

$\begin{array}{ll} x^{1/3}=5 & x^{1/3}=-1 \\ x=5^3 & x=(-1)^3 \\ x=125 & x=-1 \end{array}$

4. Answers may vary. Possible solution:
Substitute y for x^4, y^2 for x^8 which yields

$y^2-6y=0$. Next factor out y from the left side of the equation yielding $y(y-6)=0$.
Set each term equal to zero and solve for y yielding $y=0$ and $y=6$. Substitute x^4 for y and solve for x yielding $x=0$ and $x=\pm\sqrt[4]{6}$.

Use Math to Save Money

1. Gold Plan: no extra meals, $2205.
Silver Plan: four extra meals in each of the 15 weeks,
$1764 + 15 \times 4 \times \$10 = \$1764 + \$600 = \$2364$.
Bronze Plan: eight extra meals in each of the 15 weeks,
$\$1386+15\times8\times\$10 = \$1386 + \$1200 = \$2586$.
The Gold Plan is best for 20 dining-hall meals each week.

2. Gold Plan: no extra meals, $2205.
Silver Plan: two extra meals in each of the 15 weeks,
$\$1764 + 15 \times 2 \times \$10 = \$1764 + \$300 = \$2064$.
Bronze Plan: six extra meals in each of the 15 weeks,
$\$1386 + 15 \times 6 \times \$10 = \$1386 + \$900 = \$2286$.
The Silver Plan is best for 18 dining-hall meals each week.

3. Gold Plan: no extra meals, $2205.
Silver Plan: no extra meals, $1764.
Bronze Plan: four extra meals in each of the 15 weeks,
$\$1386 + 15 \times 4 \times \$10 = \$1386 + \$600 = \$1986$.
The Silver Plan is best for 16 dining-hall meals each week.

4. Gold Plan: no extra meals, $2205
    Silver Plan: no extra meals, $1764.
    Bronze Plan: two extra meals in each of the
    15 weeks,
    $1386 + 15 × 2 × $10 = $1386 + $300 = $1686.
    The Bronze Plan is best for 14 dining-hall meals
    each week.

5. If he spends $20 each week, he will spend
    15 × $20 = $300 over the course of the semester.
    The amount saved is 10% of $300:
    0.10 × $300 = $30
    He saves $30.

6. If he spends $40 each week, he will spend
    15 × $40 = $600 over the course of the semester.
    The amount saved is 10% of $600:
    0.10 × $600 = $60
    He saves $60.

7. If he spends $60 each week, he will spend
    15 × $60 = $900 over the course of the semester.
    The amount saved is 10% of $900:
    0.10 × $900 = $90
    He saves $90.

8. Gold Plan: no extra meals, $2205
    Silver Plan: one extra meal in each of the
    15 weeks,
    $1764 + 15 × 1 × $10 = $1764 + $150 = $1914.
    Bronze Plan: five extra meals in each of the
    15 weeks,
    $1386 + 15 × 5 × $10 = $1386 + $750 = $2136.
    Since he plans to spend $30 each week at other
    establishments regardless of which dining plan
    he chooses, he should choose the Silver Plan.

9. He will spend $1914 for the dining plan, and $30
    each week at other establishments less the 10%
    discount.
    $1914 + 15 × $30 − 0.10 × 15 × $30
    = $1914 + $450 − $45
    = $2319
    He will spend $2319 for food during the
    semester.

How Am I Doing? Sections 9.1–9.3
(Available online through MyMathLab.)

1. $2x^2 + 3 = 39$
    $$2x^2 = 36$$
    $$x^2 = 18$$
    $$x = \pm\sqrt{18}$$
    $$x = \pm\sqrt{9 \cdot 2}$$
    $$x = \pm 3\sqrt{2}$$

2. $(3x - 1)^2 = 28$
    $$3x - 1 = \pm\sqrt{28} = \pm 2\sqrt{7}$$
    $$3x = 1 \pm 2\sqrt{7}$$
    $$x = \frac{1 \pm 2\sqrt{7}}{3}$$

3. $$x^2 - 8x = -12$$
    $$x^2 - 8x + 16 = -12 + 16$$
    $$(x - 4)^2 = 4$$
    $$x - 4 = \pm 2$$
    $$x = 4 \pm 2$$
    $x = 2, \ x = 6$

4. $2x^2 - 4x - 3 = 0$
    $$2x^2 - 4x = 3$$
    $$x^2 - 2x = \frac{3}{2}$$
    $$x^2 - 2x + 1 = \frac{3}{2} + 1$$
    $$(x - 1)^2 = \frac{5}{2}$$
    $$x - 1 = \pm\sqrt{\frac{5}{2}}$$
    $$x - 1 = \pm\frac{\sqrt{10}}{2}$$
    $$x = 1 \pm \frac{\sqrt{10}}{2}$$
    $$x = \frac{2 \pm \sqrt{10}}{2}$$

5. $2x^2 - 8x + 3 = 0$

$a = 2, b = -8, c = 3$

$$x = \frac{-b \pm \sqrt{b^2 - 4ac}}{2a}$$

$$x = \frac{-(-8) \pm \sqrt{(-8)^2 - 4(2)(3)}}{2(2)}$$

$$x = \frac{8 \pm \sqrt{40}}{4}$$

$$x = \frac{8 \pm 2\sqrt{10}}{4}$$

$$x = \frac{2(4 \pm \sqrt{10})}{4}$$

$$x = \frac{4 \pm \sqrt{10}}{2}$$

6. $(x-1)(x+5) = 2$

$x^2 + 4x - 5 = 2$

$x^2 + 4x - 7 = 0$

$a = 1, b = 4, c = -7$

$$x = \frac{-b \pm \sqrt{b^2 - 4ac}}{2a}$$

$$x = \frac{-4 \pm \sqrt{(4)^2 - 4(1)(-7)}}{2(1)}$$

$$x = \frac{-4 \pm \sqrt{44}}{2}$$

$$x = \frac{-4 \pm 2\sqrt{11}}{2}$$

$$x = \frac{2(-2 \pm \sqrt{11})}{2}$$

$$x = -2 \pm \sqrt{11}$$

7. $4x^2 = -12x - 17$

$4x^2 + 12x + 17 = 0$

$a = 4, b = 12, c = 17$

$$x = \frac{-b \pm \sqrt{b^2 - 4ac}}{2a}$$

$$x = \frac{-12 \pm \sqrt{(12)^2 - 4(4)(17)}}{2(4)}$$

$$x = \frac{-12 \pm \sqrt{-128}}{8}$$

$$x = \frac{-12 \pm 8i\sqrt{2}}{8}$$

$$x = \frac{4(-3 \pm 2i\sqrt{2})}{8}$$

$$x = \frac{-3 \pm 2i\sqrt{2}}{2}$$

8. $5x^2 + 4x - 12 = 0$

$a = 5, b = 4, c = -12$

$$x = \frac{-b \pm \sqrt{b^2 - 4ac}}{2a}$$

$$x = \frac{-4 \pm \sqrt{(4)^2 - 4(5)(-12)}}{2(5)}$$

$$x = \frac{-4 \pm \sqrt{256}}{10}$$

$$x = \frac{-4 \pm 16}{10}$$

$$x = -2, \; x = \frac{6}{5}$$

9. $8x^2 + 3x = 3x^2 + 4x$

$5x^2 - x = 0$

$a = 5, b = -1, c = 0$

$$x = \frac{-b \pm \sqrt{b^2 - 4ac}}{2a}$$

$$x = \frac{-(-1) \pm \sqrt{(-1)^2 - 4(5)(0)}}{2(5)}$$

$$x = \frac{1 \pm 1}{10}$$

$$x = 0, \; x = \frac{1}{5}$$

10.
$$4x^2 - 3x = -6$$
$$4x^2 - 3x + 6 = 0$$
$$a = 4, b = -3, c = 6$$
$$x = \frac{-b \pm \sqrt{b^2 - 4ac}}{2a}$$
$$x = \frac{-(-3) \pm \sqrt{(-3)^2 - 4(4)(6)}}{2(4)}$$
$$x = \frac{3 \pm \sqrt{-87}}{8}$$
$$x = \frac{3 \pm i\sqrt{87}}{8}$$

11. The LCD is $x(x + 1)$.
$$\frac{18}{x} + \frac{12}{x+1} = 9$$
$$\frac{18}{x}(x)(x+1) + \frac{12}{x+1}(x)(x+1) = 9(x)(x+1)$$
$$18(x+1) + 12x = 9x(x+1)$$
$$18x + 18 + 12x = 9x^2 + 9x$$
$$9x^2 - 21x - 18 = 0$$
$$3x^2 - 7x - 6 = 0$$
$$a = 3, b = -7, c = -6$$
$$x = \frac{-b \pm \sqrt{b^2 - 4ac}}{2a}$$
$$x = \frac{-(-7) \pm \sqrt{(-7)^2 - 4(3)(-6)}}{2(3)}$$
$$x = \frac{7 \pm \sqrt{121}}{6}$$
$$x = \frac{7 \pm 11}{6}$$
$$x = 3, x = -\frac{2}{3}$$

12.
$$x^6 - 7x^3 - 8 = 0$$
$$y = x^3$$
$$y^2 - 7y - 8 = 0$$
$$(y-8)(y+1) = 0$$
$$y = 8, y = -1$$
$$x^3 = 8 \qquad\qquad x^3 = -1$$
$$x = 2 \qquad\qquad x = -1$$

13.
$$w^{4/3} - 6w^{2/3} + 8 = 0$$
$$y = w^{2/3}$$
$$y^2 - 6y + 8 = 0$$
$$(y-4)(y-2) = 0$$
$$y = 4, y = 2$$
$$w^{2/3} = 4 \qquad\qquad w^{2/3} = 2$$
$$w^2 = 64 \qquad\qquad w^2 = 8$$
$$w = \pm 8 \qquad\qquad w = \pm 2\sqrt{2}$$

14.
$$x^8 = 5x^4 - 6$$
$$x^8 - 5x^4 + 6 = 0$$
$$y = x^4$$
$$y^2 - 5y + 6 = 0$$
$$(y-2)(y-3) = 0$$
$$y = 2, y = 3$$
$$x^4 = 2 \qquad\qquad x^4 = 3$$
$$x = \pm\sqrt[4]{2} \qquad\qquad x = \pm\sqrt[4]{3}$$

15.
$$3x^{-2} = 11x^{-1} + 4$$
$$3x^{-2} - 11x^{-1} - 4 = 0$$
$$y = x^{-1}$$
$$3y^2 - 11y - 4 = 0$$
$$(3y+1)(y-4) = 0$$
$$y = -\frac{1}{3}, y = 4$$
$$x^{-1} = -\frac{1}{3}, x^{-1} = 4$$
$$x = -3, x = \frac{1}{4}$$

9.4 Exercises

1.
$$S = 16t^2$$
$$t^2 = \frac{S}{16}$$
$$t = \pm\sqrt{\frac{S}{16}} = \pm\frac{\sqrt{S}}{4}$$

3.
$$S = 9\pi r^2$$
$$r^2 = \frac{S}{9\pi}$$
$$r = \pm\sqrt{\frac{S}{9\pi}} = \pm\frac{1}{3}\sqrt{\frac{S}{\pi}}$$

5. $3N = \dfrac{2}{5}ax^2$

$x^2 = \dfrac{15N}{2a}$

$x = \pm\sqrt{\dfrac{15N}{2a}}$

7. $4(y^2 + w) - 5 = 7R$

$\qquad 4(y^2 + w) = 5 + 7R$

$\qquad 4y^2 + 4w = 5 + 7R$

$\qquad\qquad y^2 = \dfrac{7R - 4w + 5}{4}$

$\qquad\qquad y = \pm\sqrt{\dfrac{7R - 4w + 5}{4}}$

$\qquad\qquad y = \pm\dfrac{\sqrt{7R - 4w + 5}}{2}$

9. $Q = \dfrac{3mwM^2}{2c}$

$M^2 = \dfrac{2Qc}{3mw}$

$M = \pm\sqrt{\dfrac{2Qc}{3mw}}$

11. $V = \pi(r^2 + R^2)h$

$\qquad V = \pi r^2 h + \pi R^2 h$

$\qquad \pi r^2 h = V - \pi R^2 h$

$\qquad\qquad r^2 = \dfrac{V - \pi R^2 h}{\pi h}$

$\qquad\qquad r = \pm\sqrt{\dfrac{V - \pi R^2 h}{\pi h}}$

13. $7bx^2 - 3ax = 0$

$x(7bx - 3a) = 0$

$x = 0, \quad x = \dfrac{3a}{7b}$

15. $\qquad\qquad P = EI - RI^2$

$RI^2 - EI + P = 0$

$a = R, \, b = -E, \, c = P$

$I = \dfrac{-(-E) \pm \sqrt{(-E)^2 - 4RP}}{2R}$

$I = \dfrac{E \pm \sqrt{E^2 - 4RP}}{2R}$

17. $9w^2 + 5tw - 2 = 0$

$a = 9, \, b = 5t, \, c = -2$

$w = \dfrac{-(5t) \pm \sqrt{(5t)^2 - 4(9)(-2)}}{2(9)}$

$w = \dfrac{-5t \pm \sqrt{25t^2 + 72}}{18}$

19. $\qquad\qquad S = 2\pi rh + \pi r^2$

$\pi r^2 + 2\pi hr - S = 0$

$a = \pi, \, b = 2\pi h, \, c = -S$

$r = \dfrac{-2\pi h \pm \sqrt{(2\pi h)^2 - 4\pi(-S)}}{2\pi}$

$r = \dfrac{-2\pi h \pm \sqrt{4\pi^2 h^2 + 4\pi S}}{2\pi}$

$r = \dfrac{-2\pi h \pm \sqrt{4(\pi^2 h^2 + \pi S)}}{2\pi}$

$r = \dfrac{-2\pi h \pm 2\sqrt{\pi^2 h^2 + \pi S}}{2\pi}$

$r = \dfrac{-\pi h \pm \sqrt{\pi^2 h^2 + \pi S}}{\pi}$

21. $(a+1)x^2 + 5x + 2w = 0$

$a = a + 1, \, b = 5, \, c = 2w$

$x = \dfrac{-5 \pm \sqrt{5^2 - 4(a+1)(2w)}}{2(a+1)}$

$x = \dfrac{-5 \pm \sqrt{25 - 8aw - 8w}}{2a + 2}$

23. $c^2 = a^2 + b^2$

$6^2 = 4^2 + b^2$

$36 = 16 + b^2$

$b^2 = 20$

$b = \sqrt{20} = 2\sqrt{5}$

25. $\qquad c^2 = a^2 + b^2$

$\left(\sqrt{34}\right)^2 = a^2 + \left(\sqrt{19}\right)^2$

$\qquad 34 = a^2 + 19$

$\qquad a^2 = 15$

$\qquad\quad a = \sqrt{15}$

27.
$$c^2 = a^2 + b^2$$
$$12^2 = a^2 + (2a)^2$$
$$144 = 5a^2$$
$$a^2 = \frac{144}{5}$$
$$a = \sqrt{\frac{144}{5}}$$
$$a = \frac{12}{\sqrt{5}}$$
$$a = \frac{12\sqrt{5}}{5}$$
$$b = 2a = \frac{24\sqrt{5}}{5}$$

29.
$$c^2 = a^2 + b^2$$
$$10^2 = x^2 + x^2$$
$$100 = 2x^2$$
$$x^2 = 50$$
$$x = \pm\sqrt{50} = \pm 5\sqrt{2}$$
Since the length must be positive, $x = 5\sqrt{2}$.
Each leg is $5\sqrt{2}$ inches long.

31. Let x = the width of the parking lot. Then the length is $x + 0.07$. The area of a rectangle is $A = LW$.
$$x(x + 0.07) = 0.026$$
$$x^2 + 0.07x - 0.026 = 0$$
$$x = \frac{-0.07 \pm \sqrt{(0.07)^2 - 4(1)(-0.026)}}{2(1)}$$
$$x = \frac{-0.07 \pm \sqrt{0.1089}}{2} = \frac{-0.07 \pm 0.33}{2}$$
$$x = 0.13, \ x = -0.2$$
Since the width must be positive, $x = 0.13$. The width is 0.13 mile and the length is $0.13 + 0.07 = 0.2$ mile.

33. If W is the width of the barn, then $2W + 4$ is the length.
$$(2W + 4)W = 126$$
$$2W^2 + 4W - 126 = 0$$
$$W^2 + 2W - 63 = 0$$
$$(W - 7)(W + 9) = 0$$
$$W = 7, \ W = -9$$
Since the width must be positive, $W = 7$.
The width of the barn is 7 feet and the length is $2(7) + 4 = 18$ feet.

35. If b = base of triangle, then $2b + 2$ = the altitude.
$$A = \frac{1}{2}bh$$
$$\frac{1}{2}b(2b + 2) = 72$$
$$2b^2 + 2b = 144$$
$$b^2 + b - 72 = 0$$
$$(b - 8)(b + 9) = 0$$
$$b = 8, \ b = -9$$
Since the base must have a positive length, $b = 8$.
The base is 8 cm and the altitude is $2(8) + 2 = 18$ cm.

37. If v = his speed in the rain then $v + 5$ is his speed after the rain stopped.
$$\text{Time} = \frac{\text{distance}}{\text{rate}}$$
$$\frac{225}{v} + \frac{150}{v+5} = 8$$
$$225(v+5) + 150v = 8v(v+5)$$
$$225v + 1125 + 150v = 8v^2 + 40v$$
$$8v^2 - 335v - 1125 = 0$$
$$(v - 45)(8v + 25) = 0$$
$$v = 45, \ v = -3,125$$
Since his speed is positive, $v = 45$. His speed in the rain was 45 mph and his speed after the rain stopped was 50 mph.

39. t = time from home to work
$$1 \text{ hr } 16 \text{ min} = 1 + \frac{16}{60} = \frac{19}{15} \text{ hr}$$
$$\frac{19}{15} - t = \text{time from work to home}$$
The distance from home to work and the distance from work to home are the same.
Distance = (Rate)(Time)
$$50t = 45\left(\frac{19}{15} - t\right)$$
$$50t = 57 - 45t$$
$$95t = 57$$
$$t = \frac{57}{95} = \frac{3}{5}$$
$$\frac{50 \text{ mi}}{\text{hr}} \cdot \frac{3}{5} \text{ hr} = 30 \text{ mi}$$
Bob lives 30 miles from his job.

41. $N = -3.9x^2 + 70.9x + 1944$
In 2010, $x = 10$.
$N = -3.9(10)^2 + 70.9(10) + 1944 = 2263$
There were 2,263,000 incarcerated adults in 2010.

43. Let $N = 2130$.
$$2130 = -3.9x^2 + 70.9x + 1944$$
$3.9x^2 - 70.9x + 186 = 0$
$a = 3.9,\ b = -70.9,\ c = 186$
$$x = \frac{-(-70.9) \pm \sqrt{(-70.9)^2 - 4(3.9)(186)}}{2(3.9)}$$
$$= \frac{70.9 \pm 46.1}{7.8}$$
$x \approx 3.18,\ x = 15$
The number of incarcerated adults was 2,130,000 in 2003 and again in 2015.

Cumulative Review

45. $\dfrac{4}{\sqrt{3x}} = \dfrac{4}{\sqrt{3x}} \cdot \dfrac{\sqrt{3x}}{\sqrt{3x}} = \dfrac{4\sqrt{3x}}{3x}$

46. $\dfrac{5\sqrt{6}}{2\sqrt{5}} = \dfrac{5\sqrt{6}}{2\sqrt{5}} \cdot \dfrac{\sqrt{5}}{\sqrt{5}} = \dfrac{5\sqrt{30}}{10} = \dfrac{\sqrt{30}}{2}$

47. $\dfrac{3}{\sqrt{x} + \sqrt{y}} = \dfrac{3}{\sqrt{x} + \sqrt{y}} \cdot \dfrac{\sqrt{x} - \sqrt{y}}{\sqrt{x} - \sqrt{y}} = \dfrac{3\left(\sqrt{x} - \sqrt{y}\right)}{x - y}$

48. $\dfrac{2\sqrt{3}}{\sqrt{3} - \sqrt{6}} = \dfrac{2\sqrt{3}}{\sqrt{3} - \sqrt{6}} \cdot \dfrac{\sqrt{3} + \sqrt{6}}{\sqrt{3} + \sqrt{6}}$
$$= \dfrac{6 + 2\sqrt{18}}{3 - 6}$$
$$= \dfrac{6 + 6\sqrt{2}}{3 - 6}$$
$$= \dfrac{-3\left(-2 - 2\sqrt{2}\right)}{-3}$$
$$= -2 - 2\sqrt{2}$$

Quick Quiz 9.4

1. $H = \dfrac{5ab}{y^2}$
$$y^2 = \dfrac{5ab}{H}$$
$$y = \pm\sqrt{\dfrac{5ab}{H}}$$

2. $6z^2 + 7yz - 5w = 0$
$a = 6,\ b = 7y,\ c = -5w$
$$z = \dfrac{-7y \pm \sqrt{(7y)^2 - 4(6)(-5w)}}{2(6)}$$
$$z = \dfrac{-7y \pm \sqrt{49y^2 + 120w}}{12}$$

3. If $x =$ the width, then $3x + 4 =$ the length.
$$A = \text{length} \cdot \text{width}$$
$$175 = (3x + 4)x$$
$3x^2 + 4x - 175 = 0$
$(3x + 25)(x - 7) = 0$
$$x = -\dfrac{25}{3},\ x = 7$$
Since the width must be positive, $x = 7$. The width is 7 yards and the length is $3(7) + 4 = 25$ yards.

4. Answers may vary. Possible solution:
Set one leg's length $= x$, then the other leg's length $= 3x$.
Use $c^2 = a^2 + b^2$ with $c = 12$, $a = x$, and $b = 3x$.
Solve for x to find one leg length then multiply the found value of x by 3 to find the other leg's length.

9.5 Exercises

1. $f(x) = x^2 - 8x + 15$
$$\dfrac{-b}{2a} = \dfrac{-(-8)}{2(1)} = 4$$
$$f(4) = 4^2 - 8(4) + 15 = -1$$
$V(4, -1)$
$$f(0) = 0^2 - 8(0) + 15 = 15$$
y-intercept: (0, 15)
$$x^2 - 8x + 15 = 0$$
$(x - 3)(x - 5) = 0$
$x = 3,\ x = 5$
x-intercepts: (3, 0), (5, 0)

3. $g(x) = -x^2 - 8x + 9$
$$\dfrac{-b}{2a} = \dfrac{-(-8)}{2(-1)} = -4$$
$$g(-4) = -(-4)^2 - 8(-4) + 9 = 25$$
$V(-4, 25)$
$$g(0) = 0^2 - 8(0) + 9 = 9$$
y-intercept: (0, 9)

$$-x^2 - 8x + 9 = 0$$
$$x^2 + 8x - 9 = 0$$
$$(x+9)(x-1) = 0$$
$$x = -9, x = 1$$
x-intercepts: (−9, 0), (1, 0)

5. $p(x) = 3x^2 + 12x + 3$

$$\frac{-b}{2a} = \frac{-(12)}{2(3)} = -2$$

$$p(-2) = 3(-2)^2 + 12(-2) + 3 = -9$$

$V(-2, -9)$

$$p(0) = 3(0)^2 + 12(0) + 3 = 3$$

y-intercept: (0, 3)

$$3x^2 + 12x + 3 = 0$$
$$x^2 + 4x + 1 = 0$$
$$a = 1, b = 4, c = 1$$

$$x = \frac{-4 \pm \sqrt{4^2 - 4(1)(1)}}{2(1)}$$

$$x \approx -0.3, \ x \approx -3.7$$

approximate x-intercepts: (−0.3, 0), (−3.7, 0)

7. $r(x) = -3x^2 - 2x - 6$

$$\frac{-b}{2a} = \frac{-(-2)}{2(-3)} = -\frac{1}{3}$$

$$r\left(-\frac{1}{3}\right) = -3\left(-\frac{1}{3}\right)^2 - 2\left(-\frac{1}{3}\right) - 6 = -\frac{17}{3}$$

$$V\left(-\frac{1}{3}, -\frac{17}{3}\right)$$

$$r(0) = -3(0)^2 - 2(0) - 6 = -6$$

y-intercept: (0, −6)

$$-3x^2 - 2x - 6 = 0$$
$$a = -3, b = -2, c = -6$$
$$b^2 - 4ac = -68 < 0 \Rightarrow \text{no } x\text{-intercepts}$$

9. $f(x) = 2x^2 + 2x - 4$

$$\frac{-b}{2a} = \frac{-(2)}{2(2)} = -\frac{1}{2}$$

$$f\left(-\frac{1}{2}\right) = 2\left(-\frac{1}{2}\right)^2 + 2\left(-\frac{1}{2}\right) - 4 = -\frac{9}{2}$$

$$V\left(-\frac{1}{2}, -\frac{9}{2}\right)$$

$$f(0) = 2(0)^2 + 2(0) - 4 = -4$$

y-intercept: (0, −4)

$$2x^2 + 2x - 4 = 0$$
$$x^2 + x - 2 = 0$$
$$(x-1)(x+2) = 0$$
$$x = 1, x = -2$$
x-intercepts: (1, 0), (−2, 0)

11. $f(x) = x^2 - 6x + 8$

$$\frac{-b}{2a} = \frac{-(-6)}{2(1)} = 3$$

$$f(3) = 3^2 - 6(3) + 8 = -1$$

$V(3, -1)$

$$f(0) = 0^2 - 6(0) + 8 = 8$$

y-intercept: (0, 8)

$$x^2 - 6x + 8 = 0$$
$$(x-4)(x-2) = 0$$
$$x = 4, x = 2$$
x-intercepts: (4, 0), (2, 0)

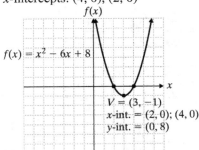

$f(x) = x^2 - 6x + 8$

$V = (3, -1)$
x-int. = (2, 0); (4, 0)
y-int. = (0, 8)

13. $g(x) = x^2 + 2x - 8$

$$\frac{-b}{2a} = \frac{-2}{2} = -1$$

$$g(-1) = (-1)^2 + 2(-1) - 8 = -9$$

$V(-1, -9)$

$$g(0) = 0^2 + 2(0) - 8 = -8$$

y-intercept: (0, −8)

$$x^2 + 2x - 8 = 0$$
$$(x+4)(x-2) = 0$$
$$x = -4, x = 2$$
x-intercepts: (−4, 0), (2, 0)

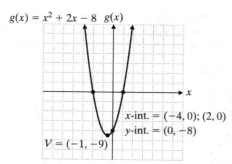

$g(x) = x^2 + 2x - 8$ $g(x)$

x-int. $= (-4, 0); (2, 0)$
y-int. $= (0, -8)$
$V = (-1, -9)$

Scale: Each unit $= 2$

15. $p(x) = -x^2 + 8x - 12$

$$\frac{-b}{2a} = \frac{-8}{2(-1)} = 4$$

$p(4) = -4^2 + 8(4) - 12 = 4$

$V = (4, 4)$

$p(0) = 0^2 + 8(0) - 12 = -12$

y-intercept: $(0, -12)$

$-x^2 + 8x - 12 = 0$

$x^2 - 8x + 12 = 0$

$(x-6)(x-2) = 0$

$x = 6, \ x = 2$

x-intercepts: $(6, 0), (2, 0)$

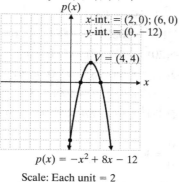

$p(x)$

x-int. $= (2, 0); (6, 0)$
y-int. $= (0, -12)$

$V = (4, 4)$

$p(x) = -x^2 + 8x - 12$

Scale: Each unit $= 2$

17. $r(x) = 3x^2 + 6x + 4$

$$\frac{-b}{2a} = \frac{-6}{2(3)} = -1$$

$r(-1) = 3(-1)^2 + 6(-1) + 4 = 1$

$V(-1, 1)$

$r(0) = 3 \cdot 0^2 + 6(0) + 4 = 4$

y-intercept: $(0, 4)$

$3x^2 + 6x + 4 = 0$

$a = 3, \ b = 6, \ c = 4$

$b^2 - 4ac = 6^2 - 4(3)(4) = -12 < 0$

x-intercepts: none

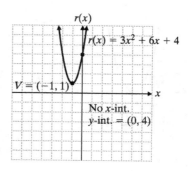

$r(x)$

$r(x) = 3x^2 + 6x + 4$

$V = (-1, 1)$

No x-int.
y-int. $= (0, 4)$

19. $f(x) = x^2 - 6x + 5$

$$\frac{-b}{2a} = \frac{-(-6)}{2(1)} = 3$$

$f(3) = 3^2 - 6(3) + 5 = -4$

$V(3, -4)$

$f(0) = 0^2 - 6(0) + 5 = 5$

y-intercept: $(0, 5)$

$x^2 - 6x + 5 = 0$

$(x-5)(x-1) = 0$

$x = 5, \ x = 1$

x-intercepts: $(5, 0), (1, 0)$

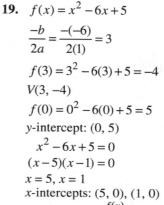

$f(x)$

$f(x) = x^2 - 6x + 5$

$V = (3, -4)$
x-int. $= (1, 0); (5, 0)$
y-int. $= (0, 5)$

21. $f(x) = x^2 - 4x + 4$

$$\frac{-b}{2a} = \frac{-(-4)}{2(1)} = 2$$

$f(2) = 2^2 - 4(2) + 4 = 0$

$V(2, 0)$

$f(0) = 0^2 - 4(0) + 4 = 4$

y-intercept: $(0, 4)$

$x^2 - 4x + 4 = 0$

$(x-2)(x-2) = 0$

$x = 2, \ x = 2$

x-intercept: $(2, 0)$

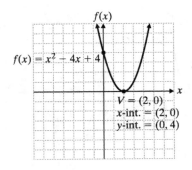

27. From the graph, $N(60) \approx 8000$. There are approximately 8,000,000 people who participate in boating and have a mean income of $60,000.

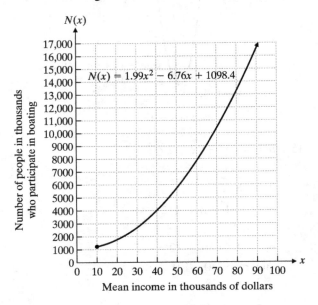

23. $f(x) = x^2 - 4$

$$\frac{-b}{2a} = \frac{-0}{2(1)} = 0$$

$$f(0) = 0^2 - 4 = -4$$

$$V(0, -4)$$

$$f(0) = 0^2 - 4 = -4$$

y-intercept: $(0, -4)$

$$x^2 - 4 = 0$$

$$x^2 = 4$$

$$x = \pm 2$$

x-intercepts: $(\pm 2, 0)$

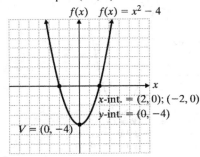

29. From the graph, $N(x) = 10,000$ when $x \approx 68$. This means 10,000,000 people who participate in boating have a mean income of $68,000.

31. $P(x) = -5x^2 + 300x - 4000$

$$P(18) = -5(18)^2 + 300(18) - 4000 = -220$$

$$P(20) = -5(20)^2 + 300(20) - 4000 = 0$$

$$P(28) = -5(28)^2 + 300(28) - 4000 = 480$$

$$P(35) = -5(35)^2 + 300(35) - 4000 = 375$$

$$P(42) = -5(42)^2 + 300(42) - 4000 = -220$$

25. $N(x) = 1.99x^2 - 6.76x + 1098.4$

$$N(10) = 1.99(10)^2 - 6.76(10) + 1098.4 = 1229.8$$

$$N(30) = 1.99(30)^2 - 6.76(30) + 1098.4 = 2686.6$$

$$N(50) = 1.99(50)^2 - 6.76(50) + 1098.4 = 5735.4$$

$$N(70) = 1.99(70)^2 - 6.76(70) + 1098.4$$
$$= 10,376.2$$

$$N(90) = 1.99(90)^2 - 6.76(90) + 1098.4 = 16,609$$

33. $\dfrac{-b}{2a} = \dfrac{-300}{2(-5)} = 30$

$$P(30) = -5(30)^2 + 300(30) - 4000 = 500$$

Making 30 tables per day will give the maximum profit of $500 per day.

35. $P(x) = 0$

$$-5x^2 + 300x - 4000 = 0$$

$$x^2 - 60x + 800 = 0$$

$$(x - 20)(x - 40) = 0$$

$x = 20, \ x = 40$

Making 20 or 40 tables per day gives a daily profit of zero dollars.

37. $d(t) = -16t^2 + 32t + 40$

$\dfrac{-b}{2a} = \dfrac{-32}{2(-16)} = 1$

$d(1) = -16(1^2) + 32(1) + 40 = 56$

$-16^2 + 32t + 40 = 0$

$t = \dfrac{-32 \pm \sqrt{32^2 - 4(-16)(40)}}{2(-16)}$

$t \approx 2.9,\ t \approx -0.9$

The maximum height is 56 feet and the ball will reach the ground about 2.9 seconds after being thrown upward.

39. $y = 2.3x^2 - 5.4x - 1.6$

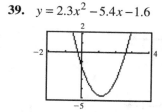

x-intercepts: $(-0.3, 0),\ (2.6, 0)$

41. From $(0, -10)$:

$-10 = a \cdot 0^2 + b \cdot 0 + c$

$-10 = c$

From $(3, 41)$:

$41 = a \cdot 3^2 + b \cdot 3 - 10$

$41 = 9a + 3b - 10$

$51 = 9a + 3b \qquad \textbf{(1)}$

From $(-1, -15)$:

$-15 = a(-1)^2 + b(-1) - 10$

$-15 = a - b - 10$

$-5 = a - b \qquad \textbf{(2)}$

Solve equation **(2)** for a.

$a = b - 5$

Substitute $b - 5$ for a in equation **(1)**.

$51 = 9(b - 5) + 3b$

$51 = 9b - 45 + 3b$

$96 = 12b$

$8 = b$

$a = b - 5 = 8 - 5 = 3$

$a = 3,\ b = 8,\ c = -10$

Cumulative Review

43. $3x - y + 2z = 12 \qquad \textbf{(1)}$

$2x - 3y + z = 5 \qquad \textbf{(2)}$

$x + 3y + 8z = 22 \qquad \textbf{(3)}$

Eliminate y by multiplying equation **(1)** by 3 and adding the result to equation **(3)**.

$9x - 3y + 6z = 36$

$\dfrac{x + 3y + 8z = 22}{10x \qquad + 14z = 58}$

$5x + 7z = 29 \qquad \textbf{(4)}$

Eliminate y by adding equation **(2)** and **(3)**.

$2x - 3y + z = 5$

$\dfrac{x + 3y + 8z = 22}{3x \qquad + 9z = 27}$

$x + 3z = 9 \qquad \textbf{(5)}$

Multiply equation **(5)** by -5 and add the result to equation **(4)**.

$5x + 7z = 29$

$\dfrac{-5x - 15z = -45}{-8z = -16}$

$z = 2$

Replace z with 2 in equation **(5)**.

$x + 3(2) = 9$

$x + 6 = 9$

$x = 3$

Replace x with 3 and z with 2 in equation **(1)**.

$3(3) - y + 2(2) = 12$

$9 - y + 4 = 12$

$-y = -1$

$y = 1$

The solution is $(3, 1, 2)$.

44. $7x + 3y - z = -2 \qquad \textbf{(1)}$

$x + 5y + 3z = 2 \qquad \textbf{(2)}$

$x + 2y + z = 1 \qquad \textbf{(3)}$

Eliminate z by multiplying equation **(1)** by 3 and adding the result to equation **(2)**.

$21x + 9y - 3z = -6$

$\dfrac{x + 5y + 3z = 2}{22x + 14y = -4}$

$11x + 7y = -2 \qquad \textbf{(4)}$

Eliminate z by adding equations **(1)** and **(3)**.

$7x + 3y - z = -2$

$\dfrac{x + 2y + z = 1}{8x + 5y = -1} \qquad \textbf{(5)}$

Multiply equation **(4)** by 5, equation **(5)** by -7, and add the results.

$55x + 35y = -10$

$\dfrac{-56x - 35y = 7}{-x = -3}$

$x = 3$

Replace x with 3 in equation **(4)**.

$$11(3) + 7y = -2$$
$$33 + 7y = -2$$
$$7y = -35$$
$$y = -5$$

Replace x with 3 and y with -5 in equation (**3**).
$$3 + 2(-5) + z = 1$$
$$3 - 10 + z = 1$$
$$z = 8$$

The solution is $(3, -5, 8)$.

Quick Quiz 9.5

1. $f(x) = -2x^2 - 4x + 6$

   $\dfrac{-b}{2a} = \dfrac{-(-4)}{2(-2)} = -1$

   $f(-1) = -2(-1)^2 - 4(-1) + 6 = -2 + 4 + 6 = 8$

   $V(-1, 8)$

2. $f(0) = -2(0)^2 - 4(0) + 6 = 6$

   y-intercept $(0, 6)$

   $-2x^2 - 4x + 6 = 0$

   $x^2 + 2x - 3 = 0$

   $(x + 3)(x - 1) = 0$

   $x = -3, x = 1$

   x-intercepts $(-3, 0)$, $(1, 0)$

3.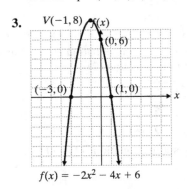

   $f(x) = -2x^2 - 4x + 6$

4. Answers may vary. Possible solution:
   The function is in standard form,
   $f(x) = ax^2 + bx + c$ with $a = 4$, $b = -9$, and
   $c = -5$. The x-coordinate of the vertex is
   $\dfrac{-b}{2a} = \dfrac{-(-9)}{2(4)} = \dfrac{9}{8}$. The y-coordinate of the vertex
   is $f\left(\dfrac{9}{8}\right)$.

9.6 Exercises

1. $3 < x$ *and* $x < 8$

2. $-4 < x$ *and* $x < 2$

5. $7 < x < 9$

7. $-2 < x \le \dfrac{1}{2}$

9. $x > 8$ *or* $x < 2$

11. $x \le -\dfrac{5}{2}$ *or* $x > 4$

13. $x \le -10$ *or* $x \ge 40$

15. $2x + 3 \le 5$ *and* $x + 1 \ge -2$
    $\quad\quad 2x \le 2 \quad\quad\quad\quad\quad x \ge -3$
    $\quad\quad\quad x \le 1$
    $-3 \le x \le 1$

17. $4x - 9 > 1$ *or* $x + 4 < -2$
    $\quad\quad 4x > 10 \quad\quad\quad\quad x < -6$
    $\quad\quad\quad x > \dfrac{10}{4}$
    $\quad\quad\quad x > \dfrac{5}{2}$

19. $x < 8$ *and* $x > 10$ do not overlap.
    No solution

21. $t < 10.9$ *or* $t > 11.2$

23. $5000 \le c \le 12{,}000$

25. $-20 \leq C \leq 11$

$-20 \leq \dfrac{5}{9}(F-32) \leq 11$

$9(-20) \leq 5(F-32) \leq 9(11)$

$-180 \leq 5F - 160 \leq 99$

$-20 \leq 5F \leq 259$

$-4° \leq F \leq 51.8°$

27. Frank will need between 60,000 yen and 70,000 yen for 2 weeks.

$60,000 \leq Y \leq 70,000$

$60,000 \leq 119(d-5) \leq 70,000$

$504.20 \leq d - 5 \leq 588.24$

$\$509.20 \leq d \leq \$593.24$

29. $x - 3 > -5$ *and* $2x + 4 < 8$

$\quad x > -2 \qquad\qquad 2x < 4$

$\quad -2 < x \qquad\qquad x < 2$

$-2 < x < 2$ is the solution.

31. $-6x + 5 \geq -1$ *and* $2 - x \leq 5$

$\quad -6x \geq -6 \qquad\qquad -x \leq 3$

$\quad\quad x \leq 1 \qquad\qquad\quad x \geq -3$

$-3 \leq x \leq 1$ is the solution.

33. $4x - 3 < -11$ *or* $7x + 2 \geq 23$

$\quad 4x < -8 \qquad\qquad 7x \geq 21$

$\quad\; x < -2 \qquad\qquad\; x \geq 3$

$x < -2$ *or* $x \geq 3$ is the solution.

35. $-0.3x + 1 \geq 0.2x$ *or* $-0.2x + 0.5 > 0.7$

Multiply by 10 on both sides of both inequalities to clear decimals.

$-3x + 10 \geq 2x$ *or* $-2x + 5 > 7$

$\quad -5x \geq -10 \qquad\qquad -2x > 2$

$\quad\quad x \leq 2 \qquad\qquad\quad x < -1$

$x \leq 2$ contains $x < -1$. $x \leq 2$ is the solution.

37. $\dfrac{5x}{2} + 1 \geq 3$ *and* $x - \dfrac{2}{3} \geq \dfrac{4}{3}$

$\quad 5x + 2 \geq 6 \qquad\qquad 3x - 2 \geq 4$

$\quad\;\; 5x \geq 4 \qquad\qquad\;\; 3x \geq 6$

$\quad\;\;\; x \geq \dfrac{4}{5} \qquad\qquad\quad x \geq 2$

$x \geq 2$ is the solution.

39. $2x + 5 < 3$ *and* $3x - 1 > -1$

$\quad 2x < -2 \qquad\qquad 3x > 0$

$\quad\; x < -1 \qquad\qquad\; x > 0$

$x < -1$ *and* $x > 0$ do not overlap.
No solution

41. $2x - 3 \geq 7$ *and* $5x - 8 \leq 2x + 7$

$\quad 2x \geq 10 \qquad\qquad 3x \leq 15$

$\quad\; x \geq 5 \qquad\qquad\;\; x \leq 5$

$x \geq 5$ *and* $x \leq 5$ overlap at $x = 5$.
$x = 5$ is the solution.

43. $\dfrac{1}{4}(x+2) + \dfrac{1}{8}(x-3) \leq 1$ *and* $\dfrac{3}{4}(x-1) > -\dfrac{1}{4}$

$\quad 2x + 4 + x - 3 \leq 8 \qquad\qquad 3x - 3 > -1$

$\quad\quad\;\; 3x + 1 \leq 8 \qquad\qquad\qquad 3x > 2$

$\quad\quad\quad\;\; 3x \leq 7 \qquad\qquad\qquad\;\; x > \dfrac{2}{3}$

$\quad\quad\quad\quad x \leq \dfrac{7}{3}$

$\dfrac{2}{3} < x \leq \dfrac{7}{3}$ is the solution.

45. The boundary points divide the number line into regions. All values of x in a given region produce results that are greater than zero, or else all the values of x in a given region produce results that are less than zero.

47. $x^2 + x - 12 < 0$

$x^2 + x - 12 = 0$

$(x + 4)(x - 3) = 0$

$x = -4$, $x = 3$

Boundary points: -4, 3

Test: $x^2 + x - 12$

Region	Test	Result
$x < -4$	-5	$8 > 0$
$-4 < x < 3$	0	$-12 < 0$
$x > 3$	4	$8 > 0$

$(x + 4)(x - 3) = 0$
$-4 < x < 3$

49. $x^2 \leq 25$

$x^2 - 25 \leq 0$

$x^2 - 25 = 0$

$(x + 5)(x - 5) = 0$

$x = -5$, $x = 5$

Boundary points: -5, 5

Test: $x^2 - 25$

Region	Test	Result
$x < -5$	-6	$11 > 0$
$-5 < x < 5$	0	$-25 < 0$
$x > 5$	6	$11 > 0$

$$(x+5)(x-5) = 0$$
$$-5 \le x \le 5$$

51. $2x^2 + x - 3 < 0$

$$2x^2 + x - 3 = 0$$
$$(2x+3)(x-1) = 0$$
$$x = -\frac{3}{2},\ x = 1$$

Boundary points: $-\frac{3}{2},\ 1$

Test: $2x^2 + x - 3$

Region	Test	Result
$x < -\frac{3}{2}$	-2	$3 > 0$
$-\frac{3}{2} < x < 1$	0	$-3 < 0$
$x > 1$	2	$7 > 0$

$$(2x+3)(x-1) = 0$$
$$-\frac{3}{2} < x < 1$$

53. $x^2 + x - 20 > 0$

$$x^2 + x - 20 = 0$$
$$(x+5)(x-4) = 0$$
$$x = -5,\ x = 4$$

Boundary points: $-5,\ 4$

Test: $x^2 + x - 20$

Region	Test	Result
$x < -5$	-6	$10 > 0$
$-5 < x < 4$	0	$-20 < 0$
$x > 4$	5	$10 > 0$

$x < -5$ or $x > 4$

55. $4x^2 \le 11x + 3$

$$4x^2 - 11x - 3 \le 0$$
$$4x^2 - 11x - 3 = 0$$
$$(4x+1)(x-3) = 0$$
$$x = -\frac{1}{4},\ x = 3$$

Boundary points: $-\frac{1}{4},\ 3$

Test: $4x^2 - 11x - 3$

Region	Test	Result
$x < -\frac{1}{4}$	-1	$12 > 0$
$-\frac{1}{4} < x < 3$	0	$-3 < 0$
$x > 3$	4	$17 > 0$

$-\frac{1}{4} \le x \le 3$

57. $6x^2 - 5x > 6$

$$6x^2 - 5x - 6 > 0$$
$$6x^2 - 5x - 6 = 0$$
$$(3x+2)(2x-3) = 0$$
$$x = -\frac{2}{3},\ x = \frac{3}{2}$$

Boundary points: $-\frac{2}{3},\ \frac{3}{2}$

Test: $6x^2 - 5x - 6$

Region	Test	Result
$x < -\frac{2}{3}$	-1	$5 > 0$
$-\frac{2}{3} < x < \frac{3}{2}$	0	$-6 < 0$
$x > \frac{3}{2}$	2	$8 > 0$

$x < -\dfrac{2}{3}$ or $x > \dfrac{3}{2}$

59.
$$-2x+30 \geq x(x+5)$$
$$-2x+30 \geq x^2+5x$$
$$0 \geq x^2+7x-30$$
$$x^2+7x-30 \leq 0$$
$$x^2+7x-30 = 0$$
$$(x+10)(x-3) = 0$$
$$x = -10, \ x = 3$$
Boundary points: $-10, 3$

Test: $x^2+7x-30$

Region	Test	Result
$x < -10$	-12	$30 > 0$
$-10 < x < 3$	0	$-30 < 0$
$x > 3$	5	$30 > 0$

$$-10 \leq x \leq 3$$

61.
$$x^2-2x \geq -1$$
$$x^2-2x+1 \geq 0$$
$$(x-1)^2 \geq 0$$
Since the square of any real number must be 0 or positive, all real numbers satisfy this inequality.

63.
$$x^2-4x \leq -4$$
$$x^2-4x+4 \leq 0$$
$$x^2-4x+4 = 0$$
$$(x-2)(x-2) = 0$$
$$x = 2$$
Boundary point: 2

Test: x^2-4x+4

Region	Test	Result
$x < 2$	0	$4 > 0$
$x > 2$	3	$1 > 0$

$$x = 2$$

65.
$$x^2-2x > 5$$
$$x^2-2x-5 > 0$$
$$x^2-2x-5 = 0$$
$$a = 1, \ b = -2, \ c = -5$$

$$x = \frac{-(-2) \pm \sqrt{(-2)^2 - 4(1)(-5)}}{2(1)}$$
$$= \frac{2 \pm \sqrt{24}}{2}$$
$$= 1 \pm \sqrt{6}$$
$$x \approx -1.4, \ x \approx 3.4$$
Approximate boundary points: $-1.4, 3.4$

Test: x^2-2x-5

Region	Test	Result
$x < 1-\sqrt{6}$	-2	$3 > 0$
$1-\sqrt{6} < x < 1+\sqrt{6}$	0	$-5 < 0$
$x > 1+\sqrt{6}$	4	$3 > 0$

$$x < 1-\sqrt{6} \ \text{ or } \ x > 1+\sqrt{6}$$
Approximately $x < -1.4$ or $x > 3.4$

67.
$$2x^2-2x < 3$$
$$2x^2-2x-3 < 0$$
$$2x^2-2x-3 = 0$$
$$a = 2, \ b = -2, \ c = -3$$
$$x = \frac{-(-2) \pm \sqrt{(-2)^2 - 4(2)(-3)}}{2(2)}$$
$$= \frac{2 \pm \sqrt{28}}{4}$$
$$= \frac{1 \pm \sqrt{7}}{2}$$
$$x \approx -0.8, \ x \approx 1.8$$
Approximate boundary points: $-0.8, 1.8$

Test: $2x^2-2x-3$

Region	Test	Result
$x < \frac{1-\sqrt{7}}{2}$	-1	$1 > 0$
$\frac{1-\sqrt{7}}{2} < x < \frac{1+\sqrt{7}}{2}$	0	$-3 < 0$
$x > \frac{1+\sqrt{7}}{2}$	2	$1 > 0$

$$\frac{1-\sqrt{7}}{2} < x < \frac{1+\sqrt{7}}{2}$$
Approximately $-0.8 < x < 1.8$

69. $2x^2 \geq x^2 - 4$

$x^2 \geq -4$

Since the square of any real number must be 0 or positive, all real numbers satisfy this inequality.

71. $5x^2 \leq 4x^2 - 1$

$x^2 \leq -1$

Since the square of any real number must be 0 or positive, no real number satisfies this inequality.

73. $s = -16t^2 + 640t$

$-16t^2 + 640t > 6000$

$0 > 16t^2 - 640t + 6000$

$0 > t^2 - 40t + 375$

$t^2 - 40t + 375 < 0$

$t^2 - 40t + 375 = 0$

$(t - 15)(t - 25) = 0$

$t = 15, t = 25$

Boundary points: 15, 25

Test: $t^2 - 40t + 375$

Region	Test	Result
$t < 15$	0	$375 > 0$
$15 < t < 25$	20	$-25 < 0$
$t > 25$	30	$75 > 0$

$15 < t < 25$

The height will be greater than 6000 feet for times greater than 15 seconds but less than 25 seconds.

75. a. Profit $= -10(x^2 - 200x + 1800)$

$-10(x^2 - 200x + 1800) > 0$

$x^2 - 200x + 1800 < 0$

$x^2 - 200x + 1800 = 0$

$a = 1, b = -200, c = 1800$

$x = \dfrac{-(-200) \pm \sqrt{(-200)^2 - 4(1)(1800)}}{2(1)}$

$x = \dfrac{200 \pm \sqrt{32,800}}{2}$

$x \approx 9.4, x \approx 190.6$

Approximate boundary points: 190.6, 9.4

Test: $x^2 - 200x + 1800$

Approximate Region	Test	Result
$x < 9.4$	0	$1800 > 0$
$9.4 < x < 190.6$	10	$-100 < 0$
$x > 190.6$	200	$1800 > 0$

Approximately $9.4 < x < 190.6$

The profit is greater than zero when more than 9.4 but fewer than 190.6 units are manufactured each day.

b. $-10(50^2 - 200(50) + 1800) = 57,000$

Daily profit is \$57,000 when 50 units are manufactured.

c. $-10(60^2 - 200(60) + 1800) = 66,000$

Daily profit is \$66,000 when 60 units are manufactured.

Cumulative Review

77. $10(18) + 14(10) + 5(16) = 400$

$10(22) + 14(12) + 5(19) = 483$

It would cost the family \$400 for the 2-hour cruise and \$483 for the 3-hour cruise.

78. There are $10 + 14 + 5 = 29$ total family members, so 23 people plan to take a cruise. Let x = the number of adults, y = the number of children, and z = the number of seniors that plan to take the cruise.

$x + y + z = 23$ **(1)**

$18x + 10y + 16z = 314$ **(2)**

$22x + 12y + 19z = 380$ **(3)**

Multiply equation **(1)** by -10 and add the result to equation **(2)** to eliminate y.

$-10x - 10y - 10z = -230$

$\underline{18x + 10y + 16z = 314}$

$8x \qquad + 6z = 84$

$4x + 3z = 42$ **(4)**

Multiply equation **(1)** by -12 and add the result to equation **(3)** to eliminate y.

$-12x - 12y - 12z = -276$

$\underline{22x + 12y + 19z = 380}$

$10x \qquad + 7z = 104$ **(5)**

Multiply equation **(4)** by 5 and equation **(5)** by -2.

$20x + 15z = 210$

$\underline{-20x - 14z = -208}$

$z = 2$

Replace z with 2 in equation **(4)**.
$$4x + 3(2) = 42$$
$$4x + 6 = 42$$
$$4x = 36$$
$$x = 9$$
Replace x with 9 and z with 2 in equation **(1)**.
$$9 + y + 2 = 23$$
$$y + 11 = 23$$
$$y = 12$$
9 adults, 12 children, and 2 seniors plan to take a cruise.

Quick Quiz 9.6

1. $3x + 2 < 8$ *and* $3x > -16$
   $3x < 6$
   $x < 2$ $x > -\dfrac{16}{3}$

   $-\dfrac{16}{3} < x < 2$ is the solution.

2. $x - 7 \le -15$ *or* $2x + 3 \ge 5$
   $x \le -8$ $2x \ge 2$
   $x \ge 1$

   $x \le -8$ *or* $x \ge 1$ is the solution.

3. $x^2 - 7x + 6 > 0$
   $x^2 - 7x + 6 = 0$
   $(x - 6)(x - 1) = 0$
   $x = 6, x = 1$
   Boundary points: 1, 6
   Test: $x^2 - 7x + 6$

Region	Test	Result
$x < 1$	0	$6 > 0$
$1 < x < 6$	2	$-4 < 0$
$x > 6$	7	$6 > 0$

   $x < 1$ or $x > 6$

4. Answers may vary. Possible solution:

   The equation $x^2 + 2x + 8 = 0$ does not have any real solutions. Thus there are no boundary points. Also, the quadratic function $f(x) = x^2 + 2x + 8$ has no x-intercepts, so the graph lies entirely above or below the x-axis. This means that the inequality is either true for all real numbers, or has no real number solutions.

9.7 Exercises

1. It will always have two solutions. One solution is when $x = b$ and one when $x = -b$. Since $b > 0$ the values of b and $-b$ are always different numbers.

3. You must first isolate the absolute value expression. To do this you add 2 to each side of the equation. The result will be $|x + 7| = 10$. Then you solve the two equations $x + 7 = 10$ and $x + 7 = -10$. The final answer is $x = 3, x = -17$.

5. $|x| = 30$
   $x = 30$ or $x = -30$
   Check: $|30| \overset{?}{=} 30$ $|-30| \overset{?}{=} 30$
   $30 = 30$ $30 = 30$

7. $|x + 7| = 15$
   $x + 7 = 15$ or $x + 7 = -15$
   $x = 8$ $x = -22$
   Check: $|8 + 7| \overset{?}{=} 15$ $|-22 + 7| \overset{?}{=} 15$
   $|15| \overset{?}{=} 15$ $|-15| \overset{?}{=} 15$
   $15 = 15$ $15 = 15$

9. $|2x - 5| = 13$
   $2x - 5 = 13$ or $2x - 5 = -13$
   $2x = 18$ $2x = -8$
   $x = 9$ $x = -4$
   Check: $|2 \cdot 9 - 5| \overset{?}{=} 13$ $|2(-4) - 5| \overset{?}{=} 13$
   $|18 - 5| \overset{?}{=} 13$ $|-8 - 5| \overset{?}{=} 13$
   $|13| \overset{?}{=} 13$ $|-13| \overset{?}{=} 13$
   $13 = 13$ $13 = 13$

11. $|1.8 - 0.4x| = 1$
    $1.8 - 0.4x = 1$ or $1.8 - 0.4x = -1$
    $-0.4x = -0.8$ $-0.4x = -2.8$
    $x = 2$ $x = 7$
    Check: $|1.8 - 0.4(2)| \overset{?}{=} 1$ $|1.8 - 0.4(7)| \overset{?}{=} 1$
    $|1.8 - 0.8| \overset{?}{=} 1$ $|1.8 - 2.8| \overset{?}{=} 1$
    $|1| \overset{?}{=} 1$ $|-1| \overset{?}{=} 1$
    $1 = 1$ $1 = 1$

13. $|x + 2| - 1 = 7$
    $|x + 2| = 8$
    $x + 2 = 8$ or $x + 2 = -8$
    $x = 6$ $x = -10$
    Check: $|6 + 2| - 1 \overset{?}{=} 7$ $|-10 + 2| - 1 \overset{?}{=} 7$
    $|8| - 1 \overset{?}{=} 7$ $|-8| - 1 \overset{?}{=} 7$
    $8 - 1 \overset{?}{=} 7$ $8 - 1 \overset{?}{=} 7$
    $7 = 7$ $7 = 7$

15. $\left|\dfrac{1}{2} - \dfrac{3}{4}x\right| + 1 = 3$

$\left|\dfrac{1}{2} - \dfrac{3}{4}x\right| = 2$

$\dfrac{1}{2} - \dfrac{3}{4}x = 2$ or $\dfrac{1}{2} - \dfrac{3}{4}x = -2$

$-\dfrac{3}{4}x = \dfrac{3}{2}$ $-\dfrac{3}{4}x = -\dfrac{5}{2}$

$x = -2$ $x = \dfrac{10}{3}$

Check: $\left|\dfrac{1}{2} - \dfrac{3}{4}(-2)\right| + 1 \overset{?}{=} 3$ $\left|\dfrac{1}{2} - \dfrac{3}{4} \cdot \dfrac{10}{3}\right| + 1 \overset{?}{=} 3$

$\left|2\right| + 1 \overset{?}{=} 3$ $\left|-2\right| + 1 \overset{?}{=} 3$

$2 + 1 \overset{?}{=} 3$ $2 + 1 \overset{?}{=} 3$

$3 = 3$ $3 = 3$

17. $\left|2 - \dfrac{2}{3}x\right| - 3 = 5$

$\left|2 - \dfrac{2}{3}x\right| = 8$

$2 - \dfrac{2}{3}x = 8$ or $2 - \dfrac{2}{3}x = -8$

$-\dfrac{2}{3}x = 6$ $-\dfrac{2}{3}x = -10$

$x = -9$ $x = 15$

Check: $\left|2 - \dfrac{2}{3}(-9)\right| - 3 \overset{?}{=} 5$

$\left|2 + 6\right| - 3 \overset{?}{=} 5$

$\left|8\right| - 3 \overset{?}{=} 5$

$8 - 3 \overset{?}{=} 5$

$5 = 5$

$\left|2 - \dfrac{2}{3}(15)\right| - 3 \overset{?}{=} 5$

$\left|2 - 10\right| - 3 \overset{?}{=} 5$

$\left|-8\right| - 3 \overset{?}{=} 5$

$8 - 3 \overset{?}{=} 5$

$5 = 5$

19. $\left|\dfrac{1 - 3x}{2}\right| = \dfrac{4}{5}$

$\dfrac{1 - 3x}{2} = \dfrac{4}{5}$ or $\dfrac{1 - 3x}{2} = -\dfrac{4}{5}$

$5 - 15x = 8$ $5 - 15x = -8$

$-15x = 3$ $-15x = -13$

$x = -\dfrac{1}{5}$ $x = \dfrac{13}{15}$

Check: $\dfrac{\left|1 - 3\left(-\frac{1}{5}\right)\right|}{2} \overset{?}{=} \dfrac{4}{5}$ $\dfrac{\left|1 - 3\left(\frac{13}{15}\right)\right|}{2} \overset{?}{=} \dfrac{4}{5}$

$\left|\dfrac{4}{5}\right| \overset{?}{=} \dfrac{4}{5}$ $\left|-\dfrac{4}{5}\right| \overset{?}{=} \dfrac{4}{5}$

$\dfrac{4}{5} = \dfrac{4}{5}$ $\dfrac{4}{5} = \dfrac{4}{5}$

21. $\left|x + 4\right| = \left|2x - 1\right|$

$x + 4 = 2x - 1$ or $x + 4 = -(2x - 1)$

$4 = x - 1$ $x + 4 = -2x + 1$

$5 = x$ $3x + 4 = 1$

$x = 5$ $3x = -3$

$x = -1$

23. $\left|\dfrac{x - 1}{2}\right| = \left|2x + 3\right|$

$\dfrac{x - 1}{2} = 2x + 3$ or $\dfrac{x - 1}{2} = -(2x + 3) = -2x - 3$

$x - 1 = 4x + 6$ $x - 1 = -4x - 6$

$-3x = 7$ $5x = -5$

$x = -\dfrac{7}{3}$ $x = -1$

25. $\left|1.5x - 2\right| = \left|x - 0.5\right|$

$1.5x - 2 = x - 0.5$ or $1.5x - 2 = -(x - 0.5)$

$15x - 20 = 10x - 5$ $15x - 20 = -10x + 5$

$5x = 15$ $25x = 25$

$x = 3$ $x = 1$

27. $\left|3 - x\right| = \left|\dfrac{x}{2} + 3\right|$

$3 - x = \dfrac{x}{2} + 3$ or $3 - x = -\left(\dfrac{x}{2} + 3\right) = -\dfrac{x}{2} - 3$

$6 - 2x = x + 6$ $6 - 2x = -x - 6$

$-3x = 0$ $-x = -12$

$x = 0$ $x = 12$

29. $\left|3(x + 4)\right| + 2 = 14$

$\left|3x + 12\right| = 12$

$3x + 12 = 12$ or $3x + 12 = -12$

$3x = 0$ $3x = -24$

$x = 0$ $x = -8$

Check: $\left|3(0 + 4)\right| + 2 \overset{?}{=} 14$ $\left|3(-8 + 4)\right| + 2 \overset{?}{=} 14$

$\left|12\right| + 2 \overset{?}{=} 14$ $\left|-12\right| + 2 \overset{?}{=} 14$

$12 + 2 \overset{?}{=} 14$ $12 + 2 \overset{?}{=} 14$

$14 = 14$ $14 = 14$

31. $\left|\dfrac{8x}{5}-2\right|=0$

$\dfrac{8x}{5}-2=0$

$\dfrac{8x}{5}=2$

$8x=10$

$x=\dfrac{10}{8}=\dfrac{5}{4}$

Check: $\left|\dfrac{8\cdot\frac{5}{4}}{5}-2\right|\stackrel{?}{=}0$

$\left|\dfrac{10}{5}-2\right|\stackrel{?}{=}0$

$\left|2-2\right|\stackrel{?}{=}0$

$\left|0\right|\stackrel{?}{=}0$

$0=0$

33. $\left|\dfrac{4}{3}x-\dfrac{1}{8}\right|=-5$

No solution, since absolute value is nonnegative.

35. $\left|\dfrac{3x-1}{3}\right|=\dfrac{2}{5}$

$\dfrac{3x-1}{3}=\dfrac{2}{5}$ or $\dfrac{3x-1}{3}=-\dfrac{2}{5}$

$5(3x-1)=3(2)$ $5(3x-1)=3(-2)$

$15x-5=6$ $15x-5=-6$

$15x=11$ $15x=-1$

$x=\dfrac{11}{15}$ $x=-\dfrac{1}{15}$

Check: $\left|\dfrac{3\cdot\frac{11}{15}-1}{3}\right|\stackrel{?}{=}\dfrac{2}{5}$ $\left|\dfrac{3\left(-\frac{1}{15}\right)-1}{3}\right|\stackrel{?}{=}\dfrac{2}{5}$

$\left|\dfrac{\frac{11}{5}-\frac{5}{5}}{3}\right|\stackrel{?}{=}\dfrac{2}{5}$ $\left|\dfrac{-\frac{1}{5}-\frac{5}{5}}{3}\right|\stackrel{?}{=}\dfrac{2}{5}$

$\left|\dfrac{6}{5}\cdot\dfrac{1}{3}\right|\stackrel{?}{=}\dfrac{2}{5}$ $\left|-\dfrac{6}{5}\cdot\dfrac{1}{3}\right|\stackrel{?}{=}\dfrac{2}{5}$

$\left|\dfrac{2}{5}\right|\stackrel{?}{=}\dfrac{2}{5}$ $\left|-\dfrac{2}{5}\right|\stackrel{?}{=}\dfrac{2}{5}$

$\dfrac{2}{5}=\dfrac{2}{5}$ $\dfrac{2}{5}=\dfrac{2}{5}$

37. $|x|\le 8$

$-8\le x\le 8$

39. $|x+4.5|<5$

$-5<x+4.5<5$

$-9.5<x<0.5$

41. $|x-3|\le 5$

$-5\le x-3\le 5$

$-2\le x\le 8$

43. $|3x+1|\le 10$

$-10\le 3x+1\le 10$

$-11\le 3x\le 9$

$-\dfrac{11}{3}\le x\le 3$

45. $|0.5-0.1x|<1$

$-1<0.5-0.1x<1$

$-1.5<-0.1x<0.5$

$15>x>-5$

$-5<x<15$

47. $\left|\dfrac{1}{4}x+2\right|<6$

$-6<\dfrac{1}{4}x+2<6$

$-24<x+8<24$

$-32<x<16$

49. $\left|\dfrac{2}{3}(x-2)\right|<4$

$-4<\dfrac{2}{3}(x-2)<4$

$-6<x-2<6$

$-4<x<8$

51. $\left|\dfrac{3x-2}{4}\right|<3$

$-3<\dfrac{3x-2}{4}<3$

$-12<3x-2<12$

$-10<3x<14$

$-\dfrac{10}{3}<x<\dfrac{14}{3}$

$-3\dfrac{1}{3}<x<4\dfrac{2}{3}$

53. $|x|>5$

$x<-5$ *or* $x>5$

55. $|x+2| > 5$

$x+2 < -5$ $\quad or \quad$ $x+2 > 5$

$\quad x < -7 \qquad\qquad x > 3$

$x < -7 \; or \; x > 3$

57. $|x-1| \geq 2$

$x-1 \leq -2$ $\quad or \quad$ $x-1 \geq 2$

$\quad x \leq -1 \qquad\qquad x \geq 3$

$x \leq -1 \; or \; x \geq 3$

59. $|4x-7| \geq 9$

$4x-7 \leq -9$ $\quad or \quad$ $4x-7 \geq 9$

$\quad 4x \leq -2 \qquad\qquad 4x \geq 16$

$\quad x \leq -\dfrac{1}{2} \qquad\qquad x \geq 4$

$x \leq -\dfrac{1}{2} \; or \; x \geq 4$

61. $|6-0.1x| > 5$

$6-0.1x > 5$ $\quad or \quad$ $6-0.1x < -5$

$\quad -0.1x > -1 \qquad\qquad -0.1x < -11$

$\quad x < 10 \qquad\qquad\qquad x > 110$

$x < 10 \; or \; x > 110$

63. $\left|\dfrac{1}{5}x - \dfrac{1}{10}\right| > 2$

$\dfrac{1}{5}x - \dfrac{1}{10} < -2$ $\quad or \quad$ $\dfrac{1}{5}x - \dfrac{1}{10} > 2$

$\quad 2x-1 < -20 \qquad\qquad 2x-1 > 20$

$\quad 2x < -19 \qquad\qquad\quad 2x > 21$

$\quad x < -\dfrac{19}{2} \qquad\qquad\quad x > \dfrac{21}{2}$

$\quad x < -9\dfrac{1}{2} \qquad\qquad\quad x > 10\dfrac{1}{2}$

$x < -9\dfrac{1}{2} \; or \; x > 10\dfrac{1}{2}$

65. $\left|\dfrac{1}{3}(x-2)\right| < 5$

$-5 < \dfrac{1}{3}(x-2) < 5$

$-15 < x-2 < 15$

$-13 < x < 17$

67. $|4x+7| < 13$

$-13 < 4x+7 < 13$

$-20 < 4x < 6$

$-5 < x < \dfrac{6}{4}$

$-5 < x < \dfrac{3}{2}$

69. $|3-8x| > 19$

$3-8x < -19$ $\quad or \quad$ $3-8x > 19$

$\quad -8x < -22 \qquad\qquad -8x > 16$

$\quad x > \dfrac{11}{4} \qquad\qquad\quad x < -2$

$x < -2 \; or \; x > 2\dfrac{3}{4}$

71. $\qquad |m-s| \leq 0.12$

$|m-18.65| \leq 0.12$

$-0.12 \leq m-18.65 \leq 0.12$

$18.53 \leq m \leq 18.77$

73. $\qquad |n-p| \leq 0.03$

$|n-17.78| \leq 0.03$

$-0.03 \leq n-17.78 \leq 0.03$

$17.75 \leq n \leq 17.81$

Cumulative Review

75. $(3x^{-3}yz^0)\left(\dfrac{5}{3}x^4y^2\right) = 5x^{-3+4}y^{1+2} \cdot 1 = 5xy^3$

76. $(3x^4y^{-3})^{-2} = 3^{-2}x^{4(-2)}y^{-3(-2)}$

$\qquad\qquad\quad = 3^{-2}x^{-8}y^6$

$\qquad\qquad\quad = \dfrac{y^6}{3^2x^8}$

$\qquad\qquad\quad = \dfrac{y^6}{9x^8}$

Quick Quiz 9.7

1. $\left|\dfrac{2}{3}x+1\right| - 3 = 5$

$\left|\dfrac{2}{3}x+1\right| = 8$

$\dfrac{2}{3}x+1 = 8$ $\quad or \quad$ $\dfrac{2}{3}x+1 = -8$

$\quad \dfrac{2}{3}x = 7 \qquad\qquad \dfrac{2}{3}x = -9$

$\quad x = \dfrac{21}{2} \qquad\qquad x = -\dfrac{27}{2}$

2. $|8x-4| \leq 20$

$-20 \leq 8x-4 \leq 20$

$-16 \leq 8x \leq 24$

$-2 \leq x \leq 3$

3. $|5x + 2| > 7$

$$5x + 2 < -7 \quad or \quad 5x + 2 > 7$$
$$5x < -9 \qquad\qquad 5x > 5$$
$$x < -\frac{9}{5} \qquad\qquad x > 1$$
$$x < -1\frac{4}{5}$$

$$x < -1\frac{4}{5} \ or \ x > 1$$

4. Answers may vary. Possible solution: Since the absolute value is always nonnegative, there is no solution. $|7x + 3|$ cannot be < -4.

Career Exploration Problems

1. $S(t) = 1.53t^2 + 2.11t + 41.06$

2011 corresponds to $t = 0$.

$S(0) = 1.53(0)^2 + 2.11(0) + 41.06 = 41.06$

2015 corresponds to $t = 4$.

$S(4) = 1.53(4)^2 + 2.11(4) + 41.06 = 73.98$

The sales in 2011 were about $41 million and the sales in 2015 were about $74 million.

2. $P(t) = -1.48t^2 + 7.02t + 63.5$

2011 corresponds to $t = 0$.

$P(0) = -1.48(0)^2 + 7.02(0) + 63.5 = 63.5$

2015 corresponds to $t = 4$.

$P(4) = -1.48(4)^2 + 7.02(4) + 63.5 = 67.9$

The sales in 2011 were about $64 million and the sales in 2015 were about $68 million.
The sales increased some, but not as much as the sales of the competitor's brand.

3. Solve $S(t) = P(t)$.

$$1.53t^2 + 2.11t + 41.06 = -1.48t^2 + 7.02t + 63.5$$
$$3.01t^2 - 4.91t - 22.44 = 0$$

$$t = \frac{-(-4.91) \pm \sqrt{(-4.91)^2 - 4(3.01)(-22.44)}}{2(3.01)}$$

$$= \frac{4.91 \pm \sqrt{294.2857}}{6.02}$$

$t \approx -2.03$, $t \approx 3.67$

Only the positive value is valid in this context. $t \approx 3.67$ corresponds to 3 years 8 months, so the sales were equal in August 2014.

4. $N(t) = -0.83t^2 + 4.15t + 47.25$

Since the model for $N(t)$ is quadratic, the maximum value will be at the vertex.

$$\frac{-b}{2a} = \frac{-(4.15)}{2(-0.83)} = 2.5$$

$$N(2.5) = -0.83(2.5)^2 + 4.15(2.5) + 47.25$$
$$= 52.4375$$

The maximum profit will be $52,437,500 and will occur 2.5 years after 2016, or midway through 2018.

You Try It

1. $3x^2 - 60 = 0$

$$3x^2 = 60$$
$$x^2 = 20$$
$$x = \pm\sqrt{20}$$
$$x = \pm 2\sqrt{5}$$

2. $2x^2 + 6x - 3 = 0$

$$2x^2 + 6x = 3$$
$$x^2 + 3x = \frac{3}{2}$$
$$x^2 + 3x + \frac{9}{4} = \frac{3}{2} + \frac{9}{4}$$
$$\left(x + \frac{3}{2}\right)^2 = \frac{15}{4}$$
$$x + \frac{3}{2} = \pm\sqrt{\frac{15}{4}}$$
$$x = -\frac{3}{2} \pm \frac{\sqrt{15}}{2}$$
$$x = \frac{-3 \pm \sqrt{15}}{2}$$

3. $x^2 - 8x = 5$

$$x^2 - 8x - 5 = 0$$
$$a = 1, \ b = -8, \ c = -5$$
$$x = \frac{-(-8) \pm \sqrt{(-8)^2 - 4(1)(-5)}}{2(1)}$$
$$x = \frac{8 \pm \sqrt{64 + 20}}{2}$$
$$x = \frac{8 \pm \sqrt{84}}{2}$$
$$x = \frac{8 \pm 2\sqrt{21}}{2}$$
$$x = 4 \pm \sqrt{21}$$

4.
$$\frac{5}{x-1}+\frac{3}{x+2}=2$$
$$(x-1)(x+2)\left(\frac{5}{x-1}\right)+(x-1)(x+2)\left(\frac{3}{x+2}\right)=(x-1)(x+2)(2)$$
$$5(x+2)+3(x-1)=2(x^2+x-2)$$
$$5x+10+3x-3=2x^2+2x-4$$
$$8x+7=2x^2+2x-4$$
$$-2x^2+6x+11=0$$
$$2x^2-6x-11=0$$

5. $x+3x^{1/2}-4=0$

$y=x^{1/2}$

$y^2+3y-4=0$
$(y+4)(y-1)=0$
$y=-4,\ y=1$
$x^{1/2}=-4,\ x^{1/2}=1$

$x=(-4)^2=16$ (extraneous)

$x=1^2=1$
The solution is $x=1$.

6. a. $2x^2-5xy-3y^2=0$
$(2x+y)(x-3y)=0$
$2x+y=0\quad$ or $\quad x-3y=0$
$$x=-\frac{y}{2}\quad\text{or}\quad x=3y$$

b. $x^2+4y^2=4a$
$$x^2=4a-4y^2$$
$$x=\pm\sqrt{4a-4y^2}$$
$$x=\pm2\sqrt{a-y^2}$$

c. $4x^2-2xw=9w$
$4x^2-2xw-9w=0$
$a=4,\ b=-2w,\ c=-9w$
$$x=\frac{-(-2w)\pm\sqrt{(-2w)^2-4(4)(-9w)}}{2(4)}$$
$$x=\frac{2w\pm\sqrt{4w^2+144w}}{8}$$
$$x=\frac{2w\pm2\sqrt{w^2+36w}}{8}$$
$$x=\frac{w\pm\sqrt{w^2+36w}}{4}$$

7.
$$c^2 = a^2 + b^2$$
$$10^2 = 5^2 + b^2$$
$$100 - 25 = b^2$$
$$75 = b^2$$
$$\sqrt{75} = b$$
$$5\sqrt{3} = b$$

8. $f(x) = x^2 + 3x - 4$

$$\frac{-b}{2a} = \frac{-3}{2(1)} = -\frac{3}{2}$$

$$f\left(-\frac{3}{2}\right) = \left(-\frac{3}{2}\right)^2 + 3\left(-\frac{3}{2}\right) - 4 = -\frac{25}{4}$$

The vertex is $\left(-\frac{3}{2}, -\frac{25}{4}\right)$.

$$f(0) = 0^2 - 3(0) - 4 = -4$$

The *y*-intercept is (0, −4).

$$x^2 + 3x - 4 = 0$$
$$(x + 4)(x - 1) = 0$$
$$x = -4, x = 1$$

The *x*-intercepts are (−4, 0) and (1, 0).

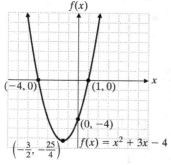

9. $x + 7 > -1$ *and* $3x + 4 < 10$

$\quad\quad x > -8 \quad\quad\quad\quad 3x < 6$

$\quad\quad\quad\quad\quad\quad\quad\quad\quad\quad x < 2$

$x > -8 \text{ and } x < 2$

10. $5x + 2 \le -8$ *or* $4x - 3 \ge 9$

$\quad\quad 5x \le -10 \quad\quad\quad 4x \ge 12$

$\quad\quad\quad x \le -2 \quad\quad\quad\quad x \ge 3$

$x \le -2 \text{ or } x \ge 3$

11.
$$2x^2 - 3x - 9 < 0$$
$$2x^2 - 3x - 9 = 0$$
$$(2x + 3)(x - 3) = 0$$
$$x = -\frac{3}{2}, \ x = 3$$

Boundary points: $-\frac{3}{2}$, 3

Test: $2x^2 - 3x - 9$

Region	Test	Result
$x < -\frac{3}{2}$	−2	5 > 0
$-\frac{3}{2} < x < 3$	0	−9 < 0
$x > 3$	4	11 > 0

$$-\frac{3}{2} < x < 3$$

12. $|3x + 5| = 11$

$\quad 3x + 5 = 11$ *or* $3x + 5 = -11$

$\quad\quad 3x = 6 \quad\quad\quad\quad 3x = -16$

$\quad\quad\quad x = 2 \quad\quad\quad\quad\quad x = -\frac{16}{3}$

13. $|2x + 7| < 17$

$$-17 < 2x + 7 < 17$$
$$-17 - 7 < 2x + 7 - 7 < 17 - 7$$
$$-24 < 2x < 10$$
$$\frac{-24}{2} < \frac{2x}{2} < \frac{10}{2}$$
$$-12 < x < 5$$

$-12 < x < 5$

14. $\left|\frac{1}{4}(x + 8)\right| > 1$

$\quad \frac{1}{4}(x + 8) < -1$ *or* $\frac{1}{4}(x + 8) > 1$

$\quad\quad \frac{1}{4}x + 2 < -1 \quad\quad\quad \frac{1}{4}x + 2 > 1$

$\quad\quad\quad x + 8 < -4 \quad\quad\quad\quad x + 8 > 4$

$\quad\quad\quad\quad x < -12 \quad\quad\quad\quad\quad x > -4$

$x < -12 \text{ or } x > -4$

Chapter 9 Review Problems

1. $6x^2 = 24$
    $x^2 = 4$
    $x = \pm\sqrt{4}$
    $x = \pm 2$

2. $(x+8)^2 = 81$
    $x + 8 = \pm\sqrt{81}$
    $x + 8 = \pm 9$
    $x = -8 \pm 9$
    $x = 1,\ x = -17$

3. $x^2 + 8x + 13 = 0$
    $x^2 + 8x = -13$
    $x^2 + 8x + 16 = -13 + 16$
    $(x+4)^2 = 3$
    $x + 4 = \pm\sqrt{3}$
    $x = -4 \pm \sqrt{3}$

4. $4x^2 - 8x + 1 = 0$
    $4x^2 - 8x = -1$
    $x^2 - 2x = -\dfrac{1}{4}$
    $x^2 - 2x + 1 = -\dfrac{1}{4} + 1$
    $(x-1)^2 = \dfrac{3}{4}$
    $x - 1 = \pm\dfrac{\sqrt{3}}{2}$
    $x = 1 \pm \dfrac{\sqrt{3}}{2}$ or $\dfrac{2 \pm \sqrt{3}}{2}$

5. $x^2 - 4x - 2 = 0$
    $a = 1,\ b = -4,\ c = -2$
    $x = \dfrac{-(-4) \pm \sqrt{(-4)^2 - 4(1)(-2)}}{2(1)}$
    $x = \dfrac{4 \pm \sqrt{24}}{2}$
    $x = \dfrac{4 \pm 2\sqrt{6}}{2} = \dfrac{2\left(2 \pm \sqrt{6}\right)}{2}$
    $x = 2 \pm \sqrt{6}$

6. $3x^2 - 8x + 4 = 0$
    $a = 3,\ b = -8,\ c = 4$
    $x = \dfrac{-(-8) \pm \sqrt{(-8)^2 - 4(3)(4)}}{2(3)}$
    $x = \dfrac{8 \pm \sqrt{16}}{6}$
    $x = \dfrac{8 \pm 4}{6}$
    $x = \dfrac{2}{3},\ x = 2$

7. $4x^2 - 12x + 9 = 0$
    $(2x - 3)^2 = 0$
    $2x - 3 = 0$
    $x = \dfrac{3}{2}$

8. $x^2 - 14 = 5x$
    $x^2 - 5x - 14 = 0$
    $(x-7)(x+2) = 0$
    $x = 7,\ x = -2$

9. $6x^2 - 23x = 4x$
    $6x^2 - 27x = 0$
    $3x(2x - 9) = 0$
    $x = 0,\ x = \dfrac{9}{2}$

10. $2x^2 = 5x - 1$
    $2x^2 - 5x + 1 = 0$
    $a = 2,\ b = -5,\ c = 1$
    $x = \dfrac{-(-5) \pm \sqrt{(-5)^2 - 4(2)(1)}}{2(2)}$
    $x = \dfrac{5 \pm \sqrt{17}}{4}$

11. $5x^2 - 10 = 0$
    $5x^2 = 10$
    $x^2 = 2$
    $x = \pm\sqrt{2}$

Copyright © 2017 Pearson Education, Inc.

12.
$$3x^2 - 2x = 15x - 10$$
$$3x^2 - 17x + 10 = 0$$
$$(3x - 2)(x - 5) = 0$$
$$x = \frac{2}{3}, \ x = 5$$

13. $6x^2 + 12x - 24 = 0$
$$x^2 + 2x - 4 = 0$$
$$a = 1, \ b = 2, \ c = -4$$
$$x = \frac{-2 \pm \sqrt{2^2 - 4(1)(-4)}}{2(1)}$$
$$x = \frac{-2 \pm \sqrt{20}}{2} = \frac{-2 \pm 2\sqrt{5}}{2} = \frac{2\left(-1 \pm \sqrt{5}\right)}{2}$$
$$x = -1 \pm \sqrt{5}$$

14. $7x^2 + 24 = 5x^2$
$$2x^2 = -24$$
$$x^2 = -12$$
$$x = \pm\sqrt{-12}$$
$$x = \pm 2i\sqrt{3}$$

15. $3x^2 + 5x + 1 = 0$
$$a = 3, \ b = 5, \ c = 1$$
$$x = \frac{-5 \pm \sqrt{5^2 - 4(3)(1)}}{2(3)}$$
$$x = \frac{-5 \pm \sqrt{13}}{6}$$

16.
$$2x(x - 4) - 4 = -x$$
$$2x^2 - 8x - 4 + x = 0$$
$$2x^2 - 7x - 4 = 0$$
$$(2x + 1)(x - 4) = 0$$
$$x = -\frac{1}{2}, \ x = 4$$

17. $9x(x + 2) + 2 = 12x$
$$9x^2 + 18x + 2 = 12x$$
$$9x^2 + 6x + 2 = 0$$
$$a = 9, \ b = 6, \ c = 2$$
$$x = \frac{-6 \pm \sqrt{6^2 - 4(9)(2)}}{2(9)}$$
$$x = \frac{-6 \pm \sqrt{-36}}{18} = \frac{-6 \pm 6i}{18} = \frac{6(-1 \pm i)}{18}$$
$$x = \frac{-1 \pm i}{3}$$

18. $\dfrac{4}{5}x^2 + x + \dfrac{1}{5} = 0$
$$4x^2 + 5x + 1 = 0$$
$$(4x + 1)(x + 1) = 0$$
$$x = -\frac{1}{4}, \ x = -1$$

19. The LCD is $6y$.
$$y + \frac{5}{3y} + \frac{17}{6} = 0$$
$$y \cdot 6y + \frac{5}{3y} \cdot 6y + \frac{17}{6} \cdot 6y = 0 \cdot 6y$$
$$6y^2 + 10 + 17y = 0$$
$$6y^2 + 17y + 10 = 0$$
$$(6y + 5)(y + 2) = 0$$
$$y = -\frac{5}{6}, \ y = -2$$

20. The LCD is y^2.
$$\frac{15}{y^2} - \frac{2}{y} = 1$$
$$\frac{15}{y^2} \cdot y^2 - \frac{2}{y} \cdot y^2 = 1 \cdot y^2$$
$$15 - 2y = y^2$$
$$y^2 + 2y - 15 = 0$$
$$(y + 5)(y - 3) = 0$$
$$y = -5, \ y = 3$$

21.
$$y(y + 1) + (y + 2)^2 = 4$$
$$y^2 + y + y^2 + 4y + 4 = 4$$
$$2y^2 + 5y = 0$$
$$y(2y + 5) = 0$$
$$y = 0, \ y = -\frac{5}{2}$$

22. The LCD is $(x + 3)(x + 1)$.

$$\frac{2x}{x+3} + \frac{3x-1}{x+1} = 3$$

$$\frac{2x}{x+3}(x+3)(x+1) + \frac{3x-1}{x+1}(x+3)(x+1) = 3(x+3)(x+1)$$

$$2x(x+1) + (3x-1)(x+3) = 3(x+3)(x+1)$$

$$2x^2 + 2x + 3x^2 + 8x - 3 = 3x^2 + 12x + 9$$

$$2x^2 - 2x - 12 = 0$$

$$x^2 - x - 6 = 0$$

$$(x-3)(x+2) = 0$$

$x = 3, x = -2$

23. $4x^2 - 5x - 3 = 0$

$a = 4, b = -5, c = -3$

$b^2 - 4ac = (-5)^2 - 4(4)(-3) = 73$

Two irrational solutions

24. $2x^2 - 7x + 6 = 0$

$a = 2, b = -7, c = 6$

$b^2 - 4ac = (-7)^2 - 4(2)(6) = 1$

Two rational solutions

25. $25x^2 - 20x + 4 = 0$

$a = 25, b = -20, c = 4$

$b^2 - 4ac = (-20)^2 - 4(25)(4) = 0$

One rational solution

26. $5, -5$

$$\begin{array}{ll} x = 5 & x = -5 \\ x - 5 = 0 & x + 5 = 0 \end{array}$$

$$(x-5)(x+5) = 0$$

$$x^2 - 25 = 0$$

27. $3i, -3i$

$$\begin{array}{ll} x = 3i & x = -3i \\ x - 3i = 0 & x + 3i = 0 \end{array}$$

$$(x-3i)(x+3i) = 0$$

$$x^2 - 9i^2 = 0$$

$$x^2 + 9 = 0$$

28. $-\dfrac{1}{4}, -\dfrac{3}{2}$

$$\begin{array}{ll} x = -\dfrac{1}{4} & x = -\dfrac{3}{2} \\ 4x = -1 & 2x = -3 \\ 4x + 1 = 0 & 2x + 3 = 0 \end{array}$$

$$(4x+1)(2x+3) = 0$$

$$8x^2 + 14x + 3 = 0$$

29. $x^4 - 6x^2 + 8 = 0$

$y = x^2$

$y^2 - 6y + 8 = 0$

$(y-4)(y-2) = 0$

$y - 4 = 0$	$y - 2 = 0$
$y = 4$	$y = 2$
$x^2 = 4$	$x^2 = 2$
$x = \pm\sqrt{4} = \pm 2$	$x = \pm\sqrt{2}$

30. $2x^6 - 5x^3 - 3 = 0$

$y = x^3$

$2y^2 - 5y - 3 = 0$

$(2y+1)(y-3) = 0$

$2y + 1 = 0$	$y - 3 = 0$
$y = -\dfrac{1}{2}$	$y = 3$
$x^3 = -\dfrac{1}{2}$	$x^3 = 3$
$x = \sqrt[3]{-\dfrac{1}{2}} = -\dfrac{\sqrt[3]{4}}{2}$	$x = \sqrt[3]{3}$

31. $x^{2/3} - 3 = 2x^{1/3}$

$x^{2/3} - 2x^{1/3} - 3 = 0$

$y = x^{1/3}$

$y^2 - 2y - 3 = 0$

$(y-3)(y+1) = 0$

$y = 3$	$y = -1$
$x^{1/3} = 3$	$x^{1/3} = -1$
$x = 27$	$x = -1$

32. $1 + 4x^{-8} = 5x^{-4}$

$4x^{-8} - 5x^{-4} + 1 = 0$

$y = x^{-4}$

$4y^2 - 5y + 1 = 0$

$(y-1)(4y-1) = 0$

$y - 1 = 0$	$4y - 1 = 0$
$y = 1$	$y = \dfrac{1}{4}$
$x^{-4} = 1$	$x^{-4} = \dfrac{1}{4}$
$x^4 = 1$	$x^4 = 4$
$x = \pm\sqrt[4]{1} = \pm 1$	$x = \pm\sqrt[4]{4} = \pm\sqrt{2}$

33. $3M = \dfrac{2A^2}{N}$

$A^2 = \dfrac{3MN}{2}$

$A = \pm\sqrt{\dfrac{3MN}{2}}$

34. $3t^2 + 4b = t^2 + 6ay$

$2t^2 = 6ay - 4b$

$t^2 = 3ay - 2b$

$t = \pm\sqrt{3ay - 2b}$

35. $yx^2 - 3x - 7 = 0$

$a = y,\ b = -3,\ c = -7$

$x = \dfrac{-(-3) \pm \sqrt{(-3)^2 - 4(y)(-7)}}{2y}$

$x = \dfrac{3 \pm \sqrt{9 + 28y}}{2y}$

36. $20d^2 - xd - x^2 = 0$

$(4d - x)(5d + x) = 0$

$4d - x = 0$	$5d + x = 0$
$d = \dfrac{x}{4}$	$d = -\dfrac{x}{5}$

37. $2y^2 + 4ay - 3a = 0$

$a = 2,\ b = 4a,\ c = -3a$

$y = \dfrac{-4a \pm \sqrt{(4a)^2 - 4(2)(-3)}}{2(2)}$

$y = \dfrac{-4a \pm \sqrt{16a^2 + 24}}{4}$

$y = \dfrac{-4a \pm \sqrt{4(4a^2 + 6)}}{4}$

$y = \dfrac{-2a \pm \sqrt{4a^2 + 6}}{2}$

38. $AB = 3x^2 + 2y^2 - 4x$

$3x^2 - 4x + 2y^2 - AB = 0$

$a = 3, b = -4, c = 2y^2 - AB$

$x = \dfrac{-(-4) \pm \sqrt{(-4)^2 - 4(3)(2y^2 - AB)}}{2(3)}$

$x = \dfrac{4 \pm \sqrt{16 - 12(2y^2 - AB)}}{6}$

$x = \dfrac{4 \pm \sqrt{16 - 24y^2 + 12AB}}{6}$

$x = \dfrac{4 \pm 2\sqrt{4 - 6y^2 + 3AB}}{6}$

$x = \dfrac{2 \pm \sqrt{4 - 6y^2 + 3AB}}{3}$

39. $c^2 = a^2 + b^2$

$c^2 = \left(3\sqrt{2}\right)^2 + 2^2$

$c^2 = 18 + 4$

$c^2 = 22$

$c = \sqrt{22}$

40. $c^2 = a^2 + b^2$

$16^2 = a^2 + 4^2$

$256 = a^2 + 16$

$a^2 = 240$

$a = \sqrt{240} = 4\sqrt{15}$

41. $c^2 = a^2 + b^2$

$6^2 = 5^2 + b^2$

$36 = 25 + b^2$

$b^2 = 11$

$b = \sqrt{11} \approx 3.3$

The car is approximately 3.3 miles from the observer.

42. Let b = the length of the base. Then $2b + 6$ = the length of the altitude.

$A = \dfrac{1}{2} ab$

$\dfrac{1}{2} b(2b + 6) = 70$

$b(2b + 6) = 140$

$2b^2 + 6b - 140 = 0$

$b^2 + 3b - 70 = 0$

$(b - 7)(b + 10) = 0$

$b - 7 = 0 \qquad\qquad b + 10 = 0$

$b = 7 \qquad\qquad\quad b = -10$

Since the length of the base must be positive, $b = 7$.

$2b + 6 = 2(7) + 6 = 20$

The base is 7 cm and the altitude is 20 cm.

43. Let W = the width of the rectangle.
Then $4W + 1$ = the length.

$(4W + 1)W = 203$

$4W^2 + W - 203 = 0$

$(W - 7)(4W + 29) = 0$

$W - 7 = 0 \qquad\qquad 4W + 29 = 0$

$W = 7 \qquad\qquad\qquad W = -\dfrac{29}{4}$

Since the width must be positive, $W = 7$.

$4W + 1 = 4(7) + 1 = 29$

The width is 7 m and the length is 29 m.

44. Let v = the faster speed. Then $v - 10$ = the slower speed.

Time $= \dfrac{\text{Distance}}{\text{Rate}}$

$5 = \dfrac{80}{v} + \dfrac{10}{v - 10}$

$5v(v - 10) = \dfrac{80}{v} v(v - 10) + \dfrac{10}{v - 10} v(v - 10)$

$5v^2 - 50v = 80v - 800 + 10v$

$5v^2 - 140v + 800 = 0$

$v^2 - 28v + 160 = 0$

$(v - 8)(v - 20) = 0$

$v - 8 = 0 \qquad\qquad v - 20 = 0$

$v = 8 \qquad\qquad\quad v = 20$

Since v must be greater than 10, $v = 20$.
The speed was 20 mph for 80 miles and 10 mph for 10 miles.

45. Let v = her speed with no rain. Then $x - 5$ = her speed in the rain.

Time $= \dfrac{\text{Distance}}{\text{Rate}}$

Copyright © 2017 Pearson Education, Inc.

$$\frac{200}{v}+\frac{90}{v-5}=6$$

$$\frac{200}{v}v(v-5)+\frac{90}{v-5}v(v-5)=6v(v-5)$$

$$200(v-5)+90v=6v(v-5)$$

$$200v-1000+90v=6v^2-30v$$

$$6v^2-320v+1000=0$$

$$(v-50)(6v-20)=0$$

$$v-50=0 \qquad 6v-20=0$$

$$v=50 \qquad\qquad v=\frac{10}{3}$$

Since v must be greater than 5, $v = 50$.
$v - 5 = 45$
The speed before the rain was 50 mph and 45 mph in the rain.

46. The length of the garden plus the walkway is $10 + 2x$, and the width is $6 + 2x$. The area of brick is the combined area minus the area of the garden.

$$(10+2x)(6+2x)-10(6)=100$$

$$60+32x+4x^2-60=100$$

$$4x^2+32x-100=0$$

$$x^2+8x-25=0$$

$$x=\frac{-8\pm\sqrt{8^2-4(1)(-25)}}{2}$$

$$x\approx-10.4,\ x\approx 2.4$$

Since the width of the walkway must be positive, $x \approx 2.4$.
The walkway should be approximately 2.4 feet wide.

47. Let x = the width of walkway in meters. The length of the pool plus the walkway is $50 + 2x$, and the width $25 + 2x$. The area of nonslip surface is the combined area minus the area of the pool.

$$76=(50+2x)(25+2x)-50(25)$$

$$76=1250+4x^2-1250$$

$$4x^2+150x-76=0$$

$$(2x+76)(2x-1)=0$$

$$2x+76=0 \qquad 2x-1=0$$

$$x=-\frac{76}{2} \qquad\quad x=\frac{1}{2}$$

Since the width of the walkway must be positive, $x=\frac{1}{2}$.
The width of the walkway is 0.5 meter.

48. $g(x)=-x^2+6x-11$

$$\frac{-b}{2a}=\frac{-6}{2(-1)}=3$$

$$g(3)=-3^2+6(3)-11=-2$$

$$V(3,-2)$$

$$g(0)=-0^2+6(0)-11=-11$$

y-intercept: $(0, -11)$

$$b^2-4ac=6^2-4(-1)(-11)=-8<0$$

x-intercepts: none

49. $f(x)=x^2+10x+25$

$$\frac{-b}{2a}=\frac{-10}{2(1)}=-5$$

$$f(-5)=(-5)^2+10(-5)+25=0$$

$$V(-5,0)$$

$$f(0)=0^2+10(0)+25=25$$

y-intercept: $(0, 25)$

$$x^2+10x+25=0$$

$$(x+5)^2=0$$

$$x=-5$$

x-intercept: $(-5, 0)$

50. $f(x)=x^2+6x+5$

$$\frac{-b}{2a}=\frac{-6}{2(1)}=-3$$

$$f(-3)=(-3)^2+6(-3)+5=-4$$

$$V(-3,-4)$$

$$f(0)=0^2+6(0)+5=5$$

y-intercept: $(0, 5)$

$$x^2+6x+5=0$$

$$(x+5)(x+1)=0$$

$$x=-5,\ x=-1$$

x-intercepts: $(-5, 0), (-1, 0)$

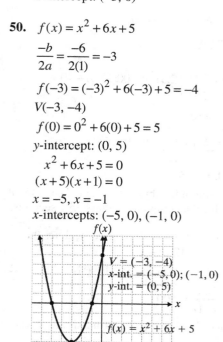

$V = (-3, -4)$
x-int. $= (-5, 0); (-1, 0)$
y-int. $= (0, 5)$

$f(x) = x^2 + 6x + 5$

51. $f(x) = -x^2 + 6x - 5$

$$\frac{-b}{2a} = \frac{-6}{2(-1)} = 3$$

$f(3) = -3^2 + 6(3) - 5 = 4$

$V(3, 4)$

$f(0) = -0^2 + 6(0) - 5 = -5$

y-intercept: $(0, -5)$

$-x^2 + 6x - 5 = 0$

$x^2 - 6x + 5 = 0$

$(x-1)(x-5) = 0$

$x = 1, x = 5$

x-intercepts: $(1, 0), (5, 0)$

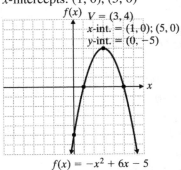

$f(x) = -x^2 + 6x - 5$

52. $h(t) = -16t^2 + 400t + 40$

$$\frac{-b}{2a} = \frac{-400}{2(-16)} = 12.5$$

$h(1.25) = -16(12.5)^2 + 400(12.5) + 40$

$h(12.5) = 2540$

$-16t^2 + 400t + 40 = 0$

$$t = \frac{-400 \pm \sqrt{400^2 - 4(-16)(40)}}{2(-16)}$$

$t \approx -0.1, t \approx 25.1$

The maximum height is 2540 feet. The amount of time for the complete flight is 25.1 seconds.

53. The revenue is the selling price, x, times the number sold, $1200 - x$.

$$R(x) = x(1200 - x) = -x^2 + 1200x$$

$$\frac{-b}{2a} = \frac{-1200}{2(-1)} = 600$$

The price should be $600 for maximum revenue.

54. $-3 \le x < 2$

55. $-8 \le x \le -4$

56. $x < -2 \text{ or } x \ge 5$

57. $x > -5 \text{ and } x < -1$

58. $x > -8 \text{ and } x < -3$

59. $x + 3 > 8 \quad \text{or} \quad x + 2 < 6$

$\qquad x > 5 \qquad\qquad x < 4$

60. $x - 2 > 7 \quad \text{or} \quad x + 3 < 2$

$\qquad x > 9 \qquad\qquad x < -1$

61. $x + 3 > 8 \quad \text{and} \quad x - 4 < -2$

$\qquad x > 5 \qquad\qquad x < 2$

Since x cannot be both > 5 and < 2, there is no solution.

62. $-1 < x + 5 < 8$

$-6 < x < 3$

63. $0 \le 5 - 3x \le 17$

$-5 \le -3x \le 12$

$$\frac{5}{3} \ge x \ge -4$$

$$-4 \le x \le \frac{5}{3}$$

$$-4 \le x \le 1\frac{2}{3}$$

64. $2x - 7 < 3 \quad \text{and} \quad 5x - 1 \ge 8$

$\qquad 2x < 10 \qquad\qquad 5x \ge 9$

$\qquad x < 5 \qquad\qquad x \ge \dfrac{9}{5}$

$$\frac{9}{5} \le x < 5$$

$$1\frac{4}{5} \le x < 5$$

65. $4x - 2 < 8 \quad \text{or} \quad 3x + 1 > 4$

$\qquad 4x < 10 \qquad\qquad 3x > 3$

$\qquad x < \dfrac{5}{2} \qquad\qquad x > 1$

The solution is all real numbers.

66. $x^2 + 7x - 18 < 0$

$x^2 + 7x - 18 = 0$

$(x + 9)(x - 2) = 0$

$x = -9, x = 2$

Boundary points: $-9, 2$

Test: $x^2 + 7x - 18$

Region	Test	Result
$x < -9$	-10	$12 > 0$
$-9 < x < 2$	0	$-18 < 0$
$x > 2$	3	$12 > 0$

$-9 < x < 2$

67. $x^2 - 9x + 20 > 0$

$x^2 - 9x + 20 = 0$

$(x - 5)(x - 4) = 0$

$x = 5, x = 4$

Boundary points: $4, 5$

Test: $x^2 - 9x + 20$

Region	Test	Result
$x < 4$	0	$20 > 0$
$4 < x < 5$	4.5	$-0.25 < 0$
$x > 5$	6	$2 > 0$

$x < 4$ or $x > 5$

68. $2x^2 - x - 6 \le 0$

$2x^2 - x - 6 = 0$

$(2x + 3)(x - 2) = 0$

$x = -\dfrac{3}{2}, x = 2$

Boundary points: $-\dfrac{3}{2}, 2$

Test: $2x^2 - x - 6$

Region	Test	Result
$x < -\frac{3}{2}$	-2	$4 > 0$
$-\frac{3}{2} < x < 2$	0	$-6 < 0$
$x > 2$	4	$22 > 0$

$-\dfrac{3}{2} \le x \le 2$

69. $3x^2 - 13x + 12 \le 0$

$3x^2 - 13x + 12 = 0$

$(3x - 4)(x - 3) = 0$

$x = \dfrac{4}{3}, x = 3$

Boundary points: $\dfrac{4}{3}, 3$

Test: $3x^2 - 13x + 12$

Region	Test	Result
$x < \frac{4}{3}$	0	$12 > 0$
$\frac{4}{3} < x < 3$	2	$-2 < 0$
$x > 3$	4	$8 > 0$

$\dfrac{4}{3} \le x \le 3$

70. $9x^2 - 4 > 0$

$9x^2 - 4 = 0$

$x^2 = \dfrac{4}{9}$

$x = \pm\sqrt{\dfrac{4}{9}} = \pm\dfrac{2}{3}$

Boundary points: $-\dfrac{2}{3}, \dfrac{2}{3}$

Test: $9x^2 - 4$

Region	Test	Result
$x < -\frac{2}{3}$	-2	$32 > 0$
$-\frac{2}{3} < x < \frac{2}{3}$	0	$-4 < 0$
$x > \frac{2}{3}$	2	$32 > 0$

$x < -\dfrac{2}{3}$ or $x > \dfrac{2}{3}$

71.
$$4x^2 - 8x \le 12 + 5x^2$$
$$0 \le x^2 + 8x + 12$$
$$x^2 + 8x + 12 \ge 0$$
$$x^2 + 8x + 12 = 0$$
$$(x+6)(x+2) = 0$$
$$x = -6, \ x = -2$$
Boundary points: −6, −2
Test: $x^2 + 8x + 12$

Region	Test	Result
$x < -6$	−8	$12 > 0$
$-6 < x < -2$	−4	$-4 < 0$
$x > -2$	0	$12 > 0$

$x \le -6$ or $x \ge -2$

72.
$$x^2 + 13x > 16 + 7x$$
$$x^2 + 6x - 16 > 0$$
$$x^2 + 6x - 16 = 0$$
$$(x+8)(x-2) = 0$$
$$x = -8, \ x = 2$$
Boundary points: −8, 2
Test: $x^2 + 6x - 16$

Region	Test	Result
$x < -8$	−10	$24 > 0$
$-8 < x < 2$	0	$-16 < 0$
$x > 2$	4	$24 > 0$

$x < -8$ or $x > 2$

73.
$$3x^2 - 12x > -11$$
$$3x^2 - 12x + 11 > 0$$
$$3x^2 - 12x + 11 = 0$$
$$a = 3, b = -12, c = 11$$
$$x = \frac{-(-12) \pm \sqrt{(-12)^2 - 4(3)(11)}}{2(3)} = \frac{12 \pm \sqrt{12}}{6}$$
$$x = \frac{6 \pm \sqrt{3}}{3}$$
$$x \approx 1.4, \ x \approx 2.6$$
Approximate boundary points: 1.4, 2.6
Test: $3x^2 - 12x + 11$

Approximate Region	Test	Result
$x < 1.4$	0	$11 > 0$
$1.4 < x < 2.6$	2	$-1 < 0$
$x > 2.6$	3	$2 > 0$

$$x < \frac{6 - \sqrt{3}}{3} \quad \text{or} \quad x > \frac{6 + \sqrt{3}}{3}$$
Approximately $x < 1.4$ or $x > 2.6$

74.
$$4x^2 + 12x + 9 < 0$$
$$(2x+3)^2 < 0$$
Since the square of any real number must be 0 or positive, there is no real solution.

75.
$$|2x - 7| + 4 = 5$$
$$|2x - 7| = 1$$
$$2x - 7 = -1 \quad \text{or} \quad 2x - 7 = 1$$
$$2x = 6 \qquad\qquad 2x = 8$$
$$x = 3 \qquad\qquad\quad x = 4$$

76.
$$\left| \frac{2}{3}x - \frac{1}{2} \right| \le 3$$
$$-3 \le \frac{2}{3}x - \frac{1}{2} \le 3$$
$$-18 \le 4x - 3 \le 18$$
$$-15 \le 4x \le 21$$
$$-\frac{15}{4} \le x \le \frac{21}{4}$$

77.
$$|2 - 5x - 4| > 13$$
$$2 - 5x - 4 > 13 \quad \text{or} \quad 2 - 5x - 4 < -13$$
$$-5x > 15 \qquad\qquad\qquad -5x < -11$$
$$x < -3 \qquad\qquad\qquad\quad x > \frac{11}{5}$$

78.
$$|x + 7| < 15$$
$$-15 < x + 7 < 15$$
$$-22 < x < 8$$

79.
$$|x + 9| < 18$$
$$-18 < x + 9 < 18$$
$$-27 < x < 9$$

80. $\left|\dfrac{1}{2}x+2\right|<\dfrac{7}{4}$

$$-\dfrac{7}{4}<\dfrac{1}{2}x+2<\dfrac{7}{4}$$
$$-7<2x+8<7$$
$$-15<2x<-1$$
$$-\dfrac{15}{2}<x<-\dfrac{1}{2}$$
$$-7\dfrac{1}{2}<x<-\dfrac{1}{2}$$

81. $|2x-1|\ge 9$

$\begin{array}{llll} 2x-1\le -9 & or & 2x-1\ge 9 \\ \quad 2x\le -8 & & \quad 2x\ge 10 \\ \quad\;\; x\le -4 & & \quad\;\; x\ge 5 \end{array}$

82. $|3x-1|\ge 2$

$\begin{array}{llll} 3x-1\le -2 & or & 3x-1\ge 2 \\ \quad 3x\le -1 & & \quad 3x\ge 3 \\ \quad\;\; x\le -\dfrac{1}{3} & & \quad\;\; x\ge 1 \end{array}$

83. $|2(x-5)|\ge 2$

$\begin{array}{llll} 2(x-5)\le -2 & or & 2(x-5)\ge 2 \\ 2x-10\le -2 & & 2x-10\ge 2 \\ \quad\; 2x\le 8 & & \quad\; 2x\ge 12 \\ \quad\;\; x\le 4 & & \quad\;\; x\ge 6 \end{array}$

How Am I Doing? Chapter 9 Test

1. $8x^2+9x=0$
$$x(8x+9)=0$$
$$x=0,\; x=-\dfrac{9}{8}$$

2. $6x^2-3x=1$
$$6x^2-3x-1=0$$
$$a=6,\, b=-3,\, c=-1$$
$$x=\dfrac{-(-3)\pm\sqrt{(-3)^2-4(6)(-1)}}{2(6)}$$
$$x=\dfrac{3\pm\sqrt{33}}{12}$$

3. The LCD is $6x$.

$$\frac{3x}{2} - \frac{8}{3} = \frac{2}{3x}$$

$$\frac{3x}{2} \cdot 6x - \frac{8}{3} \cdot 6x = \frac{2}{3x} \cdot 6x$$

$$9x^2 - 16x = 4$$

$$9x^2 - 16x - 4 = 0$$

$$(9x + 2)(x - 2) = 0$$

$$x = -\frac{2}{9}, \ x = 2$$

4. $x(x - 3) - 30 = 5(x - 2)$

$$x^2 - 3x - 30 = 5x - 10$$

$$x^2 - 8x - 20 = 0$$

$$(x - 10)(x + 2) = 0$$

$$x = 10, \ x = -2$$

5. $7x^2 - 4 = 52$

$$7x^2 = 56$$

$$x^2 = 8$$

$$x = \pm\sqrt{8} = \pm\sqrt{4 \cdot 2}$$

$$x = \pm 2\sqrt{2}$$

6. The LCD is $(2x + 1)(2x - 1) = 4x^2 - 1$.

$$\frac{2x}{2x + 1} - \frac{6}{4x^2 - 1} = \frac{x + 1}{2x - 1}$$

$$\frac{2x}{2x + 1}(2x + 1)(2x - 1) - \frac{6}{4x^2 - 1}(4x^2 - 1) = \frac{x + 1}{2x - 1}(2x + 1)(2x - 1)$$

$$2x(2x - 1) - 6 = (x + 1)(2x + 1)$$

$$4x^2 - 2x - 6 = 2x^2 + 3x + 1$$

$$2x^2 - 5x - 7 = 0$$

$$(2x - 7)(x + 1) = 0$$

$$x = \frac{7}{2}, \ x = -1$$

7. $2x^2 - 6x + 5 = 0$

$a = 2, \ b = -6, \ c = 5$

$$x = \frac{-(-6) \pm \sqrt{(-6)^2 - 4(2)(5)}}{2(2)}$$

$$x = \frac{6 \pm \sqrt{-4}}{4}$$

$$x = \frac{6 \pm 2i}{4} = \frac{2(3 \pm i)}{4}$$

$$x = \frac{3 \pm i}{2}$$

8.
$$2x(x-3) = -3$$
$$2x^2 - 6x = -3$$
$$2x^2 - 6x + 3 = 0$$
$$a = 2,\ b = -6,\ c = 3$$
$$x = \frac{-(-6) \pm \sqrt{(-6)^2 - 4(2)(3)}}{2(2)}$$
$$x = \frac{6 \pm \sqrt{12}}{4}$$
$$x = \frac{6 \pm 2\sqrt{3}}{4} = \frac{2(3 \pm \sqrt{3})}{4}$$
$$x = \frac{3 \pm \sqrt{3}}{2}$$

9. $x^4 - 11x^2 + 18 = 0$
$$y = x^2$$
$$y^2 - 11y + 18 = 0$$
$$(y-9)(y-2) = 0$$

$y - 9 = 0$	$y - 2 = 0$
$y = 9$	$y = 2$
$x^2 = 9$	$x^2 = 2$
$x = \pm 3$	$x = \pm\sqrt{2}$

10. $3x^{-2} - 11x^{-1} - 20 = 0$
$$y = x^{-1}$$
$$3y^2 - 11y - 20 = 0$$
$$(y-5)(3y+4) = 0$$

$y - 5 = 0$	$3y + 4 = 0$
$y = 5$	$y = -\dfrac{4}{3}$
$x^{-1} = 5$	$x^{-1} = -\dfrac{4}{3}$
$x = \dfrac{1}{5}$	$x = -\dfrac{3}{4}$

11. $x^{2/3} - 3x^{1/3} - 4 = 0$
$$y = x^{1/3}$$
$$y^2 - 3y - 4 = 0$$
$$(y-4)(y+1) = 0$$

$y - 4 = 0$	$y + 1 = 0$
$y = 4$	$y = -1$
$x^{1/3} = 4$	$x^{1/3} = -1$
$x = 64$	$x = -1$

12. $B = \dfrac{xyw}{z^2}$
$$z^2 = \frac{xyw}{B}$$
$$z = \pm\sqrt{\frac{xyw}{B}}$$

13. $5y^2 + 2by + 6w = 0$
$$y = \frac{-2b \pm \sqrt{(2b)^2 - 4(5)(6w)}}{2(5)}$$
$$y = \frac{-2b \pm \sqrt{4(b^2 - 30w)}}{10}$$
$$y = \frac{-2b \pm 2\sqrt{b^2 - 30w}}{10}$$
$$y = \frac{-b \pm \sqrt{b^2 - 30w}}{5}$$

14. Let x = the width of the rectangle. Then $3x + 1$ = the length.
$$x(3x+1) = 80$$
$$3x^2 + x = 80$$
$$3x^2 + x - 80 = 0$$
$$(x-5)(3x+16) = 0$$
$$x = 5,\ x = -\frac{16}{3}$$

Since the width must be positive, $x = 5$. The width is 5 miles and the length is $3(5) + 1 = 16$ miles.

15. $c^2 = a^2 + b^2$
$$c^2 = 6^2 + \left(2\sqrt{3}\right)^2$$
$$c^2 = 36 + 12 = 48$$
$$c = \sqrt{48} = \sqrt{16 \cdot 3}$$
$$c = 4\sqrt{3}$$

16. Let v = the speed for the first 6 miles. Then $v + 1$ is their speed for the other 3 miles.

$$\text{Time} = \frac{\text{Distance}}{\text{Rate}}$$

$$\frac{6}{v} + \frac{3}{v+1} = 4$$
$$6(v+1) + 3v = 4v(v+1)$$
$$6v + 6 + 3v = 4v^2 + 4v$$
$$4v^2 - 5v - 6 = 0$$
$$(v-2)(4v+3) = 0$$

$$v = 2, \ v = -\frac{3}{4}$$

Since the speed must be positive, $v = 2$.
They paddled 2 mph on the first part and 3 mph on the second part.

17. $f(x) = -x^2 - 6x - 5$

$$\frac{-b}{2a} = \frac{-(-6)}{2(-1)} = -3$$

$$f(-3) = -(-3)^2 - 6(-3) - 5$$
$$f(-3) = 4$$
$$V(-3, 4)$$

$$f(0) = -0^2 - 6(0) - 5 = -5$$

y-intercept: $(0, -5)$

$$-x^2 - 6x - 5 = 0$$
$$x^2 + 6x + 5 = 0$$
$$(x+1)(x+5) = 0$$

$x + 1 = 0 \qquad\qquad x + 5 = 0$
$\quad x = -1 \qquad\qquad\quad x = -5$

x-intercepts: $(-1, 0)$, $(-5, 0)$

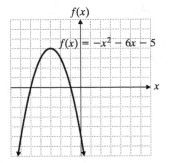

18. $-11 < 2x - 1 \le -3$
$$-10 < 2x \le -2$$
$$-5 < x \le -1$$

19. $x - 4 \le -6 \quad or \quad 2x + 1 \ge 3$
$\quad\quad x \le -2 \qquad\qquad 2x \ge 2$
$\qquad\qquad\qquad\qquad\qquad x \ge 1$

20. $\qquad 2x^2 + 3x \ge 27$
$$2x^2 + 3x - 27 \ge 0$$
$$2x^2 + 3x - 27 = 0$$
$$(2x+9)(x-3) = 0$$

$$x = -\frac{9}{2}, \ x = 3$$

Boundary points: $-\frac{9}{2}$, 3

Test: $2x^2 + 3x - 27$

Region	Test	Result
$x < -\frac{9}{2}$	-5	$8 > 0$
$-\frac{9}{2} < x < 3$	0	$-27 < 0$
$x > 3$	4	$17 > 0$

$$x \le -\frac{9}{2} \ or \ x \ge 3$$

21. $x^2 - 5x - 14 < 0$
$$x^2 - 5x - 14 = 0$$
$$(x+2)(x-7) = 0$$
$$x = -2, \ x = 7$$
Boundary points: -2, 7
Test: $x^2 - 5x - 14$

Region	Test	Result
$x < -2$	-3	$10 > 0$
$-2 < x < 7$	0	$-14 < 0$
$x > 7$	8	$10 > 0$

$$-2 < x < 7$$

22. $x^2 + 3x - 7 > 0$
$$x^2 + 3x - 7 = 0$$
$$a = 1, b = 3, c = -7$$
$$x = \frac{-3 \pm \sqrt{3^2 - 4(1)(-7)}}{2(1)} = \frac{-3 \pm \sqrt{37}}{2}$$
$$x \approx -4.5, \ x \approx 1.5$$
Approximate boundary points: -4.5, 1.5
Test: $x^2 + 3x - 7$

Approximate Region	Test	Result
$x < -4.5$	-5	$3 > 0$
$-4.5 < x < 1.5$	0	$-7 < 0$
$x > 1.5$	2	$3 > 0$

Approximately $x < -4.5$ *or* $x > 1.5$.

23. $|5x - 2| = 37$

$5x - 2 = 37$ or $5x - 2 = -37$

$5x = 39$ $5x = -35$

$x = \dfrac{39}{5}$ $x = -7$

24. $\left|\dfrac{1}{2}x + 3\right| - 2 = 4$

$\left|\dfrac{1}{2}x + 3\right| = 6$

$\dfrac{1}{2}x + 3 = 6$ or $\dfrac{1}{2}x + 3 = -6$

$x + 6 = 12$ $x + 6 = -12$

$x = 6$ $x = -18$

25. $|7x - 3| \le 18$

$-18 \le 7x - 3 \le 18$

$-15 \le 7x \le 21$

$-\dfrac{15}{7} \le x \le 3$

26. $|3x + 1| > 7$

$3x + 1 < -7$ *or* $3x + 1 > 7$

$3x < -8$ $3x > 6$

$x < -\dfrac{8}{3}$ $x > 2$

Cumulative Test for Chapters 0–9

1. $(-3x^{-2}y^3)^4 = (-3)^4 x^{-2 \cdot 4} y^{3 \cdot 4}$

$= 81x^{-8}y^{12}$

$= \dfrac{81y^{12}}{x^8}$

2. $\dfrac{1}{2}a^3 - 2a^2 + 3a - \dfrac{1}{4}a^3 - 6a + a^2 = \dfrac{1}{4}a^3 - a^2 - 3a$

3. $\dfrac{1}{3}(x-3) + 1 = \dfrac{1}{2}x - 2$

$6\left[\dfrac{1}{3}(x-3)+1\right] = 6\left(\dfrac{1}{2}x - 2\right)$

$2(x-3) + 6 = 3x - 12$

$2x - 6 + 6 = 3x - 12$

$2x = 3x - 12$

$-x = -12$

$x = 12$

4. $6x - 3y = -12$

x	y
0	4
-2	0

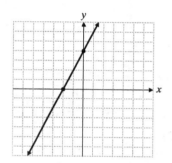

5. $2y + x = 8$

$2y = -x + 8$

$y = -\dfrac{1}{2}x + 4$

$m = -\dfrac{1}{2}$

The slope of a parallel line is $m = -\dfrac{1}{2}$.

6. $3x + 4y = -14$ **(1)**

$-x - 3y = 13$ **(2)**

Multiply equation **(2)** by 3 and add the result to equation **(1)** to eliminate x.

$3x + 4y = -14$

$\underline{-3x - 9y = 39}$

${-5y = 25}$

$y = -5$

Substitute -5 for y in equation **(2)**.

$-x - 3(-5) = 13$

$-x + 15 = 13$

$-x = -2$

$x = 2$

The solution is $(2, -5)$.

7. $125x^3 - 27y^2 = (5x)^3 - (3y)^3$
$$= (5x - 3y)(25x^2 + 15xy + 9y^2)$$

8. $\sqrt{72x^3 y^6} = \sqrt{36x^2 y^6 \cdot 2x} = 6xy^3 \sqrt{2x}$

9. $\left(5 + \sqrt{3}\right)\left(\sqrt{6} - \sqrt{2}\right) = 5\sqrt{6} - 5\sqrt{2} + \sqrt{18} - \sqrt{6}$
$$= 5\sqrt{6} - 5\sqrt{2} + \sqrt{9 \cdot 2} - \sqrt{6}$$
$$= 5\sqrt{6} - 5\sqrt{2} + 3\sqrt{2} - \sqrt{6}$$
$$= 4\sqrt{6} - 2\sqrt{2}$$

10. $\dfrac{3x}{\sqrt{6}} = \dfrac{3x}{\sqrt{6}} \cdot \dfrac{\sqrt{6}}{\sqrt{6}} = \dfrac{3x\sqrt{6}}{6} = \dfrac{x\sqrt{6}}{2}$

11. $3x^2 + 12x = 26x$
$3x^2 - 14x = 0$
$x(3x - 14) = 0$

$x = 0, \ x = \dfrac{14}{3}$

12. $\qquad 12x^2 = 11x - 2$
$12x^2 - 11x + 2 = 0$
$(4x - 1)(3x - 2) = 0$
$4x - 1 = 0 \qquad\qquad 3x - 2 = 0$
$\quad x = \dfrac{1}{4} \qquad\qquad\quad x = \dfrac{2}{3}$

13. $\qquad 44 = 3(2x - 3)^2 + 8$
$3(2x - 3)^2 = 36$
$(2x - 3)^2 = 12$
$2x - 3 = \pm\sqrt{12}$
$2x - 3 = \pm 2\sqrt{3}$
$2x = 3 \pm 2\sqrt{3}$
$x = \dfrac{3 \pm 2\sqrt{3}}{2}$

14. The LCD is x^2.
$$3 - \frac{4}{x} + \frac{5}{x^2} = 0$$
$$3x^2 - \frac{4}{x} \cdot x^2 + \frac{5}{x^2} \cdot x^2 = 0 \cdot x^2$$
$$3x^2 - 4x + 5 = 0$$

$a = 3, \ b = -4, \ c = 5$
$$x = \frac{-(-4) \pm \sqrt{(-4)^2 - 4(3)(5)}}{2(3)}$$
$$x = \frac{4 \pm \sqrt{-44}}{6}$$
$$x = \frac{4 \pm 2i\sqrt{11}}{6} = \frac{2\left(2 \pm i\sqrt{11}\right)}{6}$$
$$x = \frac{2 \pm i\sqrt{11}}{3}$$

15. $\sqrt{6x + 12} - 2 = x$
$\sqrt{6x + 12} = x + 2$
$\left(\sqrt{6x + 12}\right)^2 = (x + 2)^2$
$6x + 12 = x^2 + 4x + 4$
$0 = x^2 - 2x - 8$
$0 = (x + 2)(x - 4)$
$x = -2, \ x = 4$
Check $x = -2$: $\sqrt{6(-2) + 12} \overset{?}{=} -2 + 2$
$\sqrt{-12 + 12} \overset{?}{=} 0$
$0 = 0$ True
Check $x = 4$: $\sqrt{6(4) + 12} \overset{?}{=} 4 + 2$
$\sqrt{24 + 12} \overset{?}{=} 6$
$\sqrt{36} \overset{?}{=} 6$
$6 = 6$ True
The solutions are -2 and 4.

16. $x^{2/3} + 9x^{1/3} + 18 = 0$
$y = x^{1/3}$
$y^2 + 9y + 18 = 0$
$(y + 6)(y + 3) = 0$
$y = -6, \ y = -3$
$x^{1/3} = -6 \qquad\qquad x^{1/3} = -3$
$x = (-6)^3 = -216 \qquad x = (-3)^3 = -27$

17. $2y^2 + 5wy - 7z = 0$
$a = 2, \ b = 5w, \ c = -7z$
$$y = \frac{-5w \pm \sqrt{(5w)^2 - 4(2)(-7z)}}{2(2)}$$
$$y = \frac{-5w \pm \sqrt{25w^2 + 56z}}{4}$$

18. $3y^2 + 16z^2 = 5w$

$$3y^2 = 5w - 16z^2$$

$$y^2 = \frac{5w - 16z^2}{3}$$

$$y = \pm\sqrt{\frac{5w - 16z^2}{3}}$$

$$y = \pm\frac{\sqrt{15w - 48z^2}}{3}$$

19. $\qquad c^2 = a^2 + b^2$

$$\left(\sqrt{38}\right)^2 = 5^2 + b^2$$

$$38 = 25 + b^2$$

$$b^2 = 13$$

$$b = \sqrt{13}$$

20. Let b = the length of the base. Then $3b + 3$ = the length of the altitude.

$$A = \frac{1}{2}ab$$

$$\frac{1}{2}b(3b + 3) = 45$$

$$3b^2 + 3b = 90$$

$$3b^2 + 3b - 90 = 0$$

$$b^2 + b - 30 = 0$$

$$(b + 6)(b - 5) = 0$$

$b = -6, b = 5$

Since the length of the base must be positive, $b = 5$.

The base of the triangle is 5 meters and the altitude of the triangle is $3(5) + 3 = 18$ meters.

21. $f(x) = -x^2 + 8x - 12$

$a = -1, b = 8, c = -12$

$$x_{\text{vertex}} = \frac{-b}{2a}$$

$$x_{\text{vertex}} = \frac{-8}{2(-1)}$$

$$x_{\text{vertex}} = 4$$

$$f(x_{\text{vertex}}) = f(4) = -4^2 + 8(4) - 12 = 4$$

$V(4, 4)$

$$f(0) = -0^2 + 8(0) - 12 = -12$$

y-intercept: $(0, -12)$

$$-x^2 + 8x - 12 = 0$$

$$x^2 - 8x + 12 = 0$$

$$(x - 6)(x - 2) = 0$$

$$\begin{array}{ll} x - 6 = 0 & x - 2 = 0 \\ x = 6 & x = 2 \end{array}$$

x-intercepts: $(6, 0), (2, 0)$

22.

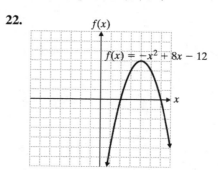

23. $x + 5 \le -4 \quad or \quad 2 - 7x \le 16$

$\qquad x \le -9 \qquad\qquad -7x \le 14$

$\qquad\qquad\qquad\qquad\qquad x \ge -2$

24. $\qquad\qquad x^2 > -3x + 18$

$$x^2 + 3x - 18 > 0$$

$$x^2 + 3x - 18 = 0$$

$$(x + 6)(x - 3) = 0$$

$x = -6, x = 3$

Boundary points: $-6, 3$

Test: $x^2 + 3x - 18$

Region	Test	Result
$x < -6$	-10	$52 > 0$
$-6 < x < 3$	0	$-18 < 0$
$x > 3$	6	$36 > 0$

$x < -6 \ or \ x > 3$

25. $|3x + 1| = 16$

$\quad 3x + 1 = 16 \quad or \quad 3x + 1 = -16$

$\qquad 3x = 15 \qquad\qquad 3x = -17$

$\qquad\quad x = 5 \qquad\qquad\qquad x = -\dfrac{17}{3}$

26. $\left|\dfrac{1}{2}x + 2\right| \le 8$

$$-8 \le \frac{1}{2}x + 2 \le 8$$

$$-16 \le x + 4 \le 16$$

$$-20 \le x \le 12$$

27. $|3x - 4| > 11$

$\quad 3x - 4 < -11 \quad or \quad 3x - 4 > 11$

$\qquad 3x < -7 \qquad\qquad 3x > 15$

$\qquad\quad x < -\dfrac{7}{3} \qquad\qquad x > 5$

$\qquad\quad x < -2\dfrac{1}{3}$

Chapter 10

1. Subtract the values of the points and use the absolute value: $|-2-4| = |-6| = 6$.

3. The equation, $(x-1)^2 + (y+2)^2 = 9 = 3^2$, is in standard form. Determine the values of h, k, and r to find the center and radius: $h = 1$, $k = -2$, and $r = 3$. The center is $(1, -2)$ and the radius is 3.

5. $d = \sqrt{(x_2 - x_1)^2 + (y_2 - y_1)^2}$
$d = \sqrt{(1-2)^2 + (6-4)^2} = \sqrt{1+4} = \sqrt{5}$

7. $d = \sqrt{(x_2 - x_1)^2 + (y_2 - y_1)^2}$
$d = \sqrt{(-2-(-4))^2 + (7-3)^2}$
$d = \sqrt{4+16} = \sqrt{20} = 2\sqrt{5}$

9. $d = \sqrt{(x_2 - x_1)^2 + (y_2 - y_1)^2}$
$d = \sqrt{(4-(-2))^2 + (-5-(-13))^2}$
$d = \sqrt{36+64} = \sqrt{100} = 10$

11. $d = \sqrt{(x_2 - x_1)^2 + (y_2 - y_1)^2}$
$d = \sqrt{\left(\dfrac{5}{4} - \dfrac{1}{4}\right)^2 + \left(-\dfrac{1}{3} - \left(-\dfrac{2}{3}\right)\right)^2}$
$d = \sqrt{1 + \dfrac{1}{9}} = \sqrt{\dfrac{10}{9}} = \dfrac{\sqrt{10}}{\sqrt{9}} = \dfrac{\sqrt{10}}{3}$

13. $d = \sqrt{(x_2 - x_1)^2 + (y_2 - y_1)^2}$
$d = \sqrt{\left(\dfrac{7}{3} - \dfrac{1}{3}\right)^2 + \left(\dfrac{1}{5} - \dfrac{3}{5}\right)^2}$
$d = \sqrt{4 + \dfrac{4}{25}} = \sqrt{\dfrac{104}{25}} = \dfrac{\sqrt{4 \cdot 26}}{\sqrt{25}} = \dfrac{2\sqrt{26}}{5}$

15. $d = \sqrt{(x_2 - x_1)^2 + (y_2 - y_1)^2}$
$d = \sqrt{(1.3 - (-5.7))^2 + (2.6 - 1.6)^2}$
$d = \sqrt{7^2 + 1^2} = \sqrt{50} = 5\sqrt{2}$

17. $d = \sqrt{(x_2 - x_1)^2 + (y_2 - y_1)^2}$
$10 = \sqrt{(1-7)^2 + (y-2)^2}$
$100 = y^2 - 4y + 40$
$y^2 - 4y - 60 = 0$
$(y-10)(y+6) = 0$
$y - 10 = 0 \qquad\qquad y + 6 = 0$
$y = 10 \qquad\qquad\quad y = -6$

19. $d = \sqrt{(x_2 - x_1)^2 + (y_2 - y_1)^2}$
$2.5 = \sqrt{(0-1.5)^2 + (y-2)^2}$
$6.25 = 2.25 + y^2 - 4y + 4$
$y^2 - 4y = 0$
$y(y-4) = 0$
$y = 0 \qquad\qquad y - 4 = 0$
$\qquad\qquad\qquad\quad y = 4$

21. $d = \sqrt{(x_2 - x_1)^2 + (y_2 - y_1)^2}$
$\sqrt{5} = \sqrt{(x-4)^2 + (5-7)^2}$
$\sqrt{5} = \sqrt{x^2 - 8x + 16 + 4}$
$5 = x^2 - 8x + 20$
$x^2 - 8x + 15 = 0$
$(x-3)(x-5) = 0$
$x - 3 = 0 \qquad\qquad x - 5 = 0$
$x = 3 \qquad\qquad\quad x = 5$

23. $d = \sqrt{(x_2 - x_1)^2 + (y_2 - y_1)^2}$
$4 = \sqrt{(2-5)^2 + (y-7)^2}$
$16 = 9 + y^2 - 14y + 49$
$y^2 - 14y + 42 = 0$
$y = \dfrac{14 \pm \sqrt{14^2 - 4(1)(42)}}{2(1)}$
$y = \dfrac{14 \pm \sqrt{28}}{2} \approx \begin{cases} 9.6 \\ 4.4 \end{cases}$
(select the shortest distance)
$y \approx 4.4$ miles
The plane can be detected at 4.4 miles.

25. $(x-h)^2 + (y-k)^2 = r^2$
$(x-(-3))^2 + (y-7)^2 = 6^2$
$(x+3)^2 + (y-7)^2 = 36$

27.
$$(x-h)^2 + (y-k)^2 = r^2$$
$$(x-(-2.4))^2 + (y-0)^2 = \left(\frac{3}{4}\right)^2$$
$$(x+2.4)^2 + y^2 = \frac{9}{16}$$

29. $(x-h)^2 + (y-k)^2 = r^2$
$$(x-0)^2 + \left(y-\frac{3}{8}\right)^2 = \left(\sqrt{3}\right)^2$$
$$x^2 + \left(y-\frac{3}{8}\right)^2 = 3$$

31.
$$x^2 + y^2 = 25$$
$$(x-0)^2 + (y-0)^2 = 5^2$$
$$C(0, 0), r = 5$$

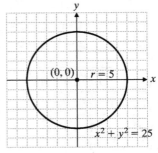

$x^2 + y^2 = 25$

33. $(x-5)^2 + (y-3)^2 = 16 = 4^2$
$$C(5, 3), r = 4$$

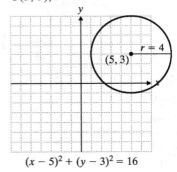

$(x-5)^2 + (y-3)^2 = 16$

35.
$$(x+2)^2 + (y-3)^2 = 25$$
$$[x-(-2)]^2 + (y-3)^2 = 5^2$$
$$C(-2, 3), r = 5$$

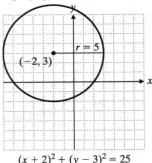

$(x+2)^2 + (y-3)^2 = 25$

37.
$$x^2 + y^2 - 4x + 10y + 4 = 0$$
$$x^2 - 4x + 4 + y^2 + 10y + 25 = -4 + 4 + 25$$
$$(x-2)^2 + (y+5)^2 = 25$$
$$C(2, -5), r = 5$$

39.
$$x^2 + y^2 - 10x + 6y - 2 = 0$$
$$x^2 - 10x + 25 + y^2 + 6y + 9 = 2 + 25 + 9$$
$$(x-5)^2 + (y+3)^2 = 36$$
$$C(5, -3), r = 6$$

41. $x^2 + y^2 + 3x - 2 = 0$
$$x^2 + 3x + y^2 = 2$$
$$x^2 + 3x + \frac{9}{4} + y^2 = 2 + \frac{9}{4} = \frac{17}{4}$$
$$\left(x+\frac{3}{2}\right)^2 + y^2 = \left(\frac{\sqrt{17}}{2}\right)^2$$
$$C\left(-\frac{3}{2}, 0\right), r = \frac{\sqrt{17}}{2}$$

43. $C(61.5, 55.8), r = 44.2, r^2 = 1953.64$
$$(x-61.5)^2 + (y-55.8)^2 = 1953.64$$

45. $(x-5.32)^2 + (y+6.54)^2 = 47.28$
$$(y+6.54)^2 = 47.28 - (x-5.32)^2$$
$$y + 6.54 = \pm\sqrt{47.28 - (x-5.32)^2}$$
$$y_1 = -6.54 + \sqrt{47.28 - (x-5.32)^2}$$
$$y_2 = -6.54 - \sqrt{47.28 - (x-5.32)^2}$$

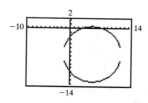

Cumulative Review

47. $4x^2 + 2x = 1$

$4x^2 + 2x - 1 = 0$

$x = \dfrac{-2 \pm \sqrt{2^2 - 4(4)(-1)}}{2(4)}$

$x = \dfrac{-2 \pm \sqrt{20}}{8} = \dfrac{-2 \pm \sqrt{4(5)}}{8} = \dfrac{-2 \pm 2\sqrt{5}}{8}$

$x = \dfrac{-1 \pm \sqrt{5}}{4}$

48. $5x^2 - 6x - 7 = 0$

$x = \dfrac{-(-6) \pm \sqrt{(-6)^2 - 4(5)(-7)}}{2(5)}$

$x = \dfrac{6 \pm \sqrt{176}}{10} = \dfrac{6 \pm \sqrt{16(11)}}{10}$

$x = \dfrac{6 \pm 4\sqrt{11}}{10} = \dfrac{3 \pm 2\sqrt{11}}{5}$

49. $V = Ah$

$V = 20 \text{ mi}^2 (150 \text{ ft}) \left(\dfrac{5280^2 \text{ ft}^2}{\text{mi}^2} \right)$

$V = 8.364 \times 10^{10} \text{ ft}^3$

There was approximately $8.364 \times 10^{10} \text{ ft}^3$ of rock and sediments that settled in this region.

50. $d = rt$

$t = \dfrac{d}{r} = \dfrac{15}{670} \text{ hr} \cdot \dfrac{3600 \text{ sec}}{\text{hr}}$

$t \approx 81 \text{ sec}$

He had approximately 81 seconds to run.

Quick Quiz 10.1

1. $d = \sqrt{(x_2 - x_1)^2 + (y_2 - y_1)^2}$

$d = \sqrt{(-2 - 3)^2 + (-6 - (-4))^2}$

$d = \sqrt{25 + 4} = \sqrt{29}$

2. $(x - h)^2 + (y - k)^2 = r^2$

$(x - 5)^2 + (y - (-6))^2 = 7^2$

$(x - 5)^2 + (y + 6)^2 = 49$

3. $x^2 + 4x + y^2 - 6y + 4 = 0$

$x^2 + 4x + 4 + y^2 - 6y + 9 = -4 + 4 + 9$

$(x + 2)^2 + (y - 3)^2 = 9$

center $(-2, 3)$, $r = 3$

4. Answers may vary. Possible solution: Use the distance formula:

$d = \sqrt{(x_2 - x_1)^2 + (y_2 - y_1)^2}$.

Fill in the known variables, and solve for the one unknown variable x.

10.2 Exercises

1. The graph of $y = x^2$ is symmetric about the y-axis. The graph of $x = y^2$ is symmetric about the x-axis.

3. Since $y = 2(x - 3)^2 + 4$ is in standard form, $y = a(x - h)^2 + k$, the vertex is $(h, k) = (3, 4)$.

5. $y = -4x^2$

$V(0, 0)$

Let $x = 0$: $y = -4(0)^2$

$\qquad\qquad\quad y = 0$

y-intercept = $(0, 0)$

x	-2	-1	0	1	2
y	-16	-4	0	-4	-16

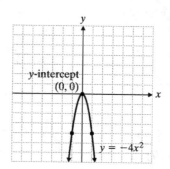

7. $y = x^2 - 2$

$V(0, -2)$

Let $x = 0$.

$y = 0^2 - 2 = -2$

y-intercept: $(0, -2)$

x	-2	-1	0	1	2
y	2	-1	-2	-1	2

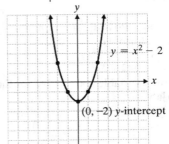

9. $y = \dfrac{1}{2}x^2 - 2$

$V(0, -2)$

Let $x = 0$.

$y = \dfrac{1}{2}(0)^2 - 2 = -2$

y-intercept: $(0, -2)$

x	-4	-2	0	2	4
y	6	0	-2	0	6

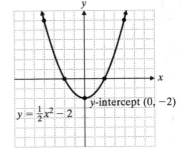

11. $y = (x - 3)^2 - 2$

$V(3, -2)$

Let $x = 0$.

$y = (0 - 3)^2 - 2 = 7$

y-intercept: $(0, 7)$

x	1	2	3	4	5
y	2	-1	-2	-1	2

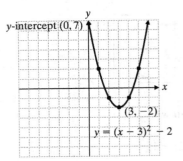

13. $y = 2(x - 1)^2 + \dfrac{3}{2}$

$V\left(1, \dfrac{3}{2}\right)$

Let $x = 0$.

$y = 2(0 - 1)^2 + \dfrac{3}{2} = \dfrac{7}{2}$

y-intercept: $\left(0, \dfrac{7}{2}\right)$

x	-1	0	1	2	3
y	$\dfrac{19}{2}$	$\dfrac{7}{2}$	$\dfrac{3}{2}$	$\dfrac{7}{2}$	$\dfrac{19}{2}$

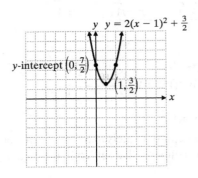

15. $y = -4\left(x + \dfrac{3}{2}\right)^2 + 5$

$V\left(-\dfrac{3}{2}, 5\right)$

Let $x = 0$.

$y = -4\left(0 + \dfrac{3}{2}\right)^2 + 5$

$y = -4$

y-intercept: $(0, -4)$

x	-3	-2	$-\dfrac{3}{2}$	-1	0
y	-4	4	5	4	-4

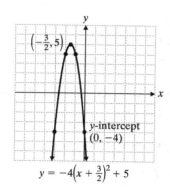

$$y = -4\left(x + \tfrac{3}{2}\right)^2 + 5$$

17. $x = \dfrac{1}{2}y^2$

$V(0, 0)$

Let $y = 0$.

$x = \dfrac{1}{2}(0)^2 = 0$

x-intercept: $(0, 0)$

x	0	2	2	4	4
y	0	2	-2	-8	8

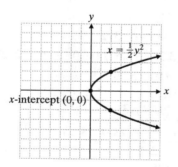

19. $x = \dfrac{1}{4}y^2 - 2$

$V(-2, 0)$

Let $y = 0$.

$x = \dfrac{1}{4}(0)^2 - 2 = -2$

x-intercept: $(-2, 0)$

x	-2	-1	-1	2	2
y	0	-2	2	-4	4

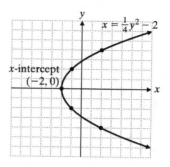

21. $x = -y^2 + 2$

$V(2, 0)$

Let $y = 0$.

$x = -0^2 + 2 = 2$

x-intercept: $(2, 0)$

x	2	-2	-2	-7	-7
y	0	-2	2	-3	3

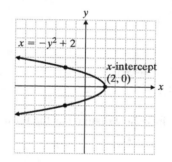

23. $x = (y - 2)^2 + 3$

$V(3, 2)$

Let $y = 0$.

$x = (0 - 2)^2 + 3 = 7$

x-intercept: $(7, 0)$

x	3	4	4	7	7
y	2	1	3	4	0

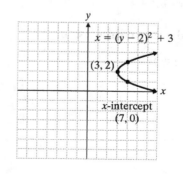

25. $x = -3(y+1)^2 - 2$

$V(-2, -1)$

Let $y = 0$.

$x = -3(0+1)^2 - 2$

$x = -5$

x-intercept: $(-5, 0)$

x	-2	-5	-5
y	-1	0	-2

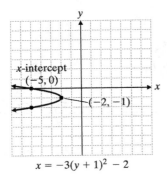

$x = -3(y+1)^2 - 2$

27. $y = x^2 - 4x - 1$

$y = (x^2 - 4x + 4) - 4 - 1$

$y = (x-2)^2 - 5$

a. Vertical since the equation has the form

   $y = a(x-h)^2 + k.$

b. Opens upward since $a > 0$.

c. Vertex: $(2, -5)$

29. $y = -2x^2 + 12x - 16$

$y = -2(x^2 - 6x + 9) + 18 - 16$

$y = -2(x-3)^2 + 2$

a. Vertical since the equation has the form

   $y = a(x-h)^2 + k$

b. Opens downward since $a < 0$.

c. Vertex: $(3, 2)$

31. $x = y^2 + 8y + 9$

$x = (y^2 + 8y + 16) - 16 + 9$

$x = (y+4)^2 - 7$

a. Horizontal since the equation has the form

   $x = a(y-k)^2 + h.$

b. Opens right since $a > 0$.

c. Vertex: $(-7, -4)$

33. $y = ax^2$

$5 = a(10)^2$

$a = \dfrac{1}{20}$

$y = \dfrac{1}{20} x^2$

35. $a = \dfrac{1}{4p} = \dfrac{1}{20} \Rightarrow p = 5$

The distance from $(0, 0)$ to the focus point is 5 inches.

37. $y = 2x^2 + 6.48x - 0.1312$

$y = 2(x^2 + 3.24x + 2.6244) - 0.1312 - 5.2488$

$y = 2(x + 1.62)^2 - 5.38$

$V(-1.62, -5.38)$

y-intercept: $(0, -0.1312)$

$\dfrac{-6.48 \pm \sqrt{6.48^2 - 4(2)(-0.1312)}}{2(2)}$

$= \begin{cases} 0.02012195 \\ -3.26012195 \end{cases}$

x-intercepts: $(0.02012195, 0), (-3.26012195, 0)$

39. $P = -2x^2 + 280x + 35,200$

$P = -2(x^2 - 140x + 4900) + 9800 + 35,200$

$P = -2(x - 70)^2 + 45,000$

Vertex: $(70, 45,000)$

The maximum monthly profit is $45,000, when 70 watches are made.

41. $E = x(900 - x) = -x^2 + 900x$

$E = -(x^2 - 900x + 202,500) + 202,500$

$E = -(x - 450)^2 + 202,500$

Vertex $(450, 202,500)$

The maximum yield is 202,500 and the number of trees per acre is 450.

Cumulative Review

43. $\sqrt{50x^3} = \sqrt{25x^2 \cdot 2x} = \sqrt{25x^2} \cdot \sqrt{2x} = 5x\sqrt{2x}$

44. $\sqrt[3]{40x^3 y^4} = \sqrt[3]{8x^3 y^3 \cdot 5y}$
$\qquad = \sqrt[3]{8x^3 y^3} \cdot \sqrt[3]{5y}$
$\qquad = 2xy\sqrt[3]{5y}$

45. $\sqrt{98x} + x\sqrt{8} - 3\sqrt{50x}$
$\quad = \sqrt{49 \cdot 2x} + x\sqrt{4 \cdot 2} - 3\sqrt{25 \cdot 2x}$
$\quad = \sqrt{49} \cdot \sqrt{2x} + x\sqrt{4} \cdot \sqrt{2} - 3\sqrt{25} \cdot \sqrt{2x}$
$\quad = 7\sqrt{2x} + 2x\sqrt{2} - 15\sqrt{2x}$
$\quad = 2x\sqrt{2} - 8\sqrt{2x}$

46. $\sqrt[3]{16x^4} + 4x\sqrt[3]{2} - 8x\sqrt[3]{54}$
$\quad = \sqrt[3]{8x^3 \cdot 2x} + 4x\sqrt[3]{2} - 8x\sqrt[3]{27 \cdot 2}$
$\quad = \sqrt[3]{8x^3} \cdot \sqrt[3]{2x} + 4x\sqrt[3]{2} - 8x\sqrt[3]{27} \cdot \sqrt[3]{2}$
$\quad = 2x\sqrt[3]{2x} + 4x\sqrt[3]{2} - 24x\sqrt[3]{2}$
$\quad = 2x\sqrt[3]{2x} - 20x\sqrt[3]{2}$

47. $8(1050)(0.88) = 7392$
Sir George can expect 7392 blooms if there is heavy rainfall this year.

48. $\dfrac{2900}{6(0.44)} \approx 1098$
There were 1098 buds on each bush.

Quick Quiz 10.2

1. $y = -3(x+2)^2 + 5$
vertex $(-2, 5)$
Let $x = 0$.
$y_{\text{int}} = -3(0+2)^2 + 5 = -7$
y-intercept $(0, -7)$

2. $x = (y-k)^2 + h$
vertex: $(3, 2)$
x-intercept $(7, 0)$
$x = (y-2)^2 + 3$
Test for point $(7, 0)$.
$7 = (0-2)^2 + 3$
$7 = 4 + 3$
$7 = 7$　True

3. $y = 2x^2 - 12x + 12$
$y = 2(x^2 - 6x) + 12$
$y = 2(x^2 - 6x + 9) - 18 + 12$
$y = 2(x-3)^2 - 6$
vertex $(3, -6)$

x	1	3	5
y	2	-6	2

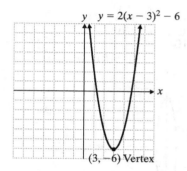

4. Answers may vary. Possible solution with $a > 0$:
$y = ax^2$, vertical, opens up
$y = -ax^2$, vertical, opens down
$x = ay^2$, horizontal, opens right
$x = -ay^2$, horizontal, opens left

10.3 Exercises

1. $\dfrac{(x+2)^2}{4} + \dfrac{(y-3)}{9} = 1$ is in the form
$\dfrac{(x-h)^2}{a^2} + \dfrac{(y-k)^2}{b^2} = 1$ where (h, k) is the center of the ellipse. In this case, the center of the ellipse is $(-2, 3)$.

3. $\dfrac{x^2}{36} + \dfrac{y^2}{4} = 1 \Rightarrow \dfrac{x^2}{6^2} + \dfrac{y^2}{2^2} = 1$
$a = 6$, $b = 2$
Intercepts: $(\pm 6, 0)$, $(0, \pm 2)$

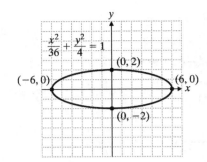

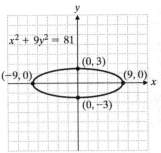

Scale: Each square $= 2$ units

5. $\dfrac{x^2}{4} + \dfrac{y^2}{100} = 1 \Rightarrow \dfrac{x^2}{2^2} + \dfrac{y^2}{10^2} = 1$

$a = 2$, $b = 10$

Intercepts: $(\pm 2, 0)$, $(0, \pm 10)$

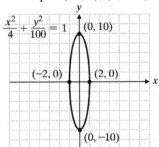

Scale: Each square $= 2$ units

7. $4x^2 + y^2 - 36 = 0$

$\dfrac{x^2}{9} + \dfrac{y^2}{36} = 1 \Rightarrow \dfrac{x^2}{3^2} + \dfrac{y^2}{6^2} = 1$

$a = 3$, $b = 6$

Intercepts: $(\pm 3, 0)$, $(0, \pm 6)$

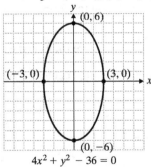

9. $x^2 + 9y^2 = 81$

$\dfrac{x^2}{81} + \dfrac{y^2}{9} = 1$

$\dfrac{x^2}{9^2} + \dfrac{y^2}{3^2} = 1$

$a = 9$, $b = 3$

Intercepts: $(\pm 9, 0)$, $(0, \pm 3)$

11. $x^2 + 12y^2 = 36$

$\dfrac{x^2}{36} + \dfrac{y^2}{3} = 1$

$\dfrac{x^2}{6^2} + \dfrac{y^2}{\sqrt{3}^2} = 1$

$a = 6$, $b = \sqrt{3}$

Intercepts: $(\pm 6, 0)$, $\left(0, \pm\sqrt{3}\right)$

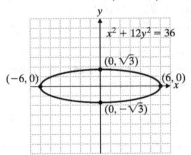

13. $\dfrac{x^2}{\frac{25}{4}} + \dfrac{y^2}{\frac{16}{9}} = 1$

$\dfrac{x^2}{\left(\frac{5}{2}\right)^2} + \dfrac{y^2}{\left(\frac{4}{3}\right)^2} = 1$

$a = \dfrac{5}{2}$, $b = \dfrac{4}{3}$

Intercepts: $\left(\pm\dfrac{5}{2}, 0\right)$, $\left(0, \pm\dfrac{4}{3}\right)$

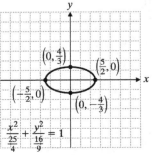

Copyright © 2017 Pearson Education, Inc.

15. $121x^2 + 64y^2 = 7744$

$$\frac{x^2}{64} + \frac{y^2}{121} = 1$$

$$\frac{x^2}{8^2} + \frac{y^2}{11^2} = 1$$

$a = 8$, $b = 11$, $C(0, 0)$
Intercepts: $(0, \pm 11)$, $(\pm 8, 0)$

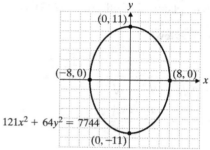

Scale: Each square = 2 units

17. $\dfrac{(x-h)^2}{a^2} + \dfrac{(y-k)^2}{b^2}$

$C(0, 0) \Rightarrow (h, k) = (0, 0)$
x-int$(13, 0) \Rightarrow a = 13$
y-int$(0, -12) \Rightarrow b = 12$

$$\frac{x^2}{13^2} + \frac{y^2}{12^2} = 1$$

$$\frac{x^2}{169} + \frac{y^2}{144} = 1$$

19. $\dfrac{(x-h)^2}{a^2} + \dfrac{(y-k)^2}{b^2} = 1$

$C(0, 0) \Rightarrow (h, k) = (0, 0)$
x-int$(6, 0) \Rightarrow a = 6$
y-int$\left(0, 4\sqrt{3}\right) \Rightarrow b = 4\sqrt{3}$

$$\frac{x^2}{6^2} + \frac{y^2}{\left(4\sqrt{3}\right)^2} = 1$$

$$\frac{x^2}{36} + \frac{y^2}{48} = 1$$

21. $\dfrac{x^2}{5013} + \dfrac{y^2}{4970} = 1 \Rightarrow a^2 = 5013$

$$a = \sqrt{5013}$$

$d = 2a = 2\sqrt{5013} \approx 142$
The largest possible distance across the ellipse is 142 million miles.

23. $\dfrac{(x-h)^2}{a^2} + \dfrac{(y-k)^2}{b^2} = 1$

$$\frac{(x-5)^2}{9} + \frac{(y-2)^2}{1} = 1$$

$$\frac{(x-5)^2}{3^2} + \frac{(y-2)^2}{1^2} = 1$$

$a = 3$, $b = 1$, $C(5, 2)$
Vertices: $(2, 2)$, $(8, 2)$, $(5, 1)$, $(5, 3)$

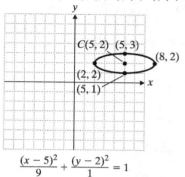

$$\frac{(x-5)^2}{9} + \frac{(y-2)^2}{1} = 1$$

25. $\dfrac{(x-h)^2}{a^2} + \dfrac{(y-k)^2}{b^2} = 1$

$$\frac{x^2}{25} + \frac{(y-4)^2}{16} = 1$$

$$\frac{x^2}{5^2} + \frac{(y-4)^2}{4^2} = 1$$

$a = 5$, $b = 4$, $C(0, 4)$
Vertices: $(5, 4)$, $(0, 8)$, $(-5, 4)$, $(0, 0)$

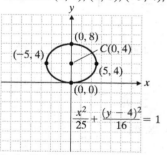

Scale: Each square = 2 units

27. $\dfrac{(x-h)^2}{a^2} + \dfrac{(y-k)^2}{b^2} = 1$

$$\frac{(x+5)^2}{16} + \frac{(y+2)^2}{36} = 1$$

$$\frac{(x+5)^2}{4^2} + \frac{(y+2)^2}{6^2} = 1$$

$a = 4$, $b = 6$, $C(-5, -2)$
Vertices: $(-1, -2)$, $(-5, 4)$, $(-9, -2)$, $(-5, -8)$

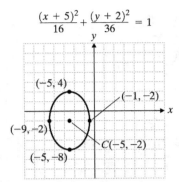

$$\frac{(x+5)^2}{16} + \frac{(y+2)^2}{36} = 1$$

$(-5, 4)$

$(-1, -2)$

$(-9, -2)$

$C(-5, -2)$

$(-5, -8)$

Scale: Each square = 2 units

29. $\left(\dfrac{4+6}{2}, \dfrac{-1+3}{2}\right) = (5, 1)$

$a = |6 - 5| = 1$
$b = |3 - 1| = 2$

$$\frac{(x-5)^2}{1^2} + \frac{(y-1)^2}{2^2} = 1$$

$$(x-5)^2 + \frac{(y-1)^2}{4} = 1$$

31. $C\left(\dfrac{0+60}{2}, \dfrac{0+40}{2}\right) = C(30, 20)$

$2a = 60$
$a = 30$
$2b = 40$
$b = 20$

$$\frac{(x-30)^2}{30^2} + \frac{(y-20)^2}{20^2} = 1$$

$$\frac{(x-30)^2}{900} + \frac{(y-20)^2}{400} = 1$$

33. $\dfrac{(x-3.6)^2}{14.98} + \dfrac{(y-5.3)^2}{28.98} = 1$

$(y-5.3)^2 = 28.98\left(1 - \dfrac{12.96}{14.98}\right)$

$y = \pm\sqrt{28.98\left(1 - \dfrac{12.967}{14.98}\right)} + 5.3$

$y = \begin{cases} 7.2768 \\ 3.3232 \end{cases}$

$$\frac{(x-3.6)^2}{14.98} + \frac{(0-5.3)^2}{28.98} = 1$$

$(x-3.6)^2 = 14.98\left(1 - \dfrac{28.09}{28.98}\right)$

$x = \pm\sqrt{14.98\left(1 - \dfrac{28.09}{28.98}\right)} + 3.6$

$x = \begin{cases} 4.2783 \\ 2.9217 \end{cases}$

x-intercepts: (4.2783, 0), (2.9217, 0)
y-intercepts: (0, 7.2768), (0, 3.3232)

Cumulative Review

35. $\dfrac{5}{\sqrt{2x} - \sqrt{y}} = \dfrac{5}{\sqrt{2x} - \sqrt{y}} \cdot \dfrac{\sqrt{2x} + \sqrt{y}}{\sqrt{2x} + \sqrt{y}}$

$= \dfrac{5\left(\sqrt{2x} + \sqrt{y}\right)}{2x - y}$

36. $\left(2\sqrt{3} + 4\sqrt{2}\right)\left(5\sqrt{6} - \sqrt{2}\right)$

$= 10\sqrt{18} - 2\sqrt{6} + 20\sqrt{12} - 8$

$= 10\sqrt{9 \cdot 2} - 2\sqrt{6} + 20\sqrt{4 \cdot 3} - 8$

$= 30\sqrt{2} - 2\sqrt{6} + 40\sqrt{3} - 8$

37. $4.5x = 102 \Rightarrow x = 22\dfrac{2}{3}$

It took $22\dfrac{2}{3}$ weeks to complete the framework.

Quick Quiz 10.3

1. $\dfrac{x^2}{a^2} + \dfrac{y^2}{b^2} = 1$

$\dfrac{x^2}{5^2} + \dfrac{y^2}{(-6)^2} = 1$

$\dfrac{x^2}{25} + \dfrac{y^2}{36} = 1$

2. $\dfrac{(x-h)^2}{a^2} + \dfrac{(y-k)^2}{b^2} = 1$

$h = x_1 + \dfrac{(x_2 - x_1)}{2} = -7 + \dfrac{1 - (-7)}{2} = -3$

$k = y_1 + \dfrac{(y_2 - y_1)}{2} = -5 + \dfrac{1 - (-5)}{2} = -2$

$a = x_2 - h = 1 - (-3) = 4$

$b = y_2 - k = 1 - (-2) = 3$

$$\frac{(x-(-3))^2}{4^2}+\frac{[y-(-2)]^2}{3^2}=1$$

$$\frac{(x+3)^2}{16}+\frac{(y+2)^2}{9}=1$$

3. $\dfrac{(x-2)^2}{9}+\dfrac{(y+1)}{25}=1$

$h=2$

$k=-1$

$a=3$

$b=5$

Center $(2, -1)$

Vertices: $(-1, -1)$, $(2, 4)$, $(5, -1)$, $(2, -6)$

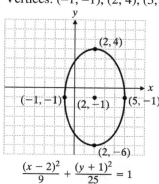

$$\frac{(x-2)^2}{9}+\frac{(y+1)^2}{25}=1$$

4. Answers may vary. Possible solution:
Add 36 to both sides of the equation, then divide both sides of the equation by 36 to put the equation in standard form. The center of the ellipse is at $(0, 0)$ so y-intercepts are $(0, \pm b)$ and x-intercepts are $(\pm a, 0)$.

Use Math to Save Money

1. Private loan: interest for 20 years.
$I = (10{,}000)(0.0465)(20) = 9300$
Total amount $= \$10{,}000 + \$9300 = \$19{,}300$
Subsidized loan: interest for
$20 - 4.5 = 15.5$ years.
$I = (10{,}000)(0.06)(15.5) = 9300$
Total amount $= \$10{,}000 + \$9300 = \$19{,}300$

2. It is surprising that the total amount is the same for both loans.

3. Monthly payment $= \dfrac{\text{Total amount}}{\text{Number of payments}}$

Private loan: $12 \times 20 = 240$ payments.

Monthly payment $= \dfrac{\$19{,}300}{240} \approx \$80.42$

Subsidized loan: $12 \times 15.5 = 186$ payments.

Monthly payment $= \dfrac{\$19{,}300}{186} \approx \$103.76$

4. The private loan has a lower monthly payment.

5.

Type of Loan	Total Amount	Number of Monthly Payments	Monthly Payment Amount	When Payments Begin
Private Loan	$19,300	240	$80.42	Immediately
Subsidized Loan	$19,300	186	$103.76	6 months after graduation

6. Answers will vary.

How Am I Doing? Sections 10.1–10.3
(Available online through MyMathLab.)

1.
$$(x-h)^2 + (y-k)^2 = r^2$$
$$(x-(-3))^2 + (y-7)^2 = \left(\sqrt{2}\right)^2$$
$$(x+3)^2 + (y-7)^2 = 2$$

2. $(x_1, y_1) = (-6, -2),\ (x_2, y_2) = (-3, 4)$
$$d = \sqrt{(x_2 - x_1)^2 + (y_2 - y_1)^2}$$
$$d = \sqrt{(-3-(-6))^2 + (4-(-2))^2}$$
$$d = \sqrt{9+36}$$
$$d = \sqrt{45} = \sqrt{9(5)}$$
$$d = 3\sqrt{5}$$

3.
$$x^2 + y^2 - 2x - 4y + 1 = 0$$
$$x^2 - 2x + 1 + y^2 - 4y + 4 = -1 + 1 + 4$$
$$(x-1)^2 + (y-2)^2 = 4 = 2^2$$
Center = (1, 2)
radius = 2

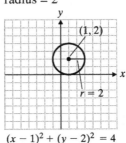

$(x-1)^2 + (y-2)^2 = 4$

4. $y = a(x-h)^2 + k$ has $x = h$ as its axis of symmetry. $y = 4(x-3)^2 + 5$ has $x = 3$ as its axis of symmetry.

5. $y = a(x-h)^2 + k$ has $V(h, k)$. Therefore, $y = \dfrac{1}{3}(x+4)^2 + 6 = \dfrac{1}{3}(x-(-4))^2 + 6$ has $V(-4, 6)$.

6. $x = (y+1)^2 + 2$

$a > 0$, opens right; horizontal

$V(2, -1)$

Let $y = 0$.

$x = (0+1)^2 + 2 = 3$

$(3, 0)$

x	2	3	3	6	6
y	−1	0	−2	1	−3

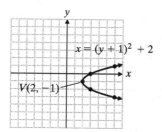

7. $x^2 = y - 4x - 1$

$y = x^2 + 4x + 1$

$y = x^2 + 4x + 4 - 3$

$y = (x+2)^2 - 3$

$a > 0$, opens up; vertical

$V(-2, -3)$

Let $x = 0$.

$y = (0+2)^2 - 3$

$y = 1$

$(0, 1)$

x	−4	0	−1	−3	−2
y	1	1	−2	−2	−3

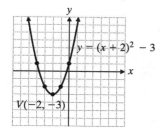

8. $C(0, 0) \Rightarrow h = 0, k = 0$

x-intercept: $(-10, 0) \Rightarrow a = 10$

y-intercept: $(0, 7) \Rightarrow b = 7$

$$\frac{(x-h)^2}{a^2} + \frac{(y-k)^2}{b^2} = 1$$

$$\frac{x^2}{10^2} + \frac{y^2}{7^2} = 1$$

$$\frac{x^2}{100} + \frac{y^2}{49} = 1$$

9. $4x^2 + y^2 - 36 = 0$

$4x^2 + y^2 = 36$

$$\frac{x^2}{9} + \frac{y^2}{36} = 1$$

$a = 3, b = 6$

Intercepts: $(0, \pm 6), (\pm 3, 0)$

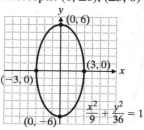

10. $\dfrac{(x+3)^2}{25} + \dfrac{(y-1)^2}{16} = 1$

$a = 5, b = 4$

Vertices: $(-8, 1), (2, 1), (-3, 5), (-3, -3)$

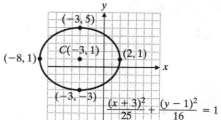

10.4 Exercises

1. The standard form of a horizontal hyperbola centered at the origin is $\dfrac{x^2}{a^2} - \dfrac{y^2}{b^2} = 1$ where a, b are real numbers and $a, b > 0$.

    311

3. $\dfrac{x^2}{16} - \dfrac{y^2}{4} = 1$ is a horizontal hyperbola centered

at the origin with vertices at (4, 0) and (−4, 0).
Draw a fundamental rectangle with corners at
(4, 2), (4, −2), (−4, 2), (−4, −2). Extend the
diagonals through the rectangle as asymptotes of
the hyperbola. Construct each branch of the
hyperbola passing through a vertex and
approaching the asymptotes.

5. $\dfrac{x^2}{4} - \dfrac{y^2}{25} = 1 \Rightarrow \dfrac{x^2}{2^2} - \dfrac{y^2}{5^2} = 1$

$a = 2, b = 5, \ V(\pm 2, 0)$

asymptotes: $y = \pm \dfrac{5}{2}x$

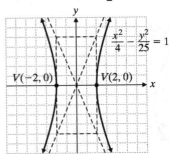

7. $\dfrac{y^2}{16} - \dfrac{x^2}{36} = 1$

$\dfrac{y^2}{4^2} - \dfrac{x^2}{6^2} = 1$

$a = 6, b = 4, \ V(0, \pm 4)$

asymptotes: $y = \pm \dfrac{4}{6}x = \pm \dfrac{2}{3}x$

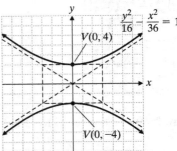

Scale: Each square = 2 units

9. $4x^2 - y^2 = 64$

$\dfrac{x^2}{16} - \dfrac{y^2}{64} = 1$

$\dfrac{x^2}{4^2} - \dfrac{y^2}{8^2} = 1$

$a = 4, b = 8, \ V(\pm 4, 0)$,

asymptotes: $y = \pm \dfrac{8}{4}x$

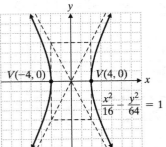

Scale: Each square = 2 units

11. $8x^2 - y^2 = 16$

$\dfrac{x^2}{2} - \dfrac{y^2}{16} = 1$

$\dfrac{x^2}{\sqrt{2}^2} - \dfrac{y^2}{4^2} = 1$

$a = \sqrt{2}, b = 4, \ V\left(\pm\sqrt{2}, 0\right)$

asymptotes: $y = \pm \dfrac{4}{\sqrt{2}}x$

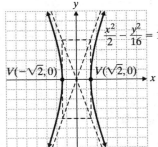

13. $4y^2 - 3x^2 = 48$

$$\frac{y^2}{12} - \frac{x^2}{16} = 1$$

$$\frac{y^2}{\left(2\sqrt{3}\right)^2} - \frac{x^2}{4^2} = 1$$

$b = 2\sqrt{3},\ a = 4$

$V\left(0,\ \pm 2\sqrt{3}\right),\ y_{\text{asymptote}} = \pm \dfrac{2\sqrt{3}}{4}x = \pm \dfrac{\sqrt{3}}{2}x$

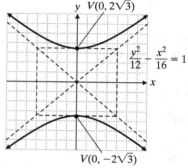

15. $V(\pm 3,\ 0) \Rightarrow a = 3,\ a^2 = 9$

asymptotes: $y = \pm \dfrac{4}{3}x$

$b = 4,\ b^2 = 16$

$$\frac{x^2}{9} - \frac{y^2}{16} = 1$$

17. $V(0,\ \pm 11)$

$b = 11,\ b^2 = 121$

asymptotes: $y = \pm \dfrac{11}{13}x$

$a = 13,\ a^2 = 169$

$$\frac{y^2}{121} - \frac{x^2}{169} = 1$$

19. $\dfrac{x^2}{a^2} - \dfrac{y^2}{b^2} = 1,$ from graph, $a = 120.$

asymptotes: $y = \dfrac{3}{1}x$

$\dfrac{b}{a} = \dfrac{3}{1}$

$\dfrac{b}{120} = \dfrac{3}{1}$

$b = 360$

$$\frac{x^2}{120^2} - \frac{y^2}{360^2} = 1$$

$$\frac{x^2}{14,400} - \frac{y^2}{129,600} = 1,$$

where x and y are measured in millions of miles.

21. $\dfrac{(x-1)^2}{4} - \dfrac{(y+2)^2}{9} = 1$

$$\frac{(x-1)^2}{2^2} - \frac{(y+2)^2}{3^2} = 1$$

$C(1,\ -2)$
$a = 2,\ b = 3$
$V(1 \pm 2,\ -2) = (3,\ -2),\ (-1,\ -2)$

asymptotes: $y = \pm \dfrac{3}{2}(x-1) - 2$

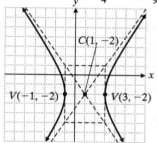

23. $\dfrac{(y+2)^2}{36} - \dfrac{(x+1)^2}{81} = 1$

$$\frac{(y+2)^2}{6^2} - \frac{(x+1)^2}{9^2} = 1$$

$C(-1,\ -2)$
$a = 6,\ b = 9$
$V(-1,\ -2 \pm 6) = (-1,\ 4),\ (-1,\ -8)$

asymptotes: $y = \pm \dfrac{6}{9}(x - (-1)) + (-2)$

$$= \pm \frac{2}{3}(x+1) - 2$$

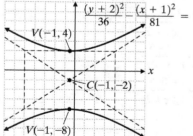

Scale: Each square = 2 units

25.
$$\frac{(x+6)^2}{7} - \frac{y^2}{3} = 1$$

$$\frac{(x-(-6))^2}{\sqrt{7}^2} - \frac{(y-0)^2}{\sqrt{3}^2} = 1$$

$C(-6, 0)$

$V\left(-6 \pm \sqrt{7}, 0\right)$

27. $C\left(4, \frac{0-14}{2}\right) = (4, -7)$ and $b = 7$

asymptote: $y = \frac{-7}{4}x = \pm\frac{b}{a}x \Rightarrow a = 4$ since $b = 7$

$$\frac{(y-(-7))^2}{7^2} - \frac{(x-4)^2}{4^2} = 1$$

$$\frac{(y+7)^2}{49} - \frac{(x-4)^2}{16} = 1$$

29. $8x^2 - y^2 = 16$
$$8(3.5)^2 - y^2 = 16$$
$$y^2 = 8(3.5)^2 - 16$$
$$y = \pm\sqrt{8(3.5)^2 - 16} = \pm9.055385138$$

Cumulative Review

31. $\frac{3}{x^2 - 5x + 6} + \frac{2}{x^2 - 4}$

$= \frac{3(x+2)}{(x-3)(x-2)(x+2)} + \frac{2(x-3)}{(x-2)(x+2)(x-3)}$

$= \frac{3x + 6 + 2x - 6}{(x-3)(x-2)(x+2)}$

$= \frac{5x}{(x-3)(x-2)(x+2)}$

32. $\frac{2x}{5x^2 + 9x - 2} - \frac{3}{5x-1}$

$= \frac{2x}{(5x-1)(x+2)} - \frac{3(x+2)}{(5x-1)(x+2)}$

$= \frac{2x - 3(x+2)}{(5x-1)(x+2)}$

$= \frac{2x - 3x - 6}{(5x-1)(x+2)}$

$= \frac{-x-6}{(5x-1)(x+2)}$ or $-\frac{x+6}{(5x-1)(x+2)}$

33. $x =$ amount grossed by the top ten movies
63.3% of $x = 0.633x = 78.1$

$$x = \frac{78.1}{0.633} \approx 123.4$$

$123.4 million was grossed by the top ten movies.

34. $x =$ number of barrels of oil (in millions) used daily in 2015
$$x + 0.19x = 105$$
$$1.19x = 105$$
$$x = \frac{105}{1.19}$$
$$x \approx 88.2$$

88.2 million barrels of oil were used daily in 2015.

Quick Quiz 10.4

1. $36y^2 - 9x^2 = 36$
$$\frac{36y^2}{36} - \frac{9x^2}{36} = \frac{36}{36}$$
$$\frac{y^2}{1} - \frac{x^2}{4} = 1$$
$b = \sqrt{1} = 1$
vertices $(0, -1), (0, 1)$

2. $\frac{x^2}{4^2} - \frac{y^2}{b} = 1$

$y = \frac{b}{a}x = \frac{5}{4}x, \; b = 5$

$$\frac{x^2}{4^2} - \frac{y^2}{5^2} = 1$$

$$\frac{x^2}{16} - \frac{y^2}{25} = 1$$

3. $\frac{y^2}{9} - \frac{x^2}{4} = 1$

$a = 2, b = 3$
Center $(0, 0)$
Vertices: $(0, 3), (0, -3)$

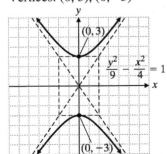

4. Answers may vary. Possible solution:
Divide both sides of the equation by 196 in order
to put the equation in standard form. This
equation describes a horizontal hyperbola, the
asymptotes of which are $y = \pm\dfrac{b}{a}x$. Substitution

yields $y = \pm\dfrac{7}{2}x$.

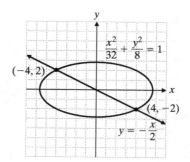

10.5 Exercises

1. $y^2 = 2x$

$y = -2x + 2$, substitute into first equation

$$(-2x+2)^2 = 2x$$
$$4x^2 - 8x + 4 = 2x$$
$$4x^2 - 10x + 4 = 0$$
$$2x^2 - 5x + 2 = 0$$
$$(2x-1)(x-2) = 0$$

$x = \begin{cases} \dfrac{1}{2}, \ y = -2\left(\dfrac{1}{2}\right)+2 = 1 \\ 2, \ y = -2(2)+2 = -2 \end{cases}$

$(2, -2)$, $\left(\dfrac{1}{2}, 1\right)$ is the solution.

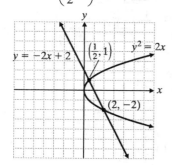

3. $x + 2y = 0$

$x = -2y$

$$x^2 + 4y^2 = 32$$
$$(-2y)^2 + 4y^2 = 32$$
$$4y^2 + 4y^2 = 8y^2 = 32$$
$$y^2 = 4$$

$y = \begin{cases} 2, \ x = -2(2) = -4 \\ -2, \ x = -2(-2) = 4 \end{cases}$

$(-4, 2)$, $(4, -2)$ is the solution.

5. $\dfrac{x^2}{1} - \dfrac{y^2}{3} = 1$

$$3x^2 - y^2 = 3$$
$$x + y = 1$$
$$y = 1 - x$$
$$3x^2 - (1-x)^2 = 3$$
$$3x^2 - 1 + 2x - x^2 = 3$$
$$2x^2 + 2x - 4 = 0$$
$$x^2 + x - 2 = 0$$
$$(x+2)(x-1) = 0$$

$\begin{array}{ll} x + 2 = 0 & x - 1 = 0 \\ x = -2 & x = 1 \end{array}$

$y = 1 - x = \begin{cases} 1-(-2) = 3 \\ 1-1 = 0 \end{cases}$

$(-2, 3)$, $(1, 0)$ is the solution.

7. $x^2 + y^2 - 16 = 0$

$$2y = x - 4$$
$$2y + 4 = x$$
$$(2y+4)^2 + y^2 - 16 = 0$$
$$4y^2 + 16y + 16 + y^2 - 16 = 0$$
$$5y^2 + 16y = 0$$
$$y(5y + 16) = 0$$

$\begin{array}{ll} y = 0 & 5y + 16 = 0 \\ & y = -\dfrac{16}{5} \end{array}$

$x = 2y + 4 = \begin{cases} 2(0)+4 = 4 \\ 2\left(-\dfrac{16}{5}\right)+4 = -\dfrac{12}{5} \end{cases}$

$(4, 0)$, $\left(-\dfrac{12}{5}, -\dfrac{16}{5}\right)$ is the solution.

9. $x^2 + 2y^2 = 4$

$y = -x + 2$

$\quad x^2 + 2(-x+2)^2 = 4$

$\quad x^2 + 2x^2 - 8x + 8 = 4$

$\quad\quad 3x^2 - 8x + 4 = 0$

$\quad\quad (3x-2)(x-2) = 0$

$3x - 2 = 0 \quad\quad\quad\quad x - 2 = 2$

$\quad x = \dfrac{2}{3} \quad\quad\quad\quad\quad x = 2$

$y = -x + 2 = \begin{cases} -\dfrac{2}{3} + 2 = \dfrac{4}{3} \\ -2 + 2 = 0 \end{cases}$

$\left(\dfrac{2}{3}, \dfrac{4}{3}\right)$, $(2, 0)$ is the solution.

11. $\dfrac{x^2}{4} - \dfrac{y^2}{4} = 1$

$\quad x^2 - y^2 = 4$

$\quad x + y - 4 = 0$

$\quad\quad x = 4 - y$

$\quad (4-y)^2 - y^2 = 4$

$16 - 8y + y^2 - y^2 = 4$

$\quad\quad\quad\quad 8y = 12$

$\quad\quad\quad\quad y = \dfrac{3}{2}$

$x = 4 - y = 4 - \dfrac{3}{2} = \dfrac{5}{2}$

$\left(\dfrac{5}{2}, \dfrac{3}{2}\right)$ is the solution.

13. $2x^2 - 5y^2 = -2 \xrightarrow{\times 2} 4x^2 - 10y^2 = -4$

$3x^2 + 2y^2 = 35 \xrightarrow{\times 5} 15x^2 + 10y^2 = 175$

$\quad\quad\quad\quad\quad\quad\quad\quad 19x^2 \quad\quad\quad = 171$

$\quad\quad\quad\quad\quad\quad\quad\quad\quad x^2 = 9$

$\quad\quad\quad\quad\quad\quad\quad\quad\quad x = \pm 3$

$y^2 = \dfrac{35 - 3x^2}{2}$

$y = \pm\sqrt{\dfrac{35 - 3x^2}{2}} = \pm\sqrt{\dfrac{35 - 3(\pm 3)^2}{2}}$

$y = \pm 2$

$(3, 2,)$, $(-3, 2)$, $(3, -2,$ $(-3, -2)$ is the solution.

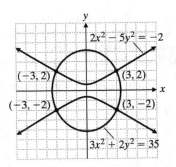

15. $x^2 + y^2 = 9$

$\dfrac{2x^2 - y^2 = 3}{3x^2 \quad\quad = 12}$

$\quad\quad x^2 = 4$

$\quad\quad x = \pm 2$

$(\pm 2)^2 + y^2 = 9$

$\quad\quad y^2 = 5$

$\quad\quad y = \pm\sqrt{5}$

$\left(2, \sqrt{5}\right), \left(-2, \sqrt{5}\right), \left(2, -\sqrt{5}\right), \left(-2, -\sqrt{5}\right)$ is the solution.

17. $x^2 + 2y^2 = 8 \longrightarrow x^2 + 2y^2 = 8$

$x^2 - y^2 = 1 \xrightarrow{\times -1} \dfrac{-x^2 + y^2 = -1}{3y^2 = 7}$

$y^2 = \dfrac{7}{3} \Rightarrow y = \pm\dfrac{\sqrt{7}}{\sqrt{3}} \cdot \dfrac{\sqrt{3}}{\sqrt{3}}$

$y = \pm\dfrac{\sqrt{21}}{3}$

$x^2 = 1 + y^2 = 1 + \dfrac{7}{3} = \dfrac{10}{3}$

$x = \pm\sqrt{\dfrac{10}{3}} = \pm\dfrac{\sqrt{30}}{3}$

$\left(\dfrac{\sqrt{30}}{3}, \dfrac{\sqrt{21}}{3}\right), \left(-\dfrac{\sqrt{30}}{3}, \dfrac{\sqrt{21}}{3}\right), \left(\dfrac{\sqrt{30}}{3}, -\dfrac{\sqrt{21}}{3}\right),$

$\left(-\dfrac{\sqrt{30}}{3}, -\dfrac{\sqrt{21}}{3}\right)$ is the solution.

19. $x^2 + y^2 = 7 \longrightarrow x^2 + y^2 = 7$

$$\dfrac{x^2}{3} - \dfrac{y^2}{9} = 1 \longrightarrow \dfrac{3x^2 - y^2 = 9}{4x^2 \qquad = 16}$$

$$x^2 = 4$$
$$x = \pm 2$$

$$y^2 = 7 - x^2 = 7 - 4 = 3$$
$$y = \pm\sqrt{3}$$

$\left(2, \sqrt{3}\right), \left(-2, \sqrt{3}\right), \left(2, -\sqrt{3}\right), \left(-2, -\sqrt{3}\right)$ is the solution.

21. $xy = -2 \Rightarrow y = -\dfrac{2}{x}$

$$x + 6y = -1$$
$$x + 6\left(-\dfrac{2}{x}\right) = -1$$
$$x - \dfrac{12}{x} = -1$$
$$x^2 - 12 = -x$$
$$x^2 + x - 12 = 0$$
$$(x + 4)(x - 3) = 0$$

$$x + 4 = 0 \qquad\qquad x - 3 = 0$$
$$x = -4 \qquad\qquad\quad x = 3$$

$$y = -\dfrac{2}{x} = \begin{cases} -\dfrac{2}{-4} = \dfrac{1}{2} \\[2mm] -\dfrac{2}{3} \end{cases}$$

$\left(-4, \dfrac{1}{2}\right), \left(3, -\dfrac{2}{3}\right)$ is the solution.

23. $xy = -6$

$$2x + y = -4 \Rightarrow y = -4 - 2x$$
$$x(-4 - 2x) = -6$$
$$2x^2 + 4x - 6 = 0$$
$$x^2 + 2x - 3 = 0$$
$$(x + 3)(x - 1) = 0$$

$$x = -3, \; x = 1$$
$$x = -3, \; y = -4 - 2(-3) = 2$$
$$x = 1, \; y = -4 - 2(1) = -6$$

$(-3, 2), (1, -6)$ is the solution.

25. $x + y = 5$

$$x = 5 - y$$
$$x^2 + y^2 = 4$$
$$(5 - y)^2 + y^2 = 4$$
$$25 - 10y + y^2 + y^2 = 4$$
$$2y^2 - 10y + 21 = 0$$

$$y = \dfrac{-(-10) \pm \sqrt{(-10)^2 - 4(2)(21)}}{2(2)}$$

$$y = \dfrac{10 \pm \sqrt{-68}}{4}$$

No real solution, line does not intersect the circle.

27. $x^2 + y^2 = 16,000,000$

$$y^2 = 16,000,000 - x^2$$
$$25,000,000x^2 - 9,000,000y^2 = 2.25 \times 10^{14}$$
$$25x^2 - 9y^2 = 2.25 \times 10^8$$
$$25x^2 - 9(16,000,000 - x^2) = 2.25 \times 10^8$$
$$34x^2 = 369,000,000$$
$$x^2 \approx 10,852,941$$
$$x \approx 3290$$

$$y^2 = 16,000,000 - x^2$$
$$y \approx \sqrt{16,000,000 - 3290^2}$$
$$y \approx 2280$$

The hyperbola intersects the circle when $(x, y) \approx (3290, 2280)$.

Cumulative Review

29. $(3x^3 - 8x^2 - 33x - 10) \div (3x + 1)$

$$\begin{array}{r} x^2 - 3x - 10 \\ 3x+1{\overline{\smash{\big)}\,3x^3 - 8x^2 - 33x - 10}} \\ \underline{3x^3 + x^2} \\ -9x^2 - 33x \\ \underline{-9x^2 - 3x} \\ -30x - 10 \\ \underline{-30x - 10} \\ 0 \end{array}$$

$(3x^3 - 8x^2 - 33x - 10) \div (3x + 1) = x^2 - 3x - 10$

30. $\dfrac{6x^4 - 24x^3 - 30x^2}{3x^3 - 21x^2 + 30x} = \dfrac{6x^2(x^2 - 4x - 5)}{3x(x^2 - 7x + 10)}$

$\qquad\qquad\qquad = \dfrac{2x(x-5)(x+1)}{(x-5)(x-2)}$

$\qquad\qquad\qquad = \dfrac{2x(x+1)}{x-2}$

Quick Quiz 10.5

1. $2x - y = 4$ (1)

$y^2 - 4x = 0$ (2)

Solve (1) for y.

$2x - y = 4$

$\quad -y = 4 - 2x$

$\quad\ y = 2x - 4$

Substitute $2x - 4$ for y in (2).

$(2x - 4)^2 - 4x = 0$

$4x^2 - 16x + 16 - 4x = 0$

$\quad 4x^2 - 20x + 16 = 0$

$\quad (4x - 16)(x - 1) = 0$

$x = 4, 1$

Solving (1) for y with $x = 4$:

$2(4) - y = 4$

$\quad -y = -4$

$\quad\ y = 4$

$(4, 4)$

Solving (1) for y with $x = 1$:

$2(1) - y = 4$

$\quad -y = 2$

$\quad\ y = -2$

$(1, -2)$

2. $y - x^2 = -4$ (1)

$x^2 + y^2 = 16$ (2)

Add the equations.

$\quad -x^2 + y = -4$

$\quad \underline{\ x^2 + y^2 = 16\ }$

$\qquad\quad y^2 + y = 12$

$\quad y^2 + y - 12 = 0$

$\quad (y + 4)(y - 3) = 0$

$y + 4 = 0 \qquad\qquad y + 3 = 0$

$\quad y = -4 \qquad\qquad\ \ y = 3$

Solving (1) for x with $y = 3$:

$3 - x^2 = -4$

$\quad -x^2 = -4 - 3$

$\qquad x^2 = 7$

$\qquad\ x = \pm\sqrt{7}$

$\left(\sqrt{7}, 3\right), \left(-\sqrt{7}, 3\right)$

Solving (1) for x with $y = -4$:

$-4 - x^2 = -4$

$\quad -x^2 = 0$

$\qquad x = 0$

$(0, -4)$

3. $(x + 2)^2 + (y - 1)^2 = 9$ (1)

$\qquad\qquad\qquad x = 2 - y$ (2)

Substitute $2 - y$ for x in (1).

$(2 - y + 2)^2 + (y - 1)^2 = 9$

$\quad 2y^2 - 10y + 8 = 0$

$\quad\ y^2 - 5y + 4 = 0$

$\quad (y - 4)(y - 1) = 0$

$y = 1, 4$

Solving (2) for x with $y = 1$:

$x = 2 - 1 = 1$

$(1, 1)$

Solving (2) for x with $y = 4$:

$x = 2 - 4 = -2$

$(-2, 4)$

4. Answers may vary. Possible solution:
Use the substitution method. Start by labeling the equations.

$y^2 + 2x^2 = 18$ (1)

$\qquad\ xy = 4$ (2)

Because the y^2 term of (1) has a coefficient of 1, choose to solve (2) for the variable y and substitute into (1).

$y = \dfrac{4}{x}$

$\left(\dfrac{4}{x}\right)^2 + 2x^2 = 18$

Solve the resulting equation for x.

$x = \pm 1, \ \pm 2\sqrt{2}$

Substitute each of the four found values of x to find corresponding y values.

$(1, 4), (-1, -4), \left(2\sqrt{2}, \sqrt{2}\right), \left(-2\sqrt{2}, -\sqrt{2}\right)$

Check for extraneous solutions.

Career Exploration Problems

1. The equation of the parabola has the form
   $y = a(x-h)^2 + k$. Since the vertex is at $(0, 0)$,
   $h = 0$ and $k = 0$. The parabola passes through the
   point $(15, -30)$.

   $$y = ax^2$$
   $$-30 = a(15)^2$$
   $$\frac{-30}{225} = a$$
   $$-\frac{2}{15} = a$$

   The equation is $y = -\frac{2}{15}x^2$.

   When $x = 8$, $y = -\frac{2}{15}(8)^2 = -\frac{128}{15} \approx -8.53$.
   The vertical distance is about 8.53 feet.

2. The equation of the ellipse is
   $\frac{(x-h)^2}{a^2} + \frac{(y-k)^2}{b^2} = 1$. Since the center of the
   ellipse is at the origin, $h = 0$ and $k = 0$. The
   ellipse has vertices at $(\pm 40, 0)$ and $(0, 30)$, so
   $a = 40$ and $b = 30$.

   $$\frac{x^2}{40^2} + \frac{y^2}{30^2} = 1$$
   $$\frac{x^2}{1600} + \frac{y^2}{900} = 1$$
   Let $x = 20$.
   $$\frac{20^2}{1600} + \frac{y^2}{900} = 1$$
   $$\frac{400}{1600} + \frac{y^2}{900} = 1$$
   $$\frac{y^2}{900} = \frac{3}{4}$$
   $$y^2 = \frac{3}{4}(900)$$
   $$y^2 = 675$$
   $$y = \pm\sqrt{675}$$
   $$y \approx \pm 26$$

   The distance is about 26 feet.

3. In $\frac{x^2}{4800} + \frac{y^2}{2500} = 1$, $a^2 = 4800$, so
   $a = \sqrt{4800} \approx 69.3$.
   $2a = 2(69.3) \approx 138.6$
   The width of the archway across the river is
   about 138.6 feet.
   Since the barges are each 25 feet wide and there
   must be 30 feet between them, the distance
   required is 2(25 feet) + (30 feet) = 80 feet.
   The archway across the river is wide enough for
   the barges to pass safely.

You Try It

1. $\sqrt{(3-7)^2 + (-1-(-4))^2} = \sqrt{(-4)^2 + (3)^2}$
   $$= \sqrt{16+9}$$
   $$= \sqrt{25}$$
   $$= 5$$

2. $(x+2)^2 + (y-4)^2 = 9$
   Center: $(-2, 4)$
   Radius: 3

   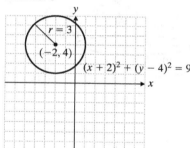

3. $y = -2(x+1)^2 + 3$
   $a = -2$, so parabola opens downward
   Vertex: $(-1, 3)$
   If $x = 0$, $y = 1$.

   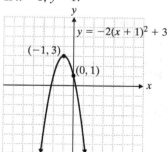

4. $x = (y-4)^2 - 1$

$a = 1$, so parabola opens to the right.
Vertex: $(-1, 4)$
If $y = 0$, $x = 15$.

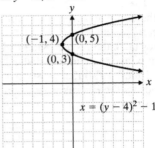

5. $\dfrac{x^2}{36} + y^2 = 1$

Center at origin.
$a^2 = 36$, $a = 6$; $b^2 = 1$, $b = 1$
Vertices: $(-6, 0)$, $(6, 0)$, $(0, 1)$, $(0, -1)$

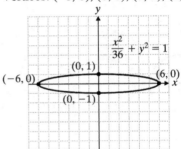

6. $\dfrac{(x-1)^2}{9} + \dfrac{(y+3)^2}{4} = 1$

Center: $(1, -3)$
$a = 3$, $b = 2$
Vertices: $(-2, -3)$, $(4, -3)$, $(1, -1)$, $(1, -5)$

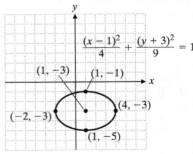

7. $\dfrac{x^2}{36} - \dfrac{y^2}{4} = 1$

Center at origin
$a = 6$, $b = 2$

Vertices: $(-6, 0)$, $(6, 0)$
Asymptotes: $y = \pm\dfrac{2}{6}x = \pm\dfrac{1}{3}x$

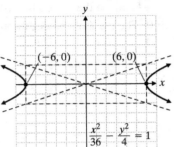

8. $\dfrac{y^2}{16} - x^2 = 1$

Center at origin
$a = 1$, $b = 4$
Vertices: $(0, 4)$, $(0, -4)$
Asymptotes: $y = \pm\dfrac{4}{1}x = \pm 4x$

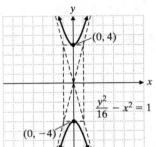

9. $\dfrac{(x-1)^2}{9} - \dfrac{(y+2)^2}{16} = 1$

Center at $(1, -2)$
$a = 3$, $b = 4$
Vertices: $(-2, -2)$, $(4, -2)$

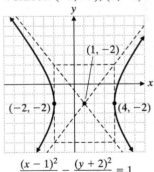

10. $\dfrac{(y+3)^2}{4} - \dfrac{(x-2)^2}{4} = 1$

Center at $(2, -3)$
$a = 2$, $b = 2$
Vertices: $(2, -1)$, $(2, -5)$

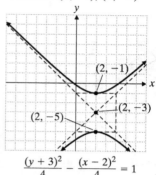

$\dfrac{(y+3)^2}{4} - \dfrac{(x-2)^2}{4} = 1$

11. $y - 3x = -2$
$\quad\quad y = 3x - 2$

$$x^2 + 3y^2 = 12$$
$$x^2 + 3(3x - 2)^2 = 12$$
$$x^2 + 3(9x^2 - 12x + 4) = 12$$
$$x^2 + 27x^2 - 36x + 12 = 12$$
$$28x^2 - 36x = 0$$
$$4x(7x - 9) = 0$$

$4x = 0 \quad\quad\quad 7x - 9 = 0$
$\ x = 0 \quad\quad\quad\quad\quad\quad x = \dfrac{9}{7}$

For $x = 0$, $y = 3(0) - 2 = -2$.

For $x = \dfrac{9}{7}$, $y = 3\left(\dfrac{9}{7}\right) - 2 = \dfrac{13}{7}$.

$(0, -2)$, $\left(\dfrac{9}{7}, \dfrac{13}{7}\right)$ is the solution.

Chapter 10 Review Problems

1. $(0, -6)$ and $(-3, 2)$

$d = \sqrt{(0 - (-3))^2 + (-6 - 2)^2} = \sqrt{9 + 64} = \sqrt{73}$

2. $(-7, 3)$ and $(-2, -1)$

$d = \sqrt{(-2 - (-7))^2 + (-1 - 3)^2} = \sqrt{25 + 16} = \sqrt{41}$

3. $\quad (x - h)^2 + (y - k)^2 = r^2$
$\quad (x - (-6))^2 + (y - 3)^2 = \sqrt{15}^2$
$\quad\quad (x + 6)^2 + (y - 3)^2 = 15$

4. $\quad (x - h)^2 + (y - k)^2 = r^2$
$\quad (x - 0)^2 + (y - (-7))^2 = 5^2$
$\quad\quad\quad x^2 + (y + 7)^2 = 25$

5. $\quad x^2 + y^2 + 2x - 6y + 5 = 0$
$\quad x^2 + 2x + 1 + y^2 - 6y + 9 = -5 + 1 + 9$
$\quad\quad\quad (x + 1)^2 + (y - 3)^2 = 5 = \sqrt{5}^2$

$C(-1, 3)$, $r = \sqrt{5}$

6. $\quad x^2 + y^2 - 10x + 12y + 52 = 0$
$\quad x^2 - 10x + 25 + y^2 + 12y + 36 = -52 + 25 + 36$
$\quad\quad\quad (x - 5)^2 + (y + 6)^2 = 9 = 3^2$

$C(5, -6)$, $r = 3$

7. $x = \dfrac{1}{3}y^2$

Let $y = 0$.

$x = \dfrac{1}{3}(0)^2 = 0$

x	y	
0	0	$V(0, 0)$
3	3	y-intercept: $(0, 0)$
3	-3	

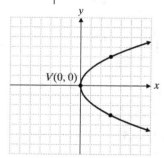

8. $x = \dfrac{1}{2}(y - 2)^2 + 4$

Let $y = 0$.

$x = \dfrac{1}{2}(0 - 2)^2 + 4$
$x = 6$

x	y	
4	2	$V(4, 2)$
6	0	y-intercept: $(6, 0)$
6	4	

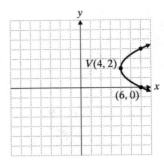

9. $y = -2(x+1)^2 - 3$

Let $x = 0$.

$y = -2(0+1)^2 - 3$

$y = -5$

x	y
-2	-5
-1	-3
0	-5

$V(-1, -3)$

y-intercept: $(0, -5)$

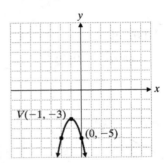

10. $x^2 + 6x = y - 4$

$y = x^2 + 6x + 9 - 9 + 4$

$y = (x+3)^2 - 5$

$V(-3, -5)$

Opens upward because $a > 0$ and it is vertical.

11. $x + 8y = y^2 + 10$

$x = y^2 - 8y + 16 - 6$

$x = (y-4)^2 - 6$

$V(-6, 4)$

Opens to right since $a > 0$ and it is horizontal.

12. $\dfrac{x^2}{4} + \dfrac{y^2}{1} = 1 \Rightarrow \dfrac{x^2}{2^2} + \dfrac{y^2}{1^2} = 1$

$a = 2$, $b = 1$, $C(0, 0)$

Vertices: $(0, 1)$, $(0, -1)$, $(-2, 0)$, $(2, 0)$

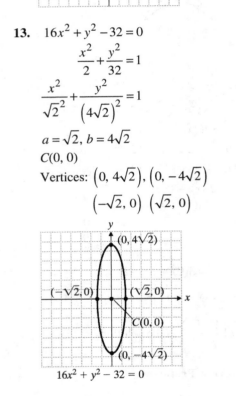

13. $16x^2 + y^2 - 32 = 0$

$\dfrac{x^2}{2} + \dfrac{y^2}{32} = 1$

$\dfrac{x^2}{\sqrt{2}^2} + \dfrac{y^2}{\left(4\sqrt{2}\right)^2} = 1$

$a = \sqrt{2}$, $b = 4\sqrt{2}$

$C(0, 0)$

Vertices: $\left(0, 4\sqrt{2}\right)$, $\left(0, -4\sqrt{2}\right)$

$\left(-\sqrt{2}, 0\right)$ $\left(\sqrt{2}, 0\right)$

14. $\dfrac{(x+5)^2}{4} + \dfrac{(y+3)^2}{25} = 1$

$\dfrac{(x-(-5))^2}{2^2} + \dfrac{(y-(-3))^2}{5^2} = 1$

$C(-5, -3)$

$a = 2$, $b = 5$

Vertices: $(-3, -3)$, $(-7, -3)$, $(-5, 2)$, $(-5, -8)$

15. $x^2 - 4y^2 - 16 = 0$

$$\frac{x^2}{16} - \frac{y^2}{4} = 1$$

$$\frac{x^2}{4^2} - \frac{y^2}{2^2} = 1$$

$C(0, 0)$

$a = 4, b = 2$

Vertices: $(-4, 0), (4, 0)$

Asymptotes: $y = \pm\frac{2}{4}x = \pm\frac{1}{2}x$

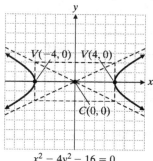

$$x^2 - 4y^2 - 16 = 0$$

16. $3y^2 - x^2 = 27$

$$\frac{y^2}{9} - \frac{x^2}{27} = 1$$

$$\frac{y^2}{3^2} - \frac{x^2}{\sqrt{27}^2} = 1$$

$a = 3, b = \sqrt{27}$

$C(0, 0)$

Vertices: $(0, 3), (0, -3)$

Asymptotes: $y = \pm\frac{3}{\sqrt{27}}x = \pm\frac{\sqrt{3}}{3}x$

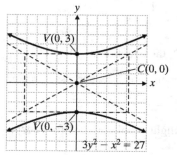

$$3y^2 - x^2 = 27$$

17.
$$\frac{(x-2)^2}{4} - \frac{(y+3)^2}{25} = 1$$

$$\frac{(x-2)^2}{2^2} - \frac{(y-(-3))^2}{5^2} = 1$$

$C(2, -3), a = 2, b = 5$

Vertices: $(0, -3), (4, -3)$

18. $x^2 + y = 9 \longrightarrow \quad x^2 + y = 9$

$\quad y - x = 3 \xrightarrow{\times -1} x \quad - y = -3$

$\qquad\qquad\qquad\qquad\overline{\quad x^2 + x = 6\quad}$

$(x + 3)(x - 2) = 0$

$\begin{array}{ll} x + 3 = 0 & x - 2 = 0 \\ x = -3 & x = 2 \end{array}$

$y = x + 3 = \begin{cases} -3 + 3 = 0 \\ 2 + 3 = 5 \end{cases}$

$(-3, 0), (2, 5)$ is the solution.

19. $x^2 + y^2 = 4$

$\quad x + y = 2$

$\qquad y = 2 - x$

$\quad x^2 + (2 - x)^2 = 4$

$x^2 + x^2 - 4x + 4 = 4$

$\qquad x^2 - 2x = 0$

$\qquad x(x - 2) = 0$

$x = 0, x = 2$

$x = 0, y = 2 - 0 = 2$

$x = 2, y = 2 - 2 = 0$

$(0, 2), (2, 0)$ is the solution.

20. $2x^2 + y^2 = 17$

$\qquad y^2 = 17 - 2x^2$

$\qquad x^2 + 2y^2 = 22$

$x^2 + 2(17 - 2x^2) = 22$

$\quad x^2 + 34 - 4x^2 = 22$

$\qquad\qquad 3x^2 = 12$

$\qquad\qquad x^2 = 4$

$\qquad\qquad x = \pm 2$

$y = \pm\sqrt{17 - 2x^2}$

$y = \pm\sqrt{17 - 2(4)}$

$y = \pm 3$

$(2, -3), (2, 3), (-2, -3), (-2, 3)$ is the solution.

21. $3x^2 - 4y^2 = 12$

$\quad 7x^2 - y^2 = 8$

$\qquad y^2 = 7x^2 - 8$

$\quad 3x^2 - 4(7x^2 - 8) = 12$

$\quad 3x^2 - 28x^2 + 32 = 12$

$\qquad\qquad 25x^2 = 20$

$\qquad\qquad\quad x^2 = \dfrac{20}{25}$

$y^2 = 7x^2 - 8 = 7 \cdot \dfrac{20}{25} - 8 = -\dfrac{12}{5}$

$y^2 > 0 \Rightarrow$ no real solution, hyperbolas do not intersect.

22. $\quad y = x^2 + 1$

$\qquad x^2 = y - 1$

$\quad x^2 + y^2 - 8y + 7 = 0$

$\quad y - 1 + y^2 - 8y + 7 = 0$

$\qquad\qquad y^2 - 7y + 6 = 0$

$\qquad\qquad (y-1)(y-6) = 0$

$\quad y - 1 = 0 \qquad\qquad y - 6 = 0$

$\qquad y = 1 \qquad\qquad\quad y = 6$

$x^2 = y - 1 = \begin{cases} 1 - 1 = 0 \\ 6 - 1 = 5 \end{cases}$

$x = \begin{cases} 0 \\ \pm\sqrt{5} \end{cases}$

$(0, 1),\ \left(\sqrt{5},\, 6\right),\ \left(-\sqrt{5},\, 6\right)$ is the solution.

23. $2x^2 + y^2 = 18$

$\quad xy = 4$

$\quad y = \dfrac{4}{x}$

$\quad 2x^2 + \left(\dfrac{4}{x}\right)^2 = 18$

$\qquad 2x^4 + 16 = 18x^2$

$\qquad x^4 - 9x^2 + 8 = 0$

$\quad (x^2 - 8)(x^2 - 1) = 0$

$\quad x^2 = 8 \qquad\qquad x^2 = 1$

$\quad x = \pm 2\sqrt{2} \qquad\quad x = \pm 1$

$y = \dfrac{4}{x} = \begin{cases} \dfrac{4}{2\sqrt{2}} = \sqrt{2} \\[4pt] \dfrac{4}{-2\sqrt{2}} = -\sqrt{2} \\[4pt] \dfrac{4}{1} = 4 \\[4pt] \dfrac{4}{-1} = -4 \end{cases}$

$\left(2\sqrt{2},\, \sqrt{2}\right),\ \left(-2\sqrt{2},\, -\sqrt{2}\right),\ (1, 4),\ (-1, -4)$ is the solution.

24. $\quad y^2 - 2x^2 = 2 \xrightarrow{\times -2} -2y^2 + 4x^2 = -4$

$\quad 2y^2 - 3x^2 = 5 \xrightarrow{} \underline{\quad 2y^2 - 3x^2 = 5 \quad}$

$\qquad\qquad\qquad\qquad\qquad\qquad x^2 = 1,\ x = \pm 1$

$y^2 = 2x^2 + 2 = 2(1) + 2 = 4$

$\quad y = \pm 2$

$(1, \pm 2),\ (-1, \pm 2)$ is the solution.

25. $y^2 = 2x$

$\quad y = \dfrac{1}{2}x + 1$

$\quad x = 2y - 2$

$\qquad\quad y^2 = 2(2y - 2) = 4y - 4$

$\quad y^2 - 4y + 4 = 0$

$\qquad (y - 2)^2 = 0$

$\qquad\qquad y = 2$

$x = 2y - 2 = 2(2) - 2 = 4 - 2 = 2$

$x = 2$

$(2, 2)$ is the solution.

26. $y = ax^2,\ a = \dfrac{1}{4p},\ \text{so } y = \dfrac{1}{4p}x^2$

Since $p = 2$, the equation is $y = \dfrac{1}{8}x^2$.

Since the opening is 5 feet across, $x = 2.5$.

$y = \dfrac{1}{8}(2.5)^2 = \dfrac{6.25}{8} = 0.78125$

The searchlight should be 0.78 foot deep.

27. $y = ax^2,\ a = \dfrac{1}{4p},\ \text{so } y = \dfrac{1}{4p}x^2$

Since the dish is 10 feet across and 4 feet deep, the point $(5, 4)$ is on the parabola.

$4 = \dfrac{1}{4p}(5)^2$

$p = \dfrac{5^2}{16} = \dfrac{25}{16} = 1.5625$

The receiver should be 1.56 feet from the center of the dish.

How Am I Doing? Chapter 10 Test

1. $(-6, -8)$ and $(-2, 5)$

   $d = \sqrt{(-2-(-6))^2 + (5-(-8))^2}$

   $d = \sqrt{16+169}$

   $d = \sqrt{185}$

2. $y^2 - 6y - x + 13 = 0$

   $x - 13 + 9 = y^2 - 6y + 9$

   $x = (y-3)^2 + 4$

   Parabola: $V(4, 3)$

   Let $y = 0$.

   $x = (0-3)^2 + 4$

   $x = 13$

   x-int: $(13, 0)$

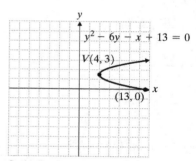

   Scale: Each square = 2 units

3. $x^2 + y^2 + 6x - 4y + 9 = 0$

   $x^2 + 6x + 9 + y^2 - 4y + 4 = -9 + 9 + 4$

   $(x+3)^2 + (y-2)^2 = 4 = 2^2$

   Circle: $C(-3, 2)$, $r = 2$

   x-int: $(-3, 0)$

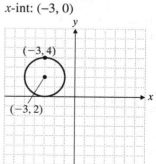

4. $\dfrac{x^2}{25} + \dfrac{y^2}{1} = 1$

   $\dfrac{x^2}{5^2} + \dfrac{y^2}{1^2} = 1$

   Ellipse: $C(0, 0)$

   $a = 5$, $b = 1$

   Vertices: $(5, 0)$, $(-5, 0)$, $(0, 1)$, $(0, -1)$

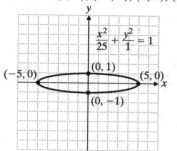

5. $\dfrac{x^2}{10} - \dfrac{y^2}{9} = 1$

   Hyperbola

   Center: $C(0, 0)$

   Vertices: $V\left(\pm\sqrt{10}, 0\right)$

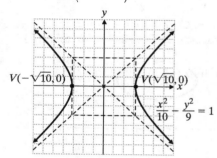

6. $y = -2(x+3)^2 + 4$

Parabola: $V(-3, 4)$

Let $x = 0$.

$y = -2(0+3)^2 + 4$

$y = -18 + 4$

$y = -14$

y-intercept: $(0, -14)$

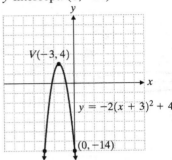

Scale: Each square = 2 units

7. $\dfrac{(x+2)^2}{16} + \dfrac{(y-5)^2}{4} = 1$

$\dfrac{(x-(-2))^2}{4^2} + \dfrac{(y-5)^2}{2^2} = 1$

Ellipse: $C(-2, 5)$, $a = 4$, $b = 2$

Vertices: $(-2, 7)$, $(-2, 3)$, $(-6, 5)$, $(2, 5)$

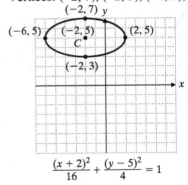

$\dfrac{(x+2)^2}{16} + \dfrac{(y-5)^2}{4} = 1$

8. $7y^2 - 7x^2 = 28$

$\dfrac{7y^2}{28} - \dfrac{7x^2}{28} = \dfrac{28}{28}$

$\dfrac{y^2}{4} - \dfrac{x^2}{4} = 1$

$\dfrac{y^2}{2^2} - \dfrac{x^2}{2^2} = 1$

Hyperbola

$C(0, 0)$

Vertices: $(0, 2)$, $(0, -2)$, $a = 2$, $b = 2$

$y_{\text{asymptote}} = \pm\dfrac{2}{2}x = \pm x$

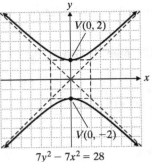

$7y^2 - 7x^2 = 28$

9. $(x-h)^2 + (y-k)^2 = r^2$

$(x-3)^2 + (y-(-5))^2 = \sqrt{8}^2$

$(x-3)^2 + (y+5)^2 = 8$

10. $C(h, k) = C(0, 0)$

$h = 0$, $k = 0$

$(3, 0) \Rightarrow a = 3$

$(0, 5) \Rightarrow b = 5$

$\dfrac{(x-h)^2}{a^2} + \dfrac{(y-k)^2}{b^2} = 1$

$\dfrac{(x-0)^2}{3^2} + \dfrac{(y-0)^2}{5^2} = 1$

$\dfrac{x^2}{9} + \dfrac{y^2}{25} = 1$

11. $x = (y-k)^2 + h$, $(h, k) = (-7, 3)$

$x = (y-3)^2 + (-7)$

$x = (y-3)^2 - 7$

Check: $(2, 0)$

$2 \overset{?}{=} (0-3)^2 - 7$

$2 \overset{?}{=} 9 - 7$

$2 = 2$

12. $C(h, k) = C(0, 0) \Rightarrow h = 0$, $k = 0$

$V(\pm 3, 0) \Rightarrow a = 3$

$y_{\text{asymptote}} = \pm\dfrac{b}{a}x = \dfrac{5}{3}x \Rightarrow b = 5$

$\dfrac{(x-h)^2}{a^2} - \dfrac{(y-k)^2}{b^2} = 1$

$\dfrac{(x-0)^2}{3^2} - \dfrac{(y-0)^2}{5^2} = 1$

$\dfrac{x^2}{9} - \dfrac{y^2}{25} = 1$

13. $-2x + y = 5$
$$y = 2x + 5$$
$$x^2 + y^2 - 25 = 0$$
$$x^2 + (2x + 5)^2 - 25 = 0$$
$$x^2 + 4x^2 + 20x + 25 - 25 = 0$$
$$5x^2 + 20x = 0$$
$$5x(x + 4) = 0$$
$x = 0, x = -4$
$y = 2x + 5 = 2(0) + 5 = 5$
$y = 2(-4) + 5 = -3$
$(0, 5), (-4, -3)$ is the solution.

14. $x^2 + y^2 = 9$
$$y = x - 3$$
$$x^2 + (x - 3)^2 = 9$$
$$x^2 + x^2 - 6x + 9 = 9$$
$$x^2 - 3x = 0$$
$$x(x - 3) = 0$$
$x = 0, x = 3$
$y = x - 3 = 0 - 3 = -3$
$y = x - 3 = 3 - 3 = 0$
$(0, -3), (3, 0)$ is the solution.

15. $4x^2 + y^2 - 4 = 0$
$$y^2 = 4 - 4x^2$$
$$9x^2 - 4y^2 - 9 = 0$$
$$9x^2 - 4(4 - 4x^2) - 9 = 0$$
$$9x^2 - 16 + 16x^2 - 9 = 0$$
$$25x^2 = 25$$
$$x^2 = 1$$
$$x = \pm 1$$
$y^2 = 4 - 4x^2 = 4 - 4(1) = 0$
$y = 0$
$(1, 0), (-1, 0)$ is the solution.

16. $x^2 + 2y^2 = 15$
$$\underline{x^2 - y^2 = 6} \text{ subtract}$$
$$3y^2 = 9 \Rightarrow y = \pm\sqrt{3}$$
$$x^2 = 6 + y^2 = 6 + 3 = 9 \Rightarrow x = \pm 3$$
$\left(3, \sqrt{3}\right), \left(-3, \sqrt{3}\right), \left(3, -\sqrt{3}\right), \left(-3, -\sqrt{3}\right)$ is the solution.

Chapter 11

1. $f(x) = 3x - 5$

$$f\left(-\frac{2}{3}\right) = 3\left(-\frac{2}{3}\right) - 5$$
$$= -2 - 5$$
$$= -7$$

3. $f(x) = 3x - 5$
$$f(a+5) = 3(a+5) - 5$$
$$= 3a + 15 - 5$$
$$= 3a + 10$$

5. $g(x) = \frac{1}{2}x - 3$

$$g(2) + g(a) = \frac{1}{2}(2) - 3 + \frac{1}{2}a - 3$$
$$= 1 - 3 + \frac{1}{2}a - 3$$
$$= \frac{1}{2}a - 5$$

7. $g(x) = \frac{1}{2}x - 3$

$$g(4a) - g(a) = \frac{1}{2}(4a) - 3 - \left[\frac{1}{2}a - 3\right]$$
$$= 2a - 3 - \frac{1}{2}a + 3$$
$$= \frac{3}{2}a$$

9. $g(x) = \frac{1}{2}x - 3$

$$g(2a-4) = \frac{1}{2}(2a-4) - 3$$
$$= a - 2 - 3$$
$$= a - 5$$

11. $g(x) = \frac{1}{2}x - 3$

$$g(a^2) - g\left(\frac{2}{5}\right) = \left(\frac{1}{2}a^2 - 3\right) - \left(\frac{1}{2} \cdot \frac{2}{5} - 3\right)$$
$$= \frac{1}{2}a^2 - 3 - \frac{1}{5} + 3$$
$$= \frac{1}{2}a^2 - \frac{1}{5}$$

13. $p(x) = 3x^2 + 4x - 2$

$$p(-2) = 3(-2)^2 + 4(-2) - 2$$
$$= 3(4) + 4(-2) - 2$$
$$= 12 - 8 - 2$$
$$= 2$$

15. $p(x) = 3x^2 + 4x - 2$

$$p\left(\frac{1}{2}\right) = 3\left(\frac{1}{2}\right)^2 + 4\left(\frac{1}{2}\right) - 2$$
$$= \frac{3}{4} + 2 - 2$$
$$= \frac{3}{4}$$

17. $p(x) = 3x^2 + 4x - 2$

$$p(a+1) = 3(a+1)^2 + 4(a+1) - 2$$
$$= 3a^2 + 6a + 3 + 4a + 4 - 2$$
$$= 3a^2 + 10a + 5$$

19. $p(x) = 3x^2 + 4x - 2$

$$p\left(-\frac{2a}{3}\right) = 3\left(-\frac{2a}{3}\right)^2 + 4\left(-\frac{2a}{3}\right) - 2$$
$$= \frac{4a^2}{3} - \frac{8a}{3} - 2$$

21. $h(x) = \sqrt{x+5}$
$$h(4) = \sqrt{4+5} = \sqrt{9} = 3$$

23. $h(x) = \sqrt{x+5}$
$$h(7) = \sqrt{7+5} = \sqrt{12} = \sqrt{4 \cdot 3} = 2\sqrt{3}$$

25. $h(x) = \sqrt{x+5}$
$$h(a^2 - 1) = \sqrt{a^2 - 1 + 5} = \sqrt{a^2 + 4}$$

27. $h(x) = \sqrt{x+5}$
$$h(-2b) = \sqrt{-2b+5}$$

29. $h(x) = \sqrt{x+5}$

$$h(4a-1) = \sqrt{4a-1+5}$$
$$= \sqrt{4a+4}$$
$$= \sqrt{4(a+1)}$$
$$= 2\sqrt{a+1}$$

31. $h(x) = \sqrt{x+5}$

$$h(b^2+b) = \sqrt{b^2+b+5}$$

33. $r(x) = \dfrac{7}{x-3}$

$$r(7) = \frac{7}{7-3} = \frac{7}{4}$$

35. $r(x) = \dfrac{7}{x-3}$

$$r(3.5) = \frac{7}{3.5-3} = \frac{7}{0.5} = 14$$

37. $r(x) = \dfrac{7}{x-3}$

$$r(a^2) = \frac{7}{a^2-3}$$

39. $r(x) = \dfrac{7}{x-3}$

$$r(a+2) = \frac{7}{a+2-3} = \frac{7}{a-1}$$

41. $r(x) = \dfrac{7}{x-3}$

$$r\left(\frac{1}{3}\right) + r(-5) = \frac{7}{\frac{1}{3}-3} + \frac{7}{-5-3}$$
$$= -\frac{21}{8} - \frac{7}{8}$$
$$= -\frac{28}{8}$$
$$= -\frac{7}{2}$$

43. $f(x) = 2x-3$

$$\frac{f(x+h)-f(x)}{h} = \frac{2(x+h)-3-(2x-3)}{h}$$
$$= \frac{2x+2h-3-2x+3}{h}$$
$$= \frac{2h}{h}$$
$$= 2$$

45. $f(x) = x^2-x$

$$\frac{f(x+h)-f(x)}{h} = \frac{(x+h)^2-(x+h)-(x^2-x)}{h}$$
$$= \frac{x^2+2xh+h^2-x-h-x^2+x}{h}$$
$$= \frac{2xh+h^2-h}{h}$$
$$= \frac{h(2x+h-1)}{h}$$
$$= 2x+h-1$$

47. $P = 2.5w^2$

   a. $P(w) = 2.5w^2$

   b. $P(20) = 2.5(20)^2 = 1000$ kilowatts

   c. $P(e) = 2.5(20+e)^2$
$$= 2.5(400+40e+e^2)$$
$$= 2.5e^2+100e+1000$$

   d. $P(2) = 2.5(2)^2+100(2)+1000$
$$= 1210 \text{ kilowatts}$$

49. The function values associated with $p(x) - 13$ would be the function values of $p(x)$ decreased by 13.

$$p(3) - 13 \approx 39 - 13 = 26$$

51. $f(x) = 3x^2-4.6x+1.23$

$$f(0.026a) = 3(0.026a)^2-4.6(0.026a)+1.23$$
$$= 0.002a^2-0.120a+1.23$$

Cumulative Review

53. $\dfrac{7}{6}+\dfrac{5}{x}=\dfrac{3}{2x}$

$7x+30=9$

$7x=-21$

$x=-3$

54.

$$\dfrac{1}{6}-\dfrac{2}{3x+6}=\dfrac{1}{2x+4}$$

$$6(x+2)\cdot\dfrac{1}{6}-6(x+2)\cdot\dfrac{2}{3(x+2)}=6(x+2)\cdot\dfrac{1}{2(x+2)}$$

$$x+2-4=3$$

$$x=5$$

55. $\dfrac{V_{\text{Earth}}}{V_{\text{Mars}}}=\dfrac{\frac{4}{3}\pi\left(\dfrac{7927}{2}\right)^3}{\frac{4}{3}\pi\left(\dfrac{4211}{2}\right)^3}$

$V_{\text{Earth}}\approx 6.67069140(V_{\text{Mars}})$

The volume of Earth is approximately 6.7 times greater than the volume of Mars.

56. $\dfrac{V_{\text{Saturn}}}{V_{\text{Neptune}}}=\dfrac{\frac{4}{3}\pi(37,366)^3}{\frac{4}{3}\pi(15,345)^3}$

$V_{\text{Saturn}}\approx 14.4387351(V_{\text{Neptune}})$

Saturn's volume is approximately 14.4 times greater than the volume of Neptune.

Quick Quiz 11.1

1. $f(x)=\dfrac{3}{5}x-4$

$f(a)=\dfrac{3}{5}a-\dfrac{20}{5}$

$f(-3)=\dfrac{3}{5}(-3)-\dfrac{20}{5}=-\dfrac{9}{5}-\dfrac{20}{5}=-\dfrac{29}{5}$

$f(a)-f(-3)=\dfrac{3a}{5}-\dfrac{20}{5}+\dfrac{29}{5}=\dfrac{3a}{5}+\dfrac{9}{5}$

2. $g(x)=2x^2-3x+4$

$g\left(\dfrac{2}{3}a\right)=2\left(\dfrac{2}{3}a\right)^2-3\left(\dfrac{2}{3}a\right)+4$

$\phantom{g\left(\dfrac{2}{3}a\right)}=2\left(\dfrac{4}{9}a^2\right)-\dfrac{6}{3}a+4$

$\phantom{g\left(\dfrac{2}{3}a\right)}=\dfrac{8}{9}a^2-2a+4$

3. $h(x) = \dfrac{3}{x+4}$

$h(a-6) = \dfrac{3}{a-6+4} = \dfrac{3}{a-2}$

4. Answers may vary. Possible solution:
   For the function $k(x) = \sqrt{3x+1}$ evaluated at
   $x = 2a - 1$, substitute $2a - 1$ for x in the function
   and solve:
   $k(2a-1) = \sqrt{3(2a-1)+1} = \sqrt{6a-2}$

11.2 Exercises

1. No, $f(x + 2)$ means to substitute $x + 2$ for x in the
   function $f(x)$. $f(x) + f(2)$ means to evaluate $f(x)$
   and $f(2)$ and then add the two results. One
   example is $f(x) = 2x + 1$
   $f(x + 2) = 2(x + 2) + 1 = 2x + 5$
   $f(x) + f(2) = 2x + 1 + 2(2) + 1 = 2x + 6$

3. To obtain the graph of $f(x) + k$ for $k > 0$, shift the
   graph of $f(x)$ <u>up</u> k units.

5. Graph fails vertical line test and does not
   represent a function.

7. Graph passes vertical line test and does represent
   a function.

9. Graph passes vertical line test and does represent
   a function.

11. Graph fails vertical line test and does not
    represent a function.

13. Graph passes vertical line test and does represent
    a function.

For Exercises 15, 17, and 19:

x	$f(x) = x^2$
-2	4
-1	1
0	0
1	1
2	4

15. $f(x) = x^2$, $h(x) = x^2 - 3$
    Shift $f(x)$ down 3 units.

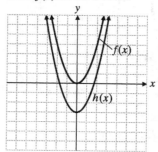

17. $f(x) = x^2$, $p(x) = (x+3)^2$
    Shift $f(x)$ left 3 units.

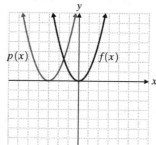

19. $f(x) = x^2$, $g(x) = (x-2)^2 + 1$
    Shift $f(x)$ right 2 units and up 1 unit.

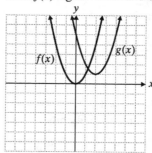

21. $f(x) = x^3$, $r(x) = x^3 - 1$
    Shift $f(x)$ down 1 unit.

x	$f(x) = x^3$
-2	-8
-1	-1
0	0
1	1
2	8

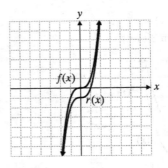

For Exercises 23 and 25:

x	$f(x)=|x|$
−2	2
−1	1
0	0
1	1
2	2

23. $f(x)=|x|$, $s(x)=|x+4|$
Shift $f(x)$ left 4 units.

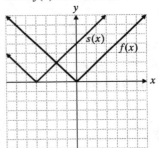

25. $f(x)=|x|$, $t(x)=|x-3|-4$
Shift right 3 units and down 4 units.

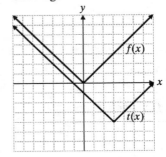

27. $f(x)=\dfrac{3}{x}$, $g(x)=\dfrac{3}{x}-2$
Shift $f(x)$ down 2 units.

x	$f(x)=\frac{3}{x}$
−3	−1
−1	−3
0	undefined
1	3
3	1

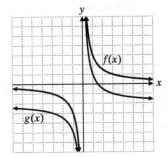

Cumulative Review

29. $\sqrt{12}+3\sqrt{50}-4\sqrt{27}=\sqrt{4\cdot3}+3\sqrt{25\cdot2}-4\sqrt{9\cdot3}$
$=2\sqrt{3}+3\cdot5\sqrt{2}-4\cdot3\sqrt{3}$
$=2\sqrt{3}+15\sqrt{2}-12\sqrt{3}$
$=15\sqrt{2}-10\sqrt{3}$

30. $\left(\sqrt{3x}-1\right)^2=\sqrt{3x}^2-2\sqrt{3x}(1)+1^2$
$=3x-2\sqrt{3x}+1$

31. $\dfrac{\sqrt{5}-2}{\sqrt{5}+1}=\dfrac{\sqrt{5}-2}{\sqrt{5}+1}\cdot\dfrac{\sqrt{5}-1}{\sqrt{5}-1}=\dfrac{5-3\sqrt{5}+2}{5-1}=\dfrac{7-3\sqrt{5}}{4}$

Quick Quiz 11.2

1. The graph of $h(x)$ is 3 units to the right of the graph of $f(x)$.

2. The graph of $g(x)$ is 4 units below the graph of $f(x)$.

3. $f(x) = |x|$, $k(x) = |x - 1| - 4$
Shift $f(x)$ to the right 1 unit and down 4 units.

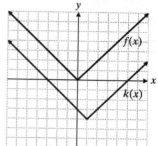

4. Answers may vary. Possible solution:
If a vertical line passes through more than one point of the graph of a relation, the relation is not a function.

Use Math to Save Money

1. December 2012: $100 \times \$2.95 = \$295$
November 2014: $100 \times \$3.55 = \$355$

2. $5 \times \$355 = \$1775$
Mark can expect to pay $1775 for the upcoming winter.

3. 4° lower = 8% savings
8% of $1775 = $0.08 \times \$1775 = \$142$
He will save $142 by setting his thermostat at 68°.

4. $\dfrac{\$355}{\$1775} = 0.20$ or 20%

20% savings = 10° lower
$72° - 10° = 62°$
Mark can save one month's heating costs by setting his thermostat at 62°.

5. Answers will vary.

6. Answers will vary.

How Am I Doing? Sections 11.1–11.2
(Available online through MyMathLab.)

1. $f(x) = 2x - 6$
$f(-3) = 2(-3) - 6 = -6 - 6 = -12$

2. $f(x) = 2x - 6$
$f(a) = 2a - 6$

3. $f(x) = 2x - 6$
$f(2a) = 2(2a) - 6 = 4a - 6$

4. $f(x) = 2x - 6$
$f(a + 2) = 2(a + 2) - 6 = 2a + 4 - 6 = 2a - 2$

5. $f(x) = 5x^2 + 2x - 3$
$f(-2) = 5(-2)^2 + 2(-2) - 3 = 20 - 4 - 3 = 13$

6. $f(x) = 5x^2 + 2x - 3$
$f(a) = 5a^2 + 2a - 3$

7. $f(x) = 5x^2 + 2x - 3$
$$\begin{aligned}
f(a - 1) &= 5(a-1)^2 + 2(a-1) - 3 \\
&= 5(a^2 - 2a + 1) + 2a - 2 - 3 \\
&= 5a^2 - 10a + 5 + 2a - 5 \\
&= 5a^2 - 8a
\end{aligned}$$

8. $f(x) = 5x^2 + 2x - 3$
$f(-2a) = 5(-2a)^2 + 2(-2a) - 3 = 20a^2 - 4a - 3$

9. $f(x) = \dfrac{3x}{x + 2}$
$$\begin{aligned}
f(a) + f(a-2) &= \frac{3a}{a+2} + \frac{3(a-2)}{a-2+2} \\
&= \frac{3a^2 + 3(a+2)(a-2)}{a(a+2)} \\
&= \frac{3a^2 + 3a^2 - 12}{a(a+2)} \\
&= \frac{6a^2 - 12}{a(a+2)} \\
&= \frac{6(a^2 - 2)}{a(a+2)}
\end{aligned}$$

10. $f(x) = \dfrac{3x}{x + 2}$
$$\begin{aligned}
f(3a) - f(3) &= \frac{3(3a)}{3a+2} - \frac{3(3)}{3+2} \\
&= \frac{9a}{3a+2} - \frac{9}{5} \\
&= \frac{45a - 9(3a+2)}{5(3a+2)} \\
&= \frac{45a - 27a - 18}{5(3a+2)} \\
&= \frac{18(a-1)}{5(3a+2)}
\end{aligned}$$

11. Graph passes vertical line test and therefore represents a function.

12. Graph does not pass vertical line test and hence does not represent a function.

13. $f(x) = |x|, s(x) = |x - 3|$
Shift $f(x)$ to the right 3 units.

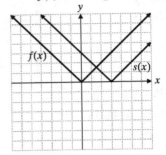

14. $f(x) = x^2, h(x) = (x+2)^2 + 3$
Shift $f(x)$ to the left 2 units and up 3 units.

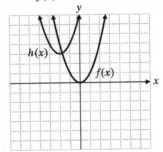

15. $f(x) = \dfrac{4}{x+2}$

x	$f(x) = \dfrac{4}{x+2}$
-5	$-\dfrac{4}{3}$
-4	-2
-3	-4
-2	undefined
0	2
1	$\dfrac{4}{3}$
2	1

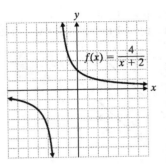

$f(x) = \dfrac{4}{x+2}$

16. The graph of $g(x) = \dfrac{4}{x+2} - 2$ is the graph of

$f(x) = \dfrac{4}{x+2}$ shifted down 2 units.

11.3 Exercises

1. $f(x) = -2x + 3, g(x) = 2 + 4x$

   a. $(f+g)(x) = f(x) + g(x)$
$$= -2x + 3 + 2 + 4x$$
$$= 2x + 5$$

   b. $(f-g)(x) = f(x) - g(x)$
$$= -2x + 3 - (2 + 4x)$$
$$= -2x + 3 - 2 - 4x$$
$$= -6x + 1$$

   c. $(f+g)(2) = 2(2) + 5 = 9$

   d. $(f-g)(-1) = -6(-1) + 1 = 7$

3. $f(x) = 2x^2 - 1, g(x) = 4x + 1$

   a. $(f+g)(x) = f(x) + g(x)$
$$= 2x^2 - 1 + 4x + 1$$
$$= 2x^2 + 4x$$

   b. $(f-g)(x) = f(x) - g(x)$
$$= 2x^2 - 1 - (4x + 1)$$
$$= 2x^2 - 1 - 4x - 1$$
$$= 2x^2 - 4x - 2$$

   c. $(f+g)(2) = 2(2)^2 + 4(2) = 8 + 8 = 16$

   d. $(f-g)(-1) = 2(-1)^2 - 4(-1) - 2$
$$= 2 + 4 - 2$$
$$= 4$$

5. $f(x) = x^3 - \dfrac{1}{2}x^2 + x, \; g(x) = x^2 - \dfrac{x}{4} - 5$

   a. $(f+g)(x) = f(x) + g(x)$
   $$= x^3 - \frac{1}{2}x^2 + x + x^2 - \frac{x}{4} - 5$$
   $$= x^3 + \frac{1}{2}x^2 + \frac{3x}{4} - 5$$

   b. $(f-g)(x) = f(x) - g(x)$
   $$= x^3 - \frac{1}{2}x^2 + x - \left(x^2 - \frac{x}{4} - 5 \right)$$
   $$= x^3 - \frac{1}{2}x^2 + x - x^2 + \frac{x}{4} + 5$$
   $$= x^3 - \frac{3}{2}x^2 + \frac{5x}{4} + 5$$

   c. $(f+g)(2) = (2)^3 + \dfrac{1}{2}(2)^2 + \dfrac{3(2)}{4} - 5 = \dfrac{13}{2}$

   d. $(f-g)(-1) = (-1)^3 - \dfrac{3}{2}(-1)^2 + \dfrac{5(-1)}{4} + 5 = \dfrac{5}{4}$

7. $f(x) = -5\sqrt{x+6}, \; g(x) = 8\sqrt{x+6}$

   a. $(f+g)(x) = f(x) + g(x)$
   $$= -5\sqrt{x+6} + 8\sqrt{x+6}$$
   $$= 3\sqrt{x+6}$$

   b. $(f-g)(x) = f(x) - g(x)$
   $$= -5\sqrt{x+6} - 8\sqrt{x+6}$$
   $$= -13\sqrt{x+6}$$

   c. $(f+g)(2) = 3\sqrt{2+6} = 3\sqrt{8} = 6\sqrt{2}$

   d. $(f-g)(-1) = -13\sqrt{-1+6} = -13\sqrt{5}$

9. $f(x) = 2x - 3, \; g(x) = -2x^2 - 3x + 1$

   a. $(fg)(x) = f(x)g(x)$
   $$= (2x-3)(-2x^2 - 3x + 1)$$
   $$= -4x^3 - 6x^2 + 2x + 6x^2 + 9x - 3$$
   $$= -4x^3 + 11x - 3$$

   b. $(fg)(-3) = -4(-3)^3 + 11(-3) - 3$
   $$= 108 - 33 - 3$$
   $$= 72$$

11. $f(x) = \dfrac{2}{x^2}, \; x \neq 0; \; g(x) = x^2 - x$

   a. $(fg)(x) = f(x)g(x)$
   $$= \frac{2}{x^2}(x^2 - x)$$
   $$= 2 - \frac{2}{x}$$
   $$= \frac{2x - 2}{x}$$
   $$= \frac{2(x-1)}{x}, \; x \neq 0$$

   b. $(fg)(-3) = \dfrac{2(-3-1)}{-3} = \dfrac{8}{3}$

13. $f(x) = \sqrt{-2x+1}, \; x \leq \dfrac{1}{2}; \; g(x) = -3x$

   a. $(fg)(x) = f(x)g(x)$
   $$= \sqrt{-2x+1}(-3x)$$
   $$= -3x\sqrt{-2x+1}$$

   b. $(fg)(-3) = -3(-3)\sqrt{-2(-3)+1} = 9\sqrt{7}$

15. $f(x) = x - 6, \; g(x) = 3x$

   a. $\left(\dfrac{f}{g} \right)(x) = \dfrac{f(x)}{g(x)} = \dfrac{x-6}{3x}, \; x \neq 0$

   b. $\left(\dfrac{f}{g} \right)(2) = \dfrac{2-6}{3(2)} = \dfrac{-4}{6} = -\dfrac{2}{3}$

17. $f(x) = x^2 - 1, \; g(x) = x - 1$

   a. $\left(\dfrac{f}{g} \right)(x) = \dfrac{f(x)}{g(x)}$
   $$= \frac{x^2 - 1}{x - 1}$$
   $$= \frac{(x-1)(x+1)}{x-1}$$
   $$= x + 1, \; x \neq 1$$

   b. $\left(\dfrac{f}{g} \right)(2) = 2 + 1 = 3$

19. $f(x) = x^2 + 10x + 25,\ g(x) = x + 5$

a. $\left(\dfrac{f}{g}\right)(x) = \dfrac{f(x)}{g(x)}$

$\qquad = \dfrac{x^2 + 10x + 25}{x + 5}$

$\qquad = \dfrac{(x+5)(x+5)}{(x+5)}$

$\qquad = x + 5,\ x \neq -5$

b. $\left(\dfrac{f}{g}\right)(2) = 2 + 5 = 7$

21. $f(x) = 4x - 1,\ g(x) = 4x^2 + 7x - 2$

a. $\left(\dfrac{f}{g}\right)(x) = \dfrac{f(x)}{g(x)} = \dfrac{4x-1}{4x^2 + 7x - 2}$

$\qquad = \dfrac{(4x-1)}{(4x-1)(x+2)}$

$\qquad = \dfrac{1}{x+2},\ x \neq -2,\ x \neq \dfrac{1}{4}$

b. $\left(\dfrac{f}{g}\right)(2) = \dfrac{1}{2+2} = \dfrac{1}{4}$

23. $f(x) = 3x + 2,\ g(x) = x^2 - 2x$

$(f - g)(x) = f(x) - g(x) = 3x + 2 - (x^2 - 2x)$

$\qquad = 3x + 2 - x^2 + 2x$

$\qquad = -x^2 + 5x + 2$

25. $g(x) = x^2 - 2x,\ h(x) = \dfrac{x-2}{3}$

$\left(\dfrac{g}{h}\right)(x) = \dfrac{g(x)}{h(x)} = \dfrac{x^2 - 2x}{\frac{x-2}{3}} = \dfrac{3x(x-2)}{x-2} = 3x,$

$x \neq 2$

27. $g(x) = x^2 - 2x,\ f(x) = 3x + 2$

$(fg)(x) = f(x)g(x)$

$\qquad = (3x + 2)(x^2 - 2x)$

$\qquad = 3x^3 - 6x^2 + 2x^2 - 4x$

$\qquad = 3x^3 - 4x^2 - 4x$

$(fg)(-1) = 3(-1)^3 - 4(-1)^2 - 4(-1)$

$\qquad = -3 - 4 + 4$

$\qquad = -3$

29. $g(x) = x^2 - 2x,\ f(x) = 3x + 2$

$\left(\dfrac{g}{f}\right)(-1) = \dfrac{g(-1)}{f(-1)} = \dfrac{(-1)^2 - 2(-1)}{3(-1) + 2} = \dfrac{3}{-1} = -3$

31. $f(x) = 2 - 3x,\ g(x) = 2x + 5$

$f[g(x)] = f[2x + 5]$

$\qquad = 2 - 3(2x + 5)$

$\qquad = 2 - 6x - 15$

$\qquad = -6x - 13$

33. $f(x) = 2x^2 + 5,\ g(x) = x - 1$

$f[g(x)] = f[x - 1]$

$\qquad = 2(x-1)^2 + 5$

$\qquad = 2(x^2 - 2x + 1) + 5$

$\qquad = 2x^2 - 4x + 7$

35. $f(x) = 8 - 5x,\ g(x) = x^2 + 3$

$f[g(x)] = f[x^2 + 3[$

$\qquad = 8 - 5(x^2 + 3)$

$\qquad = 8 - 5x^2 - 15$

$\qquad = -5x^2 - 7$

37. $f(x) = \dfrac{7}{2x - 3},\ g(x) = x + 2$

$f[g(x)] = f(x + 2)$

$\qquad = \dfrac{7}{2(x+2) - 3}$

$\qquad = \dfrac{7}{2x + 1},\ x \neq -\dfrac{1}{2}$

39. $f(x) = |x + 3|,\ g(x) = 2x - 1$

$f[g(x)] = f[2x - 1]$

$\qquad = |2x - 1 + 3|$

$\qquad = |2x + 2|$

41. $f(x) = x^2 + 2,\ g(x) = 3x + 5$

$f[g(x)] = f(3x + 5)$

$\qquad = (3x + 5)^2 + 2$

$\qquad = 9x^2 + 30x + 25 + 2$

$\qquad = 9x^2 + 30x + 27$

43. $f(x) = x^2 + 2, \; g(x) = 3x + 5$

$$g[f(x)] = g[x^2 + 2]$$
$$= 3(x^2 + 2) + 5$$
$$= 3x^2 + 6 + 5$$
$$= 3x^2 + 11$$

45. From Exercise 43, $g[f(x)] = 3x^2 + 11$.

$$g[f(0)] = 3(0)^2 + 11 = 11$$

47. $p(x) = \sqrt{x-1}, \; f(x) = x^2 + 2$

$$(p \circ f)(x) = p[f(x)]$$
$$= p[x^2 + 2]$$
$$= \sqrt{x^2 + 2 - 1}$$
$$= \sqrt{x^2 + 1}$$

49. $g(x) = 3x + 5, \; h(x) = \dfrac{1}{x}$

$$(g \circ h)\left(\sqrt{2}\right) = g\left[h\left(\sqrt{2}\right)\right]$$
$$= g\left[\frac{1}{\sqrt{2}}\right]$$
$$= 3 \cdot \frac{1}{\sqrt{2}} + 5$$
$$= 3 \cdot \frac{1}{\sqrt{2}} \cdot \frac{\sqrt{2}}{\sqrt{2}} + 5$$
$$= \frac{3\sqrt{2}}{2} + 5$$

51. $p(x) = \sqrt{x-1}, \; f(x) = x^2 + 2$

$$(p \circ f)(-3) = p[f(-3)]$$
$$= p[(-3)^2 + 2]$$
$$= p(11)$$
$$= \sqrt{11 - 1}$$
$$= \sqrt{10}$$

53. $K[C(F)] = K\left[\dfrac{5F - 160}{9}\right]$

$$= \frac{5F - 160}{9} + 273$$
$$= \frac{5F - 160 + 9(273)}{9}$$
$$= \frac{5F + 2297}{9}$$

55. $r(t) = 3t, \; a(r) = 3.14r^2$

$$a[r(t)] = a[3t] = 3.14(3t)^2 = 28.26t^2$$
$$a[r(20)] = 28.26(20)^2 = 11{,}304 \text{ ft}^2$$

Cumulative Review

56. $36x^2 - 12x + 1 = (6x)^2 - 2(6x)(1) + 1^2 = (6x - 1)^2$

57. $25x^4 - 1 = (5x^2)^2 - 1^2 = (5x^2 - 1)(5x^2 + 1)$

58. $x^4 - 10x^2 + 9 = (x^2 - 9)(x^2 - 1)$
$$= (x + 3)(x - 3)(x + 1)(x - 1)$$

59. $3x^2 - 7x + 2 = (3x - 1)(x - 2)$

Quick Quiz 11.3

1. $f(x) = 2x^2 - 4x - 8$

$$g(x) = -3x^2 + 5x - 2$$
$$(f - g)(x) = f(x) - g(x)$$
$$= (2x^2 - 4x - 8) - (-3x^2 + 5x - 2)$$
$$= 2x^2 - 4x - 8 + 3x^2 - 5x + 2$$
$$= 5x^2 - 9x - 6$$

2. $f(x) = x^2 - 3$

$$g(x) = \frac{x - 4}{2}$$
$$f[g(x)] = f\left(\frac{x - 4}{2}\right)$$
$$= \left(\frac{x - 4}{2}\right)^2 - 3$$
$$= \frac{x^2 - 8x + 16}{4} - 3$$
$$= \frac{x^2}{4} - \frac{8x}{4} + \frac{16}{4} - 3$$
$$= \frac{x^2}{4} - 2x + 1$$

3. $f(x) = x - 7$
$g(x) = 2x - 5$

$$\left(\frac{g}{f}\right)(2) = \frac{g(2)}{f(2)} = \frac{2(2) - 5}{2 - 7} = \frac{-1}{-5} = \frac{1}{5}$$

4. Answers may vary. Possible solution:
Evaluate both functions for -4, then subtract the result of $g(-4)$ from the result of $f(-4)$.

11.4 Exercises

1. A one-to-one function is a function in which no ordered pairs <u>have the same second coordinate</u>.

3. The graphs of a function f and its inverse f^{-1} are symmetric about the line <u>$y = x$</u>.

5. The graph of a horizontal line is the graph of a function because it passes the vertical line test. A horizontal line is not the graph of a one-to-one function because it fails the horizontal line test.

7. $B = \{(0, 1), (1, 0), (10, 0)\}$ is not one-to-one since two ordered pairs, $(1, 0)$ and $(10, 0)$, have the same second coordinate.

9. $F = \left\{ \left(\dfrac{2}{3}, 2 \right), \left(3, -\dfrac{4}{5} \right), \left(-\dfrac{2}{3}, -2 \right), \left(-3, \dfrac{4}{5} \right) \right\}$

   is a one-to-one function since no two ordered pairs have the same second coordinate.

11. $E = \{(2, 3.5), (-1, 8), (10, 3.5), (0, -8)\}$ is not one-to-one since two ordered pairs, $(2, 3.5)$ and $(10, 3.5)$, have the same second coordinate.

13. Graph of function passes the horizontal line test and therefore, function is one-to-one.

15. Graph of function fails the horizontal line test and therefore, function is not one-to-one.

17. Graph of function fails the horizontal line test and therefore, function is not one-to-one.

19. $J = \{(8, 2), (1, 1), (0, 0), (-8, -2)\}$

   $J^{-1} = \{(2, 8), (1, 1), (0, 0), (-2, -8)\}$

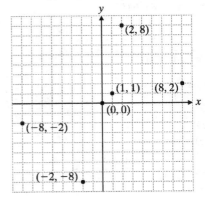

21. $f(x) = 4x - 5, \; f(x) \to y$

   $y = 4x - 5, \; x \leftrightarrow y$

   $x = 4y - 5$

   $4y = x + 5$

   $y = \dfrac{x+5}{4}, \; y \to f^{-1}(x)$

   $f^{-1}(x) = \dfrac{x+5}{4}$

23. $f(x) = x^3 - 8, \; f(x) \to y$

   $y = x^3 - 8, \; x \leftrightarrow y$

   $x = y^3 - 8$

   $y^3 = x + 8$

   $y = \sqrt[3]{x+8}, \; y \to f^{-1}(x)$

   $f^{-1}(x) = \sqrt[3]{x+8}$

25. $f(x) = -\dfrac{4}{x}, \; f(x) \to y$

   $y = -\dfrac{4}{x}, \; x \leftrightarrow y$

   $x = -\dfrac{4}{y}$

   $y = -\dfrac{4}{x}, \; y \to f^{-1}(-x)$

   $f^{-1}(x) = -\dfrac{4}{x}$

27. $f(x) = \dfrac{4}{x-5}, \; f(x) \to y$

   $y = \dfrac{4}{x-5}, \; x \leftrightarrow y$

   $x = \dfrac{4}{y-5}$

   $y - 5 = \dfrac{4}{x}$

   $y = \dfrac{4}{x} + 5, \; y \to f^{-1}(x)$

   $f^{-1}(x) = \dfrac{4}{x} + 5 \; \text{ or } \; f^{-1}(x) = \dfrac{4+5x}{x}$

29. $f(x) = \dfrac{x-3}{5}; \; f^{-1}(x) = 5x + 3$

   $f[f^{-1}(x)] = f[5x+3] = \dfrac{5x+3-3}{5} = \dfrac{5x}{5} = x$

31. $g(x) = 2x + 5$, $g(x) \rightarrow y$

$y = 2x + 5$, $x \leftrightarrow y$

$x = 2y + 5$

$2y = x - 5$

$y = \dfrac{x - 5}{2}$

$g^{-1}(x) = \dfrac{x - 5}{2}$

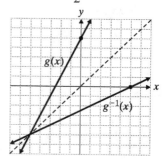

33. $h(x) = \dfrac{1}{2}x - 2$, $h(x) \rightarrow y$

$y = \dfrac{1}{2}x - 2$, $x \leftrightarrow y$

$x = \dfrac{1}{2}y - 2$

$2x = y - 4$

$y = 2x + 4$

$h^{-1}(x) = 2x + 4$

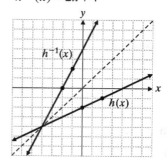

35. $r(x) = -3x - 1$, $r(x) \rightarrow y$

$y = -3x - 1$, $x \leftrightarrow y$

$x = -3y - 1$

$3y = -x - 1$

$y = -\dfrac{x + 1}{3}$, $r^{-1}(x)$

$r^{-1}(x) = -\dfrac{x + 1}{3}$

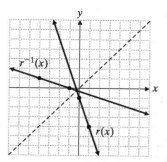

37. No; $f(x) = 2x^2 + 3$ is a vertical parabola and fails the horizontal line test; it is not one-to-one and therefore, does not have an inverse.

39. $f(x) = 2x + \dfrac{3}{2}$, $f^{-1}(x) = \dfrac{1}{2}x - \dfrac{3}{4}$

$f[f^{-1}(x)] = f\left(\dfrac{1}{2}x - \dfrac{3}{4}\right)$

$\quad\quad\quad = 2\left(\dfrac{1}{2}x - \dfrac{3}{4}\right) + \dfrac{3}{2}$

$\quad\quad\quad = x - \dfrac{3}{2} + \dfrac{3}{2}$

$\quad\quad\quad = x$

$f^{-1}[f(x)] = f^{-1}\left(2x + \dfrac{3}{2}\right)$

$\quad\quad\quad = \dfrac{1}{2}\left(2x + \dfrac{3}{2}\right) - \dfrac{3}{4}$

$\quad\quad\quad = x + \dfrac{3}{4} - \dfrac{3}{4}$

$\quad\quad\quad = x$

Cumulative Review

41. $\sqrt{20 - x} = x$

$\quad\quad\left(\sqrt{20 - x}\right)^2 = x^2$

$\quad\quad\quad\quad 20 - x = x^2$

$\quad\quad x^2 + x - 20 = 0$

$\quad\quad (x + 5)(x - 4) = 0$

$\quad\quad x + 5 = 0$ or $x - 4 = 0$

$\quad\quad\quad x = -5$ $x = 4$

$x = -5$ does not check.

The solution is $x = 4$.

42. $x^{2/3} + 7x^{1/3} + 12 = 0$

$(x^{1/3} + 4)(x^{1/3} + 3) = 0$

$x^{1/3} + 4 = 0 \qquad\qquad x^{1/3} + 3 = 0$

$\qquad x^{1/3} = -4 \qquad\qquad\quad x^{1/3} = -3$

$\qquad\qquad x = -64 \qquad\qquad\qquad x = -27$

43. x = number of people working in forest related jobs in Canada in 2013

$$x = \frac{19,446,000}{15} = 1,296,400$$

1,296,400 people worked in a job related to forests.

44. $\dfrac{45.5 - 33.4}{45.5} \approx 0.27$

The expected percent of decrease is 27%.

Quick Quiz 11.4

1. $A = \{(3, -4), (2, -6), (5, 6), (-3, 4)\}$

    a. Yes since no two ordered pairs have the same first coordinate.

    b. Yes since no two ordered pairs have the same second coordinate.

    c. $A^{-1} = \{(-4, 3), (-6, 2), (6, 5), (4, -3)\}$

2. $f(x) = 5 - 2x$

$\quad y = 5 - 2x$

$\qquad x = 5 - 2y$

$\quad x - 5 = -2y$

$\qquad y = \dfrac{-x + 5}{2}$

$\quad f^{-1}(x) = \dfrac{-x + 5}{2}$

x	$f(x)$
0	5
1	3
2	1
4	−3

x	$f^{-1}(x)$
5	0
3	1
1	2
−3	4

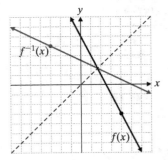

3. $f(x) = 2 - x^3$

$\quad y = 2 - x^3$

$\quad x = 2 - y^3$

$\quad x - 2 = -y^3$

$\quad 2 - x = y^3$

$\qquad y = \sqrt[3]{2 - x}$

$\quad f^{-1}(x) = \sqrt[3]{2 - x}$

4. Answers may vary. Possible solution:

To find the inverse of the function $f(x) = \dfrac{x - 5}{3}$, substitute y for $f(x)$.

$y = \dfrac{x - 5}{3}$

Interchange x and y.

$x = \dfrac{y - 5}{3}$

Solve for y in terms of x.

$x = \dfrac{y - 5}{3}$

$3x = y - 5$

$y = 3x + 5$

Replace y with $f^{-1}(x)$.

$f^{-1}(x) = 3x + 5$

Career Exploration Problems

1. $C(x) = 1.4x^2 + 8.2x + 2170$

Use $x = 0$ for 2012.

$C(0) = 1.4(0)^2 + 8.2(0) + 2170 = 2170$

Use $x = 3$ for 2015.

$C(3) = 1.4(3)^2 + 8.2(3) + 2170 = 2207.2$

There were 2170 crimes reported in 2012 and 2207 crimes reported in 2015.

2. $P(x) = 68.8x^2 + 1744.73x + 73,561$
Use $x = 0$ for 2012.

$P(0) = 68.8(0)^2 + 1744.73(0) + 73,561 = 73,561$
Use $x = 3$ for 2015.

$P(3) = 68.8(3)^2 + 1744.73(3) + 73,561$
$\qquad = 79,414.39$
The population in 2012 was 73,561; in 2015 it was 79,414.

3. the crime rate per person is

$R(x) = \dfrac{C(x)}{P(x)} = \dfrac{1.4x^2 + 8.2x + 2170}{68.8x^2 + 1744.73x + 73,561}$.

Use $x = 0$ for 2012.

$R(0) = \dfrac{C(0)}{P(0)} = \dfrac{2170}{73,561} \approx 0.0295$

Use $x = 3$ for 2015.

$R(3) = \dfrac{C(3)}{P(3)} = \dfrac{2207.2}{79,414.39} \approx 0.0278$

The crime rate per person in 2012 was 2.95%; in 2015 it was 2.78%.

4. $T(x) = 0.27x + 2028$
Use $x = 0$ for 2012.
$T(0) = 0.27(0) + 2028 = 2028$
Use $x = 3$ for 2015.
$T(3) = 0.27(3) + 2028 = 2028.81$
The petty crime rate per person is

$R(x) = \dfrac{T(x)}{P(x)} = \dfrac{0.27x + 2028}{68.8x^2 + 1744.73x + 73,561}$.

Use $x = 0$ for 2012.

$R(0) = \dfrac{T(0)}{P(0)} = \dfrac{2028}{73,561} \approx 0.0276$

Use $x = 3$ for 2015.

$R(3) = \dfrac{T(3)}{P(3)} = \dfrac{2028.81}{79,414.39} \approx 0.0255$

In 2012, there were 2028 petty crimes, a rate of 2.76%. In 2015, there were 2029 petty crimes, a rate of 2.55%.

5. The model for violent crimes is
$V(x) = C(x) - T(x)$
$\qquad = 1.4x^2 + 8.2x + 2170 - (0.27x + 2028)$
$\qquad = 1.4x^2 + 8.2x + 2170 - 0.27x - 2028$
$\qquad = 1.4x^2 + 7.93x + 142$
Use $x = 0$ for 2012.

$V(0) = 1.4(0)^2 + 7.93(0) + 142 = 142$

Use $x = 3$ for 2015.

$V(3) = 1.4(3)^2 + 7.93(3) + 142 = 178.39$
There were 142 violent crimes in 2012 and 178 violent crimes in 2015.

You Try It

1. $f(x) = -x^2 + x - 5$

    a. $f(1) = -(1)^2 + 1 - 5 = -1 + 1 - 5 = -5$

    b. $f(a) = -a^2 + a - 5$

    c. $f(a-1) = -(a-1)^2 + (a-1) - 5$
$\qquad\qquad = -(a^2 - 2a + 1) + a - 1 - 5$
$\qquad\qquad = -a^2 + 2a - 1 + a - 6$
$\qquad\qquad = -a^2 + 3a - 7$

    d. $f(4a) = -(4a)^2 + 4a - 5 = -16a^2 + 4a - 5$

2. Yes, since no vertical line intersects the graph more than once.

3. a. $f(x) = x^2$, $g(x) = x^2 - 2$
      Shift $f(x)$ down 2 units.

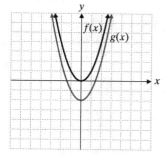

    b. $f(x) = |x|$, $g(x) = |x| + 4$
      Shift $f(x)$ up 4 units.

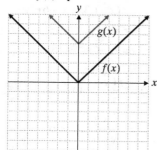

4. a. $f(x) = x^2$, $g(x) = (x+5)^2$
　　Shift $f(x)$ to the left 5 units.

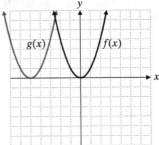

b. $f(x) = x^3$, $g(x) = (x-3)^3$
　　Shift $f(x)$ to the right 3 units.

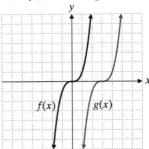

5. $f(x) = x^2 + 3x$, $g(x) = x - 6$

a. $(f+g)(x) = f(x) + g(x)$
$$= x^2 + 3x + x - 6$$
$$= x^2 + 4x - 6$$

b. $(f-g)(x) = f(x) - g(x)$
$$= x^2 + 3x - (x-6)$$
$$= x^2 + 3x - x + 6$$
$$= x^2 + 2x + 6$$

c. $(f \cdot g)(x) = f(x) \cdot g(x)$
$$= (x^2 + 3x)(x-6)$$
$$= x^3 - 6x^2 + 3x^2 - 18x$$
$$= x^3 - 3x^2 - 18x$$

d. $\left(\dfrac{f}{g}\right)(x) = \dfrac{f(x)}{g(x)}, \ g(x) \neq 0$
$$= \dfrac{x^2 + 3x}{x-6}, \ x \neq 6$$

6. $f(x) = -x + 2$, $g(x) = x^2 + 3$

a. $f[g(x)] = f[x^2 + 3]$
$$= -(x^2 + 3) + 2$$
$$= -x^2 - 3 + 2$$
$$= -x^2 - 1$$

b. $g[f(x)] = g(-x+2)$
$$= (-x+2)^2 + 3$$
$$= x^2 - 4x + 4 + 3$$
$$= x^2 - 4x + 7$$

7. $\{(-1, 0), (2, 3), (2, 5), (0, 4)\}$ is not a function since $(2, 3)$ and $(2, 5)$ have the same first coordinate. Since it is not a function, it cannot be a one-to-one function.

8. The graph does not represent a one-to-one function since it is possible to draw a horizontal line that intersects the graph in more than one point.

9. $C = \{(3, -2), (0, 4), (5, 1)\}$
Reverse the coordinates of each ordered pair.
$C^{-1} = \{(-2, 3), (4, 0), (1, 5)\}$

10. $f(x) = \dfrac{x}{2} - 5$
$$y = \dfrac{x}{2} - 5$$
$$x = \dfrac{y}{2} - 5$$
$$2x = y - 10$$
$$2x + 10 = y$$
$$f^{-1}(x) = 2x + 10$$

11. $f(x) = \dfrac{x}{3} + 2$, $f^{-1}(x) = 3x - 6$

x	$f(x)$
-6	0
0	2
3	3

x	$f^{-1}(x)$
0	-6
2	0
3	3

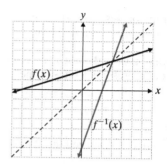

Chapter 11 Review Problems

1. $f(x) = \dfrac{1}{2}x + 3$

$f(a-1) = \dfrac{1}{2}(a-1) + 3 = \dfrac{1}{2}a - \dfrac{1}{2} + \dfrac{6}{2} = \dfrac{1}{2}a + \dfrac{5}{2}$

2. $f(x) = \dfrac{1}{2}x + 3$

$f(a-1) - f(a) = \dfrac{1}{2}(a-1) + 3 - \left(\dfrac{1}{2}a + 3\right)$

$= \dfrac{1}{2}a - \dfrac{1}{2} + 3 - \dfrac{1}{2}a - 3$

$= -\dfrac{1}{2}$

3. $f(x) = \dfrac{1}{2}x + 3$

$f(b^2 - 3) = \dfrac{1}{2}(b^2 - 3) + 3$

$= \dfrac{1}{2}b^2 - \dfrac{3}{2} + 3$

$= \dfrac{1}{2}b^2 + \dfrac{3}{2}$

4. $p(x) = -2x^2 + 3x - 1$

$p(-3) = -2(-3)^2 + 3(-3) - 1 = -18 - 9 - 1 = -28$

5. $p(x) = -2x^2 + 3x - 1$

$p(2a) + p(-2)$

$= -2(2a)^2 + 3(2a) - 1 + [-2(-2)^2 + 3(-2) - 1]$

$= -8a^2 + 6a - 1 + (-8 - 6 - 1)$

$= -8a^2 + 6a - 16$

6. $p(x) = -2x^2 + 3x - 1$

$p(a+2) = -2(a+2)^2 + 3(a+2) - 1$

$= -2(a^2 + 4a + 4) + 3(a+2) - 1$

$= -2a^2 - 8a - 8 + 3a + 6 - 1$

$= -2a^2 - 5a - 3$

7. $h(x) = |2x - 1|$

$h(0) = |2(0) - 1| = |-1| = 1$

8. $h(x) = |2x - 1|$

$h\left(\dfrac{1}{4}a\right) = \left|2\left(\dfrac{1}{4}a\right) - 1\right| = \left|\dfrac{1}{2}a - 1\right|$

9. $h(x) = |2x - 1|$

$h(2a^2 - 3a) = \left|2(2a^2 - 3a) - 1\right|$

$= \left|4a^2 - 6a - 1\right|$

10. $r(x) = \dfrac{3x}{x+4}$, $x \neq -4$

$r(5) = \dfrac{3(5)}{5+4} = \dfrac{15}{9} = \dfrac{5}{3}$

11. $r(x) = \dfrac{3x}{x+4}$, $x \neq -4$

$r(2a-5) = \dfrac{3(2a-5)}{2a-5+4} = \dfrac{6a-15}{2a-1}$

12. $r(x) = \dfrac{3x}{x+4}$, $x \neq -4$

$r(3) + r(a) = \dfrac{3(3)}{3+4} + \dfrac{3(a)}{a+4}$

$= \dfrac{9}{7} + \dfrac{3a}{a+4}$

$= \dfrac{9(a+4) + 7(3a)}{7(a+4)}$

$= \dfrac{9a + 36 + 21a}{7a + 28}$

$= \dfrac{30a + 36}{7a + 28}$

13. $f(x) = 7x - 4$

$$\frac{f(x+h)-f(x)}{h} = \frac{7(x+h)-4-(7x-4)}{h}$$

$$= \frac{7x+7h-4-7x+4}{h}$$

$$= \frac{7h}{h}$$

$$= 7$$

14. $f(x) = 6x - 5$

$$\frac{f(x+h)-f(x)}{h} = \frac{6(x+h)-5-(6x-5)}{h}$$

$$= \frac{6x+6h-5-6x+5}{h}$$

$$= \frac{6h}{h}$$

$$= 6$$

15. $f(x) = 2x^2 - 5x$

$$\frac{f(x+h)-f(x)}{h}$$

$$= \frac{2(x+h)^2 - 5(x+h) - (2x^2 - 5x)}{h}$$

$$= \frac{2x^2 + 4xh + 2h^2 - 5x - 5h - 2x^2 + 5x}{h}$$

$$= \frac{4xh + 2h^2 - 5h}{h}$$

$$= 4x + 2h - 5$$

16. a. No, the graph fails the vertical line test and therefore does not represent a function.

   b. No, unless the graph represents a function first it cannot represent a one-to-one function.

17. a. Yes, the graph passes the vertical line test and therefore represents a function.

   b. No, the graph fails the horizontal line test and therefore does not represent a one-to-one function.

18. a. No, the graph fails the vertical line test and therefore does not represent a function.

   b. No, unless the graph represents a function first it cannot represent a one-to-one function.

19. a. Yes, the graph passes the vertical line test and therefore represents a function.

   b. Yes, the graph passes the horizontal line test and therefore represents a one-to-one function.

20. $f(x) = x^2$

$g(x) = (x+2)^2 + 4$ is $f(x)$ shifted left 2 units and up 4 units.

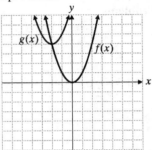

21. $f(x) = |x|$

$g(x) = |x - 4|$ is $f(x)$ shifted right 4 units.

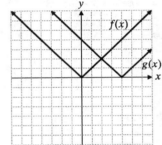

22. $f(x) = |x|$

$h(x) = |x| + 3$ is $f(x)$ shifted up 3 units.

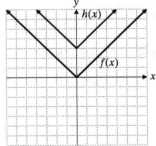

23. $f(x) = x^3$

$r(x) = (x+3)^3 + 1$ is $f(x)$ shifted left 3 units and up 1 unit.

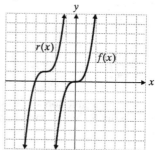

24. $f(x) = x^3$

$r(x) = (x-1)^3 + 5$ is $f(x)$ shifted right 1 unit and up 5 units.

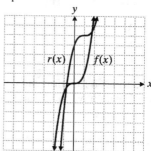

25. $f(x) = \dfrac{2}{x}$, $x \neq 0$

$r(x) = \dfrac{2}{x+3} - 2$, $x \neq -3$ is $f(x)$ shifted left 3 units and down 2 units.

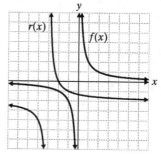

In Exercises 26–35,

$$f(x) = 3x + 5; \quad g(x) = \frac{2}{x}, \ x \neq 0$$

$$s(x) = \sqrt{x-2}, \ x \geq 2; \quad h(x) = \frac{x+1}{x-4}, \ x \neq 4$$

$$p(x) = 2x^2 - 3x + 4; \quad t(x) = -\frac{1}{2}x - 3$$

26. $(f+t)(x) = f(x) + t(x)$

$$= 3x + 5 + \left(-\frac{1}{2}x - 3\right)$$

$$= \frac{5}{2}x + 2$$

27. $(p-f)(x) = p(x) - f(x)$

$$= 2x^2 - 3x + 4 - (3x+5)$$

$$= 2x^2 - 3x + 4 - 3x - 5$$

$$= 2x^2 - 6x - 1$$

28. From Exercise 27, $(p-f)(x) = 2x^2 - 6x - 1$.

$$(p-f)(2) = 2(2)^2 - 6(2) - 1$$

$$= 2(4) - 12 - 1$$

$$= 8 - 12 - 1$$

$$= -5$$

29. $(fg)(x) = f(x)g(x) = (3x+5)\left(\dfrac{2}{x}\right) = \dfrac{6x+10}{x}$,

$x \neq 0$

30. $\left(\dfrac{g}{h}\right)(x) = \dfrac{g(x)}{h(x)} = \dfrac{\frac{2}{x}}{\frac{x+1}{x-4}} = \dfrac{2}{x} \cdot \dfrac{x-4}{x+1} = \dfrac{2x-8}{x^2+x}$,

$x \neq 0, 4, -1$

31. From Exercise 30, $\left(\dfrac{g}{h}\right)(x) = \dfrac{2x-8}{x^2+x}$,

$x \neq -1, 0, 4$.

$$\left(\frac{g}{h}\right)(-2) = \frac{2(-2)-8}{(-2)^2 + (-2)} = -6$$

32. $p[f(x)] = p[3x+5]$

$$= 2(3x+5)^2 - 3(3x+5) + 4$$

$$= 18x^2 + 60x + 50 - 9x - 15 + 4$$

$$= 18x^2 + 51x + 39$$

33. $s[p(x)] = s[2x^2 - 3x + 4]$

$$= \sqrt{2x^2 - 3x + 4 - 2}$$

$$= \sqrt{2x^2 - 3x + 2}$$

34. From Exercise 33, $s[p(x)] = \sqrt{2x^2 - 3x + 2}$.

$$s[p(2)] = \sqrt{2(2)^2 - 3(2) + 2} = 2$$

35. $f[g(x)] = f\left[\dfrac{2}{x}\right], \ x \neq 0$

$\qquad = 3\left(\dfrac{2}{x}\right) + 5$

$\qquad = \dfrac{6}{x} + 5$

$\qquad = \dfrac{6 + 5x}{x}$

$\quad g[f(x)] = g[3x + 5] = \dfrac{2}{3x + 5}$

$\quad f[g(x)] \neq g[f(x)]$

36. $f(x) = \dfrac{2}{3}x + \dfrac{1}{2}; \ f^{-1}(x) = \dfrac{6x - 3}{4}$

$\quad f^{-1}[f(x)] = f^{-1}\left[\dfrac{2}{3}x + \dfrac{1}{2}\right]$

$\qquad = \dfrac{6\left(\frac{2}{3}x + \frac{1}{2}\right) - 3}{4}$

$\qquad = \dfrac{4x + 3 - 3}{4}$

$\qquad = \dfrac{4x}{4}$

$\qquad = x$

37. $B = \{(3, 7), (7, 3), (0, 8), (0, -8)\}$

  a. $D = \{0, 3, 7\}$

  b. $R = \{-8, 3, 7, 8\}$

  c. No, the set does not define a function since two of the ordered pairs have the same first coordinate.

  d. No, since the set does not define a function it cannot define a one-to-one function.

38. $A = \{(100, 10), (200, 20), (300, 30), (400, 10)\}$

  a. $D = \{100, 200, 300, 400\}$

  b. $R = \{10, 20, 30\}$

  c. Yes, the set defines a function since no two of the ordered pairs have the same first coordinate.

  d. No, the set does not define a one-to-one function since two of the ordered pairs have the same second coordinate.

39. $D = \left\{\left(\dfrac{1}{2}, 2\right), \left(\dfrac{1}{4}, 4\right), \left(-\dfrac{1}{3}, -3\right), \left(4, \dfrac{1}{4}\right)\right\}$

  a. $D = \left\{\dfrac{1}{2}, \dfrac{1}{4}, -\dfrac{1}{3}, 4\right\}$

  b. $R = \left\{2, 4, -3, \dfrac{1}{4}\right\}$

  c. Yes, the set defines a function since no two of the ordered pairs have the same first coordinate.

  d. Yes, the set defines a one-to-one function since it is a function and no two ordered pairs have the same second coordinate.

40. $F = \{(3, 7), (2, 1), (0, -3), (1, 1)\}$

  a. $D = \{0, 1, 2, 3\}$

  b. $R = \{-3, 1, 7\}$

  c. Yes, the set defines a function since no two of the ordered pairs have the same first coordinate.

  d. No, the set does not define a one-to-one function since two of the ordered pairs have the same second coordinate.

41. $A = \left\{\left(3, \dfrac{1}{3}\right), \left(-2, -\dfrac{1}{2}\right), \left(-4, -\dfrac{1}{4}\right), \left(5, \dfrac{1}{5}\right)\right\}$

$\quad A^{-1} = \left\{\left(\dfrac{1}{3}, 3\right), \left(-\dfrac{1}{2}, -2\right), \left(-\dfrac{1}{4}, -4\right), \left(\dfrac{1}{5}, 5\right)\right\}$

42. $f(x) = -\dfrac{3}{4}x + 2, \ f(x) \rightarrow y$

$\qquad y = -\dfrac{3}{4}x + 2, \ x \leftrightarrow y$

$\qquad x = -\dfrac{3}{4}y + 2$

$\qquad -4x = 3y - 8$

$\qquad -4x + 8 = 3y$

$\qquad y = -\dfrac{4}{3}x + \dfrac{8}{3}, \ y \rightarrow f^{-1}(x)$

$\quad f^{-1}(x) = -\dfrac{4}{3}x + \dfrac{8}{3}$

43. $g(x) = -8 - 4x$, $g(x) \to y$

$y = -8 - 4x$, $x \leftrightarrow y$

$x = -8 - 4y$

$4y = -x - 8$

$y = -\frac{1}{4}x - 2$, $y \to g^{-1}(x)$

$g^{-1}(x) = -\frac{1}{4}x - 2$

44. $h(x) = \dfrac{6}{x+5}$, $h(x) \to y$

$y = \dfrac{6}{x+5}$, $x \leftrightarrow y$

$x = \dfrac{6}{y+5}$

$y + 5 = \dfrac{6}{x}$

$y = \dfrac{6}{x} - 5$ or $y = \dfrac{6 - 5x}{x}$, $y \to h^{-1}(x)$

$h^{-1}(x) = \dfrac{6}{x} - 5$ or $h^{-1}(x) = \dfrac{6 - 5x}{x}$

45. $p(x) = \sqrt[3]{x+1}$, $p(x) \to y$

$y = \sqrt[3]{x+1}$, $x \leftrightarrow y$

$x = \sqrt[3]{y+1}$

$x^3 = y + 1$

$y = x^3 - 1$, $y \to p^{-1}(x)$

$p^{-1}(x) = x^3 - 1$

46. $r(x) = x^3 + 2$, $r(x) \to y$

$y = x^3 + 2$, $x \leftrightarrow y$

$x = y^3 + 2$

$x - 2 = y^3$

$y = \sqrt[3]{x-2}$, $y \to r^{-1}(x)$

$r^{-1}(x) = \sqrt[3]{x-2}$

47. $f(x) = \dfrac{-x-2}{3}$, $f(x) \to y$

$y = \dfrac{-x-2}{3}$, $x \leftrightarrow y$

$x = \dfrac{-y-2}{3}$

$3x = -y - 2$

$y = -3x - 2$, $y \to f^{-1}(x)$

$f^{-1}(x) = -3x - 2$

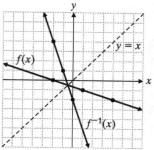

48. $f(x) = -\dfrac{3}{4}x + 1$, $f(x) \to y$

$y = -\dfrac{3}{4}x + 1$, $x \leftrightarrow y$

$x = -\dfrac{3}{4}y + 1$

$4x = -3y + 4$

$3y = -4x + 4$

$y = -\dfrac{4}{3}x + \dfrac{4}{3}$, $y \to f^{-1}(x)$

$f^{-1}(x) = -\dfrac{4}{3}x + \dfrac{4}{3}$

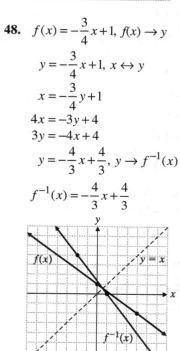

How Am I Doing? Chapter 11 Test

1. $f(x) = \dfrac{3}{4}x - 2$

$f(-8) = \dfrac{3}{4}(-8) - 2 = -6 - 2 = -8$

2. $f(x) = \dfrac{3}{4}x - 2$

$f(3a) = \dfrac{3}{4}(3a) - 2 = \dfrac{9}{4}a - 2$

3. $f(x) = \dfrac{3}{4}x - 2$

$f(a) - f(2) = \dfrac{3}{4}a - 2 - \left(\dfrac{3}{4}(2) - 2\right)$

$\qquad\qquad = \dfrac{3}{4}a - 2 - \dfrac{3}{2} + 2$

$\qquad\qquad = \dfrac{3}{4}a - \dfrac{3}{2}$

4. $f(x) = 3x^2 - 2x + 4$

$f(-6) = 3(-6)^2 - 2(-6) + 4 = 108 + 12 + 4 = 124$

5. $f(x) = 3x^2 - 2x + 4$

$f(a+1) = 3(a+1)^2 - 2(a+1) + 4$

$\qquad\quad = 3a^2 + 6a + 3 - 2a - 2 + 4$

$\qquad\quad = 3a^2 + 4a + 5$

6. $f(x) = 3x^2 - 2x + 4$

$f(a) + f(1) = 3a^2 - 2a + 4 + 3(1)^2 - 2(1) + 4$

$\qquad\qquad = 3a^2 - 2a + 4 + 3 - 2 + 4$

$\qquad\qquad = 3a^2 - 2a + 9$

7. $f(x) = 3x^2 - 2x + 4$

$f(-2a) - 2 = 3(-2a)^2 - 2(-2a) + 4 - 2$

$\qquad\qquad = 3(4a^2) + 4a + 2$

$\qquad\qquad = 12a^2 + 4a + 2$

8. a. Graph passes vertical line test and therefore represents a function.

b. Graph fails horizontal line test and does not represent a one-to-one function.

9. a. Graph passes vertical line test and therefore represents a function.

b. Graph passes horizontal line test and therefore represents a one-to-one function.

10. $f(x) = x^2$

$g(x) = (x-1)^2 + 3$ is $f(x)$ shifted right 1 unit and up 3 units.

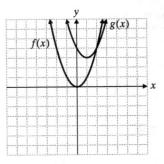

11. $f(x) = |x|$

$g(x) = |x + 1| + 2$ is $f(x)$ shifted left 1 unit and up 2 units.

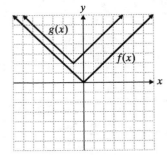

12. $f(x) = 3x^2 - x - 6$, $g(x) = -2x^2 + 5x + 7$

a. $(f+g)(x) = f(x) + g(x)$

$\qquad\qquad = 3x^2 - x - 6 + (-2x^2 + 5x + 7)$

$\qquad\qquad = 3x^2 - x - 6 - 2x^2 + 5x + 7$

$\qquad\qquad = x^2 + 4x + 1$

b. $(f-g)(x) = f(x) - g(x)$

$\qquad\qquad = 3x^2 - x - 6 - (-2x^2 + 5x + 7)$

$\qquad\qquad = 3x^2 - x - 6 + 2x^2 - 5x - 7$

$\qquad\qquad = 5x^2 - 6x - 13$

c. From b, $(f-g)(x) = 5x^2 - 6x - 13$.

$(f-g)(-2) = 5(-2)^2 - 6(-2) - 13$

$\qquad\qquad = 5(4) + 12 - 13$

$\qquad\qquad = 19$

13. $f(x) = \dfrac{3}{x}$, $x \neq 0$; $g(x) = 2x - 1$

a. $(fg)(x) = f(x)g(x) = \dfrac{3}{x}(2x - 1) = \dfrac{6x - 3}{x}$,

$x \neq 0$

b. $\left(\dfrac{f}{g}\right)(x) = \dfrac{f(x)}{g(x)} = \dfrac{\frac{3}{x}}{2x-1} = \dfrac{3}{2x^2-x}$,

$x \neq 0, \dfrac{1}{2}$

c. $f[g(x)] = f[2x-1] = \dfrac{3}{2x-1}$, $x \neq \dfrac{1}{2}$

14. $f(x) = \dfrac{1}{2}x - 3$, $g(x) = 4x + 5$

a. $(f \circ g)(x) = f[g(x)]$
$= f[4x+5]$
$= \dfrac{1}{2}(4x+5) - 3$
$= 2x + \dfrac{5}{2} - \dfrac{6}{2}$
$= 2x - \dfrac{1}{2}$

b. $(g \circ f)(x) = g[f(x)]$
$= g\left[\dfrac{1}{2}x - 3\right]$
$= 4\left(\dfrac{1}{2}x - 3\right) + 5$
$= 2x - 12 + 5$
$= 2x - 7$

c. $(f \circ g)\left(\dfrac{1}{4}\right) = 2 \cdot \dfrac{1}{4} - \dfrac{1}{2} = 0$

15. $B = \{(1, 8), (8, 1), (9, 10), (-10, 9)\}$

a. Yes, the function is one-to-one since no two ordered pairs have the same second coordinate.

b. $B^{-1} = \{(8, 1), (1, 8), (10, 9), (9, -10)\}$

16. $A = \{(1, 5), (2, 1), (4, -7), (0, 7), (-1, 5)\}$

a. The function is not one-to-one since two ordered pairs have the same second coordinate.

b. A has no inverse.

17. $f(x) = \sqrt[3]{2x-1}$, $f(x) \to y$
$y = \sqrt[3]{2x-1}$, $x \leftrightarrow y$
$x = \sqrt[3]{2y-1}$
$x^3 = 2y - 1$
$2y = x^3 + 1$
$y = \dfrac{x^3+1}{2}$, $y \to f^{-1}(x)$

$f^{-1}(x) = \dfrac{x^3+1}{2}$

18. $f(x) = -3x + 2$, $f(x) \to y$
$y = -3x + 2$, $x \leftrightarrow y$
$x = -3y + 2$
$3y = -x + 2$
$y = -\dfrac{1}{3}x + \dfrac{2}{3}$, $y \to f^{-1}(x)$

$f^{-1}(x) = -\dfrac{1}{3}x + \dfrac{2}{3}$

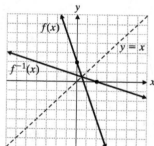

19. $f(x) = \dfrac{3}{7}x + \dfrac{1}{2}$, $f^{-1}(x) = \dfrac{14x-7}{6}$

$f^{-1}[f(x)] = f^{-1}\left[\dfrac{3}{7}x + \dfrac{1}{2}\right]$

$= \dfrac{14\left[\frac{3}{7}x + \frac{1}{2}\right] - 7}{6}$

$= \dfrac{6x + 7 - 7}{6}$

$= \dfrac{6x}{6}$

$= x$

20. $f(x) = 7 - 8x$

$\dfrac{f(x+h) - f(x)}{h} = \dfrac{7 - 8(x+h) - (7 - 8x)}{h}$

$= \dfrac{7 - 8x - 8h - 7 + 8x}{h}$

$= \dfrac{-8h}{h}$

$= -8$

Chapter 12

12.1 Exercises

1. An exponential function is a function of the form $\underline{f(x) = b^x}$, $\underline{\text{where } b > 0, b \neq 1, \text{ and } x \text{ is a real}}$ $\underline{\text{number}}$.

3. $f(x) = 3^x$

x	$y = f(x) = 3^x$
-2	$\frac{1}{9}$
-1	$\frac{1}{3}$
0	1
1	3
2	9

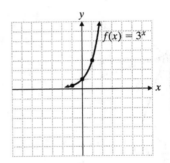

5. $f(x) = 2^{-x}$

x	$y = f(x) = 2^{-x}$
-2	4
-1	2
0	1
1	$\frac{1}{2}$
2	$\frac{1}{4}$

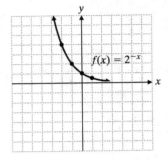

7. $f(x) = 3^{-x}$

x	$y = f(x) = 3^{-x}$
-2	9
-1	3
0	1
1	$\frac{1}{3}$
2	$\frac{1}{9}$

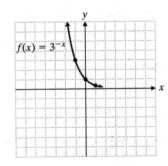

9. $f(x) = 2^{x+3}$

x	$y = f(x) = 2^{x+3}$
-3	1
-2	2
-1	4

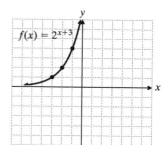

11. $f(x) = 3^{x-5}$

x	$y = f(x) = 3^{x-5}$
4	$\frac{1}{3}$
5	1
6	3

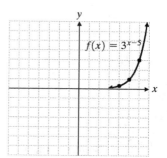

13. $f(x) = 2^x + 2$

x	$y = f(x) = 2^x + 2$
-2	$\frac{9}{4}$
-1	$\frac{5}{2}$
0	3
1	4
2	6

15. $f(x) = e^{x-1}$

x	$y = e^{x-1}$
-2	0.05
-1	0.14
0	0.37
1	1
2	2.7

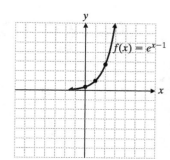

17. $f(x) = 2e^x$

x	$y = f(x) = 2e^x$
-2	0.27
-1	0.74
0	2
1	5.44
2	14.8

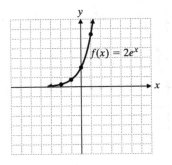

19. $f(x) = e^{1-x}$

x	$y = f(x) = e^{1-x}$
-2	20.1
-1	7.39
0	2.72
1	0.37
2	0.14

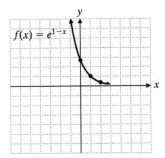

21. $2^x = 4$
$2^x = 2^2$
$x = 2$

23. $2^x = 1$
$2^x = 2^0$
$x = 0$

25. $2^x = \dfrac{1}{2} = \dfrac{1}{2^1}$
$2^x = 2^{-1}$
$x = -1$

27. $3^x = 81$
$3^x = 3^4$
$x = 4$

29. $3^x = 1$
$3^x = 3^0$
$x = 0$

31. $3^{-x} = \dfrac{1}{9} = \dfrac{1}{3^2}$
$3^{-x} = 3^{-2}$
$-x = -2$
$x = 2$

33. $4^x = 256$
$4^x = 4^4$
$x = 4$

35. $5^{x+1} = 125$
$5^{x+1} = 5^3$
$x + 1 = 3$
$x = 2$

37. $8^{3x-1} = 64$
$8^{3x-1} = 8^2$
$3x - 1 = 2$
$3x = 3$
$x = 1$

39. $A = P\left(1 + \dfrac{r}{n}\right)^{nt}$

$A = 2000\left(1 + \dfrac{0.063}{1}\right)^{1(3)} = 2402.314094$

Alicia will have \$2402.31 after 3 years.

41. $A = P\left(1 + \dfrac{r}{n}\right)^{nt}$

$A = 3000\left(1 + \dfrac{0.032}{4}\right)^{4(6)} = \$3632.34$

$A = 3000\left(1 + \dfrac{0.032}{12}\right)^{12(6)} = \$3634.08$

She will have \$3632.34 if it is compounded quarterly and \$3634.08 if it is compounded monthly.

43. $B(t) = 4000(2^t)$

$B(3) = 4000(2^3) = 32,000$

$B(9) = 4000(2^9) = 2,048,000$

At the end of 3 hours there will be 32,000 bacteria in the culture and at the end of 9 hours there will be 2,048,000 bacteria in the culture.

45. $S(f) = (1-0.18)^{f/4} = 0.82^{f/4}$

$S(20) = 0.82^{20/4} \approx 0.371$

$S(48) = 0.82^{48/4} \approx 0.092$

At a depth of 20 feet, 37.1% of the sunlight is available. Since only 9.2% of the sunlight is available at a depth of 48 feet, the spotlights will be needed.

47. $A = Ce^{-0.0004279t}$

$A = 12e^{-0.0004279(100)} \approx 11.50$

There will be approximately 11.50 mg of radium in the container after 100 years.

49. $P = 14.7e^{-0.21d}$

$P = 14.7e^{-0.21(10)} \approx 1.80$

The pressure is approximately 1.80 lb/in.^{2} on a jet flying 10 miles above sea level.

51. $A = 2.08e^{0.045t}$

Use $t = 0$ for 2000.

$A = 2.08e^{0.045(0)} = 2.08e^0 = 2.08$

Use $t = 5$ for 2005.

$A = 2.08e^{0.045(5)} \approx 2.604831$

Percent increase $= \dfrac{2.604831 - 2.08}{2.08} \approx 0.25$

The average MLB player salary was $2,080,000 in 2000 and $2,604,831 in 2005. This was about a 25% increase.

53. From the graph, the world population reached three billion people sometime in 1955. (Answers may vary.)

55. $6.07(1.021)^{25} \approx 10.2$

At a 2.1% growth rate, the world population would be about 10.2 billion people in 2025. According to the graph, the world population will be about 8 billion people in 2025.

Cumulative Review

57. $5 - 2(3 - x) = 2(2x + 5) + 1$

$5 - 6 + 2x = 4x + 10 + 1$

$2x = -12$

$x = -6$

58. $\dfrac{7}{12} + \dfrac{3}{4}x + \dfrac{5}{4} = -\dfrac{1}{6}x$

$\dfrac{3}{4}x + \dfrac{1}{6}x = -\dfrac{7}{12} - \dfrac{5}{4}$

$\dfrac{9}{12}x + \dfrac{2}{12}x = -\dfrac{7}{12} - \dfrac{15}{12}$

$\dfrac{11}{12}x = -\dfrac{11}{6}$

$6x = -12$

$x = -2$

Quick Quiz 12.1

1. $f(x) = 2^{x+4}$

x	$f(x)$
-1	8
-2	4
-3	2
-4	1

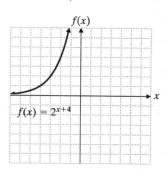

2. $A = P\left(1 + \dfrac{r}{n}\right)^{nt}$

$A = 5500\left(1 + \dfrac{0.04}{2}\right)^{2(6)} \approx 6975.33$

Dwayne will have $6975.33 after 6 years.

3. $3^{x+2} = 81$

$3^{x+2} = 3^4$

$x + 2 = 4$

$x = 2$

4. Answers may vary. Possible solution:
To solve the following equation for x:

$4^{-x} = \dfrac{1}{64}$, remember that $64 = 4^{-3}$, and replace

the right side of the equation with 4^{-3}.

$4^{-x} = 4^{-3}$

This yields the same base on both sides of the equation, and allows the use of the property of exponential functions to simplify, which states that the exponents may be isolated and compared.

$-x = -3$ or $x = 3$

12.2 Exercises

1. A logarithm is an <u>exponent</u>.

3. In the equation $y = \log_b x$, the domain (the set of permitted values of x) is <u>$x > 0$</u>.

5. $49 = 7^2 \Leftrightarrow \log_7 49 = 2$

7. $128 = 2^7 \Leftrightarrow \log_2 128 = 7$

9. $0.001 = 10^{-3} \Leftrightarrow \log_{10} 0.001 = -3$

11. $\dfrac{1}{32} = 2^{-5} \Leftrightarrow \log_2 \dfrac{1}{32} = -5$

13. $y = e^5 \Leftrightarrow \log_e y = 5$

15. $2 = \log_3 9 \Leftrightarrow 3^2 = 9$

17. $0 = \log_{17} 1 \Leftrightarrow 17^0 = 1$

19. $\dfrac{1}{2} = \log_{16} 4 \Leftrightarrow 16^{1/2} = 4$

21. $-2 = \log_{10} 0.01 \Leftrightarrow 10^{-2} = 0.01$

23. $-4 = \log_3 \left(\dfrac{1}{81}\right) \Leftrightarrow 3^{-4} = \dfrac{1}{81}$

25. $-\dfrac{3}{2} = \log_e x \Leftrightarrow e^{-3/2} = x$

27. $\log_2 x = 4$
    $2^4 = x$
    $x = 16$

29. $\log_{10} x = -3$
    $10^{-3} = x$
    $x = \dfrac{1}{1000}$

31. $\log_4 64 = y$
    $4^y = 64 = 4^3$
    $y = 3$

33. $\log_8 \left(\dfrac{1}{64}\right) = y$
    $8^y = \dfrac{1}{64} = \dfrac{1}{8^2} = 8^{-2}$
    $y = -2$

35. $\log_a 121 = 2$
    $a^2 = 121 = 11^2$
    $a = 11$

37. $\log_a 1000 = 3$
    $a^3 = 1000 = 10^3$
    $a = 10$

39. $\log_{25} 5 = w$
    $25^w = 5$
    $(5^2)^w = 5^{2w} = 5^1$
    $2w = 1$
    $w = \dfrac{1}{2}$

41. $\log_3 \left(\dfrac{1}{3}\right) = w$
    $3^w = \dfrac{1}{3} = \dfrac{1}{3^1} = 3^{-1}$
    $w = -1$

43. $\log_{15} w = 0$
    $15^0 = w$
    $w = 1$

45. $\log_w 3 = \dfrac{1}{2}$
    $w^{1/2} = 3$
    $w = 9$

47. $\log_{10} 0.001 = x$
    $10^x = 0.001 = 10^{-3}$
    $x = -3$

49. $\log_2 128 = x$
    $2^x = 128 = 2^7$
    $x = 7$

51. $\log_{23} 1 = x$

$\quad 23^x = 1 = 23^0$

$\quad x = 0$

53. $\log_6 \sqrt{6} = x$

$\quad 6^x = \sqrt{6} = 6^{1/2}$

$\quad x = \dfrac{1}{2}$

55. $\log_{57} 1 = x$

$\quad 57^x = 1 = 57^0$

$\quad x = 0$

57. $\log_3 x = y \Leftrightarrow 3^y = x$

$x = 3^y$	y
1	0
3	1
9	2

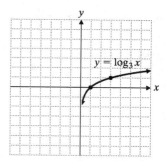

59. $\log_{1/4} x = y \Leftrightarrow \left(\dfrac{1}{4}\right)^y = x$

$y = \left(\dfrac{1}{4}\right)^y$	y
1	0
4	−1
16	−2

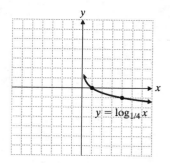

61. $\log_{10} x = y \Leftrightarrow 10^y = x$

$x = 10^y$	y
$\dfrac{1}{100}$	−2
$\dfrac{1}{10}$	−1
1	0
10	1
100	2

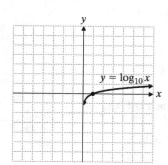

63. $f(x) = \log_3 x,\ f(x) \to y$

$\quad y = \log_3 x \Leftrightarrow 3^y = x$

$x = 3^y$	y
$\dfrac{1}{9}$	−2
$\dfrac{1}{3}$	−1
1	0
3	1
9	2

$\quad f^{-1}(x) = 3^x,\ f^{-1}(x) \to y$

$\quad y = 3^x$

x	$y = 3^x$
-2	$\frac{1}{9}$
-1	$\frac{1}{3}$
0	1
1	3
2	9

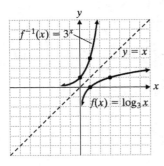

65. $\text{pH} = -\log_{10}[\text{H}^+]$

$\text{pH} = -\log_{10}[10^{-12.5}]$

$-\text{pH} = \log_{10}[10^{-12.5}]$

$10^{-\text{pH}} = 10^{-12.5}$

$-\text{pH} = -12.5$

$\text{pH} = 12.5$

The pH of bleach is 12.5.

67. $\text{pH} = -\log_{10}[\text{H}^+]$

$3.5 = -\log_{10}[\text{H}^+]$

$-3.5 = \log_{10}[\text{H}^+]$

$10^{-3.5} = \text{H}^+$

The concentration of hydrogen ions in the jelly is about $10^{-3.5}$.

69. $\text{pH} = -\log_{10}[\text{H}^+]$

$\text{pH} = -\log_{10}[1.103 \times 10^{-3}]$

$\text{pH} = -\log_{10}[0.001103]$

$\text{pH} \approx -(-2.957424488)$

$\text{pH} \approx 2.957424488$

The pH of the dressing is approximately 2.957.

71.
$$N = 1200 + 2500\log_{10} d$$
$$\log_{10} d = \frac{N - 1200}{2500}$$
$$\log_{10} 10,000 = \frac{N - 1200}{2500}$$
$$10^{\frac{N-1200}{2500}} = 10,000 = 10^4$$
$$\frac{N - 1200}{2500} = 4$$
$$N - 1200 = 10,000$$
$$N = 11,200$$

11,200 sets of software were sold.

73.
$$N = 1200 + 2500\log_{10} d$$
$$\log_{10} d = \frac{18,700 - 1200}{2500}$$
$$\log_{10} d = 7$$
$$d = 10^7$$
$$d = 10,000,000$$

$10,000,000 should be spent on advertising.

Cumulative Review

75. $y = -\frac{2}{3}x + 4$ has $m = -\frac{2}{3} \Rightarrow m_\perp = \frac{3}{2}$

$y - 1 = \frac{3}{2}(x - (-4)) = \frac{3}{2}(x + 4)$

$y - 1 = \frac{3}{2}x + 6$

$y = \frac{3}{2}x + 7$

76. $m = \frac{y_2 - y_1}{x_2 - x_1} = \frac{3 - 2}{-6 - (-1)} = -\frac{1}{5}$

77. a. $A(t) = 9000(2)^t$

$A(2) = 9000(2)^2 = 36,000$

36,000 cells will be in the culture after the first 2 hours.

b. $A(t) = 9000(2)^t$

$A(12) = 9000(2)^{12} = 36,864,00$

36,864,000 cells will be in the culture after the first 12 hours.

78. a. $C(t) = P(1.04)^t$

$C(5) = 4400(1.04)^5 = 5353.27$

In 5 years, they will charge $5353.27.

b. $C(t) = P(1.04)^t$

$C(10) = 16{,}500(1.04)^{10} = 24{,}424.03$

In 10 years, they will charge \$24,424.03.

Quick Quiz 12.2

1. $\log_3 81 = x$

$81 = 3^x$

$3^4 = 3^x$

$4 = x$

2. $\log_2 x = 6$

$x = 2^6$

$x = 64$

3. $\log_{27} 3 = w$

$3 = 27^w$

$3^1 = (3^3)^w$

$3^1 = 3^{3w}$

$1 = 3w$

$w = \dfrac{1}{3}$

4. Answers may vary. Possible solution:

To solve the equation $-\dfrac{1}{2} = \log_e x$,

write an equivalent exponential equation:

$x = e^{-1/2}$

12.3 Exercises

1. $\log_3 AB = \log_3 A + \log_3 B$

3. $\log_5 (7 \cdot 11) = \log_5 7 + \log_5 11$

5. $\log_b 9f = \log_b 9 + \log_b f$

7. $\log_9 \left(\dfrac{2}{7} \right) = \log_9 2 - \log_9 7$

9. $\log b \left(\dfrac{12}{Z} \right) = \log_b 12 - \log_b Z$

11. $\log_a \left(\dfrac{E}{F} \right) = \log_a E - \log_a F$

13. $\log_8 a^7 = 7 \log_8 a$

15. $\log_b A^{-2} = -2 \log_b A$

17. $\log_5 \sqrt{w} = \log_5 w^{1/2} = \dfrac{1}{2} \log_5 w$

19. $\log_8 x^2 y = \log_8 x^2 + \log_8 y = 2 \log_8 x + \log_8 y$

21. $\log_{11} \left(\dfrac{6M}{N} \right) = \log_{11}(6M) - \log_{11} N$

$\qquad\qquad = \log_{11} 6 + \log_{11} M - \log_{11} N$

23. $\log_2 \left(\dfrac{5xy^4}{\sqrt{z}} \right)$

$= \log_2(5xy^4) - \log_2 \sqrt{z}$

$= \log_2 5 + \log_2 x + \log_2 y^4 - \log_2 z^{1/2}$

$= \log_2 5 + \log_2 x + 4 \log_2 y - \dfrac{1}{2} \log_2 z$

25. $\log_a \sqrt[3]{\dfrac{x^4}{y}} = \log_a \left(\dfrac{x^4}{y} \right)^{1/3}$

$= \dfrac{1}{3} \log_a \left(\dfrac{x^4}{y} \right)$

$= \dfrac{1}{3} [\log_a x^4 - \log_b y]$

$= \dfrac{1}{3} [4 \log_a x - \log_b y]$

$= \dfrac{4}{3} \log_a x - \dfrac{1}{3} \log_a y$

27. $\log_4 13 + \log_4 y + \log_4 3 = \log_4(13 \cdot y \cdot 3)$

$\qquad\qquad\qquad\qquad\qquad = \log_4 39y$

29. $5 \log_3 x - \log_3 7 = \log_3 x^5 - \log_3 7 = \log_3 \left(\dfrac{x^5}{7} \right)$

31. $2 \log_b 7 + 3 \log_b y - \dfrac{1}{2} \log_b z$

$= \log_b 7^2 + \log_b y^3 - \log_b z^{1/2}$

$= \log_b 49 + \log_b y^3 - \log_b \sqrt{z}$

$= \log_b \left(\dfrac{49 y^3}{\sqrt{z}} \right)$

33. $\log_3 3 = 1$

35. $\log_e e = 1$

37. $\log_9 1 = 0$

39. $3\log_7 7 + 4\log_7 1 = 3(1) + 4(0) = 3$

41. $\log_8 x = \log_8 7$
$x = 7$

43. $\log_5(2x+7) = \log_5 29$
$2x + 7 = 29$
$2x = 22$
$x = 11$

45. $\log_3 1 = x$
$x = 0$

47. $\log_7 7 = x$
$x = 1$

49. $\log_9 27 + \log_9 x = 2$
$\log_9 27x = 2$
$27x = 9^2$
$27x = 81$
$x = 3$

51. $\log_2 7 = \log_2 x - \log_2 3$
$\log_2 7 = \log_2 \dfrac{x}{3}$
$\dfrac{x}{3} = 7$
$x = 21$

53. $3\log_5 x = \log_5 8$
$\log_5 x^3 = \log_5 8$
$x^3 = 8$
$x^3 = 2^3$
$x = 2$

55. $\log_e x = \log_e 5 + 1$
$\log_e \dfrac{x}{5} = 1$
$e^1 = \dfrac{x}{5}$
$x = 5e$

57. $\log_6(5x+21) - \log_6(x+3) = 1$
$\log_6 \dfrac{5x+21}{x+3} = 1$
$\dfrac{5x+21}{x+3} = 6$
$5x + 21 = 6x + 18$
$x = 3$

59. $5^{\log_5 4} + 3^{\log_3 2} = 4 + 2 = 6$

Cumulative Review

61. $V = \pi r^2 h = \pi(2)^2 5 \approx 62.8$
The volume is approximately 62.8 m^3.

62. $A = \pi r^2 = \pi(4)^2 \approx 50.2$
The area is approximately 50.2 m^2.

63. $5x + 3y = 9 \xrightarrow{\times 2} 10x + 6y = 18$
$7x - 2y = 25 \xrightarrow{\times 3} \underline{21x - 6y = 75}$
$\phantom{7x - 2y = 25 \xrightarrow{\times 3} 21} 31x = 93$
$\phantom{7x - 2y = 25 \xrightarrow{\times 3} 21x 6y} x = 3$
$5(3) + 3y = 9$
$ 3y = -6$
$ y = -2$
$(3, -2)$ is the solution.

64. $2x - y + z = 3$ **(1)**
$x + 2y + 2z = 1$ **(2)**
$4x + y + 2z = 0$ **(3)**
Add 2 times equation **(1)** to equation **(2)**.
$5x + 4z = 7$ **(4)**
Add equations **(1)** and **(3)**.
$6x + 3z = 3$ **(5)**
Multiply equation **(4)** by 3 and equation **(5)** by -4.
$15x + 12z = 21$
$\underline{-24x - 12z = -12}$
$-9x = 9$
$ x = -1$
Substitute $x = -1$ in equation **(4)**.
$5(-1) + 4z = 7$
$ 4z = 12$
$ z = 3$
Substitute $x = -1$, $z = 3$ in equation **(3)**.
$4(-1) + y + 2(3) = 0$
$ y = -2$
The solution is $(-1, -2, 3)$.

65. $\dfrac{1.117 \times 10^9}{1.53 \times 10^5} = \dfrac{1.117}{1.53} \times 10^{9-5}$

$$\approx 0.7301 \times 10^4$$
$$= 7301$$

The average credit card debt per household is $7301.

$$\dfrac{7301}{42,392} \times 100\% \approx 17.22\%$$

17.22% of the average yearly household income is owed to credit card companies.

66. $\dfrac{1.9 \times 10^9}{2.56 \times 10^5} = \dfrac{1.9}{2.56} \times 10^{9-5} \approx 0.7422 \times 10^4 = 7422$

The average credit card debt per household is $7422.

$$\dfrac{7422}{44,349} \times 100\% \approx 16.74\%$$

16.74% of the average yearly household income is owed to credit card companies.

Quick Quiz 12.3

1. $\log_5 \left(\dfrac{\sqrt[3]{x}}{y^4} \right) = \log_5 x^{1/3} - \log_5 y^4$

$$= \dfrac{1}{3} \log_5 x - 4 \log_5 4$$

2. $3 \log_6 x + \log_6 y - \log_6 5$

$$= \log_6 x^3 + \log_6 y - \log_6 5$$

$$= \log_6 \left(\dfrac{x^3 y}{5} \right)$$

3. $\dfrac{1}{2} \log_4 x = \log_4 25$

$$\log_4 x^{1/2} = \log_4 25$$
$$x^{1/2} = 25$$
$$x = (25)^2$$
$$x = 625$$

4. Answers may vary. Possible solution:
To simplify $\log_{10}(0.001)$, start by setting the expression equal to x.

$$\log_{10}(0.001) = x$$

$$10^x = 0.001 = 10^{-3}$$

Thus, $x = -3$, so $\log_{10} 0.001 = -3$.

Use Math to Save Money

1. $4 \times \dfrac{2}{3} = \dfrac{4}{1} \times \dfrac{2}{3} = \dfrac{8}{3}$

Lucy prepares $\dfrac{8}{3}$ cups of rice each week.

2. $\dfrac{21 \text{ ounces}}{1 \text{ cup}} \times \dfrac{8}{3} \text{ cups} = 56 \text{ ounces}$

Lucy's family consumes 56 ounces of rice each week.

3. $\dfrac{1 \text{ lb}}{16 \text{ oz}} \times 56 \text{ oz} = \dfrac{7}{2} \text{ or } 3\dfrac{1}{2} \text{ lb}$

Lucy's family eats $3\dfrac{1}{2}$ or 3.5 pounds each week.

4. $\dfrac{\$2.66}{1 \text{ lb}} \times \dfrac{3.5 \text{ lb}}{1 \text{ week}} = \$9.31 \text{ per week}$

Lucy spends $9.31 per week on name brand rice.

5. $\dfrac{\$0.88}{1 \text{ lb}} \times \dfrac{3.5 \text{ lb}}{1 \text{ week}} = \$3.08 \text{ per week}$

$9.31 - $3.08 = $6.23
Lucy could save $6.23 per week by buying the store brand rice.

6. $\dfrac{\$6.23}{\$9.31} \approx 0.67 = 67\%$

Lucy could save approximately 67% by buying the store brand.

7. $\dfrac{\$162}{1 \text{ week}} \times \dfrac{52 \text{ weeks}}{1 \text{ year}} = \$8424 \text{ per year}$

67% of $8424 = 0.67 × $8424 = $5644.08
Lucy could save $5644.08 per year by purchasing all store brand products.

8. $\dfrac{\$130}{1 \text{ week}} \times \dfrac{52 \text{ weeks}}{1 \text{ year}} = \$6760 \text{ per year}$

$8424 - $6760 = $1664
Lucy will save $1664 in the course of a year.

How Am I Doing? Sections 12.1–12.3
(Available online through MyMathLab.)

1. $f(x) = 2^{-x}$

x	$y = f(x) = 2^{-x}$
-2	4
-1	2
0	1
1	$\frac{1}{2}$
2	$\frac{1}{4}$

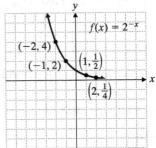

2. $3^{x+2} = 81$
$$3^{x+2} = 3^4$$
$$x + 2 = 4$$
$$x = 2$$

3. $2^x = \dfrac{1}{32} = \dfrac{1}{2^5} = 2^{-5}$
$$x = -5$$

4. $125 = 5^{3x+4}$
$$5^3 = 5^{3x+4}$$
$$3x + 4 = 3$$
$$3x = -1$$
$$x = -\frac{1}{3}$$

5. $A = P(1+r)^t$
$$A = 10{,}000(1+0.12)^4 = 15{,}735.1936$$
In 4 years Nancy will have \$15,735.19.

6. $\dfrac{1}{49} = 7^{-2}$
$$\log_7\left(\frac{1}{49}\right) = -2$$

7. $-3 = \log_{10} 0.001$
$$10^{-3} = 0.001$$

8. $\log_5 x = 3$
$$x = 5^3 = 125$$

9. $\log_x 49 = -2$
$$x^{-2} = 49$$
$$\frac{1}{x^2} = 49$$
$$x^2 = \frac{1}{49}$$
$$x = \pm\frac{1}{7}$$
Pick + since x must be > 0.
$$x = \frac{1}{7}$$

10. Let $N = \log_{10}(10{,}000)$.
$$10^N = 10{,}000$$
$$10^N = 10^4$$
$$N = 4$$

11. $\log_5\left(\dfrac{x^2 y^5}{z^3}\right) = \log_5(x^2 y^5) - \log_5 z^3$
$$= \log_5 x^2 + \log_5 y^5 - \log_5 z^3$$
$$= 2\log_5 x + 5\log_5 y - 3\log_5 z$$

12. $\dfrac{1}{2}\log_4 x - 3\log_4 w = \log_4 x^{1/2} - \log_4 w^3$
$$= \log_4 \sqrt{x} - \log_4 w^3$$
$$= \log_4\left(\frac{\sqrt{x}}{w^3}\right)$$

13. $\log_3 x + \log_3 2 = 4$
$$\log_3(2x) = 4$$
$$2x = 3^4$$
$$2x = 81$$
$$x = \frac{81}{2}$$

14. $\log_7 x = \log_7 8$
$\qquad x = 8$

15. $\log_9 1 = x$
$\qquad 9^x = 1 = 9^0$
$\qquad x = 0$

16. $\log_3 2x = 2$
$\qquad 3^2 = 2x$
$\qquad x = \dfrac{9}{2}$

17. $1 = \log_4 3x$
$\qquad 4^1 = 3x$
$\qquad x = \dfrac{4}{3}$

18. $\log_e x + \log_e 3 = 1$
$\qquad \log_e 3x = 1$
$\qquad e^1 = 3x$
$\qquad x = \dfrac{e}{3}$

12.4 Exercises

1. Error; you cannot take the logarithm of a negative number.

3. $\log 12.3 \approx 1.089905111$

5. $\log 25.6 \approx 1.408239965$

7. $\log 15 \approx 1.176091259$

9. $\log 125{,}000 \approx 5.096910013$

11. $\log 0.0123 \approx -1.910094889$

13. $\log x = 2.016$
$\qquad x = 10^{2.016} \approx 103.7528416$

15. $\log x = -2$
$\qquad x = 10^{-2} = 0.01$

17. $\log x = 3.9304$
$\qquad x = 10^{3.9304} \approx 8519.223264$

19. $\log x = 6.4683$
$\qquad x = 10^{6.4683} \approx 2{,}939{,}679.609$

21. $\log x = -3.3893$
$\qquad x = 10^{-3.3893} \approx 0.000408037$

23. $\log x = -1.5672$
$\qquad x = 10^{-1.5672} \approx 0.027089438$

25. $\text{antilog}(7.6215) \approx 41{,}831{,}168.87$

27. $\text{antilog}(-1.0826) \approx 0.0826799109$

29. $\ln 5.62 \approx 1.726331664$

31. $\ln 1.53 \approx 0.4252677354$

33. $\ln 136{,}000 \approx 11.82041016$

35. $\ln 0.00579 \approx -5.151622987$

37. $\ln x = 0.95$
$\qquad x = e^{0.95} \approx 2.585709659$

39. $\ln x = 2.4$
$\qquad x = e^{2.4} \approx 11.02317638$

41. $\ln x = -0.05$
$\qquad x = e^{-0.05} \approx 0.951229425$

43. $\ln x = -2.7$
$\qquad x = e^{-2.7} \approx 0.0672055127$

45. $\text{antilog}_e(6.1582) \approx 472.5766708$

47. $\text{antilog}_e(-2.1298) \approx 0.1188610637$

49. $\log_3 9.2 = \dfrac{\log 9.2}{\log 3} \approx 2.020006063$

51. $\log_7 7.35 = \dfrac{\log 7.35}{\log 7} \approx 1.025073184$

53. $\log_6 0.127 = \dfrac{\log 0.127}{\log 6} \approx -1.151699337$

55. $\log_{15} 12 = \dfrac{\log 12}{\log 15} \approx 0.917599921$

57. $\log_4 0.07733 = \dfrac{\ln 0.07733}{\ln 4} \approx -1.84641399$

59. $\log_{21} 436 = \dfrac{\ln 436}{\ln 21} \approx 1.996254706$

61. $\ln 1537 \approx 7.337587744$

63. $\text{antilog}_e(-1.874) \approx 0.153508399$

65. $\log x = 8.5634$
$\quad x = 10^{8.5634} \approx 3.65931672 \times 10^8$

67. $\log_4 x = 0.8645$
$\quad x = 4^{0.8645}$
$\quad x \approx 3.314979618$

69. $y = \log_6 x$

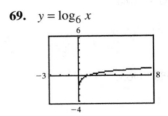

71. $y = \log_{0.4} x$

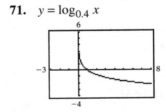

73. $N = 32.53 + 1.55 \ln x$
Use $x = 10$ for 2000.
$N = 32.53 + 1.55 \ln 10 \approx 36.10$
Use $x = 20$ for 2010.
$N = 32.53 + 1.55 \ln 20 \approx 37.17$

$\text{Percent increase} = \dfrac{37.17 - 36.10}{36.10}$

$\qquad\qquad = \dfrac{1.07}{36.10}$

$\qquad\qquad \approx 0.030$

The median age of the U.S. population in 2000 was 36.10; it was 37.17 in 2010. This was an increase of 3.0%.

75. $R = \log x$
$R = \log 56{,}000$
$R \approx 4.75$

77. $R = \log x$
$\quad 7.8 = \log x$
$\quad x = 10^{7.8} \approx 63{,}095{,}734$
The shock wave was about 63,095,734 times greater than the smallest detectable shock wave.

Cumulative Review

79. $3x^2 - 11x - 5 = 0$
$a = 3,\, b = -11,\, c = -5$
$$x = \frac{-(-11) \pm \sqrt{(-11)^2 - 4(3)(-5)}}{2(3)} = \frac{11 \pm \sqrt{181}}{6}$$

80. $2y^2 + 4y - 3 = 0$
$a = 2,\, b = 4,\, c = -3$
$$y = \frac{-4 \pm \sqrt{4^2 - 4(2)(-3)}}{2(2)}$$
$$= \frac{-4 \pm \sqrt{40}}{4}$$
$$= \frac{-4 \pm 2\sqrt{10}}{4}$$
$$= \frac{-2 \pm \sqrt{10}}{2}$$

81. Let x, y, z, w, and s be the distances between adjacent exits. So the distance between Exit 1 and Exit 2 is x, the distance between Exit 1 and Exit 3 is $x + y$, etc.
$$
\begin{aligned}
x + y + z + w + s &= 36 \quad (1) \\
x + y &= 12 \quad (2) \\
y + z &= 15 \quad (3) \\
z + w &= 12 \quad (4) \\
w + s &= 15 \quad (5)
\end{aligned}
$$
From equations (2) and (4) substitute 12 for $x + y$ and 12 for $z + w$ into equation (1).
$12 + 12 + s = 36$
$\qquad\qquad s = 12$
Substitute 12 for s in equation (5).
$w + 12 = 15$
$\qquad w = 3$
Substitute 3 for w in equation (4).
$z + 3 = 12$
$\qquad z = 9$
Substitute 9 for z in equation (3).
$y + 9 = 15$
$\qquad y = 6$
Substitute 6 for y in equation (2).
$x + 6 = 12$
$\qquad x = 6$

Thus, $x = 6$, $y = 6$, $z = 9$, $w = 3$, and $s = 12$.
The distance between Exit 1 and Exit 2 is x, or 6 miles.
The distance between Exit 1 and Exit 3 is $x + y = 6 + 6$, or 12 miles.

82. Let x, y, z, w, and s be the distances between adjacent exits. So the distance between Exit 1 and Exit 2 is x, the distance between Exit 1 and Exit 3 is $x + y$, etc.

$$\begin{aligned} x + y + z + w + s &= 36 \quad (1)\\ x + y &= 12 \quad (2)\\ y + z &= 15 \quad (3)\\ z + w &= 12 \quad (4)\\ w + s &= 15 \quad (5) \end{aligned}$$

From equations (2) and (4) substitute 12 for $x + y$ and 12 for $z + w$ into equation (1).

$$12 + 12 + s = 36$$
$$s = 12$$

Substitute 12 for s in equation (5).

$$w + 12 = 15$$
$$w = 3$$

Substitute 3 for w in equation (4).

$$z + 3 = 12$$
$$z = 9$$

Substitute 9 for z in equation (3).

$$y + 9 = 15$$
$$y = 6$$

Substitute 6 for y in equation (2).

$$x + 6 = 12$$
$$x = 6$$

Thus, $x = 6$, $y = 6$, $z = 9$, $w = 3$, and $s = 12$.
The distance between Exit 1 and Exit 4 is $x + y + z = 6 + 6 + 9$, or 21 miles.
The distance between Exit 1 and Exit 5 is $x + y + z + w = 6 + 6 + 9 + 3$, or 24 miles.

Quick Quiz 12.4

1. $\log 9.36 \approx 0.9713$

2.
$$\log x = 0.2253$$
$$\log_{10} x = 0.2253$$
$$x = 10^{0.2253}$$
$$x \approx 1.68$$

3. $\log_5 8.26 = \dfrac{\ln 8.26}{\ln 5} \approx 1.3119$

4. Answers may vary. Possible solution:
In order to solve for x in $\ln x = 1.7821$, take the antilog of both sides:
$$x = e^{1.7821}$$

12.5 Exercises

1.
$$\log_7\left(\frac{2}{3}x + 3\right) + \log_7 3 = 2$$
$$\log_7 3\left(\frac{2}{3}x + 3\right) = 2$$
$$\log_7(2x + 9) = 2$$
$$2x + 9 = 7^2$$
$$2x + 9 = 49$$
$$2x = 40$$
$$x = 20$$

Check: $\log_7\left(\frac{2}{3}\cdot 20 + 3\right) + \log_7 3 \overset{?}{=} 2$

$$\log_7 \frac{49}{3} + \log_7 3 \overset{?}{=} 2$$
$$\log_7 49 \overset{?}{=} 2$$
$$2 = 2$$

3.
$$\log_9 3 + \log_9(2x + 1) = 1$$
$$\log_9(3(2x + 1)) = 1$$
$$\log_9(6x + 3) = 1$$
$$6x + 3 = 9^1$$
$$6x + 3 = 9$$
$$6x = 6$$
$$x = 1$$

Check: $\log_9 3 + \log_9(2\cdot 1 + 1) \overset{?}{=} 1$
$$\log_9 3 + \log_9 3 \overset{?}{=} 1$$
$$\log_9 9 \overset{?}{=} 1$$
$$1 = 1$$

5.
$$\log_2\left(x + \frac{4}{3}\right) = 5 - \log_2 6$$
$$\log_2\left(x + \frac{4}{3}\right) + \log_2 6 = 5$$
$$\log_2(6x + 8) = 5$$
$$6x + 8 = 2^5$$
$$6x = 24$$
$$x = 4$$

Check: $\log_2\left(4 + \frac{4}{3}\right) \overset{?}{=} 5 - \log_2 6$
$$\log_2\left(\frac{16}{3}\right) \overset{?}{=} 5 - \log_2(3\cdot 2)$$
$$\log_2 16 - \log_2 \overset{?}{=} 5 - \log_2 3 - \log_2 2$$
$$4 - \log_2 3 \overset{?}{=} 5 - \log_2 3 - 1$$
$$4 - \log_2 3 = 4 - \log_2 3$$

7.
$$\log(30x+40)=2+\log(x-1)$$
$$\log(30x+40)-\log(x-1)=2$$
$$\log\left(\frac{30x+40}{x-1}\right)=2$$
$$\frac{30x+40}{x-1}=10^2$$
$$30x+40=100x-100$$
$$70x=140$$
$$x=2$$
Check: $\log(30(2)+40)\overset{?}{=}2+\log(2-1)$
$$\log 100 \overset{?}{=} 2+\log(1)$$
$$2 \overset{?}{=} 2+\log(1)=2+0$$
$$2=2$$

9.
$$2+\log_6(x-1)=\log_6(12x)$$
$$\log_6(x-1)-\log_6(12x)=-2$$
$$\log_6\left(\frac{x-1}{12x}\right)=-2$$
$$\frac{x-1}{12x}=6^{-2}$$
$$36x-36=12x$$
$$24x=36$$
$$x=\frac{3}{2}$$
Check: $2+\log_6\left(\frac{3}{2}-1\right)\overset{?}{=}\log_6\left(12\cdot\frac{3}{2}\right)$
$$2+\log_6\frac{1}{2}\overset{?}{=}\log_6\frac{36}{2}$$
$$2+\log_6 1-\log_2 6\overset{?}{=}\log_6\frac{36}{2}$$
$$2+0-\log_6 2\overset{?}{=}\log_6 36-\log_6 2$$
$$2-\log_6 2=2-\log_6 2$$

11. $\log(75x+50)-\log x=2$
$$\log\left(\frac{75x+50}{x}\right)=2$$
$$\frac{75x+50}{x}=10^2$$
$$100x=75x+50$$
$$25x=50$$
$$x=\frac{50}{25}=2$$
Check: $\log(75(2)+50)-\log(2)\overset{?}{=}2$
$$\log(200)-\log(2)\overset{?}{=}2$$
$$\log\left(\frac{200}{2}\right)\overset{?}{=}2$$
$$\log 100 \overset{?}{=} 2$$
$$2=2$$

13. $\log_3(x+6)+\log_3 x=3$
$$\log_3(x(x+6))=3$$
$$x(x+6)=3^3$$
$$x^2+6x-27=0$$
$$(x+9)(x-3)=0$$
$$x=-9 \text{ gives } \log_3(\text{negative})$$
$$x=3 \text{ is the solution.}$$
Check: $\log_3(3+6)+\log_3 3\overset{?}{=}3$
$$\log_3 9+\log_3 3\overset{?}{=}3$$
$$2+1\overset{?}{=}3$$
$$3=3$$

15.
$$1+\log(x-2)=\log(6x)$$
$$\log(6x)-\log(x-2)=1$$
$$\log\frac{6x}{x-2}=1$$
$$\frac{6x}{x-2}=10^1$$
$$6x=10x-20$$
$$4x=20$$
$$x=5$$
Check: $1+\log(5-2)\overset{?}{=}\log(6\cdot 5)=\log 30$
$$1+\log 3\overset{?}{=}\log(10\cdot 3)$$
$$1+\log 3\overset{?}{=}\log 10+\log 3$$
$$1+\log 3=1+\log 3$$

17.
$$\log_2(x+5)-2=\log_2 x$$
$$\log_2(x+5)-\log_2 x=2$$
$$\log_2\frac{x+5}{x}=2$$
$$\frac{x+5}{x}=2^2=4$$
$$x+5=4x$$
$$3x=5$$
$$x=\frac{5}{3}$$
Check: $\log_2\left(\frac{5}{3}+5\right)-2\overset{?}{=}\log_2\frac{5}{3}$
$$\log_2\frac{20}{3}-2\overset{?}{=}\log_2\frac{5}{3}$$
$$\log_2\left(4\cdot\frac{5}{3}\right)-2\overset{?}{=}\log_2\frac{5}{3}$$
$$\log_2 4+\log_2\frac{5}{3}-2\overset{?}{=}\log_2\frac{5}{3}$$
$$2+\log_2\frac{5}{3}-2\overset{?}{=}\log_2\frac{5}{3}$$
$$\log_2\frac{5}{3}=\log_2\frac{5}{3}$$

19.

$$2\log_7 x = \log_7(x+4) + \log_7 2$$
$$\log_7 x^2 - \log_7(x+4) = \log_7 2$$
$$\log_7 \frac{x^2}{x+4} = \log_7 2$$
$$\frac{x^2}{x+4} = 2$$
$$x^2 = 2x + 8$$
$$x^2 - 2x - 8 = 0$$
$$(x-4)(x+2) = 0$$

$x = 4$, $x = -2$, reject, gives $\log_7$ (negative)

Check: $2\log_7 4 \stackrel{?}{=} \log_7(4+4) + \log_7 2$

$$\log_7 4^2 \stackrel{?}{=} \log_7(8) + \log_7 2$$
$$\log_7 16 \stackrel{?}{=} \log_7(8 \cdot 2)$$
$$\log_7 16 = \log_7 16$$

21.

$$\ln 10 - \ln x = \ln(x-3)$$
$$\ln \frac{10}{x} = \ln(x-3)$$
$$\frac{10}{x} = x - 3$$
$$x^2 - 3x - 10 = 0$$
$$(x-5)(x+2) = 0$$

$x = 5$, $x = -2$, reject, gives $\ln$(negative)

Check: $\ln 10 - \ln 5 \stackrel{?}{=} \ln(5-3)$

$$\ln \frac{10}{5} \stackrel{?}{=} \ln 2$$
$$\ln 2 = \ln 2$$

23.

$$7^{x+3} = 12$$
$$\log 7^{x+3} = \log 12$$
$$(x+3)\log 7 = \log 12$$
$$x + 3 = \frac{\log 12}{\log 7}$$
$$x = \frac{\log 12}{\log 7} - 3$$
$$x = \frac{\log 12 - 3\log 7}{\log 7}$$

25.

$$2^{3x+4} = 17$$
$$\log 2^{3x+4} = \log 17$$
$$(3x+4)\log 2 = \log 17$$
$$3x + 4 = \frac{\log 17}{\log 2}$$
$$3x = \frac{\log 17}{\log 2} - 4 \cdot \frac{\log 2}{\log 2}$$
$$x = \frac{\log 17}{3\log 2} - 4 \cdot \frac{\log 2}{3\log 2}$$
$$x = \frac{\log 17 - 4\log 2}{3\log 2}$$

27.

$$8^{2x-1} = 90$$
$$\log 8^{2x-1} = \log 90$$
$$(2x-1)\log 8 = \log 90$$
$$2x - 1 = \frac{\log 90}{\log 8}$$
$$2x = 1 + \frac{\log 90}{\log 8}$$
$$x = \frac{1}{2}\left(1 + \frac{\log 90}{\log 8}\right)$$
$$x \approx 1.582$$

29.

$$5^x = 4^{x+1}$$
$$\log 5^x = \log 4^{x+1}$$
$$x\log 5 = (x+1)\log 4$$
$$x\log 5 = x\log 4 + \log 4$$
$$x(\log 5 - \log 4) = \log 4$$
$$x = \frac{\log 4}{\log 5 - \log 4}$$
$$x \approx 6.213$$

31.

$$28 = e^{x-2}$$
$$\ln 28 = \ln e^{x-2}$$
$$\ln 28 = (x-2)\ln e$$
$$2 + \ln 28 = x$$
$$5.332 \approx x$$

33.

$$88 = e^{2x+1}$$
$$\ln 88 = \ln e^{2x+1}$$
$$\ln 88 = (2x+1)\ln e$$
$$\ln 88 = (2x+1)(1)$$
$$\ln 88 = 2x + 1$$
$$2x = -1 + \ln 88$$
$$x = \frac{-1 + \ln 88}{2}$$
$$x \approx 1.739$$

35.
$$A = P(1+r)^t$$
$$5000 = 1500(1+0.08)^t$$
$$1.08^t = \frac{10}{3}$$
$$\ln 1.08^t = \ln \frac{10}{3}$$
$$t = \frac{\ln \frac{10}{3}}{\ln 1.08} \approx 15.64$$
It will take approximately 16 years.

37.
$$A = P(1+r)^t$$
$$3P = P(1+0.06)^t$$
$$1.06^t = 3$$
$$\ln 1.06^t = \ln 3$$
$$t \ln 1.06 = \ln 3$$
$$t = \frac{\ln 3}{\ln 1.06}$$
$$t \approx 18.85$$
It will take approximately 19 years.

39.
$$A = P(1+r)^t$$
$$6500 = 5000(1+r)^6$$
$$1.3 = (1+r)^6$$
$$\ln 1.3 = \ln(1+r)^6 = 6\ln(1+r)$$
$$\ln(1+r) = \frac{\ln 1.3}{6}$$
$$e^{\ln(1+r)} = e^{\frac{\ln 1.3}{6}}$$
$$1+r = e^{\frac{\ln 1.3}{6}}$$
$$r = e^{\frac{\ln 1.3}{6}} - 1$$
$$r \approx 0.044698$$
The rate is approximately 4.5%.

41.
$$A = A_0 e^{rt}$$
$$7 = 3e^{0.021t}$$
$$\frac{7}{3} = e^{0.021t}$$
$$\ln \frac{7}{3} = \ln e^{0.021t}$$
$$\ln 7 - \ln 3 = 0.021t$$
$$t = \frac{\ln 7 - \ln 3}{0.021} \approx 40.35$$
It would take approximately 40 years.

43.
$$A = A_0 e^{rt}$$
$$9 = 6e^{0.013t}$$
$$\frac{9}{6} = e^{0.013t}$$
$$\ln 1.5 = \ln e^{0.013t}$$
$$\ln 1.5 = 0.013t$$
$$t = \frac{\ln 1.5}{0.013} \approx 31.12$$
It would take approximately 31 years.

45. $N = 120,500(1.04)^x$
Use $x = 2$ for 2014.
$$N = 120,500(1.04)^2 \approx 130,333$$
There were approximately 130,333 physical therapy assistants in 2014.

47.
$$N = 120,500(1.04)^x$$
$$175,000 = 120,500(1.04)^x$$
$$\frac{175,000}{120,500} = 1.04^x$$
$$\frac{350}{241} = 1.04^x$$
$$\ln\left(\frac{350}{241}\right) = \ln 1.04^x$$
$$\ln\left(\frac{350}{241}\right) = x \ln 1.04$$
$$x = \frac{\ln\left(\frac{350}{241}\right)}{\ln 1.04} \approx 9.5$$
$$2012 + 9.5 = 2021.5$$
The number of physical therapy assistants will reach 175,000 during the year 2021.

49.
$$A = A_0 e^{rt}$$
$$120,000 = 80,000e^{0.015t}$$
$$e^{0.015t} = 1.5$$
$$\ln e^{0.015t} = \ln 1.5$$
$$0.015t = \ln 1.5$$
$$t = \frac{\ln 1.5}{0.015}$$
$$t \approx 27.03$$
It will take approximately 27 years.

51.
$$A = A_0 e^{rt}$$
$$1800 = 200 e^{0.04t}$$
$$e^{0.04t} = 9$$
$$0.04t = \ln 9$$
$$t = \frac{\ln 9}{0.04} \approx 54.93$$
It will take approximately 55 hours.

53. $A = A_0 e^{rt}$
$$A = 29{,}959 e^{0.06(6)} \approx 42{,}941.12$$
By the end of 2015 approximately 42,941 people are expected to be infected.

55. $R = \log\left(\dfrac{I}{I_0}\right)$
$$7.1 = \log\left(\frac{I_{LP}}{I_0}\right) = \log I_{LP} - \log I_0$$
$$8.2 = \log\left(\frac{I_{KI}}{I_0}\right) = \log I_{KI} - \log I_0$$
Subtracting the two equations gives
$$-1.1 = \log I_{LP} - \log I_{KI} = \log \frac{I_{LP}}{I_{KI}}$$
$$10^{\log\frac{I_{LP}}{I_{KI}}} = 10^{-1.1}$$
$$\frac{I_{LP}}{I_{KI}} = 10^{-1.1}$$
$$I_{KI} = 10^{1.1} I_{LP} \approx 12.6 I_{LP}$$
The Kurile earthquake was about 12.6 times as intense as the Loma Prieta earthquake.

57. $R = \log\left(\dfrac{I}{I_0}\right)$
$$8.3 = \log\left(\frac{I_S}{I_0}\right) = \log I_S - \log I_0$$
$$6.8 = \log\left(\frac{I_J}{I_0}\right) = \log I_J - \log I_0$$
Subtracting the two equations gives
$$1.5 = \log I_S - \log I_J = \log \frac{I_S}{I_J}$$
$$10^{\log\frac{I_S}{I_J}} = 10^{1.5}$$
$$\frac{I_S}{I_J} = 10^{1.5}$$
$$I_S = 10^{1.5} I_J \approx 31.6 I_J$$

The intensity of the San Francisco earthquake was approximately 31.6 times greater than the intensity of the Japanese earthquake.

Cumulative Review

59. $\left(\sqrt{3} + 2\sqrt{2}\right)\left(\sqrt{6} - \sqrt{2}\right) = \sqrt{18} - \sqrt{6} + 2\sqrt{12} - 4$
$$= 3\sqrt{2} - \sqrt{6} + 4\sqrt{3} - 4$$

60. $\sqrt{98x^3 y^2} = \sqrt{49x^2 y^2 \cdot 2x} = 7xy\sqrt{2x}$

61. Total students: 285 (given)
9 years old: 3 (given)
10 years old: 11 (given)
11 years old: 11 + 24 = 35
12 years old: 70 (given)
13 years old: 70 + 18 = 88
14 or 15 years old: 78 (given)
Thus, the number of students in each age category were:
9 years old: 3
10 years old: 11
11 years old: 35
12 years old: 70
13 years old: 88
14 or 15 years old: 78

62. $\dfrac{3.5 \times 10^9 \text{ pounds}}{16 \text{ kilometers}} \cdot \dfrac{1 \text{ kilometer}}{0.62 \text{ mile}} \cdot \dfrac{1 \text{ U.S. dollar}}{0.64 \text{ pound}}$
$$\approx 0.551285282 \times 10^9 \text{ dollars/mi}$$
$$= \frac{\$551{,}285{,}282}{\text{mile}}$$
The extension cost approximately $551,285,282 per mile.

Quick Quiz 12.5

1.
$$2 - \log_6 x = \log_6(x + 5)$$
$$2 = \log_6 x + \log_6(x + 5)$$
$$2 = \log_6[x(x + 5)]$$
$$2 = \log_6(x^2 + 5x)$$
$$6^2 = x^2 + 5x$$
$$x^2 + 5 - 36 = 0$$
$$(x + 9)(x - 4) = 0$$
$$x + 9 = 0 \text{ or } x - 4 = 0$$
$$x = -9, 4$$
-9 is extraneous. 4 is the only solution.

2. $\log_3(x-5) + \log_3(x+1) = \log_3 7$

$\log_3[(x-5)(x+1)] = \log_3 7$

$x^2 + x - 5x - 5 = 7$

$x^2 - 4x - 12 = 0$

$(x-6)(x+2) = 0$

$x - 6 = 0$ or $x + 2 = 0$

$x = 6, -2$

−2 is extraneous. 6 is the only solution.

3. $6^{x+2} = 9$

$\log 6^{x+2} = \log 9$

$(x+2)\log 6 = \log 9$

$x + 2 = \dfrac{\log 9}{\log 6}$

$x = \dfrac{\log 9}{\log 6} - 2$

$x \approx -0.774$

4.   Answers may vary. Possible solution:

To solve $26 = 52e^{3x}$, start by dividing both sides by 52.

$\dfrac{1}{2} = e^{3x}$

Then take the natural log of both sides.

$\ln\left(\dfrac{1}{2}\right) = \ln e^{3x}$

$\ln\left(\dfrac{1}{2}\right) = 3x \ln e$

$\ln\left(\dfrac{1}{2}\right) = 3x$

Isolate the x.

$\dfrac{\ln\left(\frac{1}{2}\right)}{3} = x$

Solve with a calculator.

$x \approx -0.231$

Career Exploration Problems

1. $N = 14,000(1.032)^x$

Use $x = 4$ for 2017.

$N = 14,000(1.032)^4 \approx 15,879.9$

There are expected to be 15,880 data information security analysts in 2017.

2. $N = 14,000(1.032)^x$

$18,000 = 14,000(1.032)^x$

$\dfrac{18,000}{14,000} = 1.032^x$

$\dfrac{9}{7} = 1.032^x$

$\ln\left(\dfrac{9}{7}\right) = \ln 1.032^x$

$\ln\left(\dfrac{9}{7}\right) = x \ln 1.032$

$x = \dfrac{\ln\left(\frac{9}{7}\right)}{\ln 1.032} \approx 7.98$

The number of data information security analysts is expected to reach 18,000 in 2021 (2013 + 8).

3.   Use $A = A_0 e^{rt}$ with $A = 18,000$, $A_0 = 14,000$, and $r = 3.2\% = 0.032$.

$18,000 = 14,000e^{0.032t}$

$\dfrac{18,000}{14,000} = e^{0.032t}$

$\dfrac{9}{7} = e^{0.032t}$

$\ln\left(\dfrac{9}{7}\right) = \ln e^{0.032t}$

$\ln\left(\dfrac{9}{7}\right) = 0.032t \ln e$

$\ln\left(\dfrac{9}{7}\right) = 0.032t$

$t = \dfrac{\ln\left(\frac{9}{7}\right)}{0.032} \approx 7.85$

Using the continuous growth model, the number of data information security analysts is expected to reach 18,000 in 2021 (2013 + 8). This is the same year predicted by the previous model.

4.   Use $A = A_0(1+r)^t$ with $A_0 = 45,000$, $r = 0.8\% = 0.008$, and $t = 2016 - 2013 = 3$.

$A = 45,000(1 + 0.008)^3$

$= 45,000(1.008)^3$

$\approx 46,089$

The average amount of debt for four-year school graduates is expected to be \$46,089 in 2016.

You Try It

1. $f(x) = \left(\dfrac{1}{4}\right)^x$

x	$y = f(x) = \left(\frac{1}{4}\right)^x$
-2	16
-1	4
0	1
1	$\frac{1}{4}$
2	$\frac{1}{16}$

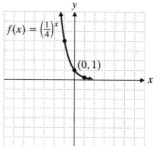

2. $3^x = \dfrac{1}{81} = \dfrac{1}{3^4}$

$3^x = 3^{-4}$

$x = -4$

3. a. $\log_4 9 = 0.5x \Leftrightarrow 4^{0.5x} = 9$

    b. $28 = 5^x \Leftrightarrow \log_5 28 = x$

    c. $\log_4\left(\dfrac{1}{64}\right) = x$

$4^x = \dfrac{1}{64} = \dfrac{1}{4^3}$

$4^x = 4^{-3}$

$x = -3$

4. a. $\log_2\left(\dfrac{a\sqrt{b}}{c^2}\right) = \log_2\left(a\sqrt{b}\right) - \log_2 c^2$

$= \log_2 a + \log_2 \sqrt{b} - \log_2 c^2$

$= \log_2 a + \dfrac{1}{2}\log_2 b - 2\log_2 c$

b. $3\log_5 a + \dfrac{2}{3}\log_5 b - 2\log_5 c$

$= \log_5 a^3 + \log_5 b^{2/3} - \log_5 c^2$

$= \log_5(a^3 b^{2/3}) - \log_5 c^2$

$= \log_5\left(\dfrac{a^3 b^{2/3}}{c^2}\right)$

c. $\log_8 1 - \log 10^3 + \log_4 4 = 0 - 3 + 1 = -2$

5. a. $\log 16.5 \approx 1.2174839$

    b. $\ln 32.7 \approx 3.4873751$

6. a. $\log x = 2.075$

$x = 10^{2.075}$

$x \approx 118.8502227$

    b. $\ln x = 1.528$

$x = e^{1.528}$

$x \approx 4.6089497$

7. $\log_8 2.5 = \dfrac{\ln 2.5}{\ln 8} \approx 0.4406427$

8. $-\log_6 3 + \log_6 8x = \log_6(x^2 - 1)$

$\log_6 8x = \log_6(x^2 - 1) + \log_6 3$

$\log_6 8x = \log_6[3(x^2 - 1)]$

$8x = 3(x^2 - 1)$

$0 = 3x^2 - 8x - 3$

$0 = (3x + 1)(x - 3)$

$3x + 1 = 0 \qquad\qquad x - 3 = 0$

$x = -\dfrac{1}{3} \qquad\qquad x = 3$

Discard the negative value. $x = 3$ is the solution.

9. $3^{x+2} = 11$

$\log 3^{x+2} = \log 11$

$(x + 2)\log 3 = \log 11$

$x\log 3 + 2\log 3 = \log 11$

$x\log 3 = \log 11 - 2\log 3$

$x = \dfrac{\log 11 - 2\log 3}{\log 3}$

$x \approx 0.1826583$

Chapter 12 Review Problems

1. $f(x) = 4^{3+x}$

x	$y = f(x) = 4^{3+x}$
-5	$\frac{1}{16}$
-4	$\frac{1}{4}$
-3	1
-2	4
-1	16

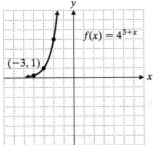

2. $f(x) = e^{x-3}$

x	$y = f(x) = e^{x-3}$
1	0.14
2	0.37
3	1
4	2.72
5	7.39

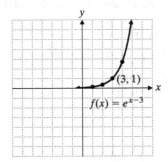

3. $3^{3x+1} = 81$
    $3^{3x+1} = 3^4$
    $3x+1 = 4$
    $3x = 3$
    $x = 1$

4. $2^{x+9} = 128$
    $2^{x+9} = 2^7$
    $x+9 = 7$
    $x = -2$

5. $-2 = \log_{10}(0.01) \Leftrightarrow 10^{-2} = 0.01$

6. $8 = 4^{3/2} \Leftrightarrow \log_4 8 = \frac{3}{2}$

7. $\log_w 16 = 4$
    $w^4 = 16 = 2^4$
    $w = 2$

8. $\log_8 x = 0$
    $8^0 = x$
    $x = 1$

9. $\log_7 w = -1$
    $7^{-1} = w$
    $w = \frac{1}{7}$

10. $\log_w 64 = 3$
    $w^3 = 64 = 4^3$
    $w = 4$

11. $\log_{10} w = -1$
    $10^{-1} = w$
    $w = 0.1$ or $\frac{1}{10}$

12. $\log_{10} 1000 = x$
    $10^x = 1000 = 10^3$
    $x = 3$

13. $\log_2 64 = x$
    $2^x = 64 = 2^6$
    $x = 6$

14. $\log_2 \dfrac{1}{4} = x$

$2^x = \dfrac{1}{4} = 2^{-2}$

$x = -2$

15. $y = \log_3 x \Leftrightarrow 3^y = x$

$x = 3^y$	y
$\dfrac{1}{9}$	-2
$\dfrac{1}{3}$	-1
1	0
3	1
9	2

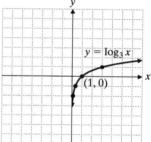

16. $\log_2 \left(\dfrac{5x}{\sqrt{w}} \right) = \log_2 (5x) - \log_2 \sqrt{w}$

$= \log_2 (5x) - \log_2 w^{1/2}$

$= \log_2 5 + \log_2 x - \dfrac{1}{2} \log_2 w$

17. $\log_2 x^3 \sqrt{y} = \log_2 x^3 + \log_2 \sqrt{y}$

$= 3 \log_2 x + \log_2 y^{1/2}$

$= 3 \log_2 x + \dfrac{1}{2} \log_2 y$

18. $\log_3 x + \log_3 w^{1/2} - \log_3 2$

$= \log_3 x + \log_3 \sqrt{w} - \log_3 2$

$= \log_3 \left(x\sqrt{w} \right) - \log_3 2$

$= \log_3 \dfrac{x\sqrt{w}}{2}$

19. $4 \log_8 w - \dfrac{1}{3} \log_8 z = \log_8 w^4 - \log_8 z^{1/3}$

$= \log_8 w^4 - \log_8 \sqrt[3]{z}$

$= \log_8 \dfrac{w^4}{\sqrt[3]{z}}$

20. $\log_e e^6 = 6 \log_e e = 6(1) = 6$

21. $\log 23.8 \approx 1.376576957$

22. $\log 0.0817 \approx -1.087777943$

23. $\ln 3.92 \approx 1.366091654$

24. $\ln 803 \approx 6.688354714$

25. $\log n = 1.1367$

$n = 10^{1.1367}$

$n \approx 13.69935122$

26. $\ln n = 1.7$

$n = e^{1.7} \approx 5.473947392$

27. $\log_8 2.81 = \dfrac{\ln 2.81}{\ln 8} \approx 0.4968567101$

28. $\log_4 72 = \dfrac{\ln 72}{\ln 4} \approx 3.084962501$

29. $\log_5 100 - \log_5 x = \log_5 4$

$\log_5 \dfrac{100}{x} = \log_5 4$

$\dfrac{100}{x} = 4$

$x = 25$

Check: $\log_5 100 - \log_5 25 \overset{?}{=} \log_5 4$

$\log_5 \dfrac{100}{25} \overset{?}{=} \log_5 4$

$\log_5 4 = \log_5 4$

30. $\log_8 x + \log_8 3 = \log_8 75$

$\log_8 (3x) = \log_8 75$

$3x = 75$

$x = 25$

Check: $\log_8 25 + \log_8 3 \overset{?}{=} \log_8 75$

$\log_8 (25 \cdot 3) \overset{?}{=} \log_8 75$

$\log_8 75 = \log_8 75$

31. $\log_{11}\left(\dfrac{4}{3}x+7\right)+\log_{11}3=2$

$\log_{11}\left[3\left(\dfrac{4}{3}x+7\right)\right]=2$

$\log_{11}(4x+21)=2$

$4x+21=11^2$

$4x+21=121$

$4x=100$

$x=25$

Check: $\log_{11}\left(\dfrac{4}{3}\cdot 25+7\right)+\log_{11}3\overset{?}{=}2$

$\log_{11}\left(\left(\dfrac{4}{3}\cdot 25+7\right)\cdot 3\right)\overset{?}{=}2$

$\log_{11}(100+21)\overset{?}{=}2$

$\log_{11}(121)\overset{?}{=}2$

$2=2$

32. $\log_8(x-3)=-1+\log_8 6x$

$\log_8(x-3)-\log_8 6x=-1$

$\log_8\dfrac{x-3}{6x}=-1$

$\dfrac{x-3}{6x}=8^{-1}=\dfrac{1}{8}$

$8x-24=6x$

$2x=24$

$x=12$

Check: $\log_8(12-3)\overset{?}{=}-1+\log_8(6\cdot 12)$

$\log_8(9)\overset{?}{=}-\log_8 8+\log_8(72)$

$\log_8(9)\overset{?}{=}\log_8\left(\dfrac{72}{8}\right)$

$\log_8(9)=\log_8(9)$

33. $\log(2t+3)+\log(4t-1)=2\log 3$

$\log((2t+3)(4t-1))=\log 3^2=\log 9$

$(2t+3)(4t-1)=9$

$8t^2+10t-3=9$

$8t^2+10t-12=0$

$4t^2+5t-6=0$

$(4t-3)(t+2)=0$

$t=\dfrac{3}{4}$, $t=-2$, reject -2 since it gives

log(negative).

$t=\dfrac{3}{4}$ is the solution.

Check: $\log\left(2\cdot\dfrac{3}{4}+3\right)+\log\left(4\cdot\dfrac{3}{4}-1\right)\overset{?}{=}2\log 3$

$\log\left(\dfrac{9}{2}\right)+\log(2)\overset{?}{=}2\log 3$

$\log\left(\dfrac{9}{2}\cdot 2\right)\overset{?}{=}2\log 3$

$\log(9)\overset{?}{=}2\log 3$

$\log(3^2)\overset{?}{=}2\log 3$

$2\log 3=2\log 3$

34. $\log(2t+4)-\log(3t+1)=\log 6$

$\log\left(\dfrac{2t+4}{3t+1}\right)=\log 6$

$\dfrac{2t+4}{3t+1}=6$

$2t+4=18t+6$

$16t=-2$

$t=-\dfrac{1}{8}$

Check: $\log\left(2\cdot\dfrac{-1}{8}+4\right)-\log\left(3\cdot\dfrac{-1}{8}+1\right)\overset{?}{=}\log 6$

$\log\left(\dfrac{15}{4}\right)-\log\left(\dfrac{5}{8}\right)\overset{?}{=}\log 6$

$\log\left(\dfrac{\frac{15}{4}}{\frac{5}{8}}\right)\overset{?}{=}\log 6$

$\log\left(\dfrac{\frac{15}{4}}{\frac{5}{8}}\right)\overset{?}{=}\log 6$

$\log 6=\log 6$

35. $3^x=14$

$\log 3^x=\log 14$

$x\log 3=\log 14$

$x=\dfrac{\log 14}{\log 3}$

36. $5^{x+3}=130$

$\log 5^{x+3}=\log 130$

$(x+3)\log 5=\log 130$

$x+3=\dfrac{\log 130}{\log 5}$

$x=\dfrac{\log 130}{\log 5}-3$

$x=\dfrac{\log 30-3\log 5}{\log 5}$

37.
$$e^{2x-1} = 100$$
$$\ln(e^{2x-1}) = \ln 100$$
$$(2x-1)\ln e = \ln 100$$
$$2x-1 = \ln 100$$
$$2x = 1 + \ln 100$$
$$x = \frac{1 + \ln 100}{2}$$

38.
$$2^{3x+1} = 5^x$$
$$\ln 2^{3x+1} = \ln 5^x$$
$$(3x+1)\ln 2 = x\ln 5$$
$$3x\ln 2 + \ln 2 = x\ln 5$$
$$x(3\ln 2 - \ln 5) = -\ln 2$$
$$x = \frac{\ln 2}{\ln 5 - 3\ln 2}$$
$$x \approx -1.4748$$

39.
$$e^{3x-4} = 20$$
$$\ln e^{3x-4} = \ln 20$$
$$(3x-4)\ln e = \ln 20$$
$$3x-4 = \ln 20$$
$$x = \frac{\ln 20 + 4}{3}$$
$$x \approx 2.3319$$

40.
$$1.03^x = 20$$
$$\ln 1.03^x = \ln 20$$
$$x\ln 1.03 = \ln 20$$
$$x = \frac{\ln 20}{\ln 1.03}$$
$$x \approx 101.3482$$

41.
$$A = P(1+r)^t$$
$$2P = P(1+0.08)^t$$
$$2 = 1.08^t$$
$$\ln 2 = \ln 1.08^t$$
$$\ln 2 = t\ln 1.08$$
$$t = \frac{\ln 2}{\ln 1.08} \approx 9$$
It will take about 9 years to double money in the account.

42. $A = P(1+r)^t$
$$A = 5000(1+0.06)^4 = 6312.38$$
He would have $6312.38 in the account after 4 years.

43.
$$A = A_0 e^{rt}$$
$$16 = 7e^{0.02t}$$
$$\ln\frac{16}{7} = \ln e^{0.02t}$$
$$\ln\frac{16}{7} = 0.02t\ln e$$
$$t = \frac{\ln\frac{16}{7}}{0.02} \approx 41.33$$
It will take approximately 41 years.

44.
$$A = A_0 e^{rt}$$
$$2600 = 2000e^{0.03t}$$
$$\ln 1.3 = \ln e^{0.03t}$$
$$\ln 1.3 = 0.03t$$
$$t = \frac{\ln 1.3}{0.03} \approx 8.75$$
It will take approximately 9 years.

45. $W = p_0 V_0 \ln\left(\dfrac{V_1}{V_0}\right)$

   a. $W = 40(15)\ln\left(\dfrac{24}{15}\right) \approx 282$

      The work done is approximately 282 lb.

   b. $100 = p_0(8)\ln\left(\dfrac{40}{8}\right)$

      $p_0 = \dfrac{100}{(8)\ln\left(\frac{40}{8}\right)} \approx 7.77$

      The pressure is approximately 7.77 lb/in.³.

46. $M = \log\left(\dfrac{I}{I_0}\right)$

$$8.4 = \log\left(\frac{I_A}{I_0}\right) = \log I_A - \log I_0$$
$$-\left(6.7 = \log\left(\frac{I_T}{I_0}\right) = \log I_T - \log I_0\right)$$
$$\overline{1.7 = \log I_A - \log I_T}$$
$$\log\frac{I_A}{I_T} = 1.7$$
$$10^{\log\frac{I_A}{I_T}} = 10^{1.7}$$
$$\frac{I_A}{I_T} \approx 50.12$$
$$I_A \approx 50.12 I_T$$
The Alaska earthquake was about 50.1 times more intense than the Turkey earthquake.

How Am I Doing? Chapter 12 Test

1. $f(x) = 3^{x+1}$

x	$y = f(x) = 3^{x+1}$
−1	1
0	3

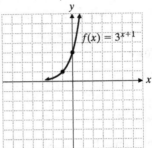

2. $f(x) = \log_2 x$

x	$y = f(x) = \log_2 x$
$\frac{1}{2}$	−1
1	0
2	2
4	2

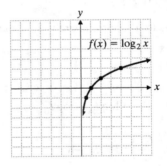

3. $4^{x+3} = 64 = 4^3$
$x + 3 = 3$
$x = 0$

4. $\log_w 125 = 3$
$w^3 = 125 = 5^3$
$w = 5$

5. $\log_8 x = -2 \Leftrightarrow x = 8^{-2} = \dfrac{1}{64}$

6. $2\log_7 x + \log_7 y - \log_7 4 = \log_7 x^2 + \log_7 \dfrac{y}{4}$

$$= \log_7 \dfrac{x^2 y}{4}$$

7. $\ln 5.99 \approx 1.7901$

8. $\log 23.6 \approx 1.3729$

9. $\log_3 1.62 = \dfrac{\log 1.62}{\log 3} \approx 0.4391$

10. $\log x = 3.7284$
$x = 10^{3.7284}$
$x \approx 5350.569382$

11. $\ln x = 0.14$
$x = e^{0.14} \approx 1.150273799$

12. $\log_8 (x+3) - \log_8 2x = \log_8 4$
$\log_8 \left(\dfrac{x+3}{2x} \right) = \log_8 4$
$\dfrac{x+3}{2x} = 4$
$x + 3 = 8x$
$7x = 3$
$x = \dfrac{3}{7}$

Check: $\log_8 \left(\dfrac{3}{7} + 3 \right) - \log_8 \left(2 \cdot \dfrac{3}{7} \right) \overset{?}{=} \log_8 4$

$\log_8 \left(\dfrac{24}{7} \right) - \log_8 \left(\dfrac{6}{7} \right) \overset{?}{=} \log_8 4$

$\log_8 \left(\dfrac{\frac{24}{7}}{\frac{6}{7}} \right) \overset{?}{=} \log_8 4$

$\log_8 4 = \log_8 4$

13. $\log_8 2x + \log_8 6 = 2$
$\log_8 ((2x)(6)) = 2$
$12x = 8^2$
$12x = 64$
$x = \dfrac{16}{3}$

Check: $\log_8 2 \cdot \frac{16}{3} + \log_8 6 \overset{?}{=} 2$

$\log_8\left(2 \cdot \frac{16}{3} \cdot 6\right) \overset{?}{=} 2$

$\log_8 64 \overset{?}{=} 2$

$\log_8 8^2 \overset{?}{=} 2$

$2 = 2$

14. $e^{5x-3} = 57$

$\ln e^{5x-3} = \ln 57$

$(5x-3)\ln e = \ln 57$

$5x = 3 + \ln 57$

$x = \frac{3 + \ln 57}{5}$

15. $5^{3x+6} = 17$

$\ln 5^{3x+6} = \ln 17$

$(3x+6)\ln 5 = \ln 17$

$3x + 6 = \frac{\ln 17}{\ln 5}$

$x = \frac{-6 + \frac{\ln 17}{\ln 5}}{3} \approx -1.4132$

16. $A = P(1+r)^t$

$A = 2000(1+0.08)^5 \approx 2938.656$

Henry will have \$2938.66.

17. $A = P(1+r)^t$

$2P = P(1+0.05)^t$

$2 = (1.05)^t$

$\ln 2 = \ln 1.05^t$

$\ln 2 = t\ln 1.05$

$t = \frac{\ln 2}{\ln 1.05} \approx 14.21$

It will take about 14 years to double her money.

Practice Final Examination

1. $3\dfrac{1}{4}+2\dfrac{3}{5}=\dfrac{13}{4}+\dfrac{13}{5}=\dfrac{65}{20}+\dfrac{52}{20}=\dfrac{117}{20}=5\dfrac{17}{20}$

2. $\left(1\dfrac{1}{6}\right)\left(2\dfrac{2}{3}\right)=\left(\dfrac{7}{6}\right)\left(\dfrac{8}{3}\right)=\dfrac{7}{2\cdot3}\times\dfrac{2\cdot4}{3}=\dfrac{28}{9}=3\dfrac{1}{9}$

3. $\dfrac{15}{4}\div\dfrac{3}{8}=\dfrac{15}{4}\times\dfrac{8}{3}=\dfrac{3\cdot5}{4}\times\dfrac{4\cdot2}{3}=\dfrac{10}{1}=10$

4.
$$
\begin{array}{r}
1.63 \\
\times\ 3.05 \\
\hline
815 \\
4\,890 \\
\hline
4.9715
\end{array}
$$

5. $0.0006\!\!\wedge\overline{)120.0000\!\!\wedge}$ quotient $20\,0000$
$$
\begin{array}{r}
\underline{12} \\
0
\end{array}
$$

$12\div0.0006=200{,}000$

6. 7% of $64{,}000=0.07\times64{,}000=4480$

7. $(4-3)^2+\sqrt{9}\div(-3)+4=1^2+3\div(-3)+4$
$$
\begin{aligned}
&=1+(-1)+4 \\
&=4
\end{aligned}
$$

8. $5a-2ab-3a^2-6a-8ab+2a^2$
$$
\begin{aligned}
&=(5a-6a)+(-2ab-8ab)+(-3a^2+2a^2) \\
&=-a-10ab-a^2
\end{aligned}
$$

9. $-2x+3y\{7-2[x-(4x+y)]\}$
$$
\begin{aligned}
&=-2x+3y[7-2(x-4x-y)] \\
&=-2x+3y[7-2(-3x-y)] \\
&=-2x+3y(7+6x+2y) \\
&=-2x+21y+18xy+6y^2
\end{aligned}
$$

10. $2x^2-3xy-4y,\ x=-2,\ y=3$

$2(-2)^2-3(-2)(3)-4(3)=8+18-12=14$

11. $F=\dfrac{9}{5}C+32$

$F=\dfrac{9}{5}(-35)+32$

$F=-31$

The temperature is $-31°F$.

12. $\dfrac{1}{3}x-4=\dfrac{1}{2}x+1$

$\dfrac{1}{6}x=-5$

$x=-30$

13. $-4x+3y=7$

$3y=4x+7$

$y=\dfrac{4x+7}{3}$

14. $5x+3-(4x-2)\le6x-8$

$5x+3-4x+2\le6x-8$

$-5x\le-13$

$x\ge\dfrac{13}{5}=2.6$

15. $P=2L+2W=1760$

$L+W=880$

$2W-200+W=880$

$3W=1080$

$W=360$

$L=2W-200=520$

The width is 360 meters and the length is 520 meters.

16. $r=\dfrac{d}{2}=\dfrac{6}{2}=3$

$V=\dfrac{4}{3}\pi r^3\approx\dfrac{4}{3}(3.14)(3)^3=\dfrac{4}{3}(3.14)(27)=113.04$

The volume is approximately 113.04 cubic feet.

17. $7x-2y=-14$

Let $x=0$.

$7(0)-2y=-14$

$y=7$

Let $y=0$.

$7x-2(0)=-14$

$x=-2$

x	y
0	7
-2	0

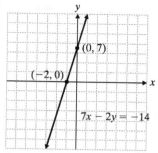

Scale: Each unit = 2

18. $3x - 4y \le 6$

Graph $3x - 4y = 6$ with a solid line.
Test point: (0, 0)
$3(0) - 4(0) \le 6$
$0 \le 6,$ True
Shade the region containing (0, 0).

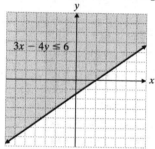

19. $m = \dfrac{y_2 - y_1}{x_2 - x_1} = \dfrac{-3 - 5}{-2 - 1} = \dfrac{8}{3}$

20. $y = -\dfrac{2}{3}x + 4;\ m = -\dfrac{2}{3}$

The slope of a parallel line is $m = -\dfrac{2}{3}$.

21. $f(x) = 3x^2 - 4x - 3$
$f(3) = 3(3^2) - 4(3) - 3$
$f(3) = 27 - 12 - 3$
$f(3) = 12$

22. $f(x) = 3x^2 - 4x - 3$
$f(-2) = 3((-2)^2) - 4(-2) - 3$
$f(-2) = 12 + 8 - 3$
$f(-2) = 17$

23. {(2, −7), (−1, 1), (3, 2)}
Domain: {2, −1, 3}
Range: {−7, 1, 2}

24. $\dfrac{1}{2}x + \dfrac{2}{3}y = 1$ **(1)**

$\dfrac{1}{3}x + y = -1$ **(2)**

Multiply equation **(1)** by 6 and equation **(2)** by 3 to clear fractions.
$3x + 4y = 6$ **(3)**
$x + 3y = -3$ **(4)**

Multiply equation **(4)** by −3 and add to equation **(3)**.

$$\begin{array}{r} 3x + 4y = 6 \\ -3x - 9y = 9 \\ \hline -5y = 15 \\ y = -3 \end{array}$$

Replace y with −3 in equation **(4)**.
$x + 3(-3) = -3$
$x - 9 = -3$
$x = 6$
The solution is (6, −3).

25. $4x - 3y = 12$ **(1)**
$3x - 4y = 2$ **(2)**
Multiply equation **(1)** by 3 and equation **(2)** by −4.

$$\begin{array}{r} 12x - 9y = 36 \\ -12x + 16y = -8 \\ \hline 7y = 28 \\ y = 4 \end{array}$$

Replace y with 4 in equation **(1)**.
$4x - 3(4) = 12$
$4x - 12 = 12$
$4x = 24$
$x = 6$
The solution is (6, 4).

26. Solve the system.
$2x + 3y - z = 16$ **(1)**
$x - y + 3z = -9$ **(2)**
$5x + 2y - z = 15$ **(3)**
Multiply equation **(1)** by 3 and add to equation **(2)**.

$$\begin{array}{r} 6x + 9y - 3z = 48 \\ x - y + 3z = -9 \\ \hline 7x + 8y = 39 \end{array}\ \textbf{(4)}$$

Multiply equation **(3)** by 3 and add to equation **(2)**.

$$\begin{array}{r} x - y + 3z = -9 \\ 15x + 6y - 3z = 45 \\ \hline 16x + 5y = 36 \end{array}\ \textbf{(5)}$$

Multiply equation **(4)** by 5, equation **(5)** by −8, and add the results.

$$35x + 40y = 195$$
$$\underline{-128x - 40y = -288}$$
$$-93x \qquad = -93$$
$$x = 1$$

Replace x with 1 in equation (**4**).
$$7(1) + 8y = 39$$
$$7 + 8y = 39$$
$$8y = 32$$
$$y = 4$$

Replace x with 1 and y with 4 in equation (**1**).
$$2(1) + 3(4) - z = 16$$
$$2 + 12 - z = 16$$
$$14 - z = 16$$
$$-z = 2$$
$$z = -2$$

The solution is $(1, 4, -2)$.

27. $y + z = 2$ (**1**)
$\quad x + \quad + z = 5$ (**2**)
$\quad x + y \quad = 5$ (**3**)

Subtract equation (**3**) from equation (**2**).
$$x \quad + z = 5$$
$$\underline{-x - y \quad = -5}$$
$$-y + z = 0 \quad (\mathbf{4})$$

Add equations (**1**) and (**4**).
$$y + z = 2$$
$$\underline{-y + z = 0}$$
$$2z = 2$$
$$z = 1$$

Replace z with 1 in equation (**1**).
$$y + 1 = 2$$
$$y = 1$$

Replace z with 1 in equation (**2**).
$$x + 1 = 5$$
$$x = 4$$

The solution is $(4, 1, 1)$.

28. $3y \geq 8x - 12$ $\qquad$ $2x + 3y \leq -6$
Test point: $(0, 0)$ $\qquad$ Test point: $(0, 0)$
$3(0) \geq 8(0) - 12$ $\qquad$ $2(0) + 3(0) \leq -6$
$\quad 0 \geq -12$, true $\qquad\quad$ $0 \leq -6$, false

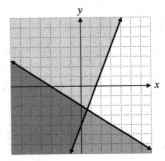

29. $(-3x^2 y)(-6x^3 y^4) = -3(-6)x^{2+3}y^{1+4} = 18x^5 y^5$

30. $(3x - 1)(2x + 5) = 6x^2 + 15x - 2x - 5$
$$= 6x^2 + 13x - 5$$

31.
$$\begin{array}{r} x + 4 \\ x+3 \overline{)\; x^2 + 7x + 12} \\ \underline{x^2 + 3x} \\ 4x + 12 \\ \underline{4x + 12} \\ 0 \end{array}$$

$(x^2 + 7x + 12) \div (x + 3) = x + 4$

32. $9x^2 - 30x + 25 = (3x)^2 - 2(3x)(5) + 5^2$
$$= (3x - 5)^2$$

33. $x^3 + 2x^2 - 4x - 8 = x^2(x + 2) - 4(x + 2)$
$$= (x + 2)(x^2 - 4)$$
$$= (x + 2)(x^2 - 2^2)$$
$$= (x + 2)(x + 2)(x - 2)$$

34. $2x^3 + 15x^2 - 8x = x(2x^2 + 15x - 8)$
$$= x(2x - 1)(x + 8)$$

35. $x^2 + 15x + 54 = 0$
$$(x + 9)(x + 6) = 0$$
$$x + 9 = 0 \qquad\qquad x + 6 = 0$$
$$x = -9 \qquad\qquad\quad x = -6$$

36. $\dfrac{9x^3 - x}{3x^2 - 8x - 3} = \dfrac{x(9x^2 - 1)}{(3x + 1)(x - 3)}$
$$= \dfrac{x(3x + 1)(3x - 1)}{(3x + 1)(x - 3)}$$
$$= \dfrac{x(3x - 1)}{(x - 3)}$$

37. $\dfrac{x^2 - 9}{2x^2 + 7x + 3} \div \dfrac{x^2 - 3x}{2x^2 + 11x + 5}$
$$= \dfrac{(x + 3)(x - 3)}{(2x + 1)(x + 3)} \cdot \dfrac{(2x + 1)(x + 5)}{x(x - 3)}$$
$$= \dfrac{x + 5}{x}$$

38. $\dfrac{3x}{x+5} - \dfrac{2}{x^2+7x+10}$

$= \dfrac{3x(x+2)}{(x+5)(x+2)} - \dfrac{2}{(x+5)(x+2)}$

$= \dfrac{3x(x+2)-2}{(x+5)(x+2)}$

$= \dfrac{3x^2+6x-2}{(x+5)(x+2)}$

39. $\dfrac{\frac{3}{2x+1}+2}{1-\frac{2}{4x^2-1}} = \dfrac{(2x+1)(2x-1)\left(\frac{3}{2x+1}+2\right)}{(2x+1)(2x-1)\left(1-\frac{2}{(2x+1)(2x-1)}\right)}$

$= \dfrac{3(2x-1)+2(2x+1)(2x-1)}{(2x+1)(2x-1)-2}$

$= \dfrac{6x-3+2(4x^2-1)}{4x^2-1-2}$

$= \dfrac{6x-3+8x^2-2}{4x^2-3}$

$= \dfrac{8x^2+6x-5}{4x^2-3}$

40.

$$\dfrac{x-1}{x^2-4} = \dfrac{2}{x+2} + \dfrac{4}{x-2}$$

$$\dfrac{x-1}{(x+2)(x-2)} = \dfrac{2}{x+2} + \dfrac{4}{x-2}$$

$$(x+2)(x-2)\left(\dfrac{x-1}{(x+2)(x-1)}\right) = (x+2)(x-2)\cdot\dfrac{2}{x+2} + (x+2)(x-2)\cdot\dfrac{4}{x-2}$$

$$x-1 = 2(x-2)+4(x+2)$$

$$x-1 = 2x-4+4x+8$$

$$5x = -5$$

$$x = -1$$

41. $\dfrac{5x^{-4}y^{-2}}{15x^{-1/2}y^3} = \dfrac{5}{15}x^{-4-(-1/2)}y^{-2-3}$

$= \dfrac{1}{3}x^{-7/2}y^{-5}$

$= \dfrac{1}{3x^{-7/2}y^5}$

42. $\sqrt[3]{40x^4y^7} = \sqrt[3]{8x^3y^6\cdot 5xy}$

$= \sqrt[3]{8x^3y^6}\cdot\sqrt[3]{5xy}$

$= 2xy^2\sqrt[3]{5xy}$

43. $5\sqrt{2}-3\sqrt{50}+4\sqrt{98} = 5\sqrt{2}-3\sqrt{25\cdot 2}+4\sqrt{49\cdot 2}$

$= 5\sqrt{2}-15\sqrt{2}+28\sqrt{2}$

$= 18\sqrt{2}$

44. $\dfrac{2\sqrt{3}+1}{3\sqrt{3}-\sqrt{2}} = \dfrac{2\sqrt{3}+1}{3\sqrt{3}-\sqrt{2}} \cdot \dfrac{3\sqrt{3}+\sqrt{2}}{3\sqrt{3}+\sqrt{2}}$

$\qquad = \dfrac{6\sqrt{9}+2\sqrt{6}+3\sqrt{3}+\sqrt{2}}{\left(3\sqrt{3}\right)^2 - \left(\sqrt{2}\right)^2}$

$\qquad = \dfrac{6\cdot 3 + 2\sqrt{6}+3\sqrt{3}+\sqrt{2}}{9\cdot 3 - 2}$

$\qquad = \dfrac{18+2\sqrt{6}+3\sqrt{3}+\sqrt{2}}{25}$

45. $i^3 + \sqrt{-25} + \sqrt{-16} = -i + 5i + 4i = 8i$

46. $\qquad \sqrt{x+7} = x+5$

$\qquad \left(\sqrt{x+7}\right)^2 = (x+5)^2$

$\qquad\qquad x+7 = x^2 + 10x + 25$

$\qquad x^2 + 9x + 18 = 0$

$\qquad (x+6)(x+3) = 0$

$\qquad x+6 = 0 \qquad\qquad x+3 = 0$

$\qquad\quad x = -6 \qquad\qquad\quad x = -3$

Check: $\sqrt{-6+7} \overset{?}{=} -6+5$

$\qquad\qquad\quad \sqrt{1} \overset{?}{=} -1$

$\qquad\qquad\qquad 1 \ne -1$

$\sqrt{-3+7} \overset{?}{=} -3+5$

$\qquad \sqrt{4} \overset{?}{=} -3+5$

$\qquad\quad 2 = 2$

$x = -3$ is the solution.

47. $y = kx^2$

$\qquad 15 = k(2)^2$

$\qquad k = \dfrac{15}{4}$

$\qquad y = \dfrac{15}{4}x^2$

$\qquad y = \dfrac{15}{4}(3)^2 = 33.75$

48. $\qquad 5x(x+1) = 1 + 6x$

$\qquad 5x^2 + 5x = 1 + 6x$

$\qquad 5x^2 - x - 1 = 0$, use quadratic formula

$\qquad x = \dfrac{-(-1) \pm \sqrt{(-1)^2 - 4(5)(-1)}}{2(5)}$

$\qquad x = \dfrac{1 \pm \sqrt{21}}{10}$

49. $5x^2 - 9x = -12x$

$\qquad 5x^2 + 3x = 0$

$\qquad x(5x+3) = 0$

$\qquad x = 0$ or $5x+3 = 0$

$\qquad\qquad\qquad x = -\dfrac{3}{5}$

50. $x^{2/3} + 5x^{1/3} - 14 = 0$, let $x^{1/3} = w$, $x^{2/3} = w^2$

$\qquad w^2 + 5w - 14 = 0$

$\qquad (w-2)(w+7) = 0$

$\qquad w-2 = 0 \qquad\qquad w+7 = 0$

$\qquad\quad w = 2 \qquad\qquad\qquad w = -7$

$\qquad x^{1/3} = 2 \qquad\qquad x^{1/3} = -7$

$\qquad\quad x = 8 \qquad\qquad\qquad x = -343$

51. $3x^2 - 11x - 4 \ge 0$

$\qquad 3x^2 - 11x - 4 = 0$

$\qquad (3x+1)(x-4) = 0$

$\qquad 3x+1 = 0 \qquad$ or $\quad x-4 = 0$

$\qquad\quad x = -\dfrac{1}{3} \qquad\qquad\qquad x = 4$

Region I: Test $x = -1$

$3(-1)^2 - 11(-1) - 4 = 10 > 0$

Region II: Test $x = 0$

$3(0)^2 - 11(0) - 4 = -4 < 0$

Region III: Test $x = 5$

$3(5)^2 - 11(5) - 4 = 16 > 0$

$x \le -\dfrac{1}{3}$ or $x \ge 4$

52. $f(x) = -x^2 - 4x + 5$

parabola, opening downward

$\qquad f(0) = -0^2 - 4(0) + 5 = 5 \Rightarrow y\text{-int: } (0, 5)$

$\qquad -x^2 - 4x + 5 = 0$

$\qquad\quad x^2 + 4x - 5 = 0$

$\qquad (x+5)(x-1) = 0$

$\qquad x = -5, \ x = 1$

$\qquad x\text{-int: } (-5, 0), \ (1, 0)$

$\qquad -\dfrac{b}{2a} = -\dfrac{-4}{2(-1)} = -2$

$\qquad f(-2) = -(-2)^2 - 4(-2) + 5 = 9$

$\qquad V(-2, 9)$

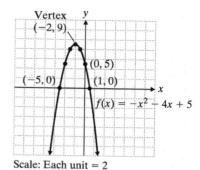

Scale: Each unit = 2

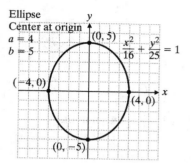

53. $L = 3W + 1$

$A = 52 \text{ cm}^2$

$A = LW = (3W + 1)W = 52$

$3W^2 + W - 52 = 0$

$(W - 4)(2W + 13) = 0$

$W = 4$ or $W = -\dfrac{13}{3}$ reject, $W > 0$

$L = 3W + 1 = 13$

The width is 4 cm and the length is 13 cm.

54. $\left| \dfrac{2}{3}x - 4 \right| = 2$

$\dfrac{2}{3}x - 4 = 2$ or $\dfrac{2}{3}x - 4 = -2$

$2x - 12 = 6$ $\qquad$ $2x - 12 = -6$

$2x = 18$ $\qquad\quad$ $2x = 6$

$x = 9$ $\qquad\qquad$ $x = 3$

55. $|2x - 5| < 10$

$-10 < 2x - 5 < 10$

$-5 < 2x < 15$

$-\dfrac{5}{2} < x < \dfrac{15}{2}$

56. $\qquad x^2 + y^2 + 6x - 4y = -9$

$x^2 + 6x + 9 + y^2 - 4y + 4 = -9 + 9 + 4 = 4$

$\qquad (x + 3)^2 + (y - 2)^2 = 2^2$

$C(-3, 2),\ r = 2$

57. $\dfrac{x^2}{16} + \dfrac{y^2}{25} = 1$

$\dfrac{x^2}{4^2} + \dfrac{y^2}{5^2} = 1$, ellipse: $C(0, 0)$

$a = 4,\ b = 5$, x-int: $(\pm 4, 0)$

y-int: $(0, \pm 5)$

58. $\dfrac{x^2}{4} - \dfrac{y^2}{9} = 1$

$\dfrac{x^2}{2^2} - \dfrac{y^2}{3^2} = 1$, hyperbola

$C(0, 0),\ a = 2,\ b = 3$

x-int: $(\pm 2, 0)$

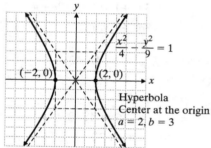

Hyperbola
Center at the origin
$a = 2,\ b = 3$

59. $x = (y - 3)^2 + 5$

parabola opening right, $V(5, 3)$

$x = (0 - 3)^2 + 5 = 14$, x-int: $(14, 0)$

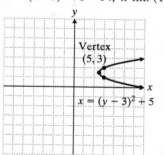

Scale: Each unit = 2

60. $x^2 + y^2 = 16$

$x^2 - y = 4 \Rightarrow y = x^2 - 4$

$x^2 + (x^2 - 4)^2 = 16$

$x^2 + x^4 - 8x^2 + 16 = 16$

$x^4 - 7x^2 = 0$

$x^2(x^2 - 7) = 0$

$x^2 = 0$ $x^2 = 7$

$x = 0$ $x = \pm\sqrt{7}$

$y = x^2 - 4 = 0^2 - 4 = -4$

$y = x^2 - 4 = \left(\pm\sqrt{7}\right)^2 - 4 = 3$

$(0, -4)$, $\left(\pm\sqrt{7}, 3\right)$ is the solution.

61. $f(x) = 3x^2 - 2x + 5$

    a. $f(-1) = 3(-1)^2 - 2(-1) + 5 = 3 + 2 + 5 = 10$

    b. $f(a) = 3a^2 - 2a + 5$

    c. $f(a+2) = 3(a+2)^2 - 2(a+2) + 5$
               $= 3a^2 + 12a + 12 - 2a - 4 + 5$
               $= 3a^2 + 10a + 13$

62. $f(x) = 5x^2 - 3$, $g(x) = -4x - 2$
    $f[g(x)] = f(-4x - 2)$
           $= 5(-4x - 2)^2 - 3$
           $= 5(16x^2 + 16x + 4) - 3$
           $= 80x^2 + 80x + 20 - 3$
           $= 80x^2 + 80x + 17$

63. $f(x) = \dfrac{1}{2}x - 7$, $f(x) \rightarrow y$

    $y = \dfrac{1}{2}x - 7$, $x \leftrightarrow y$

    $x = \dfrac{1}{2}y - 7$

    $\dfrac{1}{2}y = x + 7$

    $y = 2x + 14$

    $y \rightarrow f^{-1}(x)$

    $f^{-1}(x) = 2x + 14$

64.

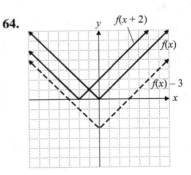

65. $f(x) = 2^{1-x}$

x	$y = f(x) = 2^{1-x}$
-1	4
0	2
1	1

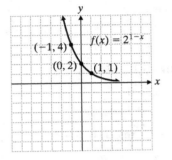

66. $\log_5 x = -4$
    $5^{-4} = x$
    $x = \dfrac{1}{5^4}$
    $x = \dfrac{1}{625}$ or 0.0016

67. $\log_4(3x + 1) = 3$
    $3x + 1 = 4^3 = 64$
    $3x + 1 = 64$
    $3x = 63$
    $x = 21$

68. $\log_{10} 0.01 = y$
    $10^y = 0.01$
    $10^y = 10^{-2}$
    $y = -2$

69.
    $\log_2 6 + \log_2 x = 4 + \log_2(x - 5)$
       $\log_2(6x) = 4 + \log_2(x - 5)$
  $\log_2(6x) - \log_2(x - 5) = 4$
          $\log_2 \dfrac{6x}{x - 5} = 4$
            $\dfrac{6x}{x - 5} = 2^4 = 16$
         $6x = 16(x - 5) = 16x - 80$
      $16x - 80 = 6x$
         $10x = 80$
           $x = 8$